Advances in Geographic Information Science

The ***Advances in Geographic Information Science*** series aims to: present current and emerging innovations in GIScience; describe new and robust GIScience methods for use in transdisciplinary problem solving and decision making contexts, illustrate GIScience case studies for use in teaching and training situations, analyze GIScience tools at the forefront of scientific research, and examine the future of GIScience in an expanding knowledge-based economy. The scope of the series is broad and will encompass work from all subject disciplines that develop and use an explicitly spatial perspective for analysis and problem-solving.

Md. Omar Sarif · Ayyoob Sharifi
Editors

Advanced Geospatial Intelligence and AI for Environmental Resilience and Sustainable Development

Springer

Editors
Md. Omar Sarif
Center for Peaceful and Sustainable Futures (CEPEAS), The IDEC Institute
Hiroshima University
Higashihiroshima, Hiroshima, Japan

Department of Geography
Lovely Professional University
Phagwara, Punjab, India

Ayyoob Sharifi
The IDEC Institute and Network for Education and Research on Peace and Sustainability (NERPS)
Hiroshima University
Higashihiroshima, Hiroshima, Japan

ISSN 1867-2434 ISSN 1867-2442 (electronic)
Advances in Geographic Information Science
ISBN 978-3-032-03713-8 ISBN 978-3-032-03714-5 (eBook)
https://doi.org/10.1007/978-3-032-03714-5

This Springer imprint is published by the registered company Springer Nature Switzerland AG
The registered company address is: Gewerbestrasse 11, 6330 Cham, Switzerland

Preface

The book, '*Advanced Geospatial Intelligence and AI for Environmental Resilience and Sustainable Development*', emerges at a critical juncture when climate change, rapid urbanization, and ecosystem degradation pose escalating challenges to environmental and societal stability. This volume provides a comprehensive examination of how advanced geospatial technologies and artificial intelligence (AI) are transforming environmental monitoring, analysis, and decision-making. By integrating cutting-edge geospatial intelligence with machine learning and AI-driven methods, this book provides interdisciplinary solutions aimed at promoting resilience and advancing sustainable development goals across urban, rural, coastal, and ecologically sensitive regions.

The volume comprises eighteen chapters, systematically grouped into five thematic sections. These sections are: (1) Urban Intelligence and Sustainable Cities, (2) Geospatial Analytics for Climate Change and Cryospheric Studies, (3) Geospatial and AI Solutions for Agriculture and Forest Management, (4) Coastal and Environmental Risk Management, and (5) Environmental Prediction and Resilience through Geospatial Intelligence. Each chapter has been meticulously selected and peer-reviewed to ensure scientific rigor and relevance to contemporary geospatial and environmental challenges.

The thematic organization of this book reflects the multidimensional nature of geospatial AI applications in the present era. Section-1 highlights urban sustainability, heat island mitigation, and infrastructure planning. Section-2 addresses the pressing impacts of climate change and glacier dynamics. Section-3 focuses on agriculture, forestry, and watershed prioritization using AI-enhanced geospatial tools. Section-4 delves into coastal zone management, groundwater quality, and natural hazard mitigation. Finally, Section-5 explores AI-powered environmental forecasting and air pollution modeling. Collectively, these sections underscore the transformative power of GeoAI in shaping resilient and sustainable environments.

We firmly believe that this volume will serve as a valuable resource for researchers, practitioners, policymakers, and students engaged in geospatial science, AI applications, and environmental resilience studies. We are confident that the insights offered in this book will foster cross-disciplinary collaborations and inspire innovative approaches to achieving a sustainable future.

Md. Omar Sarif
Center for Peaceful and Sustainable Futures (CEPEAS), The IDEC Institute
Hiroshima University
Higashihiroshima, Hiroshima, Japan

Department of Geography
Lovely Professional University
Phagwara, Punjab, India

Ayyoob Sharifi
The IDEC Institute and Network for Education and Research on Peace and Sustainability (NERPS)
Hiroshima University
Higashihiroshima, Hiroshima, Japan

Contents

Urban Intelligence and Sustainable Cities

Using Geospatial Technologies and Machine Learning to Map and Mitigate Urban Heat Islands and Improve Urban Livability 3
Md. Omar Sarif and Ayyoob Sharifi
1 Introduction 3
2 Data and Methods 6
3 Results 11
4 Discussion 18
5 Conclusion 20
References 20

Urban Expansion and Thermal Environment in a Medium Size City: Role of Urban Green and Blue Spaces in Mitigating Surface Urban Heat Island Intensity 25
Shahfahad, Pratik Kumar Mishra, Md. Rejaul Islam, Rambooshan Tiwari, Vinod Kumar Tripathi, Ahmad Mujtaba Siddiqui, and Atiqur Rahman
1 Introduction 25
2 Study Area 27
3 Data Sources and Methods 27
4 Results and Discussion 31
5 Conclusion 38
References 39

Identification of On-Road and Roadside Objects Using Random Forest-Based Prediction Model 43
Parvej Khan
1 Introduction 43
2 Literature Review 44
3 Dataset 48
4 Methodology 48
5 Experiment 59
6 Conclusion 66
References 66

Analyzing the Role of Geospatial Technologies and AI in Urban Infrastructure Planning and the Development of Smart Cities, Including Transportation Systems, Utilities, and Public Services 69
Vedprakash Maralapalle, Jayatheja Muktinutalapati, Bogireddy Chandra, Gangadhara Reddy Narala, and Rajiv Iyer
1 Introduction 69
2 Geospatial Technologies in Urban Infrastructure 73
3 Artificial Intelligence in Smart Cities 77
4 Integration of Geospatial Technologies and AI 85
5 Challenges and Limitations in Implementing Geospatial Technologies and AI in Smart Cities 89
6 Future Directions 94
7 Conclusion 97
References 98

Geospatial Analytics for Climate Change and Cryospheric Studies

Assessing the Impact of Night-Time Light as Proxy for Climate Change: A Geospatial Approach in Himachal Pradesh 103
Sumalya Ray and Md. Omar Sarif
1 Introduction 103
2 Study Area 104
3 Datasets and Methodologies 105
4 Results 111
5 Discussion 121
6 Conclusion 123
References 124

Decadal Study on Supraglacial Debris Cover of Kaer Glacier, Eastern Himalayas 127
Sivaranjani Sivalingam and Saravanan Subbarayan
1 Introduction 127
2 Recent Developments 129
3 Study Area and Its Rationale 130
4 Materials and Methods 132
5 Results and Discussion 135
6 Study Limitations Resolution Constraints 141
7 Uncertainty Estimation 142
8 Future Implications 143
9 Conclusion 144
References 144

Assessing the Impact of Climate Change on Glaciers Through Region Wise Mass Balance and AI Tools, Contributing to Understanding of Environmental Changes in Upper Indus Basin and Kashmir Himalayas 149
Suhail Ahmad Dar, Md. Omar Sarif, and Sumalya Ray
1 Introduction 149
2 Study Area 150
3 Methodology and Data Used 155
4 Results and Discussion 159
5 Discussion 167
6 Conclusion 169
References 170

Climate Change Impacts on Various Ecosystems: A Geospatial and AI Perspective 175
Mallika Bhuyan, Preeti Sardar, C. Jeganathan, and Girish S. Pujar
1 Introduction 175
2 Forest Ecosystems 177
3 Coastal Ecosystems 187
4 Conclusion 198
References 199

Geospatial and AI Solutions for Agriculture and Forest Management

Early Identification and Pathogenesis of Leaf Diseases in *Magnifera Indica* Using Transfer Learning 207
Arshad Husain and Subhojit Bhowmick
1 Introduction 207
2 Data Acquisition and Methodology 208
3 Result and Discussion 210
4 Conclusion 213
References 214

Examining the Role of Geospatial Data and AI in Conducting Comprehensive Forestry Inventories and Mapping to Support Sustainable Forest Management 215
Manohar Yadav and Mukesh Kumar Verma
1 Introduction 215
2 Overview of Geospatial Data in Forestry Management 218
3 Role of Artificial Intelligence in Forestry Applications 220
4 Comprehensive Forestry Inventories: Methodologies and Workflow 223
5 Mapping for Sustainable Forest Management 225
6 Challenges and Limitations 226
7 Future Perspectives and Recommendations 228
8 Conclusion 230
References 231

Prioritization of Erosion Prone Watersheds of the Song River Basin Using Morphometric Indices Coupled with AHP-TOPSIS and Geospatial Technology 237
Sunil Singh, Kabysruti Bora, Sanju Purohit, and Jitendra Binwal
1 Introduction 237
2 Data and Methodology 239
3 Results and Discussion 244
4 Conclusion 259
References 260

Coastal and Environmental Risk Management

Application of Geospatial Technologies and AI to Detect and Analyze the Shoreline Change of Visakhapatnam, a Coastal District 265
Sravanthi Karnata and Namrata Kumari
1 Introduction 265
2 Recent Developments 266
3 Study Rationale 269
4 Materials and Methods 270
5 Results and Discussions 280
6 Perspectives 287
7 Conclusion 287
References 288

A Systematic and Structured Review on Groundwater Contamination, Sources, Prevalence, Health Risk and Mitigation Strategies in Relation with Land Use/Land Cover in Bihar State of India 291
Sushmita Kumari and Md. Omar Sarif
1 Introduction 291
2 Data and Methods 296
3 Explore Significance of Land Use/Land Cover Mapping and Dynamics in the Context of Groundwater Contamination 299
4 Assessment of Groundwater Quality 303
5 Quantification Approach for Arsenic in Ground Water 306
6 The Impact of Arsenic Contamination on Human Health 312
7 The Available Techniques to Minimize and Mitigate Arsenic Contamination 321
8 Limitations 323
9 Need of This Study 325
10 Discussion 326
11 Conclusion 326
References 327

Examining the Role of Geospatial Technologies and AI in Disaster Risk Management and Response, Including Hazard Mapping, Vulnerability Assessment, and Real-Time Monitoring 335
Sumedha Surbhi Singh and C. Jeganathan
1 Introduction 335
2 Recent Developments 338
3 Use of AI in Different Disaster Risk Management 339
4 Discussion 357
5 Conclusion 359
References 360

GIS-Based Landslide Susceptibility Assessment for Coonoor Taluk, Nilgiris District South India Tamil Nadu Using Machine Learning Algorithm 367
D. Padmanathan, R. Jegankumar, K. Prakash, and D. Shankar Raja
1 Introduction 367
2 Significance of the Study 371
3 Study Area 371
4 Database and Methodology 373
5 Results and Discussion 374
6 Evaluation of Model Training Accuracy 387
7 Conclusion 388
References 389

Environmental Prediction and Resilience through Geospatial Intelligence

Forecasting Drought from Satellite Imagery: Comparing Stochastic, Classical Machine Learning, and Deep Learning Approaches ... 395
Mesfin F. Woldmariam
1 Introduction ... 395
2 Recent Developments ... 396
3 Machine Learning and Drought Modeling ... 399
4 Statistical, Machine Learning, and Deep Learning Techniques ... 400
5 Materials and Methods ... 402
6 Data Preparation and Modeling ... 403
7 Conclusion and Limitation ... 414
References ... 414

Smart Geospatial Intelligence and AI for a Resilient Environment: A Review ... 417
Arshad Husain and Subhojit Bhowmick
1 Introduction ... 417
2 Related Work ... 418
3 AI and Geospatial Intelligence (GEOINT) ... 418
4 Applications of Geospatial Intelligence in Environmental Resilience ... 422
5 Applications of Geospatial Intelligence in Environmental Resilience ... 426
6 Conclusion ... 429
References ... 430

Uncovering Air Pollution Risks in Indian Scenario with Explainable AI: A Case Study Perspective ... 433
Shallu Sharma, Jagriti Saini, and Sumit Kumar
1 Introduction ... 433
2 Recent Developments ... 435
3 Material and Methods ... 437
4 Results and Discussion ... 441
5 Perspectives: Study Limitations and Future Implications ... 446
6 Conclusion ... 446
References ... 450

About the Editors

Asst. Prof. (Dr.) Md. Omar Sarif is Assistant Professor (Special Appointment) at the Center for Peaceful and Sustainable Futures (CEPEAS), The IDEC Institute, Hiroshima University, Japan, since April 2025. Prior to this role, he served as Assistant Professor in the Department of Geography at Lovely Professional University (LPU), Punjab, India (2022–2024), and concurrently in the Department of Research Impact and Outcome, Research and Development Cell of LPU (2023–2024). He received his Ph.D. in Engineering (GIS and Remote Sensing) from Motilal Nehru National Institute of Technology (MNNIT) Allahabad, India, in 2022, where his doctoral research explored urban sprawl, land use/land cover change, urban heat island dynamics, ecological vulnerability, and urban future trajectory prediction. He also holds an M.Tech. in Remote Sensing from Birla Institute of Technology, Mesra, in 2016; an M.A. in Geography from Jamia Millia Islamia, New Delhi, in 2014 (where he served as Vice President of the Jamia Geography Association during 2013-2014); and a B.A. (Hons.) in Geography from Aligarh Muslim University in 2012.

Dr. Sarif's research interests span a wide spectrum of geospatial domains, including urban morphological change, LULC simulation, SUHI analysis, carbon sequestration, groundwater contamination, glacier dynamics, machine learning applications, and climate change modeling. He has guided several postgraduate and undergraduates for Master Dissertation and Bachelor Capstone project. He also has been supervising several Ph.D. scholars in Geography and Geospatial Science domains since July 2022.

As Associate Editor of a section titled, *Environmental Informatics and Remote Sensing* in *Frontiers in Environmental Science*, he continues to contribute to scholarly discourse. He has served as Reviewer for over 25 SCI/Scopus-indexed journals. He has published over 25 peer-reviewed research articles in reputed SCI/Scopus-indexed journals, presented at national and international conferences, and contributed multiple book chapters.

Prof. (Dr.) Ayyoob Sharifi is Professor at the IDEC Institute, Hiroshima University, Japan, and the University of Queensland, Australia. He currently serves as Director of the Center for Peaceful and Sustainable Futures (CEPEAS) at Hiroshima University. His academic journey includes a Doctor of Engineering degree from Nagoya University, Japan, where he also completed a Postgraduate Certificate in Integrated Environmental Studies. Prior to that, he earned his M.Sc. in Urban and Regional Planning from Iran University of Science and Technology (IUST), Tehran, and a B.Sc. in Civil Engineering–Geomatics—from Tabriz University, Iran.

Professor Sharifi is Internationally Recognized Scholar in the domains of urban sustainability, climate change adaptation and mitigation, and resilient urban systems. He was Lead Author of the Intergovernmental Panel on Climate Change (IPCC) Sixth Assessment Report (AR6) for Working Group III and Contributing Lead Author of the Second Assessment Report on Climate Change and Cities. He also serves on the advisory committee of the Future Earth Urban Knowledge Action Network (KAN).

His academic contributions span over 300 peer-reviewed journal articles in prestigious SCI/Scopus-indexed journals such as Sustainable Cities and Society, Journal of Environmental Management, Urban Climate, Cities, and Renewable and Sustainable Energy Reviews. He is also Associate Editor for Urban Climate, Journal of Environmental Management, and Climate Resilience and Sustainability and Section Editor of Frontiers in Climate.

In addition to his research, Professor Sharifi actively collaborates with international organizations and academic networks to advance the integration of sustainability science with policy and planning. His interdisciplinary approach and prolific academic output make him one of the leading voices in GeoAI, environmental planning, and sustainable urban development.

Urban Intelligence and Sustainable Cities

Using Geospatial Technologies and Machine Learning to Map and Mitigate Urban Heat Islands and Improve Urban Livability

Md. Omar Sarif and Ayyoob Sharifi

1 Introduction

In recent decades, urbanization has accelerated at an unprecedented pace, reshaping landscapes and transforming natural ecosystems into densely built environments (Shukla et al. 2019). While urban expansion is often associated with economic development and infrastructural growth, it also gives rise to a range of environmental challenges (Hui et al. 2023; Mansour et al. 2020). One of the most pressing among these is the Urban Heat Island (UHI) effect, a phenomenon where urban areas experience higher temperatures than their rural surroundings due to anthropogenic modifications in land use, increased impervious surfaces, and reduced vegetative cover (Bokaie et al. 2016). As global temperatures rise and urban populations swell, understanding and mitigating UHI effects becomes critical for enhancing the sustainability and livability of cities (Sarif et al. 2024).

The UHI effect significantly impacts urban energy demand (Suppa and Ballarini 2023), air quality (Hui et al. 2023), human health (Abutaleb et al. 2015), and overall environmental resilience (Moss et al. 2019). Elevated temperatures in cities can lead to increased use of air conditioning, thus raising electricity consumption and exacerbating greenhouse gas emissions (Jain et al. 2019). Furthermore, heat exposure disproportionately affects vulnerable populations, including the elderly, children, and

Md. O. Sarif (✉)
Center for Peaceful and Sustainable Futures (CEPEAS), The IDEC Institute, Hiroshima University, Higashihiroshima City, Hiroshima, Japan
e-mail: sarif@hiroshima-u.ac.jp

A. Sharifi
The IDEC Institute and Network for Education and Research On Peace and Sustainability (NERPS), Hiroshima University, Higashihiroshima City, Hiroshima, Japan
e-mail: sharifi@hiroshima-u.ac.jp

M. O. Sarif and A. Sharifi (eds.), *Advanced Geospatial Intelligence and AI for Environmental Resilience and Sustainable Development*, Advances in Geographic Information Science, https://doi.org/10.1007/978-3-032-03714-5_1

those with pre-existing health conditions, thereby intensifying urban health disparities (Macharia and Kiage 2025). In this context, mapping UHI-prone areas, identifying their underlying drivers, and implementing mitigation strategies are essential components of sustainable urban planning and climate adaptation.

Emerging technologies in the fields of geospatial analysis and machine learning have revolutionized the way we study and manage urban environments (Alaoui et al. 2024). Remote sensing provides multi-temporal satellite imagery that enables the quantification of land surface temperature (LST) (Mal et al. 2020), vegetation indices (Ambrosino et al. 2023), built-up areas (Chakraborty et al. 2021), and albedo (He et al. 2018) with a high degree of spatial accuracy. Geographic Information Systems (GIS) facilitate the integration, visualization, and spatial analysis of these datasets to understand spatial patterns and relationships (Johny and Mary 2016). Meanwhile, machine learning techniques enhance the predictive capacity of models, enabling data-driven insights into complex urban systems (Pham et al. 2023). Together, these tools offer a powerful framework for UHI detection, analysis, and mitigation.

In the context of India, industrial cities represent unique urban ecosystems where the interplay between economic activity, land use patterns, and environmental degradation is particularly complex (Sharma and Ghuge 2024). Ludhiana, located in the northern state of Punjab, exemplifies such a setting. Often referred to as the "Manchester of India," Ludhiana is a major industrial and manufacturing hub characterized by high population density, widespread informal industrial operations, unplanned urban growth, and inadequate green infrastructure (Shankar et al. 2024). Over the years, the city has witnessed rapid urban expansion, which has led to significant changes in its land use and land cover (LULC) configuration (Singh and Kalota 2019). These transformations have contributed to localized thermal anomalies, resulting in heightened UHI effects.

Despite Ludhiana's economic significance and the environmental stresses, it faces, there is a surprising lack of focused research on the spatial dynamics of heat stress in the city. Existing urban studies in India have primarily centered around megacities such as Delhi (Naikoo et al. 2020), Mumbai (Matloob et al. 2021), and Bengaluru (Govind and Ramesh 2019), often overlooking medium-sized industrial cities. Furthermore, most UHI-related studies have used traditional GIS methods or statistical models that may not fully capture the nonlinear and dynamic relationships between urban form, land indices, and thermal patterns (Abdullahi et al. 2015; Guha and Govil 2021). Therefore, there is a need for an integrated and intelligent approach that combines geospatial technologies with machine learning to understand and address urban thermal environments in a more holistic and actionable manner.

Despite the availability of high-resolution satellite imagery and spatial modeling tools, several research gaps persist in the study of urban heat in Indian industrial cities (Gumma et al. 2017; Pandey et al. 2012; Shahfahad et al. 2020). Firstly, there is a lack of integrated analysis that combines urban form metrics, environmental indices, and temperature patterns in a single analytical framework, especially using machine learning techniques. Most studies either focus on LST mapping or land use change detection independently, without exploring their interlinkages. Secondly, the role of industrial activity in shaping UHI patterns has not been sufficiently studied in the Indian context. While large urban agglomerations have received scholarly attention, mid-sized industrial cities like Ludhiana remain underrepresented in the literature. Addressing these gaps is crucial for developing localized and actionable urban climate policies.

By focusing on Ludhiana city, this study contributes to the broader discourse on sustainable urban development in industrial settings. The integration of geospatial technologies and machine learning not only enhances the accuracy and efficiency of UHI mapping but also offers scalable solutions that can be replicated in similar urban contexts. The findings of this research are expected to inform urban planners, policy-makers, and environmental stakeholders in devising effective heat mitigation strategies.

The primary intention of this study is grounded in the understanding that built-up density is a primary driver of the UHI effect. High-density urban areas, especially those with minimal green cover and extensive concrete surfaces, tend to absorb and retain more heat, thereby contributing to elevated LST. Mapping built-up density will provide insights into urban growth patterns and help identify potential thermal risk zones. The study further focuses on quantifying the current thermal state of the city, particularly in industrial clusters where emissions and land sealing are significant. Thermal profiling using remote sensing data will enable a comprehensive understanding of how different land uses contribute to temperature variation. Then, this research emphasizes the importance of biophysical indicators, viz., vegetation, imperviousness, and reflectivity in modulating urban temperatures. By evaluating these indices, the study will assess the city's environmental resilience and identify potential leverage points for UHI mitigation. This study aims to bridge this gap by investigating how geospatial technologies and machine learning can be employed to map and mitigate UHIs in Ludhiana. The research is guided by the following key objectives: (i) To extract a built-up density map to analyze how urbanization is spatially spreading across Ludhiana and to identify high-density zones; (ii) To evaluate the intensification of the thermal state in the city using satellite-derived LST data, with a specific focus on how industrial configurations influence the heat landscape; (iii) To identify the locations of spatial UHI and Hot spots in shaping the city's thermal resilience and its ability to achieve urban sustainability.

This study underscores the potential of modern geospatial and machine learning-driven approaches in addressing one of the most urgent urban environmental issues—urban heat islands. By mapping hotspots, understanding their underlying causes, and identifying data-driven mitigation strategies, the research aims to support Ludhiana's transition towards a more sustainable, livable, and climate-resilient urban future.

2 Data and Methods

2.1 Study Area

Ludhiana, one of the largest and most industrialized cities in the northern Indian state of Punjab, serves as the study area for the present investigation. Geographically, Ludhiana city is situated between latitudes 30° 50′2″ N to 30° 57′18″ N and longitudes 75° 45′52″ E to 75° 57′40″ E, considered 1600 km^2 of area as a study area in this research form city centre (Fig. 1). Located on the banks of the Sutlej River, the city lies within the Indo-Gangetic alluvial plains, characterized by flat topography and fertile soils, which have historically contributed to both its agricultural productivity and urban expansion (Singh and Kalota 2019). The city experiences a semi-arid climate with marked seasonal variations (Majumder et al. 2021). Summers are extremely hot, with temperatures often soaring above 40 °C, while winters are relatively cool and dry (Singh and Kalota 2019). According to the India Meteorological Department (IMD), Ludhiana receives an average annual rainfall of about 700 mm, primarily during the monsoon months of July to September (IMD 2010). The city's geographical position and climate dynamics make it particularly susceptible to urban heat island (UHI) effects, especially in densely built-up and industrial zones.

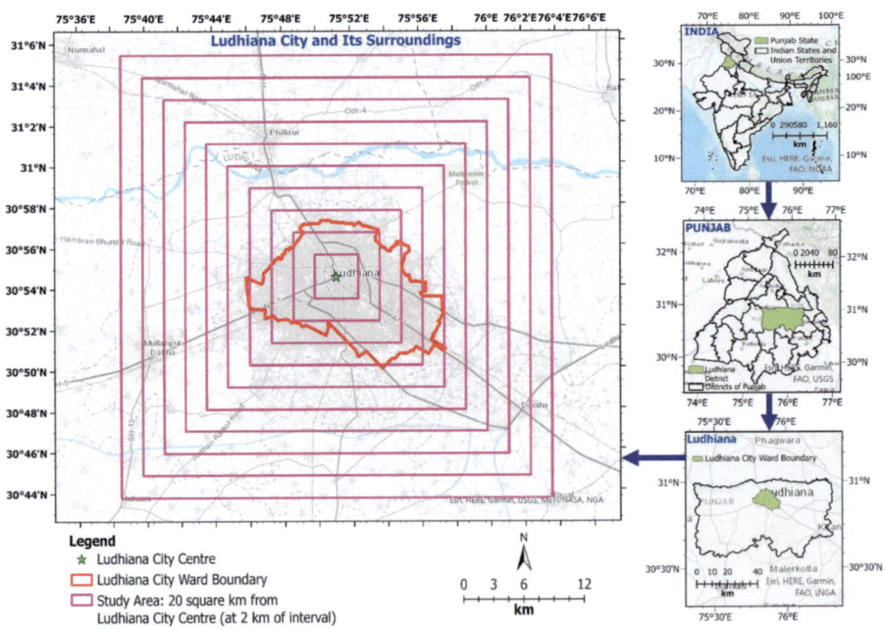

Fig. 1 Study area map: Ludhiana city of India

Ludhiana's urban morphology is shaped significantly by its role as an economic and industrial hub (Singh and Kalota 2019). Referred to as the "Manchester of India," Ludhiana hosts a diverse range of manufacturing sectors, including textiles, bicycle production, hosiery, and machine tools (Shankar et al. 2024). The city has witnessed rapid urban growth, particularly in the post-liberalization era of the 1990s, driven by both in-migration and the spatial expansion of urban functions (Singh and Kalota 2019). According to the Census of India, Ludhiana's population rose from approximately 1.1 million in 2001 to nearly 1.6 million in 2011 (Singh and Kalota 2019), with current estimates placing it above 2 million (UN 2018). The spatial expansion of the city has led to an increase in impervious surfaces and a corresponding decline in vegetation and open spaces (Majumder et al. 2021). This change in land cover configuration has significantly altered the city's microclimate and thermal environment, making it a critical case for examining the urban heat island phenomenon in the context of industrialization and urbanization.

Administratively, Ludhiana is governed by the Ludhiana Municipal Corporation and divided into various wards and planning zones (DECC 2019). The city is also part of the Ludhiana Metropolitan Region (LMR), which includes peri-urban and rural hinterlands experiencing urban spillover effects (DECC 2019). This broader urban footprint is essential for understanding the socio-environmental implications of land use transitions and heat exposure patterns across the city.

Ludhiana's landscape comprises a heterogeneous mix of land uses including residential colonies, industrial estates, commercial zones, institutional establishments, agricultural lands, and scattered green spaces such as parks and tree-lined avenues (DECC 2019). Despite several greening initiatives by the local administration, the overall vegetation density remains inadequate in many urban cores and industrial belts, contributing to the intensification of land surface temperatures (LST).

2.2 Data Used

This study has utilized Landsat 5 Thematic Mapper and Landsat 8 Operational Land Imager (OLI) imagery, both offering 30 m spatial resolution, to analyze Urban dynamics. Two summer-period time points: 1994 and 2024, were selected (see Table 1), maintaining a consistent dry-season window to maximize acquisition of cloud-free scenes and ensure optimal spectral quality for comparative analysis. This 30-year interval effectively captures long-term urbanization trends and their impacts on land surface characteristics.

2.3 Methods

The whole framework has been presented in Fig. 2. Figure 2 shows that all methods have been linked and executed in this research through this flowchart.

Table 1 Data used in the study area

Sl. no.	Name of the data	Acquisition date	Spatial resolution	Source
1	Landsat TM	08th June 1994	30 m	USGS earth explorer (https://earthexplorer.usgs.gov/)
2	Landsat OLI/TIRS	18th June 2024	30 m	
3	Boundary layers	April 2025	–	Survey of India (https://onlinemaps.surveyofindia.gov.in/Digital_Product_Show.aspx)
4	City ward boundary layer	April 2025	–	Municipal corporation of Ludhiana (https://mcludhiana.gov.in/pdf/MCL_Wards_gismap.pdf

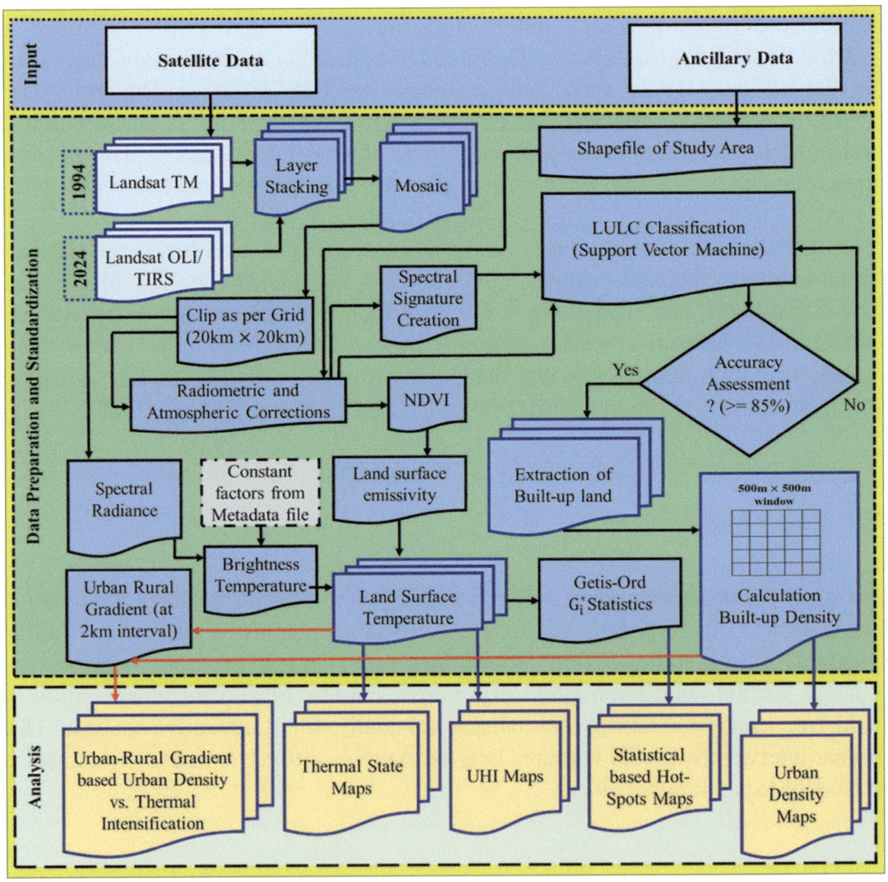

Fig. 2 Flowchart of the research study

2.3.1 Urban Density Mapping

Built-up density mapping is a technique used to quantify the concentration of built-up (impervious) surfaces within a defined spatial unit, typically in the form of a grid or buffer. This allows for spatial characterization of urbanization intensity, providing insight into how cities expand and which zones become densified over time. In this study, built-up density is calculated using classified Built-up land based on Land Use/Land Cover (LULC) maps derived from remote sensing imagery based on Anderson Scheme (Anderson et al. 1976). This LULC has been carried out using supervised classification where Support Vector Machine technique has been incorporated (Sarif et al. 2020). The LULC maps have found $>=85\%$ of overall accuracy. This study has used grid or window (500 m × 500 m) to calculate built-up density (BD) using the following formula:

$$BD_i = \left(\frac{A_{BU_i}}{A_{T_i}}\right) \times 100 \tag{1}$$

where, BD_i represents built-up density (%) in grid or window i, A_{BU_i} represents area of built-up land (km^2) within grid i, A_{T_i} represents total area (km^2) of grid i. The result is multiplied by 100 to express it as a percentage. Built-up density map, classifying it into rural area category (e.g., $<=10\%$ is a threshold value for rural or non-urban area) using reclassification.

To understand urban expansion trends, you can generate built-up density maps for multiple years and then compute the change:

$$\Delta BD_i = BD_i^{t2} - BD_i^{t1} \tag{2}$$

where, BD_i^{t1} and BD_i^{t2} are built-up densities for grid i at time $t1$ and $t2$, respectively.

2.3.2 Thermal State Intensification Mapping

Urban thermal state intensification refers to the amplification of surface temperature anomalies within urban areas over time, primarily driven by changes in land cover, industrial expansion, and anthropogenic heat emissions. Ludhiana, being a major industrial hub in northern India, exhibits significant spatial heterogeneity in surface temperature distribution. To understand the evolution of this thermal state, three core analytical stages were adopted: (i) Urban Heat Island (UHI) mapping, and (ii) Hotspot analysis.

UHI Mapping

UHI is characterized by elevated LST in urbanized or industrial zones compared to surrounding rural areas (Halder et al. 2021). The UHI Intensity (UHII) can be quantified by comparing average LST of urban and rural zones through Eq. 3:

$$UHII = LST_{urban} - \overline{LST}_{rural} \quad (3)$$

where, LST_{urban} represents LST of the particular pixel of city landscape, $\overline{LST}_{rural}$ represents mean LST of rural or non-urban area. This study considered those areas as rural areas where urban density was found less than equal to 10% as per aforesaid Eq. 1.

The LST (°C) is extracted from thermal bands (e.g., Band 6 of Landsat 5 and Band 10 of Landsat 8) using the radiative transfer model by Eq. 4:

$$LST = \frac{T_B}{1 + \left(\frac{\lambda \times T_B}{\rho}\right) \times \ln(\varepsilon)} - 273.15 \quad (4)$$

where, T_B is Brightness temperature (kelvin) derived from radiance by Eq. 5, λ is Central wavelength ($\approx 11.5\ \mu$m for Band 6 of Landsat 5, and $\approx 10.8\ \mu$m for band 10 of Landsat 8), $\rho = h \times c/\sigma \approx 1.438 \times 10^{-2}$m.K, ε is Surface emissivity estimated from NDVI.

$$T_B = \frac{K_2}{\ln\left(\frac{K_1}{L_\lambda} + 1\right)} \quad (5)$$

where, L_λ is spectral radiance (in W $\times$ m^{-2} $\times$ sr^{-1} $\times$ μm^{-1}), derived from DN values by calculating Eqs. 6 or 7 according to the Landsat sensors; K_1 and K_2 are calibration constants specific to the sensor (come from MTL.txt metadata file). L_λ is calculated using Eq. (6) for Landsat 8 and Eq. (7) for Landsat 5.

$$L_\lambda = M_L \times Q_{cal} + A_L \quad (6)$$

$$L_\lambda = \left(\frac{L_{MAX} - L_{MIN}}{Q_{CALMAX} - Q_{CALMIN}}\right) \times (Q_{cal} - Q_{CALMIN}) + L_{MIN} \quad (7)$$

where, M_L is Radiance multiplicative scaling factor, A_L is Radiance multiplicative scaling factor, and Q_{cal} is Pixel value (DN). L_{MAX} is spectral radiance for DN of 255, L_{MIN} is Spectral radiance for DN of 0, Q_{CALMAX} is 255, and Q_{CALMIN} is 0.

Hotspots Analysis

To detect localized high-temperature zones, spatial statistics were employed on the LST raster to identify thermal hotspots (Habibi and Alesheikh 2017). The Getis-Ord G_i^* statistic is used to assess whether high LST values cluster spatially using Eq. 8:

$$G_i^* = \frac{\sum_{j=1}^{n} w_{ij}x_j - \overline{X}\sum_{j=1}^{n} w_{ij}}{S\sqrt{\left[\frac{n\sum_{j=1}^{n} w_{ij}^2 - \left(\sum_{j=1}^{n} w_{ij}\right)^2}{n-1}\right]}} \tag{8}$$

where, x_j represents LST at location j, w_{ij} represents spatial weight between locations i and j, $\overline{X}$ mean LST, S represents standard deviation of LST, n number of spatial units. Statistically significant high G_i^* scores (Z > 1.96) denote hotspots, while low scores (Z < −1.96) indicate cold spots.

3 Results

3.1 Analysis of Spatial Urbanization Pattern

Urban density analysis over ten zones in Ludhiana (Fig. 3 and Table 2) indicates a consistent increase from 1994 to 2024. Densification is most evident in Zones 1, 2, 3, and 4. Zone 3 shows the highest change (+ 38.25%), increasing from 42.18 to 80.43%. Zones 1–3 represent older, more urbanized cores where densities already exceeded 40% in 1994 and these zones further exceeded 80% in 2024. Peripheral Zones (7–10), though showing lower absolute densities (from 14.94 to 19.91%) during 1994–2024, demonstrate significant relative growth, indicating suburban expansion. However, zones 1–6 experienced more than 23% of growth during 1994–2024.

Figure 4 shows that zone-1 has the highest urban density (%) in both 1994 and 2024 by 62.09% and 88.97% respectively. Also, zone-10 has the lowest urban density (%)

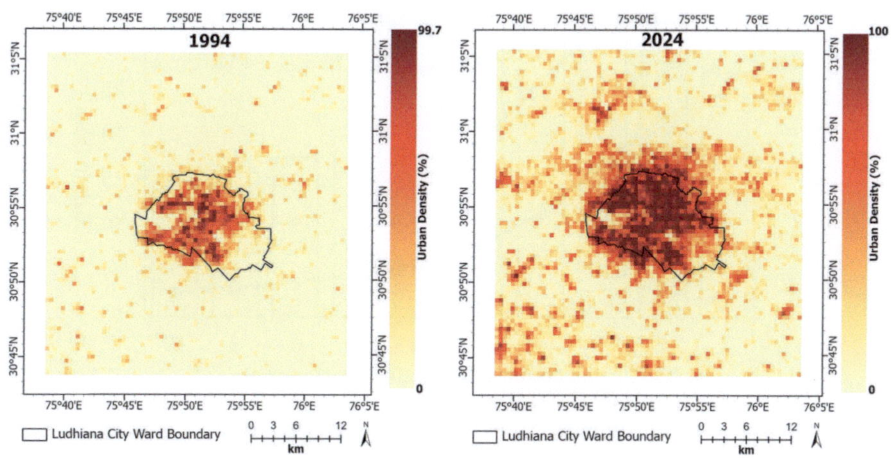

Fig. 3 Urban density maps of 1994 and 2024 for Ludhiana

Table 2 Distribution of urban density (%) in Ludhiana during 1994–2024

Zones	Urban density (%)		Difference of urban density (%)
	1994	2024	Δ 1994–2024
1	62.09	88.97	26.88
2	54.76	87.30	32.54
3	42.18	80.43	38.25
4	29.03	65.05	36.02
5	19.97	48.55	28.58
6	14.48	37.67	23.19
7	11.25	31.16	19.91
8	9.23	26.77	17.54
9	7.66	23.61	15.95
10	6.62	21.56	14.94

in both 1994 and 2024 by 6.62% and 21.56% respectively. However, the industrial areas have been observed and expanded in zones 2–5 (especially in south-east directions) during 1994–2024. Overall, it is evident that rapid urban growth has occurred from the city centre to the city periphery in the study area. Each zone crosses 20% of urban densities in 2024, which may be a concern for other LULC, especially for forest coverage and agriculture. Therefore, special attention should be given to zones 4–8, which have urban densities ranging from 26.77 to 65.05% in 2024, as these zones are undergoing significant densification.

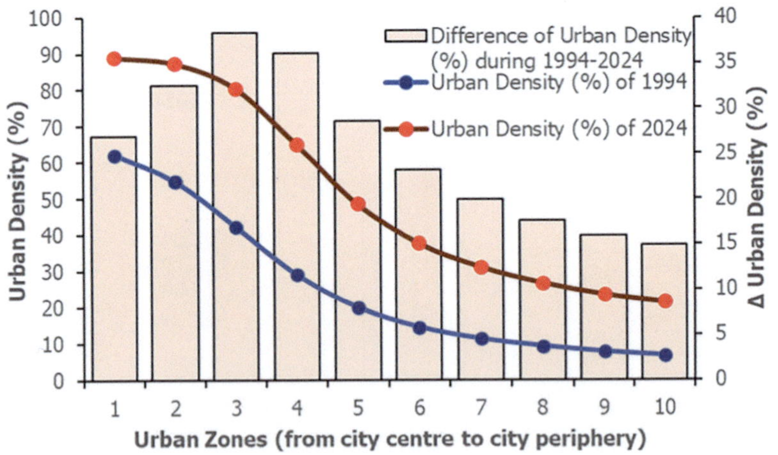

Fig. 4 Distribution of urban densities (%) and its magnitude over Ludhiana city at 2 km of intervals

3.2 Evaluation of Thermal State Intensification

Thermal state analysis over ten zones in Ludhiana (Fig. 4 and Table 3) shows a consistent intensification from 1994 to 2024. Thermal intensification is most evident in Zones 1–4 in 1994, but it shifts to zones 3–8. Zone 10 shows the highest mean LST change (+ 2.52 °C), increasing from 35.43 °C to 37.95 °C. Zones 1–4 represent intensified, more urbanized cores where mean LST already exceeded 36 °C in 1994 and these zones further exceeded 37.8 °C in 2024. Peripheral Zones (8–10), though showing lower LST intensification (nearly 2.5 °C) during 1994–2024, demonstrate significant relative intensification, indicating suburban thermal hotness. However, zones 4–10 experienced more than > 2 °C of intensification during 1994–2024.

Figure 5 shows that zone-5 has the highest mean LST intensification in 2024 by 38.17 °C but this zone was observed with 35.70 °C of mean LST in 1994. This is because of the presence of industrial areas in this zone (especially in south-east directions). Meanwhile, zone-1 has the lowest mean LST intensification in 2024 because of a significant amount of forest coverage. Overall, it is evident that a swift mean LST intensification has been observed from the city centre to the city periphery in the study area by 0.55 °C–2.52 °C during 1994–2024. Each zone crosses 37.7 °C of mean LST, which is very concerning for human livelihood and even for biodiversity.

Figure 6a depicts that urban density and mean LST have been decreasing from the city centre to the city periphery (except zone 1 where mean LST is slightly less than zone 2). This shows clear evidence of Urban Heat Island phenomena. However, Fig. 6b shows the same UHI evidence (except zone 1–3 where a significant amount of forest coverage found (especially in the east side of the zones)). Even Figs. 3 and 4 clearly depict that the lower LST is experienced in the non-urban areas (river and forest).

Table 3 Distribution of mean LST (°C) in Ludhiana city during 1994–2024

Zones	Mean LST (°C)		Difference of Mean LST (°C)
	1994	2024	Δ 1994–2024
1	37.19	37.80	0.61
2	37.41	37.96	0.55
3	36.89	38.02	1.13
4	36.07	38.15	2.08
5	35.70	38.17	2.47
6	35.71	38.10	2.39
7	35.65	38.09	2.44
8	35.54	38.03	2.49
9	35.49	37.98	2.49
10	35.43	37.95	2.52

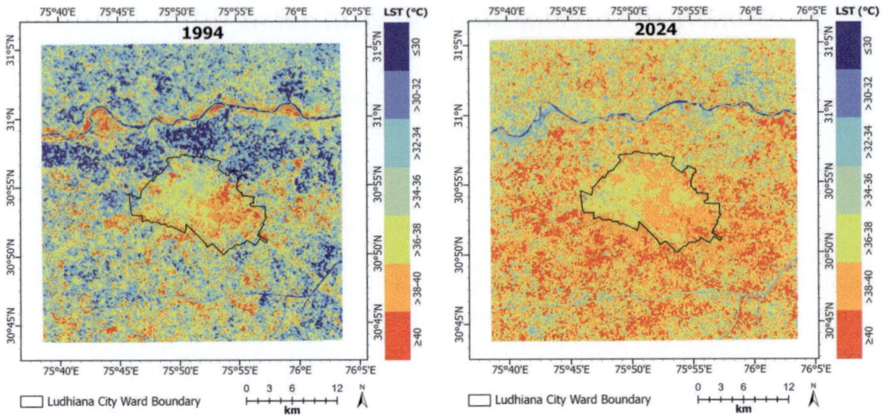

Fig. 5 Thermal state maps of 1994 and 2024 for Ludhiana city

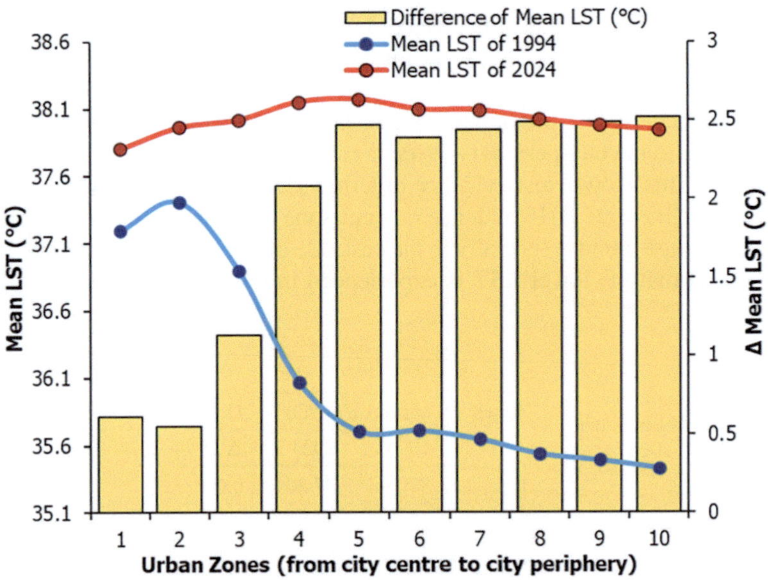

Fig. 6 Distribution of mean LST (°C) over urban zones along with its magnitude of Ludhiana city at 2 km of intervals

3.3 *Identification of UHI Locations*

The spatial distribution of UHI-affected areas in Ludhiana (Figs. 7, and 8; Table 3) reveals a consistent intensification of UHI phenomena between 1994 and 2024. UHI zones are identified based on land surface temperature (LST) anomalies, where values greater than 0 °C represent regions experiencing higher LST than rural areas and thus stronger UHI effects. Conversely, values less than or equal to 0 °C denote cooler areas where UHI effects are minimal or absent. The variation in UHI impact is closely linked to land cover configurations, with impervious surfaces and vegetative cover playing a significant role.

To facilitate a clearer understanding, UHI intensity has been categorized into six distinct classes: Very Cold (≤ − 2 °C) that typically associated with dense forests, water bodies, and wetlands; Cold (> − 2 °C to ≤ − 1 °C) that generally comprising moderately forested areas and moist soils; Cool (> − 1 °C to ≤ 0 °C) that includes grasslands and scrublands; Warm (> 0 °C to ≤ 1 °C) that often linked to transitional zones with limited vegetation; Hot (> 1 °C to ≤ 2 °C) that mostly impervious surfaces and fallow agricultural lands; and Very Hot (> 2 °C) that primarily barren land, sand, and heavily built-up areas.

Analysis over the 30-year period shows a stark decline of 222.30 km^2 (26.15%) in the Very Cold category, signaling a substantial loss of natural cooling sources such as forests and water bodies. Similarly, the Cold class decreased by 31.55 km^2 (1.97%), reflecting degradation in moderately vegetated and wet soil areas. These

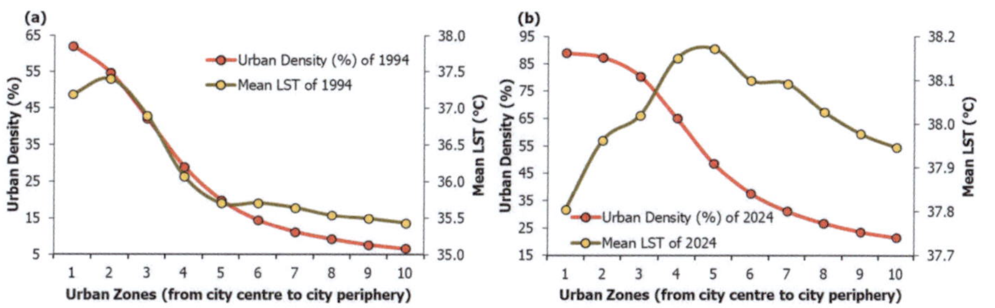

Fig. 7 Comparison of urban density with mean LST (°C): **a** 1994, and **b** 2024

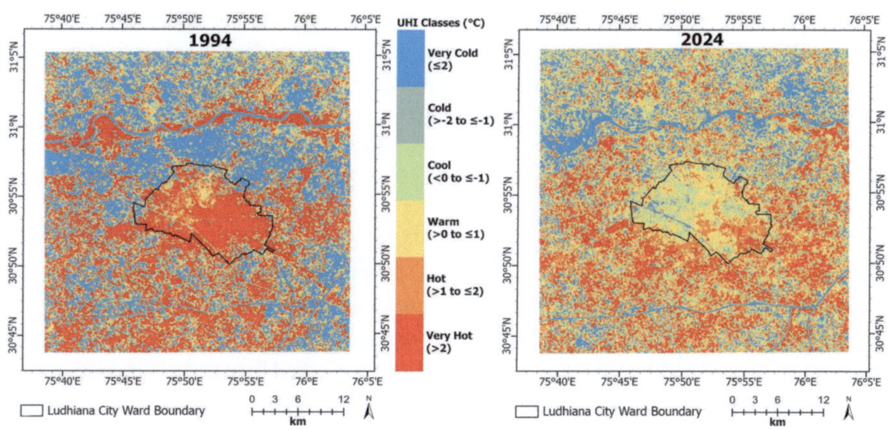

Fig. 8 Spatial distribution of UHI over Ludhiana

reductions are alarming, as they indicate a shrinking presence of natural elements critical for temperature regulation. On a different note, the Cool class expanded by 144.09 km^2 (9.01%). Although this increase appears positive at first glance, it primarily represents grasslands and scrublands, which offer limited thermal comfort compared to forested or water-covered areas.

Interestingly, the Very Hot class showed a notable decrease of 233.76 km^2 (14.61%), which is a promising development, suggesting a reduction in extreme heating surfaces like bare soil and impervious materials. However, this positive trend is tempered by the rise in the Hot and Warm classes. The Hot category increased by 178.17 km^2 (11.14%), comprising mainly impervious surfaces and abandoned agricultural lands—both significant contributors to UHI. Likewise, the Warm class expanded by 165.35 km^2 (10.33%), further confirming the spatial growth of moderately heat-affected zones, largely due to urbanization and land surface modification.

This UHI analysis highlights a concerning expansion of thermally uncomfortable areas across Ludhiana, driven by increasing impervious surface coverage (e.g., urban built-up areas, industrial sprawl, and concrete infrastructure), ongoing deforestation, and the proliferation of dry, non-vegetated lands. These land use transformations are exacerbating UHI intensity and diminishing the city's natural cooling capacity, posing significant challenges for urban sustainability and public health.

3.4 Extractions of Hotspots Assessment

The hotspot assessment using Getis-Ord G_i^* statistics is presented in Figs. 9 and 10. The results reveal that the areas classified as Very Cold Spot, Cold Spot, and Cool Spot have significantly decreased by 55.95 km^2 (3.50%), 15.55 km^2 (0.97%), and 16.05 km^2 (1.00%), respectively, between 1994 and 2024 (Table 4). This decline provides further evidence of the UHI effect, likely driven by the extensive loss of natural cooling elements such as forests and water bodies. These findings are concerning and need immediate attention. Additionally, the analysis shows that the Hot Spot area increased by 35.38 km^2 (2.21%), indicating the emergence of new heat-concentrated zones. On the other hand, it is somewhat reassuring that the Warm Spot and Very Hot Spot categories have decreased by 22.80 km^2 (1.43%) and 59.81 km^2 (3.74%), respectively.

The results clearly indicate that the "Not Significant" class increased by 134.79 km^2 (8.42%) between 1994 and 2024. This highlights that a larger proportion of land has shifted into this category, suggesting a reduction in the spatial clustering of extreme temperatures and possible climate homogenization or increased variability. This transition points toward a warming trend accompanied by a redistribution, rather than an amplification, of temperature extremes (Table 5 and Fig. 11).

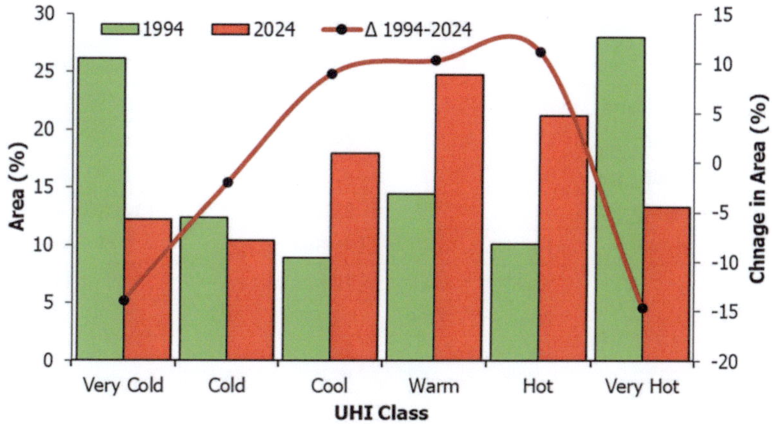

Fig. 9 Distribution and magnitude of classes of UHI in Ludhiana

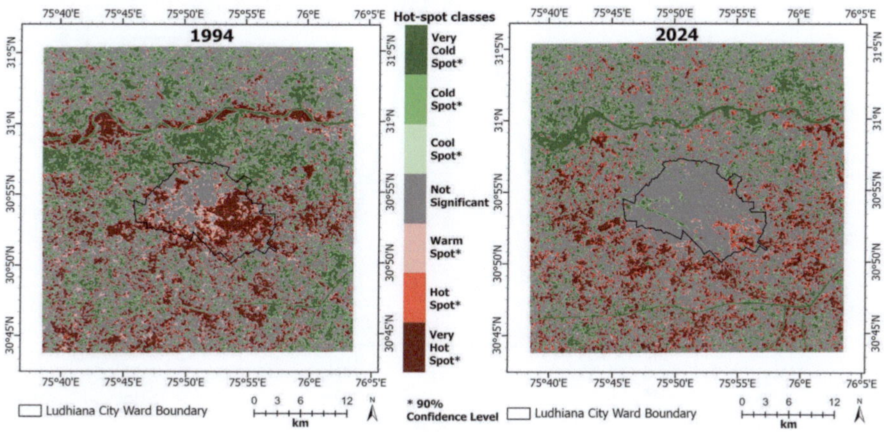

Fig. 10 Spatial distribution of hotspot maps using Getis-Ord G_i^* of Ludhiana

Table 4 Classes of UHI to understand its criticality in Ludhiana

Sl. no.	Class	1994	2024	Δ 1994–2024
1	Very cold (< = − 2)	418.55 (26.16%)	196.25 (12.27%)	− 222.30 (− 13.89%)
2	Cold (> − 2 to < = − 1)	198.18 (12.39%)	166.63 (10.41%)	− 31.55 (− 1.97%)
3	Cool (> − 1 to < = 0)	143.34 (8.96%)	287.43 (17.96%)	144.09 (9.01%)
4	Warm (> 0 to < = 1)	231.10 (14.44%)	396.45 (24.78%)	165.35 (10.33%)
5	Hot (> 1 to < = 2)	161.67 (10.10%)	339.84 (21.24%)	178.17 (11.14%)
6	Very hot (> 2)	447.16 (27.95%)	213.40 (13.34%)	− 233.76 (− 14.61%)
Total		1600 (100%)	1600 (100%)	–

Table 5 Different classes of Hotspot maps using Getis-Ord G_i^* of Ludhiana

Sl. no	Class	1994	2024	Δ 1994–2024
1	Very cold spot	245.32 (15.33%)	189.37 (11.84%)	– 55.95 (– 3.50%)
2	Cold spot	89.69 (5.61%)	74.14 (4.63%)	– 15.55 (– 0.97%)
3	Cool spot	63.07 (3.94%)	47.02 (2.94%)	– 16.05 (1.00%)
4	Not significant	796.53 (49.78%)	931.32 (58.21%)	134.79 (8.42%)
4	Warm spot	108.16 (6.76%)	85.36 (5.33%)	– 22.80 (– 1.43%)
5	Hot spot	93.91 (5.87%)	129.28 (8.08%)	35.38 (2.21%)
6	Very hot spot	203.31 (12.71%)	143.50 (8.97%)	– 59.81 (– 3.74%)
Total		1600 (100%)	1600 (100%)	–

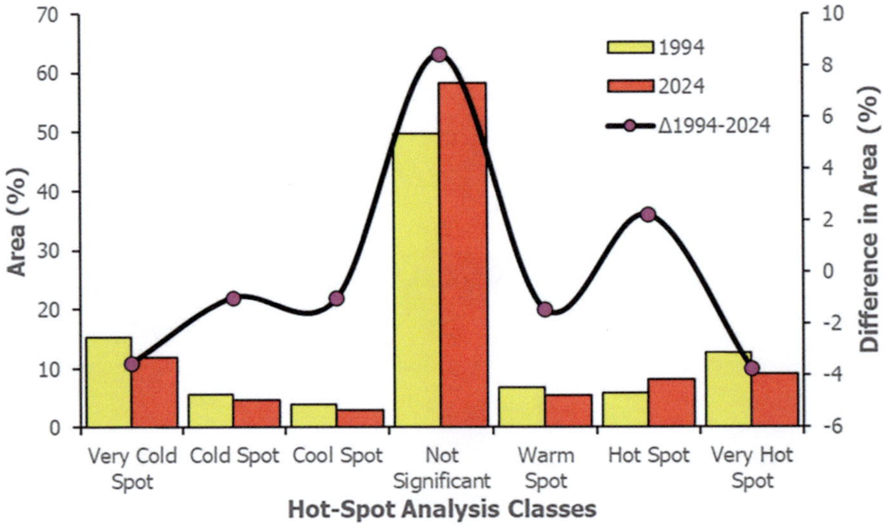

Fig. 11 Hotspot classes using Getis-Ord G_i^* over Ludhiana

4 Discussion

4.1 Urbanization's Role in UHI Intensification

The findings from this study clearly demonstrate that urbanization has been a major driver of UHI intensification in Ludhiana over the past three decades. The built-up density analysis indicates a substantial increase in impervious surfaces across all urban zones, particularly in Zones 1–5, where urban densities exceeded 48% by 2024. This urban expansion, often occurring at the expense of vegetated and agricultural lands, has led to an increase in LST, contributing to intensified UHI effects.

Industrialization in Ludhiana, concentrated in Zones 2–5 (especially in north-east directions), further exacerbates the UHI phenomenon through heat emissions, sealed surfaces, and a lack of green infrastructure. As seen in the spatial LST trends, zones with higher built-up densities consistently experienced greater thermal intensification, with Zone 5 registering the highest mean LST in 2024. The transformation of vegetated land into industrial and residential developments reduces surface albedo and evapotranspiration capacity, both of which are critical for natural cooling.

The significant decline in cool spot, cold spot and very cold spot areas (a cumulative decrease of 87.55 km^2 (5.47%)), along with the rise in hot spot regions, illustrates the thermal consequences of unregulated urban expansion. Moreover, hotspot and UHI analyzes confirm the emergence of new thermal risk zones in peripheral and industrial neighborhoods, suggesting that heat stress is no longer confined to core city areas. Collectively, these patterns reveal how rapid, poorly managed urbanization directly fuels the intensification and spatial spread of UHI.

4.2 Strategies of Urban Planning Approach to Improve Urban Livability

The growing evidence of UHI expansion in Ludhiana underscores an urgent need to rethink urban planning frameworks to enhance urban livability and climate resilience. As the city's urban density and industrial footprint continue to expand, strategic, evidence-based urban planning is essential to mitigate the health, ecological, and energy impacts of rising urban temperatures (Obateru et al. 2024; Sen et al. 2019).

Urban heat mitigation must be embedded into city development strategies through the promotion of green infrastructure, such as urban forests, green roofs, and parks (Kumar et al. 2024). Given the significant loss of natural cooling zones identified in this study, restoration of ecological (blue-green infrastructure) buffers, particularly in areas with high thermal intensity, can help rebalance the urban thermal environment (Islam et al. 2024). Additionally, incorporating urban form metrics, such as building density, orientation, vegetation, and open space ratios, into land-use planning can facilitate airflow and reduce localized heat buildup (Zhang et al. 2025). These strategies can be strictly implemented in zones 4–8, as they are undergoing significant densification, with urban density ranging from 26.77 to 65.05% as of 2024. These zones have also experienced a higher mean LST increase of 2.08 °C–2.49 °C over the past 30 years, from 1994 to 2024.

Zoning regulations must prioritize compact, mixed-use development interspersed with adequate green and blue spaces. The planning of industrial zones should include green buffers and reflective materials to minimize heat emissions. Furthermore, the insights into hotspot locations demonstrated in this research can support proactive planning by identifying future heat hotspots and enabling early interventions.

By integrating climate-sensitive planning tools and geospatial-AI technologies into municipal governance, Ludhiana can move towards a more sustainable, equitable, and thermally comfortable urban future (Alastal and Shaqfa 2022). Ultimately, a multi-disciplinary and anticipatory planning approach is critical to balance urban growth with environmental resilience and public health.

5 Conclusion

This study presents a comprehensive spatiotemporal evaluation of UHI dynamics in Ludhiana, a mid-sized industrial city in northern India, over a 30-year period (1994–2024). By integrating remote sensing, GIS, and machine learning techniques, the research captures how rapid urban expansion and land surface modifications have transformed the city's thermal landscape. The urban density analysis reveals a significant increase in impervious surface coverage (> 20% in all zones), particularly in Zones 1–3, where urban density exceeded 80% by 2024. Parallel to this, a notable rise in mean LST was observed, ranging from 0.55 °C to 2.52 °C across various urban zones (Zone 1–Zone 10).

A key novelty of this study lies in its integrated methodological approach by combining urban density mapping, LST-based UHI profiling, and Getis-Ord G_i^* hotspot analysis, to identify and spatially characterize thermal stress patterns. The results indicate a pronounced shift in thermal hotspots from central urban zones to newly urbanized industrial and peripheral areas, especially in zones 4–8. The inclusion of UHI modeling using historical LST and spatial urban density adds a future-oriented perspective, enabling the identification of potential thermal risk zones that may emerge due to ongoing urban and industrial expansion.

The findings underscore the urgent need for climate-sensitive urban planning in Ludhiana. The loss of natural cooling zones, such as forests and water bodies, and the spatial growth of high-temperature zones call for targeted interventions. Strategies such as green infrastructure development, heat-resilient urban design, and the regulation of industrial sprawl are essential to curb further UHI intensification and improve urban livability. This study provides not only a replicable analytical framework for UHI assessment but also actionable insights to guide sustainable and thermally comfortable urban development.

References

Abdullahi S, Pradhan B, Mansor S, Shariff ARM (2015) GIS-based modeling for the spatial measurement and evaluation of mixed land use development for a compact city. Giscience Remote Sens 52(1):18–39. https://doi.org/10.1080/15481603.2014.993854

Abutaleb K, Ngie A, Darwish A, Ahmed M, Arafat S, Ahmed F (2015) Assessment of urban heat island using remotely sensed imagery over Greater Cairo Egypt. Adv Remote Sens 4:35–47. https://doi.org/10.4236/ars.2015.41004

Alaoui ME, EL Amraoui K, Masmoudi L, Ettouhami A, Rouchdi M (2024) Unleashing the potential of IoT, artificial intelligence, and UAVs in contemporary agriculture: a comprehensive review. J Terrramech 115:100986. https://doi.org/10.1016/j.jterra.2024.100986

Alastal AI, Shaqfa AH (2022) GeoAI technologies and their application areas in urban planning and development: concepts, opportunities and challenges in smart city (Kuwait, study case). J Data Anal Inf Process 10(02):110–126. https://doi.org/10.4236/jdaip.2022.102007

Ambrosino A, Di Benedetto A, Fiani M (2023) LiDAR data and HRSI to evaluate the mitigating effect of forests into rockfall risk analysis using SOM: Mt San Liberatore case study. Remote Sensing 15(18):1–22. https://doi.org/10.3390/rs15184523

Anderson JR, Hardy EE, Roach JT, Witmer RE (1976) A land use and land cover classification system for use with remote sensor data. In: A revision of the land use classification system as presented in U.S. Geological Survey Circular 671. Washington.

Bokaie M, Zarkesh MK, Arasteh PD, Hosseini A (2016) Assessment of urban heat island based on the relationship between land surface temperature and land use/land cover in Tehran. Sustain Cities Soc 23:94–104. https://doi.org/10.1016/j.scs.2016.03.009

Chakraborty S, Pramanik S, Follmann A, Giri B, Mondal B, Patel PP et al (2021) Dominant urban form and its relation to nighttime land surface temperature in the rapidly urbanizing National Capital Region of India. Urban Climate 40(101002):1–15. https://doi.org/10.1016/j.uclim.2021.101002

DECC (2019) Action plan for cleaner air. https://cpcb.nic.in/Actionplan/Ludhiana.pdf

Govind NR, Ramesh H (2019) The impact of spatiotemporal patterns of land use land cover and land surface temperature on an urban cool island: a case study of Bengaluru. Environ Monit Assess 191(283):1–20

Guha S, Govil H (2021) Annual assessment on the relationship between land surface temperature and six remote sensing indices using landsat data from 1988 to 2019. Geocarto Int 1–20. https://doi.org/10.1080/10106049.2021.1886339

Gumma MK, Mohammad I, Nedumaran S, Whitbread A, Lagerkvist CJ (2017) Urban sprawl and adverse impacts on agricultural land: A case study on Hyderabad India. Remote Sens 9(11):1–16. https://doi.org/10.3390/rs9111136

Habibi R, Alesheikh AA (2017) An assessment of spatial pattern characterization of air pollution: a case study of CO and PM2.5 in Tehran, Iran. ISPRS Int J Geo-Inf 6(9):270. https://doi.org/10.3390/ijgi6090270

Halder B, Bandyopadhyay J, Banik P (2021) Monitoring the effect of urban development on urban heat island based on remote sensing and geo-spatial approach in Kolkata and adjacent areas, India. Sustain Cities Soc 74. https://doi.org/10.1016/j.scs.2021.103186

He T, Liang S, Wang D, Caȯ Y, Gao F, Yu Y, Feng M (2018) Remote sensing of environment evaluating land surface albedo estimation from landsat MSS, TM, ETM +, and OLI data based on the uni fi ed direct estimation approach. Remote Sens Environ 204:181–196. https://doi.org/10.1016/j.rse.2017.10.031

Hui CX, Dan G, Alamri S, Toghraie D (2023) Greening smart cities: An investigation of the integration of urban natural resources and smart city technologies for promoting environmental sustainability. Sustain Cities Soc 99:104985. https://doi.org/10.1016/j.scs.2023.104985

IMD (2010) Climatological tables of observatories in India (1981–2010). https://imdpune.gov.in/library/public/1981-2010%20CLIM%20NORMALS%20%28STATWISE%29.pdf

Islam MR, Shahfahad Talukdar S, Rihan M, Rahman A (2024) Evaluating cooling effect of blue-green infrastructure on urban thermal environment in a metropolitan city: using geospatial and machine learning techniques. Sustain Cities Soc 113. https://doi.org/10.1016/j.scs.2024.105666

Jain S, Sannigrahi S, Sen S, Bhatt S, Chakraborti S, Rahmat S (2019) Urban heat island intensity and its mitigation strategies in the fast-growing urban area. J Urban Manag 1–14. https://doi.org/10.1016/j.jum.2019.09.004

Johny A, Mary M (2016) Geoinformatics in sprawl modelling—an overview, pp 79–86

Kumar P, Debele SE, Khalili S, Halios CH, Sahani J, Aghamohammadi N et al (2024) Urban heat mitigation by green and blue infrastructure: drivers, effectiveness, and future needs. Innovation 5(2). https://doi.org/10.1016/j.xinn.2024.100588

Macharia CW, Kiage L (2025) Examining the nexus of social vulnerability, land cover dynamics, and heat exposure in Reno, Nevada, USA. Remote Sens Appl: Soc Environ 37:1–15. https://doi.org/10.1016/j.rsase.2024.101400

Majumder A, Setia R, Kingra PK, Sembhi H, Singh SP, Pateriya B (2021) Estimation of land surface temperature using different retrieval methods for studying the spatiotemporal variations of surface urban heat and cold islands in Indian Punjab. Environ Dev Sustain 23(11):15921–15942. https://doi.org/10.1007/s10668-021-01321-3

Mal S, Rani S, Maharana P (2020) Estimation of spatio-temporal variability in land surface temperature over the Ganga River Basin using MODIS data. Geocarto Int 1–23. https://doi.org/10.1080/10106049.2020.1869331

Mansour S, Al-belushi M, Al-awadhi T (2020) Land use policy monitoring land use and land cover changes in the mountainous cities of Oman using GIS and CA-Markov modelling techniques. Land Use Policy 91:1–15. https://doi.org/10.1016/j.landusepol.2019.104414

Matloob A, Sarif MO, Um J-S (2021) Exploring correlation between OCO-2 XCO_2 and DMSP/OLS nightlight imagery signature in four selected locations in India. Spat Inf Res 29(1):123–135. https://doi.org/10.1007/s41324-021-00381-x

Moss JL, Doick KJ, Smith S, Shahrestani M (2019) Influence of evaporative cooling by urban forests on cooling demand in cities. Urban For Urban Green 37:65–73. https://doi.org/10.1016/j.ufug.2018.07.023

Naikoo MW, Rihan M, Ishtiaque M, Shahfahad (2020) Analyses of land use land cover (LULC) change and built-up expansion in the suburb of a metropolitan city: spatio-temporal analysis of Delhi NCR using landsat datasets. J Urban Manag 9(3):347–359. https://doi.org/10.1016/j.jum.2020.05.004

Obateru RO, Okhimamhe AA, Fashae OA, Aweda E, Dragovich D, Conrad C (2024) Community-based assessment of the dynamics of urban landscape characteristics and ecosystem services in the rainforest and guinea savanna ecoregions of Nigeria. J Environ Manag 360. https://doi.org/10.1016/j.jenvman.2024.121191

Pandey P, Kumar D, Prakash A, Masih J, Singh M, Kumar S et al (2012) A study of urban heat island and its association with particulate matter during winter months over Delhi. Sci Total Environ 414:494–507. https://doi.org/10.1016/j.scitotenv.2011.10.043

Pham HV, Dal Barco MK, Cadau M, Harris R, Furlan E, Torresan S et al (2023) Multi-model chain for climate change scenario analysis to support coastal erosion and water quality risk management for the Metropolitan city of Venice. Sci Total Environ 904:166310. https://doi.org/10.1016/j.scitotenv.2023.166310

Sarif MO, Gupta RD, Sharifi A (2024) Predicting Prayagraj's urbanization trajectory using CA-ANN Modelling: population pressures and land use dynamics. J Environ Manage 370:1–5. https://doi.org/10.1016/j.jenvman.2024.122469

Sarif MO, Rimal B, Stork NE (2020) Assessment of changes in land use/land cover and land surface temperatures and their impact on surface urban heat island phenomena in the Kathmandu Valley (1988–2018). ISPRS Int J Geo Inf 9(12):1–29. https://doi.org/10.3390/ijgi9120726

Sen S, Roesler J, Ruddell B, Middel A (2019) Cool pavement strategies for urban heat island mitigation in suburban Phoenix Arizona. Sustainability 11(16):4452. https://doi.org/10.3390/su11164452

Shahfahad Kumari B, Tayyab M, Ahmed IA, Baig MRI, Khan MF, Rahman A (2020) Longitudinal study of land surface temperature (LST) using mono- and split-window algorithms and its relationship with NDVI and NDBI over selected metro cities of India. Arab J Geosci 13(1040):1–19. https://doi.org/10.1007/s12517-020-06068-1

Shankar H, Singh D, Chauhan P (2024) Investigation of groundwater induced land subsidence in Ludhiana city using InSAR and Sentinel-1 data. Quat Sci Adv 13:1–12. https://doi.org/10.1016/j.qsa.2023.100151

Sharma GK, Ghuge VV (2024) How urban growth dynamics impact the air quality? A case of eight Indian metropolitan cities. Sci Total Environ 930. https://doi.org/10.1016/j.scitotenv.2024.172399

Shukla PR, Skea J, Buendia EC, Masson-Delmotte V, Pörtner H-O, Roberts DC et al (2019) Climate change and land: an IPCC special report on climate change, desertification, land degradation, sustainable land management, food security, and greenhouse gas fluxes in terrestrial ecosystems. Available at https://www.ipcc.ch/site/assets/uploads/20. IPCC (AR6). https://doi.org/10.4337/9781784710644

Singh R, Kalota D (2019) Urban sprawl and its impact on generation of urban heat island: a case study of Ludhiana city. J Indian Soc Remote Sens 47(9):1567–1576. https://doi.org/10.1007/s12524-019-00994-8

Suppa AR, Ballarini I (2023) Supporting climate-neutral cities with urban energy modeling: a review of building retrofit scenarios, focused on decision-making, energy and environmental performance, and cost. Sustain Cities Soc 98:104832. https://doi.org/10.1016/j.scs.2023.104832

UN (2018) The world's cities in 2018. Departmentof economic and social affairs, population division (2018). In: The world's cities in 2018—data booklet (ST/ESA/ SER.A/417). UN

Zhang L, Bagan H, Chen C, Yoshida T (2025) Exploring the impact of urban morphology on river cooling effects: a case study of the Arakawa river in Tokyo. Ecological Indicators 172. https://doi.org/10.1016/j.ecolind.2025.113288

Urban Expansion and Thermal Environment in a Medium Size City: Role of Urban Green and Blue Spaces in Mitigating Surface Urban Heat Island Intensity

Shahfahad, Pratik Kumar Mishra, Md. Rejaul Islam, Rambooshan Tiwari, Vinod Kumar Tripathi, Ahmad Mujtaba Siddiqui, and Atiqur Rahman

1 Introduction

The rise in the percentage of people residing in towns and cities is known as urbanization. At the same time, urban expansion is the outward growth of city or town into surrounding areas, often accompanied by a rising population, infrastructure

Shahfahad (✉) · P. K. Mishra · R. Tiwari · V. K. Tripathi
Department of Geography, Institute of Science, Banaras Hindu University, Varanasi, Uttar Pradesh, India
e-mail: sfahad@bhu.ac.in

P. K. Mishra
e-mail: pratik26@bhu.ac.in

R. Tiwari
e-mail: rambo_bhu@rediffmail.com

V. K. Tripathi
e-mail: vktri@bhu.ac.in

Md. R. Islam · A. Rahman
Department of Geography, Faculty of Sciences, Jamia Millia Islamia, New Delhi, India
e-mail: mdrejaulislam23@gmail.com

A. Rahman
e-mail: arahman2@jmi.ac.in

A. M. Siddiqui
Department of Geography, Faculty of Sciences, Aligarh Muslim University, Aligarh, Uttar Pradesh, India
e-mail: ahmad.Mujtaba73@gmail.com

M. O. Sarif and A. Sharifi (eds.), *Advanced Geospatial Intelligence and AI for Environmental Resilience and Sustainable Development*, Advances in Geographic Information Science, https://doi.org/10.1007/978-3-032-03714-5_2

development and economic growth (Angel et al. 2011; Naikoo et al. 2024). The urban expansion is a crucial aspect of the modern-day urbanization which results in reshaping the urban landscape and influencing thermal environment of cities and towns. As cities expand outwards, natural vegetation, wetlands, and agricultural lands gets replaced by impervious surfaces such as roads, buildings, and other infrastructures (Novotný et al. 2022; Shahfahad et al. 2024). This transformation leads to profound environmental consequences, including rising land surface temperature (LST) and the intensifying the surface urban heat island (SUHI) effect (Sarif et al. 2020; Arshad et al. 2022).

Due to persistent urban expansion, cities are witnessing significant rise in LST which ultimately led to intensification of SUHI phenomenon (Yang and Zhao 2023; Peng et al. 2024). This is because the urban areas, with dense built-up surfaces, absorb and retain more heat than natural landscapes, leading to rising LST and SUHI (Hidalgo-García and Arco-Díaz 2022; Talukdar et al. 2024). Especially during heatwaves, this phenomenon worsens thermal discomfort, raises the energy required for cooling, and has a negative impact on health (Mohammad and Weng 2025). In medium and small-sized cities, the effect of SUHI can be severe because of rapid and unregulated urban expansion and limited green infrastructure (Mentaschi et al. 2022). Hence, identifying the effective mitigation strategies are essential for reducing urban heat stress and enhancing thermal comfort in these growing urban areas (Wang and Aktas 2024; Cuce et al. 2025).

Urban green and blue spaces (UGBS) such as parks, urban forests, water bodies, and wetlands, are widely regarded as crucial solutions to mitigate SUHI effects (Ebi and Bowen 2023; Islam et al. 2024). Urban green spaces like parks and dense vegetated areas regulates urban microclimate by providing shade, increasing evapotranspiration, and reducing heat accumulation in built-up areas (Bobo Merga et al. 2024). Similarly, the water bodies in urban areas helps in mitigating SUHI through evaporative cooling, acting as thermal buffers that moderate surrounding temperatures (Marquez-Torres et al. 2025). Several studies have evaluated the mitigating effects of UGBS in urban environment. For instance, Murtinová et al. (2022) noted that the densely urbanized areas have about 3.5 °C high LST than the UGBS in a medium sized town of Slovakia. Islam et al. (2024) evaluated the cooling effect of UGBS and noted that UGBS significantly mitigate SUHI although the spatial configuration of UGBS the landscape pattern within and surrounding areas also plays important role. Qiu et al. (2023) examined the cooling effect of UGBS in urban environment and noted that the size of UGBS have significant impact on their cooling intensity. Similarly, Gunawardena et al. (2017) noted that the magnitude of cooling effects of urban green spaces depends on their geometry, size and spread.

Although, studies have examined the effectiveness of UGBS in mitigating SUHI effect in urban areas, most of these studies have been done for large metropolitan cities while medium and small sized cities being less studied. Moreover, in India, none of the studies have been done to examine the role of UGBS in mitigating SUHI effect for medium sized cities like Darbhanga. The medium and small sized cities of developing nations like India are witnessing very fast urban expansion (Dhanaraj and Angadi 2021; Shahfahad et al. 2025) which may have significant impacts on the

SUHI. Therefore, identifying the underlying factors of rising SUHI and its mitigation measures is crucial for sustainable urban planning and heat mitigation. Thus, this study aims to examine the spatio-temporal pattern of urban expansion and its impact on UGBS and SUHI effect in Darbhanga city of Bihar.

2 Study Area

Darbhanga city is a medium sized town located in northern part of Bihar on 26.17° N latitude and 85.9° E longitude (Fig. 1). The city is located on the bank of Baghmati River, an important tributary of Ganga. The city is the headquarter of Darbhanga district which is known as the cultural capital of Bihar state. Darbhanga has a unique historical and cultural significance since ancient times. It was home of Darbhanga Raj, a princely state which ruled most of northern Bihar during medieval period and British rule in India. Darbhanga city has an area of about 46 km^2 and is home of more than 5 lakh population as per latest estimate. Darbhanga has witnessed very fast increase in population during last few decades which has resulted in significant alteration of its LULC pattern. The surroundings of city are mostly agrarian and is drained by a number of small and large rivers which makes the land fertile by depositing alluvium. The climate of region is humid sub-tropical type (Koppen-Cfa) with annual rainfall averaging between 150 and 200 cm. The winter temperature of Darbhanga ranges between 10 and 25 °C while summer temperature averaging between 35 and 40 °C which sometimes reaches up to 45 °C. Due to this Darbhanga city witnesses extreme heat-waves although, city have several small and large sized artificial lakes which significantly mitigate the heat-waves.

3 Data Sources and Methods

3.1 Data Source

In this study, Landsat 5 (TM) data for 1991, 2001 and 2011 while Landsat 8 (OLI/TIRS) for 2021 have been used for LULC mapping and LST retrieval. The Landsat datasets were downloaded from the USGS Earth Explorer website. The datasets were selected for the month of March and April to avoid any king of error caused by cloud cover or haze as the region have tropical humid type climate and hence clouds and haze are frequent. The details of satellite data used are given in Table 1.

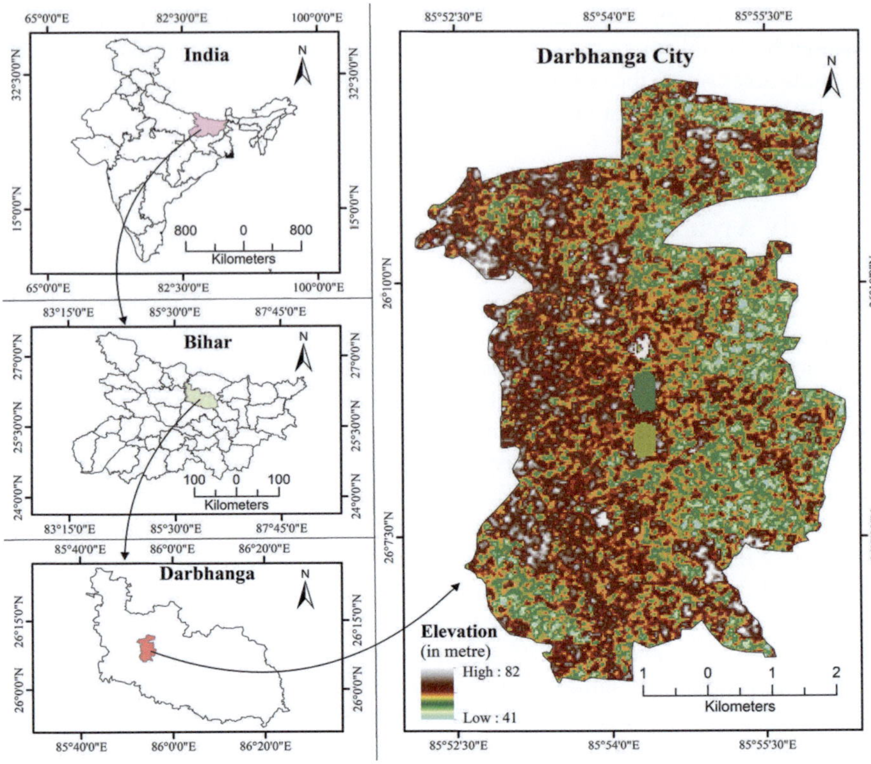

Fig. 1 Location map of study area: Darbhanga city

Table 1 Details of the Landsat data used in the study

S. no	Satellite	Sensor	Date of Acquisition	Path/row	Spatial resolution (m)	Cloud cover (%)
1	Landsat 5	TM	20-03-1991	140/42	30 (120*)	0.00
2	Landsat 5	TM	07-04-2001	140/42	30 (120*)	0.00
3	Landsat 5	TM	27-03-2011	140/42	30 (120*)	1.00
4	Landsat 8	OLI/TIRS	22-03-2021	140/42	30 (100*)	0.16

3.2 Satellite Data Pre-processing

The Landsat datasets downloaded from USGS website were subject to pre-processing which includes making of composite images, taking subset of study area, as well as eliminating the errors caused due to atmospheric and radiometric calibration errors. Thus, firstly, layer stacking of individual bands of Landsat datasets were done to create false colour composite (FCC) and then subset of study area was taken in ERDAS Imagine software. For eliminating the atmospheric and radiometric errors from satellite data, Atmospheric and Topographic Correction (ATCOR) was used which was integrated in ERDAS Imagine software. ATCOR module significantly improves the radiometric fidelity of satellite imageries and ensures consistency across spatio-temporal scales of the imageries for a reliable comparison over space and time. ATCOR module effectively reduce the inaccuracies and errors caused due to sensor and atmospheric factors (Shahfahad et al. 2025).

3.3 LULC Classification and Accuracy Assessment

Studies have used several techniques for LULC classification from satellite including supervised and unsupervised techniques (Talukdar et al. 2020; Shahfahad et al. 2023a). Traditionally, the LULC classification from satellite data were done using pixel-based classification techniques which were time consuming and relied on spectral signature of individual pixels. These techniques include supervised classification techniques like Maximum Likelihood Classifier (MLC) and Mahalanobis Distance (MD) and unsupervised techniques like ISODATA and K-means clustering. Although, techniques were time consuming, they produced reliable outcomes with satisfactory accuracy. Later, advanced techniques were introduced for LULC classification such as Object-Based Image Analysis (OBIA) as well as artificial intelligence (AI) based machine learning (ML) and deep learning (DL) techniques. These techniques produce high accuracy LULC maps in quick time and hence, have become popular in a short period of time. In this study, Spectral Angle Mapper (SAM) techniques have been used for LULC classification which is ML based supervised classification technique. SAM computes the angle between a reference spectra and spectral signature of a pixel. A slight angle specifies a sophisticated comparation, leading to class alignment. The SAM is effective in LULC classification from multi-spectral datasets as it is less sensitive to radiance variation and can classify LULC based on spectral shape rather than absolute reflectance values.

3.4 LST Retrieval and SUHII Mapping

To retrieve LST from thermal bands of Landsat data, a number of steps have been performed such as conversion of DN number to spectral radiance, emissivity correction through vegetation index, and conversion to spectral radiance to brightness temperature (Fan et al. 2017; Al Shawabkeh et al. 2025). In this study, firstly, the DN number in thermal band of Landsat 5 was converted into spectral radiance using Eq. 1.

$$L_\lambda = \left\{ \frac{L_{MAX\lambda} - L_{MIN\lambda}}{Q_{CalMAX} - Q_{CalMIN}} \right\} \times (Q_{Cal} - Q_{CalMIN}) + L_{MIN\lambda} \quad (1)$$

where, L_λ denotes the spectral radiance, while $L_{MAX\lambda}$ and $L_{MIN\lambda}$ correspond to sensor's maximum and minimum spectral radiance, respectively, scaled according to Q_{CalMAX} and Q_{CalMIN}. Q_{Cal} refers to the calibrated quantized pixel value.

From Landsat 8, the DN number to spectral radiance was done using Eq. 2.

$$L_\lambda = M_L \times Q_{Cal} + A_L \tag{2}$$

where, M_L represents the multiplicative rescaling factor associated with thermal band, Q_{Cal} denotes calibrated quantized value of pixel, and A_L denotes additive rescaling factor associated with thermal band.

Additionally, the retrieval of brightness temperature employing the calibration constant at the Top-of-Atmosphere was conducted using Eq. 3.

$$BT = \frac{K_2}{\ln\left(\frac{K_1}{L_\lambda} + 1\right)} \tag{3}$$

where, BT denotes the brightness temperature (Kelvin), L_λ signifies radiance, while K_1 and K_2 are the thermal calibration constants given in metadata.

The method for emissivity correction for LST retrieval was proposed by Sobrino et al. (2004). In this, firstly, we calculate proportion of vegetation ($P_{v)}$ as given in Eq. 4.

$$P_v = \left[\frac{NDVI - NDVI_{min}}{NDVI_{max} - NDVI_{min}}\right]^2 \tag{4}$$

Finally, emissivity (ε) correction is done using Eq. 5.

$$\varepsilon = 0.004P_V + 0.986 \tag{5}$$

Using corrected emissivity (ε) and brightness temperature (BT), the LST calculation is done using Eq. 6 given by Artis and Carnahan (1982).

$$T = \left[\frac{BT}{1 + \left(\lambda \times \frac{BT}{P}\right)\ln(\varepsilon)}\right] \tag{6}$$

where, T represents retrieved LST, BT denkotes brightness temperature, λ refers to wavelength of radiance emitted, $P = h \times c/s$; in which h represents Planck's constant (6.624×10^{-34} J s), c denotes velocity of the light (2.998×10^8 m/s), while s denotes Boltzmann constant (1.38×10^{-23} J/K) and ε is the corrected emissivity.

Traditionally, the rural–urban temperature difference is used for modeling the SUHII in cities. However, in recent years, several have used LST deviation technique for modeling SUHII (Filho and Karam, 2016; Hidalgo-García and Arco-Díaz 2022; Shahfahad et al. 2023b). The LST deviation technique uses mean LST of urban area as a limit for marking SUHI and non-SUHI zones as well as determining the SUHII.

In this research, the LST deviation technique is used for the modeling of as given in Eq. 7.

$$SUHII = LST(°C) - LST_{mean}(°C) \quad (7)$$

where, LST(°C) is the LST of a given pixel and LST_{mean} (°C) is the mean LST of all the pixels.

3.5 Method for Assessing Linkages Between LULC Types and SUHII

To establish linkages between SUHII and LULC pattern, we used violin plots in this study. Violin plots are data visualization techniques which combines the features of box plot and kernal density plot which makes it very effective in examining the distribution and variation among variables. The width of violin plots shows the frequency of data points which helps in identifying if the distribution in bimodal, skewed, or unimodal. Unlike traditional box plots, violin plots give a complete insight, especially for the large datasets by showing the complete distribution. In this research, the violin plots were created using python Matplotlib feature for visualizing the distribution and variation in SUHII across LULC types for the years 1991, 2001, 2011, and 2021. For this, the data of SUHII and LULC were taken for the respective years and then structured with LULC being classified in six categories. The violin plots visualize the distribution and density of SUHII values for each LULC types, depicting the median, interquartile and density of SUHII values. The plots allowed for a comparative investigation of SUHII variation across LULC categories, highlighting role of vegetation cover and water bodies in mitigation of SUHII.

4 Results and Discussion

4.1 Spatio-Temporal Variation in LULC Pattern

Over the last three decades (1991–2021), Darbhanga city has experienced an identifiable urban expansion, with built-up areas gradually displacing open spaces, farmland, and shrubland (Fig. 2). The city's core witnessed the maximum growth at first, but as time passed, the central, western, and southern regions have witnessed the highest growth in built-up areas. In the meantime, only slight changes have occurred in the northern and the periphery regions. The conversion of open land, shrubland, and cropland into built-up areas has been the main trend in the whole region. In contrast, LULC changes in the northern and northeastern regions have mostly involved the conversion of shrubland and cropland into open and built-up areas. In 1991 open land was present all over the area. But after 10 years the area transformed into an agriculture and shrubland. Whereas the two biggest changes among non-built-up areas from 2001 to 2011 were the transformation of shrubland into open land and dense vegetation into shrubland. Lastly, the area gets encroached upon by built-up areas. Open land is available only on the east side in the sporadic part.

During 1991 the dense vegetation spread all over the area and continued to decline. In 2021, which is available only in small patches. The area of dense vegetation between 1991 and 2021 declined sharply and was replaced by the built-up areas, while the scenario for open land has completely changed. Studies have replacement of vegetation cover, open land and other natural surfaces by built-up surface is common in the process of urban expansion in different parts of world (Moisa et al. 2025; Salam et al. 2024; Rahman and Rahman 2023; Maskooni et al. 2021). It has pre-occupied a significant area in the north, east, and western parts of the city. But after 30 years the open land has been converted into a built-up area. In the case of cropland, a significant transformation can be observed in the eastern, northern, and southern areas of the city. The area under the cropland has witnessed a consistent decline over the decades. From 2001 to 2021 the cropland area has been encroached by the built-up area. Waterbodies are mainly available in the central part of the city and remain constant. This may be due to the government's protected zone. However, the most notable change from 1991 to 2021 observed the conversion of cropland to the built-up area, which was followed by the conversion of shrubland to the built-up area and open land to the built-up area. The trend between 1991 and 2021 of ongoing urban expansion is witnessed at the expense of natural land cover.

Figure 3 shows the spatio-temporal variation in area under different LULC categories in Darbhanga city during 1991–2021. The built-up area has increased consistently in Darbhanga city from about 13% of the total city area to more than 44% while open land and cropland has declined consistently (from 8.5 to 2.3% and 23.6 to 17.7%, respectively). At the same time, dense vegetation has declined to more

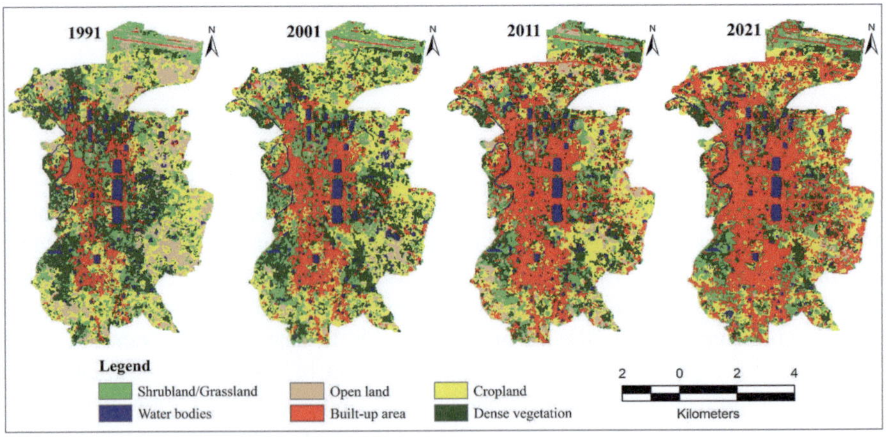

Fig. 2 LULC pattern in Darbhanga city

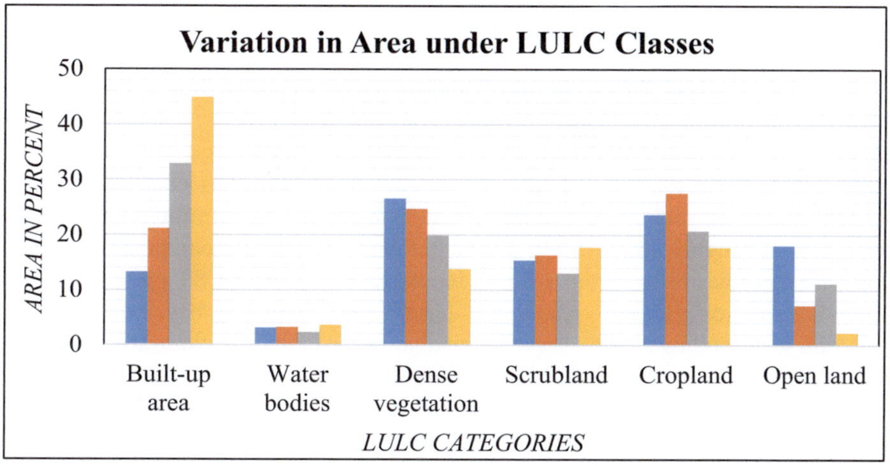

Fig. 3 Temporal variation in area under LULC types in Darbhanga city

than half of its total area of 1991 (i.e. from about 26% to 13%). This is because during last three decades, the city's population has increased to more than double (Chaturvedi and Avishek 2025; Shahfahad et al. 2025) and hence, the demand for residential, commercial, and other built infrastructures have increased dramatically. As a result, large-scale encroachment of urban vegetation and agricultural land has been noted in the city and its surroundings. Shrubland has showed a fluctuating pattern although it has overall witnessed a minor increase from about 15 to 17.5%. Most interesting finding is water bodies did not witnessed a decline in Darbhanga city during 1991–2021. This is because most of the water bodies are either protected or in the form of man-made lakes which have economic and spiritual significance for the local dwellers.

4.2 *Spatio-Temporal Variation in LST*

Study shows a clear trend of rising LST over time in the Darbhanga city during 1991–2021. It demonstrates significant spatiotemporal variations over three decades. This pattern is consistent with research on LST dynamics in Indian cities, which emphasize how anthropogenic activities, LULC changes, and urbanization, affects the local thermal environments (Bera et al. 2022; Shahfahad et al. 2023b). In 1991, LST was very high in the northern and southeastern parts of the city, while low LST was observed in scattered patches, mainly in central and southern parts, where minimum LST was 23.49 °C, and the maximum was 33.50 °C. This is due to the presence of vegetation and water bodies, which help to maintain lower temperatures. However, the northern and western regions show the first signs of increased warming in 2001. It may be due to the expansion of the city center in terms of built-up area (Fig. 4).

The minimum LST falls to 22.58 °C, indicating seasonal or climatic fluctuations, while the maximum LST (33.42 °C) stays constant. In the 2011, higher LST values are more widely distributed throughout the city, with urbanized areas exhibiting more intense heating. The overall temperature distribution has slightly changed, with the maximum temperature remaining at 33.43 °C and the minimum temperature remaining at roughly 23.36 °C. In 2021, there is widespread urban heat concentration, and there has been a considerable increase in maximum LST up to 34.54 °C. There are still some cooler areas in the eastern section, mostly because of vegetation, and waterbodies as they serve as heat sinks, reducing the warming effects of city. However, the general pattern points towards increasing heat stress brought on by urbanization. The rapid urbanization has caused a major shift in LST patterns, highlighting the necessity of sustainable urban planning to lessen heat stress and enhance environmental resilience.

4.3 Variation in SUHII

Figure 5 shows the spatio-temporal pattern of SUHII in Darbhanga city during 1991–2021. Results revealed that the city have witnessed a gradual intensification of SUHII with 2021 seeing a marked rise in high and extreme SUHII zones in comparison to previous decades. In 1991, only a few isolated moderate and high-intensity zones were present. Mainly in the northern part of city. The majority of the city had low or very low SUHII values. But by 2001, moderate and high SUHII zones had clearly expanded, especially in the city's central region. Due to increased built-up surfaces that absorb and retain more heat, resulting in higher SUHII values, highlighting rapid urbanization and land cover changes contributing to increasing surface temperatures. In 2011, a minor variation in SUHII pattern can be witnessed while in 2021, most notable increase in high and very high SUHII areas can be witnessed especially in the central and northern parts of the city, suggesting a rise in urban heat stress. This is because Darbhanga city has undergone significant urbanization, leading to increased built-up surfaces that absorb and retain more heat, resulting in higher SUHII values (Chaturvedi and Avishek 2025). The replacement of natural land cover with asphalt, concrete, and rooftops reduces natural cooling processes, leading to higher SUHII (Bokaie et al. 2016). Urban greening initiatives such as increasing tree cover, rooftop gardens, and parks can help mitigate UHI effects. Also, maintaining existing water bodies and integrating water-sensitive urban designs can enhance cooling effects in urban areas.

Table 2 shows that area under different SUHII zones have significantly varied during 1991–2021 in Darbhanga city. Although, the area under no SUHII zones have not witnessed a significant change during study period, but area under very low SUHII zone have witnessed a decline of about 1 km^2 while area under moderate and high SUHII have declined by about 0.12 and 0.44 km^2 respectively. On the

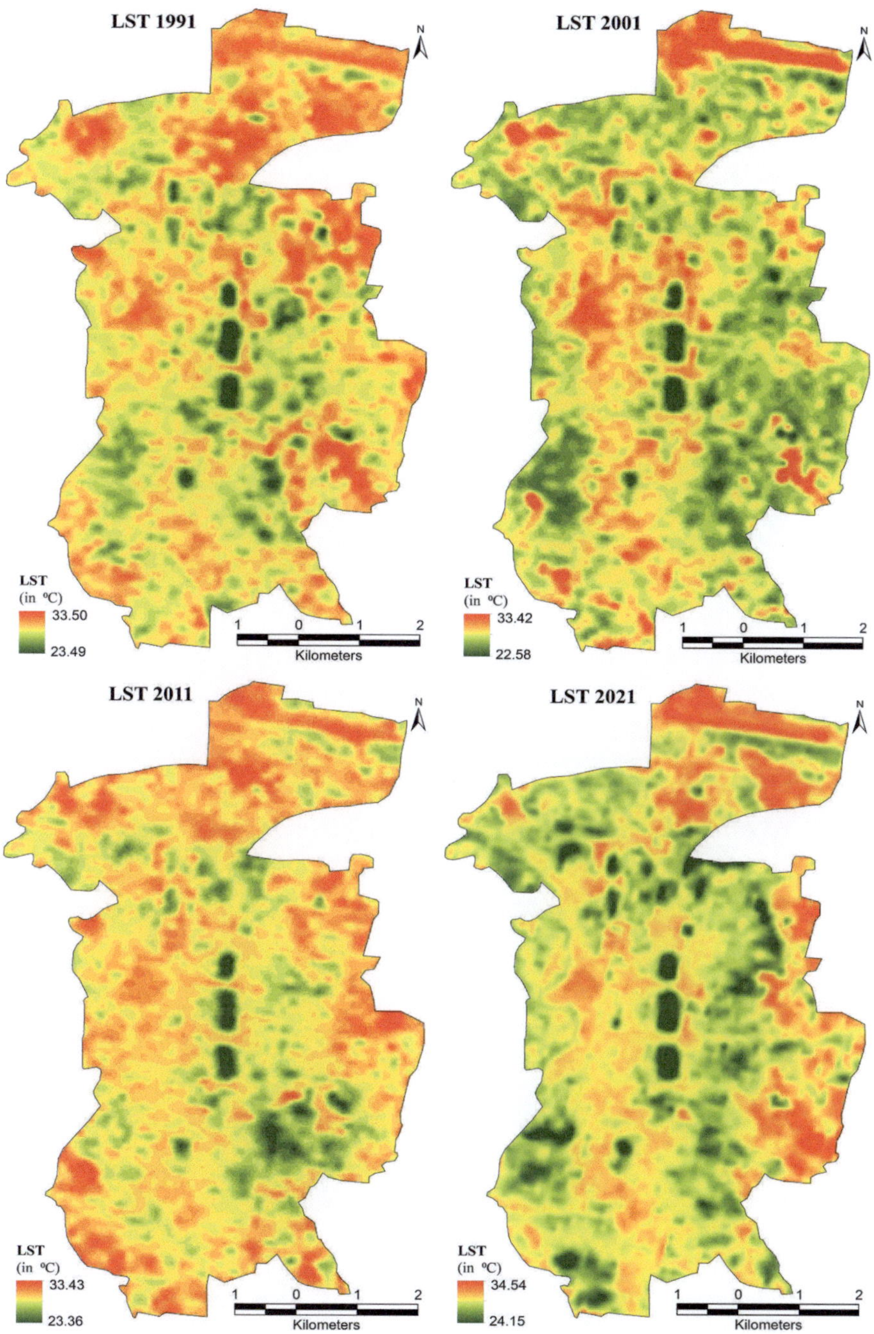

Fig. 4 Spatio-temporal pattern of LST

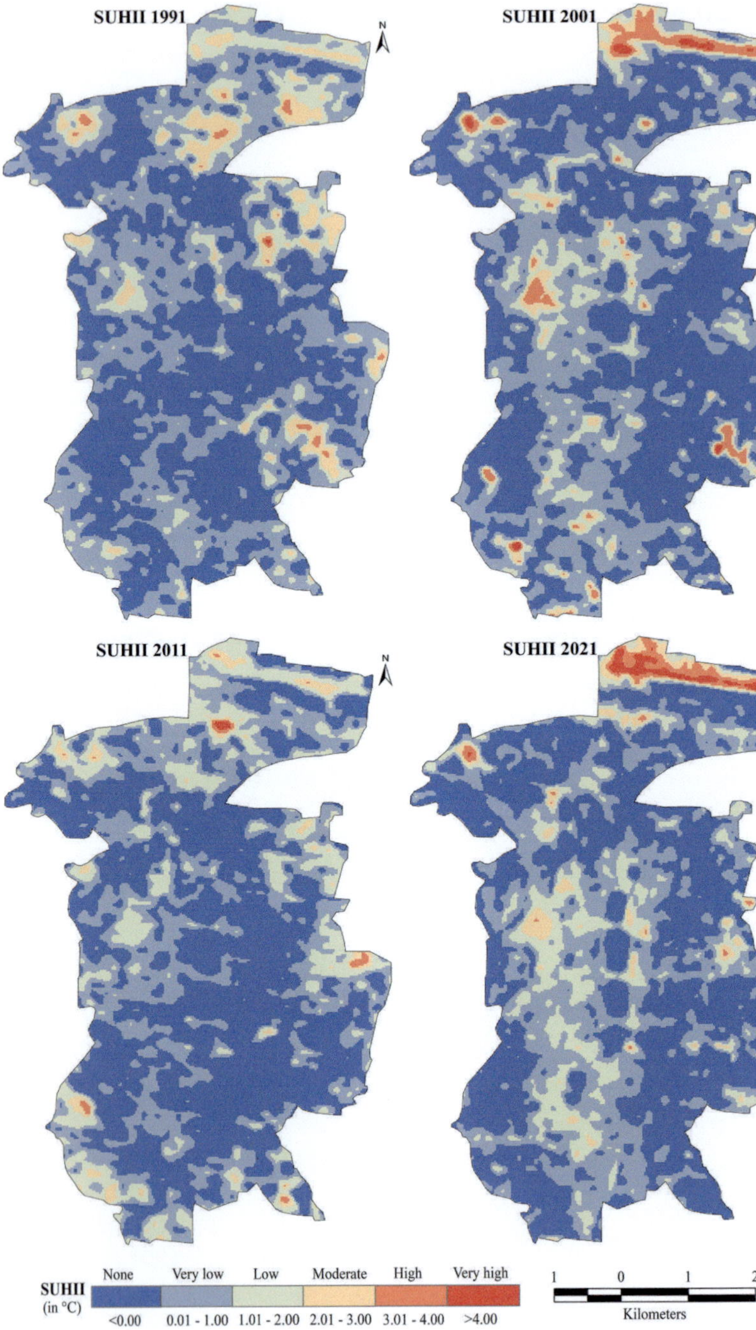

Fig. 5 Pattern of SUHII during 1991–2021

Table 2 Variation in area under SUHII zones in Darbhanga city

SUHII zones/year	Area (in km^2)			
	1991	2001	2011	2021
None	23.22	21.37	24.23	23.26
Very low	15.92	17.35	14.47	14.99
Low	4.48	4.96	6.60	5.74
Moderate	1.40	2.40	0.82	1.28
High	1.08	0.20	0.14	0.64
Very high	0.18	0.01	0.04	0.38

other hand, area under very high SUHII zone have increased by 0.20 km^2 and low SUHII zone have increased by 1.26 km^2. This illustrates that the vulnerability to heat waves have increased in city while thermal comfort has declined gradually. The heat related illness and mortality have been noted from several cities of world (Cleland et al. 2023; Sharma et al. 2024; Huang et al. 2025). Hence, it is important to frame the strategies to mitigate rising SUHII and LST in the city.

4.4 Variation in LST with LULC Types

Figure 6 illustrates the relationship between SUHII and LULC pattern in Darbhanga city using violin plots. The probability density of SUHII in violin plots describes the distribution and variation in SUHII values across LULC categories. Results describes that built-up area have consistent high SUHII during all four study years. Further, the median value of SUHII have increased over built-up area in the city, which indicates the increasing urban heating and declining thermal comfort over time as a consequence of built-up expansion. At the same time, water bodies and dense vegetation have low SUHII values, suggesting the role of water bodies and vegetation areas in mitigating SUHII in the city. This result is similar to a number of studies which noted increasing SUHII as a consequence of urbanization led built-up area growth and the mitigation of SUHII through green and blue infrastructures (He et al. 2023; Islam et al. 2024; Su et al. 2025). Cropland, shrubland, and open land shows moderate to low SUHII with a significant variation in distribution of mean and median values. This may be due to the seasonal thermal anomaly and the LULC transformation in the city and its surroundings. This consistent rise in SUHII, particularly over built-up area in the city indicates the impact of urban expansion on SUHII and thermal environment of the city. This highlights the urgent need of effective urban planning strategies to mitigate the intensifying SUHII and reducing heat stress in fast growing cities like Darbhanga.

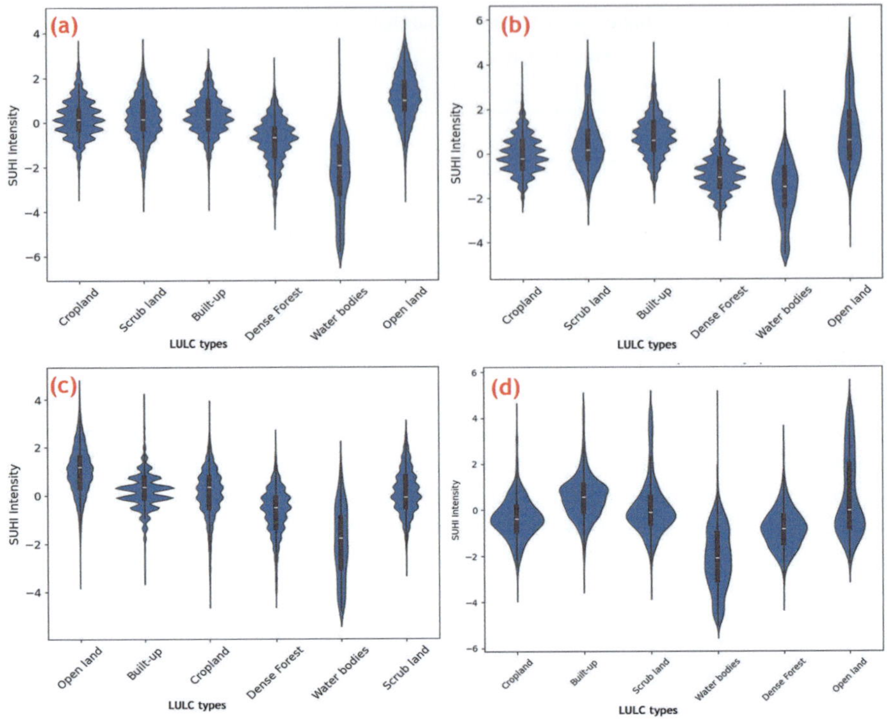

Fig. 6 Violin diagram shows variation in SUHII over different LULC types for **a** 1991, **b** 2001, **c** 2011, and **d** 2021

5 Conclusion

Study presents detailed investigation of the impact of urban expansion on urban green and blue spaces and its impact on SUHII in a medium sized city of India using geospatial techniques. Results show that during last three decades, Darbhanga city have undergone significant LULC transformation, mostly in terms of encroachment of built-up area over open land, cropland, and dense vegetation. The built-up area has increase by about 15 km^2 (239.35%) in the city during 1991–2021. At the same time dense vegetation, scrubland, open land, and cropland have declined significantly during same period. At the same time water bodies remained consistent and have not witnessed a significant change as most water bodies are either protected by government or are in the form of artificial man-made lakes for economic and cultural significance. The expansion of built-up area has resulted in significant rise in LST and SUHII in the city. The minimum and maximum LST have witnessed an increase of about 1 °C in the city while area under higher SUHII values have significantly increased. This highlights the impact of urban expansion and decline in urban green and blue spaces in enhancing SUHII phenomenon in cities like Darbhanga. Study highlights the need of effective urban planning strategies to combat the increasing

SUHII in cities like Darbhanga and protecting urban green and blue spaces for this. Moreover, future studies may integrate socio-economic and demographic data in studying the SUHII pattern across Darbhanga city to identify the impact of urban expansion on thermal environment of city and may aid in dealing with rising SUHII more effectively.

References

Al Shawabkeh R, Al-Hawwari L, Al-Fugara A, Saleem K, Smerat A, Abualigah L (2025) Analyzing urban landscapes and land surface temperature for sustainable development in industrial zones. Int J Environ Sci Technol 1–24

Angel S, Parent J, Civco DL, Blei A, Potere D (2011) The dimensions of global urban expansion: estimates and projections for all countries, 2000–2050. Prog Plan 75(2):53–107

Arshad S, Ahmad SR, Abbas S, Asharf A, Siddiqui NA, Ul Islam Z (2022) Quantifying the contribution of diminishing green spaces and urban sprawl to urban heat island effect in a rapidly urbanizing metropolitan city of Pakistan. Land Use Policy 113:105874

Bera D, Chatterjee ND, Ghosh S, Dinda S, Bera S (2022) Recent trends of land surface temperature in relation to the influencing factors using Google earth engine platform and time series products in megacities of India. J Clean Prod 379:134735

Bobo Merga B, Tabor KW, Melka GA (2024) Analysis of the cooling effects of urban green spaces in mitigating micro-climate change using geospatial techniques in Adama city Ethiopia. Sustain Environ 10(1):2350806

Bokaie M, Zarkesh MK, Arasteh PD, Hosseini A (2016) Assessment of urban heat island based on the relationship between land surface temperature and land use/land cover in Tehran. Sustain Cities Soc 23:94–104

Chaturvedi G, Avishek K (2025) Trend analysis of the indicators of urban wetland change and its implications using geospatial indices and statistical analysis. Environ Sustain Indic 100659

Cleland SE, Steinhardt W, Neas LM, West JJ, Rappold AG (2023) Urban heat island impacts on heat-related cardiovascular morbidity: a time series analysis of older adults in US metropolitan areas. Environ Int 178:108005

Cuce PM, Cuce E, Santamouris M (2025) Towards sustainable and climate-resilient cities: mitigating urban heat islands through green infrastructure. Sustainability 17(3):1303

Dhanaraj K, Angadi DP (2021) Urban expansion quantification from remote sensing data for sustainable land-use planning in Mangaluru India. Remote Sens Appl Soc Environ 23:100602

Ebi KL, Bowen K (2023) Green and blue spaces: crucial for healthy, sustainable urban futures. Lancet 401(10376):529–530

Fan C, Myint SW, Kaplan S, Middel A, Zheng B, Rahman A, Blumberg DG (2017) Understanding the impact of urbanization on surface urban heat islands—a longitudinal analysis of the oasis effect in subtropical desert cities. Remote Sens 9(7):672

Filho AJP, Karam HA (2016) Estimation of long term low resolution surface urban heat island intensities for tropical cities using MODIS remote sensing data. Urban Clim 17:32–66

Gunawardena KR, Wells MJ, Kershaw T (2017) Utilising green and bluespace to mitigate urban heat island intensity. Sci Total Environ 584:1040–1055

He T, Zhou R, Ma Q, Li C, Liu D, Fang X, Gao J (2023) Quantifying the effects of urban development intensity on the surface urban heat island across building climate zones. Appl Geogr 158:103052

Hidalgo-García D, Arco-Díaz J (2022) Modeling the Surface Urban Heat Island (SUHI) to study of its relationship with variations in the thermal field and with the indices of land use in the metropolitan area of Granada (Spain). Sustain Cities Soc 87:104166

Huang Y, Mao F, Xu Y, Qi J, Zhang W, Li C, Zhou M (2025) Association between urban heat islands and heat-related mortality in 70 pairs of adjacent urban-rural counties among 8 topographic regions in China. Sustain Cities Soc 106182

Islam MR, Shahfahad, Talukdar S, Rihan M, Rahman A (2024) Evaluating cooling effect of blue-green infrastructure on urban thermal environment in a metropolitan city: using geospatial and machine learning techniques. Sustain Cities Soc 113:105666 (2024)

Marquez-Torres A, Kumar S, Aznarez C, Jenerette GD (2025) Assessing the cooling potential of green and blue infrastructure from twelve US cities with contrasting climate conditions. Urban Urban Green 104:128660

Maskooni EK, Hashemi H, Berndtsson R, Daneshkar Arasteh P, Kazemi M (2021) Impact of spatiotemporal land-use and land-cover changes on surface urban heat islands in a semiarid region using Landsat data. Int J Digit Earth 14(2):250–270

Mentaschi L, Duveiller G, Zulian G, Corbane C, Pesaresi M, Maes J, Feyen L (2022) Global long-term mapping of surface temperature shows intensified intra-city urban heat island extremes. Glob Environ Change 72:102441

Mohammad P, Weng Q (2025) Asian heat stress variations in a changing climate: Implications for disproportionate urban and rural population exposure. Habitat Int 156:103294

Moisa MB, Negash D, Reta Roba Z, Debesa Wedajo G, Tsegay Deribew K, Benti F, Obsi Gemeda D (2025) Spatial analysis of urban expansion and its effects on land surface temperature using geospatial techniques in Addis Ababa and Shaggar Cities Ethiopia. All Earth 37(1):1–23

Naikoo MW, Bindajam AA, Talukdar S, Tayyab M, Mallick J, Ishtiaq M, Rahman A (2024) Monitoring dynamics of urban expansion using time series Landsat imageries and machine learning in Delhi NCR. Environ Dev Sustain 1–34

Novotný J, Chakraborty S, Maity I (2022) Urban expansion of the 43 worlds' largest megacities: a search for unified macro-patterns. Habitat Int 129:102676

Peng J, Qiao R, Wang Q, Yu S, Dong J, Yang Z (2024) Diversified evolutionary patterns of surface urban heat island in new expansion areas of 31 Chinese cities. npj Urban Sustain 4(1):14

Qiu X, Kil SH, Jo HK, Park C, Song W, Choi YE (2023) Cooling effect of urban blue and green spaces: a case study of Changsha, China. Int J Environ Res Public Health 20(3):2613

Rejaur Rahman M, Rahman A (2023) Urban green and blue spaces dynamics—a geospatial analysis using remote sensing, machine learning and landscape metrics in Rajshahi metropolitan city, Bangladesh. In: Advancements in urban environmental studies: application of geospatial technology and artificial intelligence in urban studies, pp 137–159. Springer, Cham

Salam M, Islam MK, Jahan I, Chowdhury MA (2024) Assessing the impacts of vegetation loss and land surface temperature on surface urban heat island (SUHI) in Gazipur district Bangladesh. Comput Urban Sci 4(1):24

Sarif MO, Rimal B, Stork NE (2020) Assessment of changes in land use/land cover and land surface temperatures and their impact on surface urban heat island phenomena in the Kathmandu valley (1988–2018). ISPRS Int J Geo-Inf 9(12):726

Shahfahad, Talukdar S, Naikoo MW, Rahman A, Gagnon AS, Islam ARMT, Mosavi A (2023a) Comparative evaluation of operational land imager sensor on board landsat 8 and landsat 9 for land use land cover mapping over a heterogeneous landscape. Geocarto Int 38(1):2152496

Shahfahad, Bindajam AA, Naikoo MW, Horo JP, Mallick J, Rihan M, Rahman A (2023b) Response of soil moisture and vegetation conditions in seasonal variation of land surface temperature and surface urban heat island intensity in sub-tropical semi-arid cities. Theor Appl Climatol 153(1):367–395

Shahfahad, Talukdar S, Naikoo MW, Rahman A (2024) Urban expansion and vegetation dynamics: the role of protected areas in preventing vegetation loss in a growing mega city. Habitat Int 150:103129

Shahfahad, Rihan M, Islam MR, Ansari I, Talukdar S, Siddiqui AM, Rahman A (2025) Decadal pattern of built-up expansion and its consequences on urban green and blue space fragmentation. J Indian Soc Remote Sens 1–16

Sharma A, Dutta P, Shah P, Iyer V, He H, Sapkota A, Wang YC (2024) Characterizing the effects of extreme heat events on all-cause mortality: a case study in Ahmedabad city of India, 2002–2018. Urban Clim 54:101832

Su W, Zhang W, Chang Q (2025) Tailored green and blue infrastructure for heat mitigation under renewal planning of urban blocks in Beijing. Sci Total Environ 967:178759

Talukdar S, Singha P, Mahato S, Pal S, Liou YA, Rahman A (2020) Land-use land-cover classification by machine learning classifiers for satellite observations—a review. Remote Sens 12(7):1135

Talukdar S, Naikoo MW, Rihan M, Mohammad P, Rahman A (2024) Seasonal dynamics of land surface temperature and urban thermal comfort with land use land cover pattern in semi-arid Indian cities: Insights for sustainable urban management. Urban Clim 57:102105

Wang K, Aktas YD (2024) Mitigation of heat exposure in urban areas. In: Lackner M, Sajjadi B, Chen WY (eds) Handbook of climate change mitigation and adaptation. Springer, New York, NY. https://doi.org/10.1007/978-1-4614-6431-0_197-1.

Yang C, Zhao S (2023) Diverse seasonal hysteresis of surface urban heat islands across Chinese cities: patterns and drivers. Remote Sens Environ 294:113644

Identification of On-Road and Roadside Objects Using Random Forest-Based Prediction Model

Parvej Khan

1 Introduction

Pole like features, trees, vehicles and building alongside the road corridor are the crucial components of the road environment. Their identification and classifications form the scene serve many applications such as City management (Gonzalez-Barbosa et al. 2022; Wang et al. 2019), street and traffic control (Javanmardi et al. 2019; Ranzinger and Gleixner 1997), autonomous vehicles (Chiang et al. 2020), disaster management (He et al. 2016), infrastructure management (Soheilian et al. 2010), change detection (Xu et al. 2015), land cover (Wu et al. 2018). Additionally, creating and updating accurate maps and spatial databases has been demanded by various applications such as urban planning, and intelligent transportation systems. The recent development in light detection and ranging (LiDAR) have exposed its huge potential in creating detailed, high density three dimensional (3D) point cloud data. Which is very imperative for many GIS base applications. There are four types of Laser scanning system satellite-based laser scanning, airborne laser scanning (ALS), terrestrial laser scanning (TLS), and vehicle-borne laser scanning (VLS) based on platform on which it is installed (Wu et al. 2013).

ALS is used to extract the features by many researchers (Zhang et al. 2014). As ALS data is collected from top of the scene so points from sides of the features (such as facades of building, stem and branches of tree) are missing (Puttonen et al. 2011). Thus, ALS do not form complete 3D structure. Unlike ALS, TLS and MLS acquire the data with relatively high density (Pfeifer and Mandlburger 2008). TLS collects data inefficiently (Li et al. 2017) compared to other LiDAR data acquisition technique but covers exhaustive detail. It is unwise using TLS for large corridor mapping. MLS overcomes the challenges of detailed vertical feature and high-density data

P. Khan (✉)
SimDaaS Autonomy Pvt. Ltd., Kanpur, India
e-mail: parvej.khan@simdaas.com

M. O. Sarif and A. Sharifi (eds.), *Advanced Geospatial Intelligence and AI for Environmental Resilience and Sustainable Development*, Advances in Geographic Information Science, https://doi.org/10.1007/978-3-032-03714-5_3

acquisition in case of ALS and speedy data acquisition in case TLS. We used MLS data for classification of the features.

2 Literature Review

The object identification and classification using mobile LiDAR data have several roadways and urban planning-related applications, for which many methods have been developed. These methods can be grouped into three categories: (i) point-based, (ii) segment-based, and (iii) object-based methods, which are summarized in Table 1. This categorization is based on the raw point cloud segmentation and generation of segmented data, where the unit entity used for the processing with chosen classifier is one of the entities: point, group of points as the segment, and a group of points as the object. The point-based classification methods analyze each point with its neighbors and label it with a class. In the case of segment-based and object-based methods, the segments are generated from the input point clouds, which represent, parts of the object or whole objects, respectively and then they are classified into the object class (Che et al. 2019). In general, The point cloud segmentation methods can be categorized as (a) hierarchical clustering, (b) region growing, (c) model fitting, (d) edge-based approach, and (e) hybrid methods (Park et al. 2019), based on the approach adopted to generate unit entity, that is the point, segment, object, and their processing. The hierarchical clustering methods compute geometrical and radiometric feature variables, such as Euclidean distance between points, point density, locally estimated surface normal, residual of best-fitting surface, point's reflectance, etc. (Grilli et al. 2017). The cluster formation, merging, and splitting are performed, so that each cluster uniquely represents an object. In Shkaberina et al. (2022), the greedy strategy was used for cluster combination and the proximity measures were used for combining the similar clusters.

The segmented data/training data are used to compute the feature descriptors to train the prediction models or the classifiers for the object's identification and classification. There are several feature descriptors discussed in the literature, which have been used for the classifiers training and object classification. These feature descriptors are categorized into four categories: (a) geometric, (b) radiometric, (c) colorimetric, and (d) contextual features. The geometric feature descriptors are often used in the literature for the training of the classifier and segmented data classification. In Yang and Dong (2013) the point clusters are generated in the neighborhood of a point as linear, planar, or spherical-shaped points cluster using geometric features, such as eigenvalues, dimensionality features, normal vector, and principal direction. Many other geometric features, such as height histogram, point density, length, and area of the object, were used for the classification of urban roadway corridor objects (Fan et al. 2013). Tilt, height, and width were also employed as geometric features for characterizing the objects in mobile LiDAR point cloud data for the classification of the urban scene (Balado et al. 2018). The discriminative properties of the classes such as elevation of the segments, planarity, space projection, and shape feature of the

Table 1 Summary of object identification and classification methods for mobile LiDAR data

Test site	Segmentation approach	Number of feature descriptors	Number of classes	Classifier	Accuracy metrics	Reference
Point-based						
Urban scenes	Region growing	12 (geometric)	4 (lamps, traffic signs, trees, others)	SVM	Overall accuracy: 95%	Yang and Dong (2013)
Segment-based						
Urban	Region growing	36 (geometric)	9 (Façade, ground, vegetation, car, motorcycle, pedestrian, traffic sign, streetlamp, bollard)	High-order graphical model	Overall accuracy: 97.13% and 95.79% (for two datasets)	Li et al. (2019)
Urban	Hierarchical clustering	9 (structure), 3 (reflectance)	3 (car, street lamp, traffic sign)	Hough forest	N/A	Yang et al. (2015)
Urban road	Hierarchical clustering	3 (structural), 2 (reflectance)	4 (street lamp, palm tree, car, and traffic sign)	Hough forest	N/A	Wang et al. (2014)
Urban corridor	Vertical layering along Z-axis	4 (geometric)	6 (buildings, roads, vehicles, trees, poles, and traffic signs)	Geometrical fitting	Overall Accuracy: 92.37%	Fan et al. (2013)
Street scenes	Super-voxels	4 (geometric), 1 (radiometric), texture, semantic correlations	4 (vehicle, building, pole-like object, street tree)	Probabilistic graphical model (Bayesian network and Markov random fields)	Overall accuracy: > 96% (for different objects)	Kang and Yang (2018)
Urban	PCA	3 (geometric)	5 (road, sidewalk, tread, riser, curb)	Topologic	Overall accuracy: 97%	Balado et al. (2018)

(continued)

Table 1 (continued)

Test site	Segmentation approach	Number of feature descriptors	Number of classes	Classifier	Accuracy metrics	Reference
Urban road environment	Octree	U-Net	7 (buildings, vegetation (i.e., trees, shrubs, and grass), fences, lamps, cars, pedestrians, and others)	FCN	Overall accuracy: 82.9–94% (for different datasets)	Xiang et al. (2019)
Road scenes	Octree algorithm	8 (geometric), 2 (contextual), 3 (color)	7 (palm tree, cycas, brushwood, vehicle, light pole, grass, road)	3D patch-based match graph structure	Average precision: 0.84	Luo et al. (2016)
Urban	3D modified Fisher vector	N/A	14 (truck, bus, car, 4wd, building, tree, pedestrian, pole, traffic light, traffic sign, pillar, Trunk, Ute, Van)	CNN (3DmFV-Net)	Different for model's variant	Ben-Shabat et al. (2018)
Objects-based						
Urban	Pairwise linkage (P-linkage)	8 (geometric), 10 (Local features)	6 (buildings, utility poles, cars, trees, ground, and others)	Evaluated for: SVM, random forests, Naïve Bayesian, extreme learning machine	Precision: 80.8−92.9%	Xiang et al. (2018)
Suburban road environment	Connected component	17 (geometric)	6 (tree, lamp post, traffic pole, car, pedestrian, hoarding)	SVM	Accuracy: 87.9%	Lehtomäki et al. (2015)

(continued)

Table 1 (continued)

Test site	Segmentation approach	Number of feature descriptors	Number of classes	Classifier	Accuracy metrics	Reference
Urban road	Euclidean distance clustering and VNCut	Template matching	3 (light poles, traffic signposts, and bus stations)	3-D object matching framework	Completeness: 0.94 Correctness: 0.97	Yu et al. (2015)
Urban scene	Hierarchical clustering	4 (geometric), 3 (colorimetric)	8 (cars trees ground buildings light poles grass shrubs)	RF/multilabel CRF	Overall Accuracy: 98.3% and 97% (for two datasets)	Li et al. (2017)
Urban Road	Voxelization	SigVox descriptors	2 (street lamp poles and 4 traffic signs)	SigVox cluster matching	Overall accuracy: 95%	Wang et al. (2017)
Urban Scene	Link-chain method	4 (geometrical)	5 (buildings, trees, cars, pedestrians, and poles)	Geometrical models and local descriptor matching	Overall Accuracy: 90%	Aijazi et al. (2013)
Urban road	Connected component	N/A	14 (truck, bus, car, 4wd, building, tree, pedestrian, pole, traffic light, traffic sign, pillar, Trunk, Ute, Van)	VoxNet, Multiclass TrAdaBoos	NA	He et al. (2020)
Urban road	Super-voxel	Bag-of-Word (BoW)	5 (light pole, light pole with the flag, guideboard, cycas, and traffic sign)	Subgraph matching	Correctness: 85% Completeness: 80%	Guo et al. (2021)
Urban and suburban scenes	Flood-fill algorithm	N/A	3 (pedestrians, cars, and street clutter)	DRF-Net	Accuracy: 97.92%	Chiang et al. (2020)

SVM support vector machine, *CNN* convolution neural network, *FCN* fully convolution network, *CRF* conditional random field

cylindrical models were also used in the classification of point cloud data (Li et al. 2017). The radiometric features such as the intensity recorded for the laser beam reflectance from the object were used in many methods. But, sometimes unavailability of intensity values (Xiang et al. 2018) in the raw point cloud data restricts the usage of intensity-dependent methods. Color information, such as RGB values, of an object in the point cloud data, is commonly represented as the colorimetric features. Colorimetric detail with the point cloud provides texture information and in addition to the geometric information, it is very helpful for image-based semantic labeling (Luo et al. 2016). The RGB values are considered one of the important factors for discriminating the objects (Aijazi et al. 2013) and thus can help in achieving higher accuracy in the classification. But its uses become limited due to the high computational cost and in many cases, the RGB values are not available with the raw mobile LiDAR data. The appearance consistency, spatial adjacency, and inter-category color similarities make the classification more challenging thus contextual information is required in such cases (Luo et al. 2016).

3 Dataset

The proposed method is tested on a multiclass MLS point cloud dataset. Occlusion, complex environment scenes, irregular pattern of the feature distribution are the major challenges of the LiDAR data post-processing. Considering these challenges test dataset is selected so to assess the proposed method against these challenges. The dataset was acquired from the peri-urban localities along the state highway (SH) 85, on the outskirts of Bengaluru city. In this, low-lying plants and shrub of varying heights were grown on the ground (see Fig. 1). The terrain was flat to gently slope away from the road and heavily vegetated.

4 Methodology

See Fig. 2.

4.1 Pre-processing

For preprocessing, we observe MLS data in two major group of points i.e. ground and non-ground points. To classify the objects, we removed the ground points first and for this we adopted the approach of Yadav et al. (2021). The remaining non-ground points represented as P_{ng} are clustered into multi-scale clusters using density-based spatial clustering of applications with noise (DBSCAN)-based clustering method (Ester et al. 1996). Each cluster is represented by m_i and all the clusters are grouped

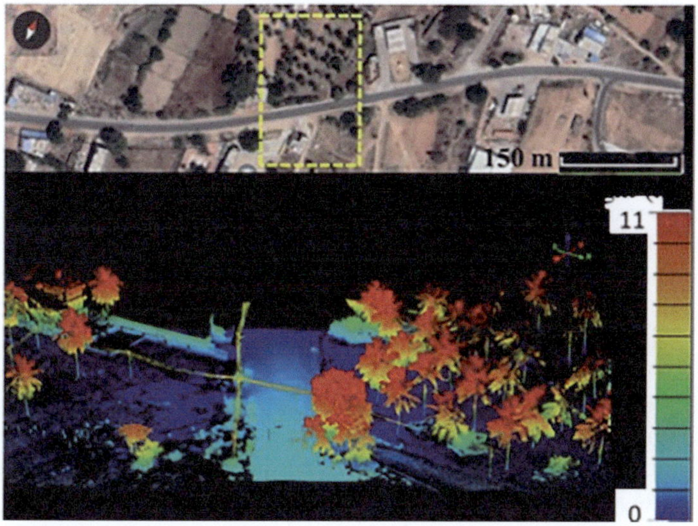

Fig. 1 Input dataset

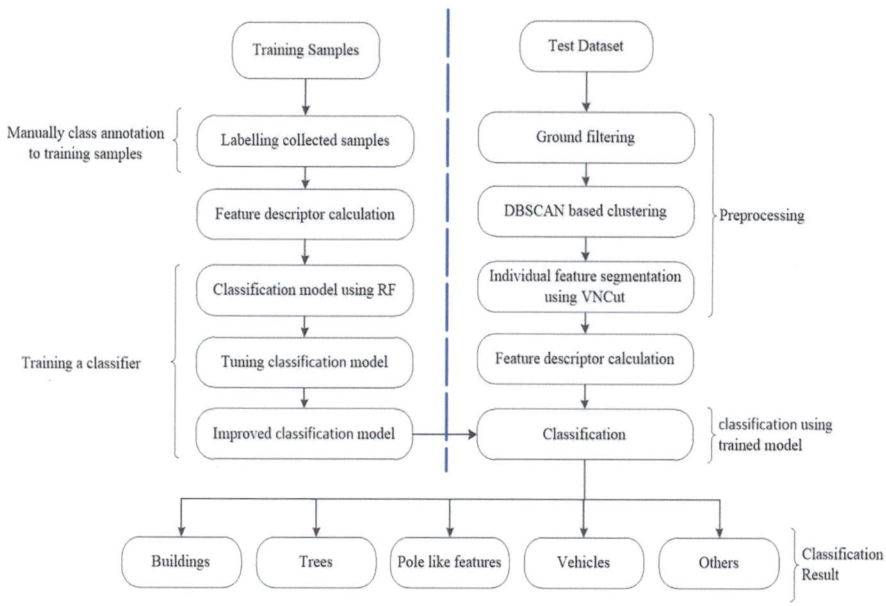

Fig. 2 Methodology

into set $M = \{m_i\}$, where i is the index value that varies from 1 to N_c. The N_c is the total number of clusters. P_{ng} is now divided into clusters, let say P_{ng} has total N_f objects, where each object is represented by f_j and all objects are grouped into set $F = \{f_j\}$, where j is the index value that varies from 1 to N_f. The N_f is the total number of objects, let's say if it is 100, then it can be made with 20 trees, 15 buildings, 10 vehicles, 20 pole-like features, and 35 other objects. If these non-ground objects are non-overlapping and isolated, then in such cases, the clusters generated by DBSCAN-based clustering satisfy $N_c = N_f$, where each cluster represents only one object. But in the roadway scenes, where objects are overlapped, the DBSCAN-based clustering does not ensure $N_c = N_f$. . It implies that it is not necessary that the DBSCAN-based clustering always generates a separate cluster for each object, mainly in dense roadway environments, where objects are overlapped and connected with other objects. In such cases, data clustering suffers from an: (1) over-segmentation: $N_c > N_f$, and (2) under-segmentation: $N_c < N_f$, even after DBSACN parameters tuning to generate the best clusters. Thus, further processing of cluster set M is required to achieve the desired condition, that is $N_c = N_f$, because for accurate object classification, a cluster should represent only one object and vice versa, and satisfy $\forall m_i : |F| = 1$.

In the case of under-segmentation, among the clusters formed by DBSCAN, some of the clusters are made up of multiple objects corresponding to overlapping and connected objects. In this experiment, the under-segmentation of the clusters is preferred over over-segmentation. It is aimed to further split the clusters made up of multiple objects into the clusters that represent the individual object. Therefore, in DBSACN-based clustering, the parameters are chosen in such a way that, the mobile LiDAR data is clustered as under-segmentation.

A voxel-based normalized cut (VNCut) algorithm (Yu et al. 2015) is adopted to split the cluster made up of the overlapping and connected objects into the multiple clusters, which uniquely represent an object. The normalized cut (Ncut) is used in 2D image segmentation by partitioning the weighted connected graph. The VNCut is the extended version of Ncut into voxelized 3D space and it ensures the optimal bipartition, where the segmented clusters represent the individual object.

Now the set of clusters M is revised after the splitting of clusters from overlapped and connected objects in such a way that each cluster represents only a complete object. Further, the clusters having a height of less than 1 m, such as shrubs, small boundaries, fragmented power lines, and other low-lying objects are removed as the aim of this experiment is to identify the building, pole-like features, trees, and vehicles. Finally, the newly generated clusters are given as input to the trained classification model for the object labeling.

4.2 Training a Prediction Model for Object Identification

Collection and Labeling of Training Samples

In this chapter, the one of objectives is to identify on-road and roadside objects, such as pole-like features, trees, vehicles, and buildings from mobile LiDAR data. This objective is achieved by the classification and labeling of mobile LiDAR data into four object classes: pole-like features, trees, vehicles, and buildings. In addition to these four classes, the other objects are grouped into a common fifth class namely other. A set of training samples $S = \{s_1, s_2, \ldots s_{ns}\}$, where i is the index value varies from 1 to n_s. n_s is the total samples, which are collected from mobile LiDAR data of various types of roadway scenes to train a prediction model for object identification in the mobile LiDAR data.

After the collection, the training samples s_i are labeled to their respective object classes and a descriptive model is formed. For the training samples collection, the different types of roadway settings having inter-and intra-class variability among the objects are selected, such that the training samples become representative of various types of roadway environments, such as urban, peri-urban, rural, mountainous, etc. (Chehata et al. 2009; Ni et al. 2018). The scene cases like complex objects with non-uniform point distribution, partially scanned objects, data acquisition from different MLS systems at variable point density and level of geometrical details, etc. are considered while training samples collection. For the training data collection, literature (Ni et al. 2018; Vosselman et al. 2017) suggests using the input mobile LiDAR dataset itself. As the training samples are collected from a specific roadway environment, the classifier will also be able to classify similar scenes accurately. Thus, the classifier becomes more scene-specific. To overcome this, the training samples are collected from a variety of roadway scenes having inter- and intra-class variability, different shapes, and sizes of objects, etc. using mobile LiDAR data. Thus, the classifier which is trained with such a comprehensive set of training samples and selected feature descriptors is able to perform satisfactorily to classify the mobile LiDAR data irrespective of types of roadway environment from where the data is collected. The number of points formed in the object cluster affects the classification, but there is a need for a robust method independent on the number of points in the object cluster. It can be achieved by selecting the feature descriptors for training, which are independent of the number of points formed the objects. Further in this chapter, several complex cases from all five classes are highlighted and discussed.

The labeled training samples of four object classes are building (c_b), tree (c_t), pole-like feature (c_p) and vehicle (c_v) are grouped into a set of interested classes C_i, while the training samples from classes other than the previously mentioned four classes are grouped in the fifth class that is denoted by a notation C_{other}. These object classes $C = \{C_i, C_{other}\}$ can be distinguished based on the selected geometric descriptors.

Derivation of Feature Descriptors Metrics

In this section, the feature descriptors, which are used to define the set of object classes C and their significance in differentiating the objects in general as well as complex roadway environments having inter- and intra-class variability, data gaps, occlusions, etc. are evaluated. For this evaluation, the training samples s_i, each labeled to one of the object classes C are used. The mathematical expression for each feature descriptor is represented by v_j is derived, where j varies from 1 to n and n is the total number of descriptors, which are combined into a set $V = \{v_j\}$. The feature descriptors are categorized into height-derived, area-derived, and statistically-derived descriptors, which are summarized in Table 2 and explained below:

4.3 Geometrical Analysis of Building, Tree, Pole-Like Features, and Vehicle

The selection of feature descriptors is one of the most important analytical parts of the proposed methodology. It is performed in two different stages: (1) feature descriptors are selected based on analyzing the geometry and point distribution of four classes (i.e. building, tree, pole-like features, vehicles) from two different views: top and side views, (2) further, the feature descriptors selected in the previous stage are experimentally examined and ranked based on their performance in the training stage of the prediction model for the classification. The first stage of feature descriptors is based on the object's points distribution and its geometry analysis for differentiating the object classes from each other. It is discussed in this section and the experimental part for feature descriptors ranking is discussed in Sect. 5.2. For geometrical analysis of the object classes and feature descriptors selection, multiple training samples having different sizes and shapes with intra-class variability are selected from the set of training samples: $S = \{s_i\},$. The selected samples are geometrically analyzed from multiple views, i.e. top and side views for selecting the best descriptors, which can successfully differentiate the various object classes.

In the case of trees as one of the object classes in mobile LiDAR data: (1) the shape of the points distribution of tree trunk observed from the top view (see Fig. 3) is compact and approximately circular (Wu et al. 2013); (2) Point cloud of a tree have one linear segment perpendicular to the ground terrain (i.e. trunk) and one volumetric segment (i.e. crown) supported by the linear segment (Yang et al. 2015); (3) Tree trunk is a vertical linear structure which has a circular cross-section, while tree crown has a volumetric shape with a circular cross-section. The crown cross-sectional area is more than the trunk; (4) Ratio of the geometric center and center of mass of the tree's points computed along the z-axis is less than 0.

In the case of pole-like features, the bare poles support various attachments, such as street lights, power lines, traffic lights, road signs advertisement boards, transformers, etc.

The geometric observations of pole-like feature's point cloud from different views (i.e. top and side) (see Fig. 4) conclude: (1) the main component of pole-like feature is the pole part, which is linear in shape and vertically oriented towards z-axis of the coordinate system used to store mobile LiDAR data. (2) the height of the pole is

Table 2 The description and analysis of the feature descriptors for differentiating the object classes

Feature descriptor	Definition	Description
H	Height	Differentiates the objects which have a higher value of height (such as trees, pole-like features (utility poles, light poles, bare poles and other poles), buildings, etc.) than the other smaller objects (such as pedestrians, two-wheelers: motorbikes, cycles, etc.)
Hmean	Mean height	Buildings that have uniform point distribution on their facades can be differentiated from trees due to their volumetric crown and linear trunk
R_ HmeanHhalf	The ratio of mean height to half of height	Differentiates the objects for which, the points are concentrated in either half of the object along the z-axis
Ap	The projection area of the objects is computed on the object's points projected on the horizontal plane, that is, x–y plane	Objects having a larger projection area on the horizontal plane (such as buildings, trees, etc.) are differentiated from other objects, which have relatively smaller projection areas (such as utility poles, light poles, traffic signs, etc.)
R_Apl_Apu	The projection area of two profiles of the object that is, the ratio of the projection area of the lower segment to the upper segment on the horizontal plane	Differentiates objects which have a projection area of the lower segment either larger or smaller than the projection area of the upper segment from other objects which have projection areas nearly equal or very close for both segments
Cm	Circularity measure the polygon formed by an object from the top view, i.e. over x–y plane	Differentiates objects based on compactness and circular shape of object's points projected on the horizontal plane, for example, poles-like features, and trees are differentiated from buildings and vehicles
Evmax	Highest Eigenvalue	To evaluate the spread along the direction of the highest points spread
Evbtw	Second highest Eigenvalue that lies between the highest and lowest eigenvalues	To evaluate the spread along the direction of average points spread
Evmin	Lowest eigenvalue	To evaluate the spread along the direction of the lowest points spread
Evmax_angle	Orientation of highest eigenvalue	The objects which have the maximum spread along perpendicular and parallel directions to the z-axis can be differentiated
Linearity	Ratio of Evmax-Evmin to Evmax	The objects which form linear shapes such as pole-like features and narrow trees are differentiated from other objects like buildings
Sphericity	The ratio of Evbtw-Evmin to Evmax	The objects such as buildings, and trees, are differentiated from pole-like features and vehicles

(continued)

Table 2 (continued)

Feature descriptor	Definition	Description
HHR	The ratio of the actual area of point cloud outline to the area of point cloud's convex hull that is computed over the horizontal plane. It is, called as horizontal hollow ratio (HHR)	The buildings can be differentiated from trees and vehicles
Zkurt	Kurtosis of z values	Trees generally have less value of kurtosis than pole-like features, thus they can be differentiated

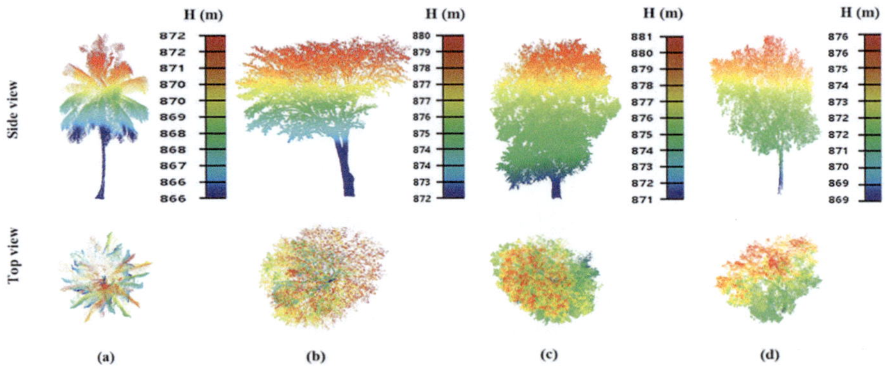

Fig. 3 Training samples represent different shapes and sizes of trees with intra-class variability: (I) side view and (II) top view

greater than a threshold height (3) the area of the pole computed using the projected pole's points on the horizontal plane (x–y plane) is very small (Cabo et al. 2014) compared with other objects such as a tree, building, vehicles, etc. (4) the shape of the poles computed using projected pole's points on the horizontal plane is circular.

In the case of buildings, the geometry of the building structures varies with each other in terms of the numbers of stories width, length, building footprint, etc.

Through the geometrical analysis of many building training samples (see Fig. 5) some common properties are derived despite the huge variability among buildings: (1) A building structure has at least two planer segments, i.e. facades (Yang et al. 2015); (2) The planer segment has large height and it is considered greater than a threshold height; (3) In mobile LiDAR data, the side looking scanners do not see the building roof in their direct view, therefore only building facades data is captured, thus in the top view building looks as hollow structure, which is bounded by at least two sides.

The absolute height and other height-derived parameters can easily differentiate vehicles from most non-vehicle objects, such as trees, buildings, and pole-like

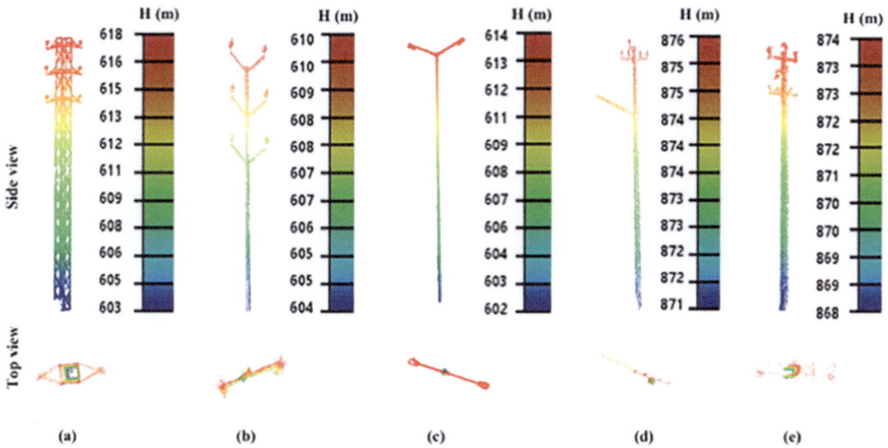

Fig. 4 Training samples represent different shapes and sizes of pole-like features with intra-class variability: (I) side view, and (II) top view

features. Based on the geometrical analysis of the vehicle's training samples (see Fig. 6), it is concluded that: (1) Vehicle in mobile LiDAR data has at least 3 planer segments (Yang et al. 2015); (2) Height of the vehicle lies within a range defined by threshold heights. (3) Vehicle points form hollow structures (4) Vehicle points satisfy rectangularity when points are projected on a horizontal plane and shape is computed in 2D (i.e., x–y plane) (Zhang et al. 2014).

4.4 Random Forest-Based Prediction Model

RF is an ensemble of decision trees that are grown independently of each other using the randomized feature descriptors (Breiman 2001). The selected feature descriptors are computed on the training samples of various object classes. When the model is trained, the model checks the votes generated by the decision trees, and a final prediction is made based on the most voted object class.

4.5 Tuning of the Prediction Model

The prediction accuracy of the random forest-based model for identifying the object classes is influenced by many parameters such as the size of the forest *ntree*, a number of feature descriptors used at the tree node for the training samples splitting *mtry*, the performance of feature descriptors, complexity present in the dataset, etc. The performance of the trained model is needed to be optimized based on these parameters to

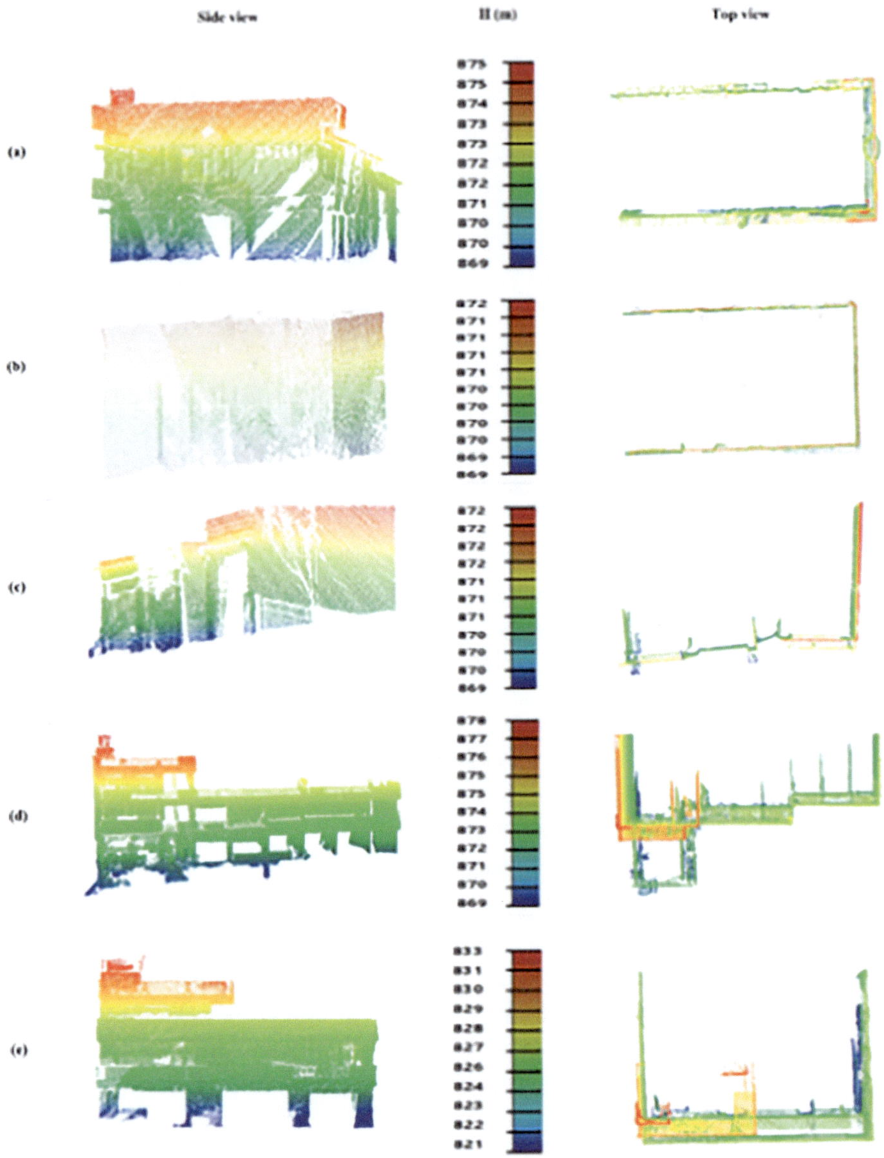

Fig. 5 Training samples represent different shapes and sizes of buildings with intra-class variability: (I) side view and (II) top view

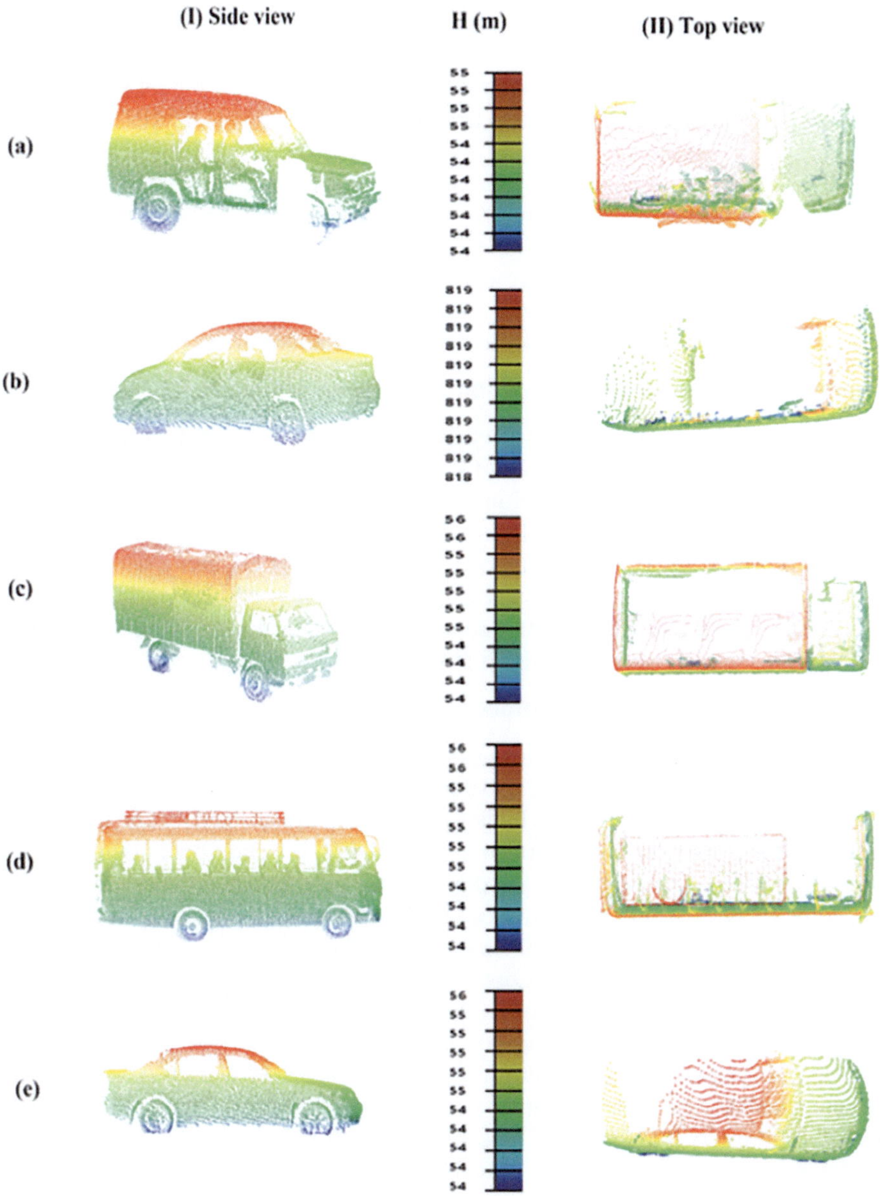

Fig. 6 Training samples represent different shapes and sizes of vehicles with intra-class variability: (I) side view, and (II) top view

achieve the highest possible object class identification and classification accuracy. In this experiment, *ntree* and *mtry* were tuned by testing the model on the tuning dataset. The following sections discuss the approaches adopted for the model performance optimization based on the parameters tuning.

Size of the Forest

To select an optimized size of the forest two different accuracy metrics; overall accuracy (OA) and kappa coefficient (k) are used. For the model to be optimized, the variations of OA and k with respect to the *ntree* varies from 100 to 1000 are investigated and variations are plotted: (a) between *ntree* and OA, and (b) between *ntree* and k for selecting the tuned value of *ntree*. relation;

Number of Feature Descriptors at the Tree Node

The number of feature descriptors *mtry* used at the node for sample splitting affects the object class prediction accuracy of the individual decision tree. As the random forest is an ensemble of *ntree* decision trees, therefore the performance of the overall model for object class prediction is affected by the selection of *mtry* value. Similar to *ntree*, *mtry* are also tuned by analyzing the variation of OA and k with respect to the *mtry* For the model to be optimized, the value of *mtry* is varied from 1 to 9 and the corresponding OA and k are plotted and analyzed for the *mtry* tuning. The *mtry* which gives the highest value for both OA and k is selected.

Performance of Feature Descriptors

In the first stage (Sect. 4.2) the feature descriptors used in the training of an object class prediction model are selected based on the geometry of the object class which may be present in the mobile LiDAR data. In the second stage, the selected feature descriptors are reassessed based on their performance in the object class prediction i.e., how effective a feature descriptor is for the accurate prediction. For this descriptor's performance assessment, the Mean Decrease Accuracy for each descriptor is plotted. This plot represents how much the accuracy of the object class prediction model reduces if a particular feature descriptor is removed from the overall processing.

In addition to this, the ability of the feature descriptors to discriminate an object from others is also analyzed using box plot-based approach (see Sect. 4.2). In the box plot, the maximum, minimum, and median values of feature descriptors are presented and used for the descriptor performance assessment. While selecting the feature descriptors it should also be considered that the computational time of the model increases with the number of feature descriptors. Thus, selecting only a few but the effective feature descriptor is recommended.

5 Experiment

The methodology developed for object identification was tested on a mobile LiDAR dataset selected from the semi-urban test site (see Sect. 3) and the performance evaluation of the methodology was performed. In the selected test sites, challenging cases such as occlusions, intra- and inter-class variations, complex roadway settings, and irregularly distributed objects were present (Table 3). These challenging cases are reported in the literature and it is mentioned that the methodology performance becomes limited in the presence of these challenging cases (Li et al. 2017). Therefore, the dataset having all these challenging cases is selected for testing the performance of the proposed methodology. The characteristics of the dataset are given in Table 3.

5.1 *Result*

Prediction Model Training and Parameter Tuning

A mobile LiDAR dataset of semi-urban environment was used to test the performance of the developed methodology and the results of clusters labeling to the object class were evaluated using accuracy metrics: precision, recall, and F1-score. Further overall accuracy and kappa coefficient were calculated. The object class identification was performed through a prediction model based on RF. For training of the prediction model, 250 training samples were collected from different types of roadway environment mobile LiDAR datasets. Table 4 shows the training sample counts for different object classes, viz. building (c_b), tree (c_t), pole-like feature (c_p), vehicle (c_v) and fifth class that is denoted by a notation C_{other}. Training samples for these classes were collected by considering the varying object geometry, partially scanned objects, occlusions, etc.

In the training data, the samples were collected from the object classes, where the object class was represented by the object point cluster having complete geometry of the object. In addition to these samples, object point clusters that were from the partially acquired object geometry due to the main occlusion were also chosen as training samples. A total of 54 training samples for the object class buildings that have

Table 3 Overall characteristics of input mobile LiDAR dataset acquired from two different test sites

Test site (length)	Number of points	Terrain slope	Partially scanned objects	Objects overlapping	Intra-class dissimilarity
Semi-Urban (800 m)	46,192,791	Gentle	***	**	**

The relative properties of the dataset are specified by symbols: * indicates *low*, ** indicates *medium*, and *** indicates *high*

Table 4 Statistics of the training samples of object class, which were collected from various types of roadway environments

						Total
Class name	Building	Tree	Pole-like features	Vehicle	Other	5
Number of training samples	54	82	52	29	33	250

at least two facades were collected. The samples having partial geometry, non-planar, and tilted facades were also included. Sometimes for the low height buildings, the MLS system can scan the points from the rooftop, which are generally missing due to the oblique geometry of MLS's laser scanners, thus this type of building sample was also included in the training samples. The trees with varying geometry from large to small size of the crown with varying point density were included in the training data for tree class. The trunks of some of the trees are tilted, sometimes the trunk is almost absent in the mobile LiDAR data due to occlusion, such tree datasets were also included in the training data for the robust model training. The object points clusters represent complete pole-like features that were chosen. The pole-like features with various complexities, such as partially scanned, different types of attachments fitted with the pole, and tilted poles were included in the training data. In the vehicle training samples collection, passenger cars, trucks, large buses, and other vehicles were considered. These vehicles have different dimensions, and some of the vehicles were partially scanned due to the obstructed sides of the vehicle, which were not in direct line of sight of the MLS's laser scanners. The training samples from the fifth class named as other class included 33 training samples from the object class such as pedestrians, two-wheelers, large size hoardings, boundary walls, and other objects which form unique geometrical structures. Further, the selected 14 feature descriptors were calculated on each training sample and utilized in the training of the RF-based prediction model. In RF the bootstrap datasets were created from these observations and total *ntree* decision trees were generated. These trained decision trees vote to classify the objects and using these votes RF predicts the class of an object.

The threshold parameters, viz *ntree* and *mtry* were selected based on several preliminary experiments (models are tested on the tuning dataset) and analysis of results in terms of accuracy metrics, i.e., overall accuracy and kappa coefficients (see Fig. 7). The variation of OA and k were assessed at different combinations of *ntree* and *mtry*. The OA and k for different combinations of *ntree* varies from 100 to 1000 and *mtry* varies from 1 to 9 were computed. Figure 7a and b show the variation plot of OA and k respectively for different combinations of *ntree* and *mtry* using the trained RF prediction model. The graph shows that the combination of $ntree = 300$ and $mtry = 4$ gives the maximum values of OA and k. thus these values are selected to train the RF-based prediction model.

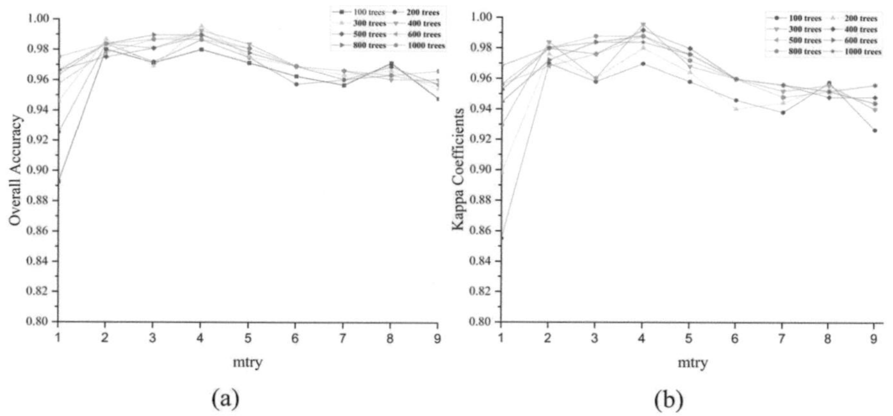

Fig. 7 Graphs showing the variation of **a** overall accuracy (OA) versus *mtry*, and **b** kappa coefficient versus *mtry*, considering *ntree* varies from 100 to 1000

Quantitative Analysis

The object-based performance analysis of the proposed methodology is shown in Table 5. Figure 8 shows the object identification results obtained. The overall accuracy and kappa coefficients were 97.92% and 96.98%, respectively. The accuracy metric, i.e. precision was better for object classes: building, tree, and pole-like features compared with the vehicles and other classes. The highest precision was achieved for the class trees which signifies that our trained model was able to classify the tree most accurately in comparison to other classes. For pole-like features and vehicles, the recall value was maximum. On the scale accuracy metric, i.e. F1-score, the performance of the methodology was highest for pole-like feature identification.

Table 5 Accuracy metrics for the object class identification (object-based)

R/P	Building	Tree	Pole like feature	Vehicle	Others	Total	Precision (%)
Building	39	1	0	0	0	40	97.50
Tree	1	127	0	0	0	128	99.21
Pole Like Feature	0	0	81	0	1	82	98.78
Vehicle	1	0	0	4	0	5	80
Others	0	2	0	0	32	34	94.11
Total	41	130	81	4	33	289	
Recall (%)	95.12	97.69	100	100	96.96		
F1-score (%)	98.44	98.44	99.38	88.88	95.52		

OA: 97.92%, k: 96.98%

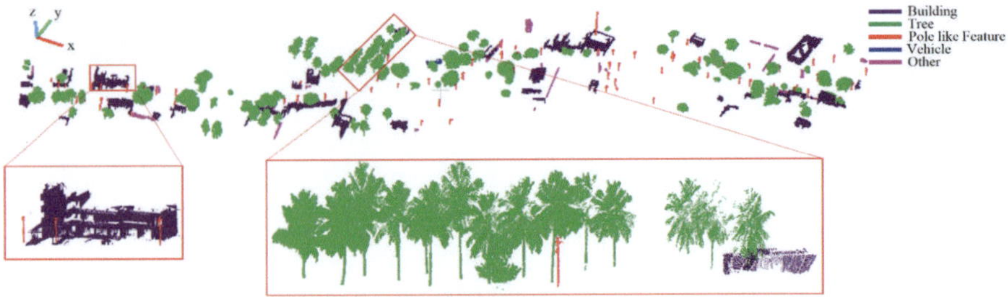

Fig. 8 Identified on-road and roadside objects

Table 6 Accuracy metrics for the object class identification (object point's-based)

R/P	Building	Tree	Pole like feature	Vehicle	Others	Total	Precision (%)
Building	5,115,517	122,703	0	0	0	5,238,220	97.65
Tree	61,094	7,047,785	0	0	0	7,108,879	99.14
Pole like feature	0	0	230,769	0	6849	237,618	97.11
Vehicle	6670	0	0	43,798	0	50,468	86.78
Others	0	6420	0	0	743,806	750,226	99.14
Total	5,183,281	7,176,908	230,769	43,798	750,655	13,385,411	
Recall (%)	98.69	98.20	100	100	99.08		
F1-score (%)	98.66	98.66	98.53	92.92	99.11		

OA: 98.47%, k: 97.28%

In Table 6, the point-based accuracy analysis of the proposed methodology is presented. In the confusion matrix, the OA and k were reported as 98.47% and 97.28%, respectively.

5.2 *Discussion*

Performance Evaluation of Feature Descriptors

In addition to *ntree* and *mtry* the feature descriptors used at the training stage are to be selected for the best performance of the proposed methodology. In this research work, the random forest package in MATLAB software is used to evaluate the performance of feature descriptors. The performance of each feature descriptor is evaluated by analyzing its contribution to the object class prediction. Figure 9 shows the importance of feature descriptors based on their object identification and classification

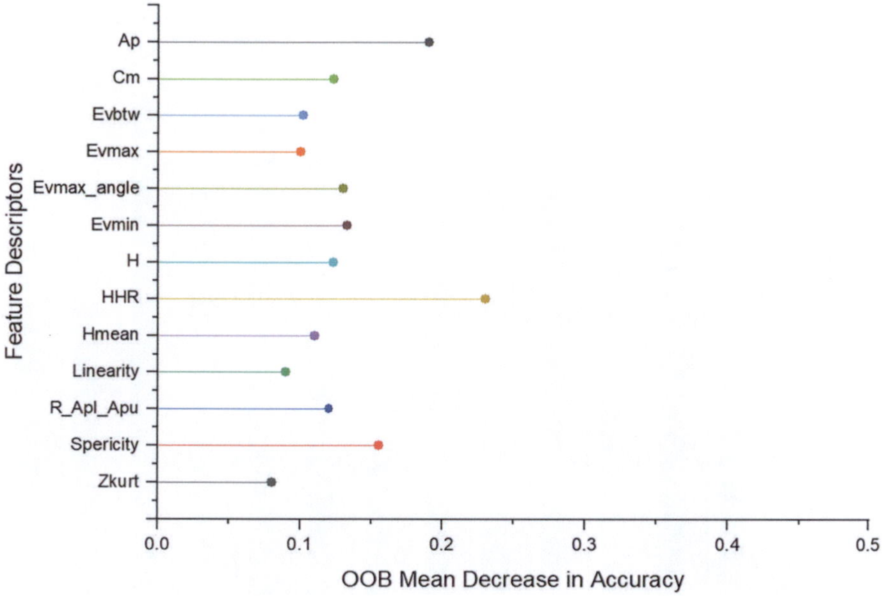

Fig. 9 Variable importance in terms of OOB mean decrease in accuracy

performance. From Fig. 9 the *HHR* performance is best on the scale of Mean Decrease Accuracy while Ap is the second most effective feature descriptor. Though *Zkurt* is found as the lowest performing feature descriptor but its exclusion reduced the model performance. In the proposed methodology total of fourteen feature descriptors are considered based on their overall performance in the model-based object class identification process.

In Fig. 10, bar graphs show the change in OA and k for different combinations of feature descriptors at a time. Initially, the model was trained using the best performer feature descriptors i.e. HHR, then OA and k were plotted. Again, the model was trained using the feature descriptor used previously with the next best performer feature descriptor. Similarly, multiple models were trained and both OA and k were plotted. The graphs show the increasing tendency of accuracy parameters with the increase in the number of feature descriptors in training, but get almost saturated after 10 feature descriptors. Still, it shows a very gradual increase in OA and k till the use of all the 14 feature descriptors.

The Capability of Feature Descriptors for Classification

The performance of the top 4 ranked feature descriptors, each differentiating the five object classes, is discussed through the box plot (see Fig. 11). The ability to discriminate an object class varies among the descriptors. Buildings have a high value of the HHR compared to the other four object classes, while among these four classes vehicles have the highest value of HHR (Fig. 11a). For the remaining three classes, the values are very close for HHR. Therefore, the HHR is highly efficient

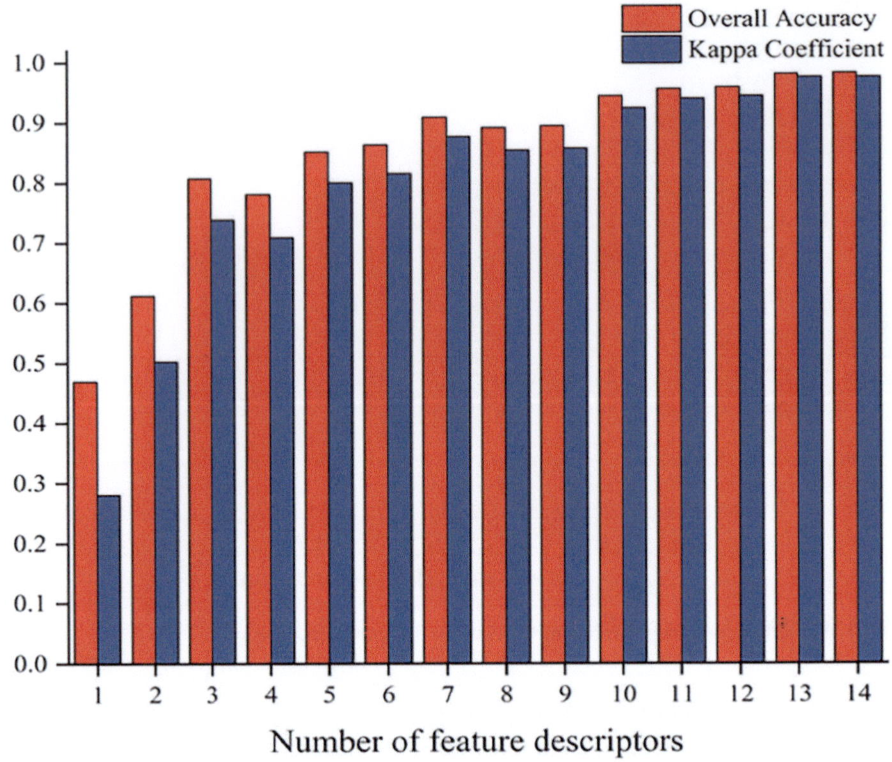

Fig. 10 Bar graph of overall accuracy and kappa coefficient for different number of feature descriptors at a time

for discriminating buildings from vehicles. It can be observed in Fig. 11b that the area of the projected 2D point cloud (A_p) is highest in the case of buildings, while for trees it is lower than the buildings but higher than the rest of the three classes. Due to the relatively larger areas, the buildings and trees could be effectively differentiated from the remaining three object classes, especially the pole-like features. Figure 11c shows a distinct difference in minimum Eigen value among all five object classes, where the difference is highest between trees and pole-like features. A similar pattern can be observed for the sphericity in Fig. 11d.

Sensitivity Analysis of Random Forest-Based Prediction Model

The sensitivity of the proposed methodology on parameters *ntree* and *mtry* was evaluated by examining the variations of OA and k with the change in *ntree* and *mtry*. In Fig. 12a the OA and k are plotted with respect to *mtry* varies from 1 to 9 while *ntree* is fixed to 300. Figure 12b shows the variation of OA and k when *ntree* varies from 100 to 1000 and *mtry* is fixed to 4. It is concluded from the above figure that changes in *mtry* and *ntree* have the least effect on the overall accuracy and kappa coefficient. For the different values of *ntree* the variations of OA and k are almost

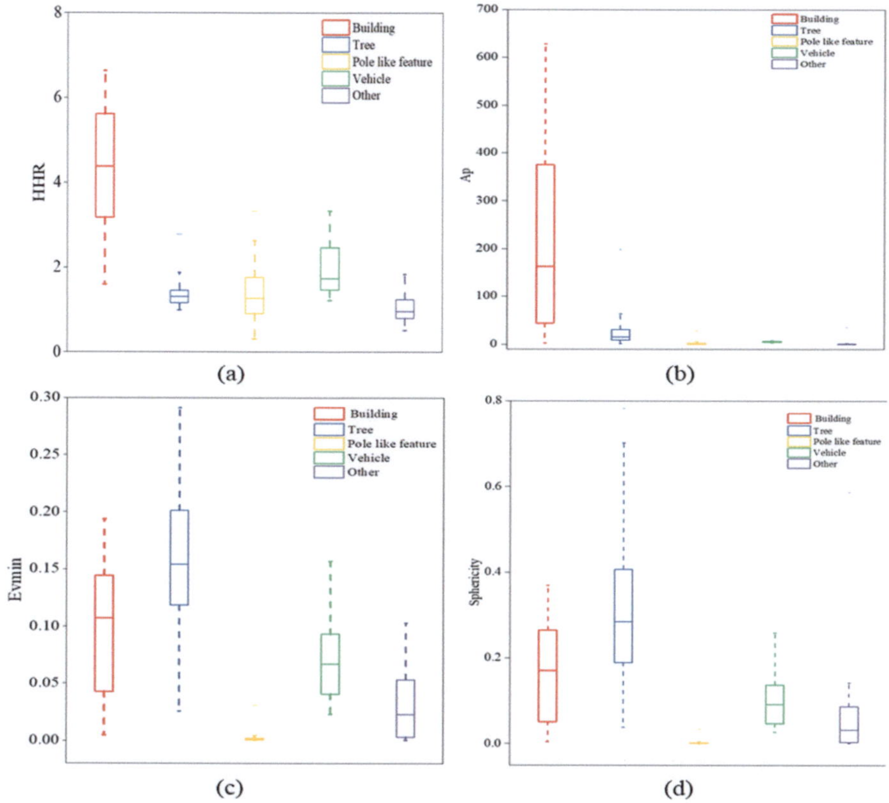

Fig. 11 Box plots of top 4 ranked feature descriptors; **a** horizontal hollow ratio (HHR), **b** area of feature's 2D projected point cloud (A_p), **c** minimum Eigenvalue (Evmin), and **d** sphericity

linear and approximately parallel to the *x*-axis. Thus, variation in the *ntree* value has a very negligent effect on the proposed method performance. The variation in *mtry* is also low and it is less than 7% and 5% for OA and k, respectively.

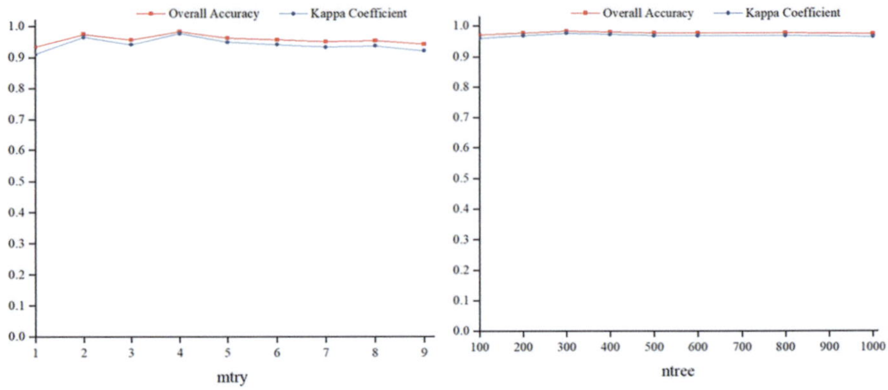

Fig. 12 Graph plot of overall accuracy and kappa coefficient for different values of *ntree* and *mtry*

6 Conclusion

This chapter aimed to present a robust and accurate method for the identification of various objects along the roadway environment using mobile LiDAR data. In the proposed methodology, first MLS data was preprocessed by filtering the non-ground points, and clustering. Further, an object class prediction model based on RF was trained using fourteen feature descriptors which were selected after a comprehensive geometric analysis of the object classes. After tuning the object class prediction model the test dataset was classified into five object classes buildings, pole-like features, trees, vehicles, and others. This chapter also presents the performance evaluation of feature descriptors and sensitivity analysis. The results show that the developed methodologies can label the mobile LiDAR clustered non-ground data into various object classes with OA of 97.92% and κ of 96.98% for semi-urban test sites.

References

Aijazi AK, Checchin P, Trassoudaine L (2013) Segmentation based classification of 3D urban point clouds: a super-voxel based approach with evaluation. Remote Sens 5:1624–1650

Balado J, Díaz-Vilariño L, Arias P, González-Jorge H (2018) Automatic classification of urban ground elements from mobile laser scanning data. Autom Constr 86:226–239

Ben-Shabat Y, Lindenbaum M, Fischer A (2018) 3dmfv: three-dimensional point cloud classification in real-time using convolutional neural networks. IEEE Rob Autom Lett 3:3145–3152

Breiman L (2001) Random For. Mach Learn 45:5–32. https://doi.org/10.1023/A:1010933404324

Cabo C, Ordoñez C, García-Cortés S, Martínez J (2014) An algorithm for automatic detection of pole-like street furniture objects from mobile laser scanner point clouds. ISPRS J Photogramm Remote Sens 87:47–56

Che E, Jung J, Olsen MJ (2019) Object recognition, segmentation, and classification of mobile laser scanning point clouds: a state of the art review. Sensors 19:810

Chehata N, Guo L, Mallet C (2009) Airborne Lidar feature selection for urban classification using random forests. In: Laserscanning. Paris

Chiang CH, Kuo CH, Lin CC, Chiang HT (2020) 3D point cloud classification for autonomous driving via dense-residual fusion network. IEEE Access 8(163775):163783

Ester M, Kriegel HP, Sander J, Xu X (1996) A density-based algorithm for discovering clusters in large spatial databases with noise. In Kdd 96(226):231

Fan H, Yao W, Tang L (2013) Identifying man-made objects along urban road corridors from mobile LiDAR data. IEEE Geosci Remote Sens Lett 11:950–954

Gonzalez-Barbosa J-J, Flores-Rodrıguez KL, Ornelas-Rodrıguez FJ, Trujillo-Romero F, Gonzalez-Barbosa EA, Hurtado-Ramos JB (2022) Using mobile laser scanner and imagery for urban management applications. IAES Int J Rob Autom 11:89–110. https://doi.org/10.11591/ijra.v11i2.pp89-110

Grilli E, Menna F, Remondino F (2017) A Review of point clouds segmentation and classification algorithms. Int Arch Photogramm Remote Sens Spatial Inform Sci 42:339

Guo W, Chen J, Wang W, Luo H, Wang S (2021) Three-dimensional object co localization from mobile LiDAR point clouds. IEEE Trans Intell Transp Syst 22:1996–2007

He M, Zhu Q, Du Z, Hu H, Ding Y, Chen M (2016) A 3D shape descriptor based on contour clusters for damaged roof detection using airborne LiDAR point clouds. Remote Sens 8:189

He H, Khoshelham K, Fraser C (2020) A multiclass TrAdaBoost transfer learning algorithm for the classification of mobile Lidar data. ISPRS J Photogramm Remote Sens 166:118–127

Javanmardi M, Song Z, Qi X (2019) Automated traffic sign and light pole detection in mobile LiDAR scanning data. IET Intel Transp Syst 13:803–815. https://doi.org/10.1049/iet-its.2018.5360

Kang Z, Yang J (2018) A probabilistic graphical model for the classification of mobile LiDAR point clouds. ISPRS J Photogramm Remote Sens 143:108–123. https://doi.org/10.1016/j.isprsjprs.2018.04.018

Lehtomäki M, Jaakkola A, Hyyppä J, Lampinen J, Kaartinen H, Kukko A, Puttonen E, Hyyppä H (2015) Object classification and recognition from mobile laser scanning point clouds in a road environment. IEEE Trans Geosci Remote Sens 54:1226–1239

Li Y, Li L, Li D, Yang F, Liu Y (2017) A density-based clustering method for urban scene mobile laser scanning data segmentation. Remote Sens 9:331

Li Y, Wu B, Ge X (2019) Structural segmentation and classification of mobile laser scanning point clouds with large variations in point density. ISPRS J Photogramm Remote Sens 153:151–165

Luo H, Wang C, Wen C, Cai Z, Chen Z, Wang H, Yu Y, Li J (2016) Patch-based semantic labeling of road scene using colorized mobile LiDAR point clouds. IEEE Trans Intell Transp Syst 17:1286–1297. https://doi.org/10.1109/TITS.2015.2499196

Ni H, Lin X, Zhang J, Chen D, Peethambaran J (2018) Joint clusters and iterative graph cuts for ALS point cloud filtering. IEEE J Selected Topics Appl Earth Observ Remote Sens 11:990–1004

Park SY, Lee DG, Yoo EJ, Lee DC, Pfeifer N, Stadler P, Briese C, Pingel TJ, Clarke KC, McBride WA (2019) Segmentation of Lidar data using multilevel cube code. In: Proceedings of OEEPE workshop on airborne laser scanning and interferometric SAR for detailed digital terrain models, vol 3612, p 156

Pfeifer N, Mandlburger G (2008) LiDAR data filtering and DTM generation. ISBN 978-1-4200-5142-1

Puttonen E, Jaakkola A, Litkey P, Hyyppä J (2011) Tree classification with fused mobile laser scanning and hyperspectral data. Sensors 11:5158–5182

Ranzinger M, Gleixner G (1997) GIS datasets for 3D urban planning. Comput Environ Urban Syst 21:159–173. https://doi.org/10.1016/S0198-9715(97)10005-9

Shkaberina G, Verenev L, Tovbis E, Rezova N, Kazakovtsev L (2022) Clustering algorithm with a greedy agglomerative heuristic and special distance measures. Algorithms 15:191. https://doi.org/10.3390/a15060191

Soheilian B, Paparoditis N, Boldo D (2010) 3d road marking reconstruction from street-level calibrated stereo pairs. ISPRS J Photogramm Remote Sens 65:347–359

Vosselman G, Coenen M, Rottensteiner F (2017) Contextual segment-based classification of airborne laser scanner data. ISPRS J Photogramm Remote Sens 128:354–371

Wang H, Wang C, Luo H, Li P, Cheng M, Wen C, Li J (2014) Object detection in terrestrial laser scanning point clouds based on hough forest. IEEE Geosci Remote Sens Lett 11:1807–1811

Wang J, Lindenbergh R, Menenti M (2017) SigVox–A 3D feature matching algorithm for automatic street object recognition in mobile laser scanning point clouds. ISPRS J Photogramm Remote Sens 128:111–129

Wang Y, Chen Q, Zhu Q, Liu L, Li C, Zheng D (2019) A survey of mobile laser scanning applications and key techniques over urban areas. Remote Sens 11:1540. https://doi.org/10.3390/rs11131540

Wu B, Yu B, Yue W, Shu S, Tan W, Hu C, Huang Y, Wu J, Liu H (2013) A voxel-based method for automated identification and morphological parameters estimation of individual street trees from mobile laser scanning data. Remote Sens 5:584–611

Wu Q, Zhong R, Zhao W, Song K, Du L (2018) Land-cover classification using GF-2 images and airborne lidar data based on random forest. Int J Remote Sens 40:1–17. https://doi.org/10.1080/01431161.2018.1483090

Xiang B, Tu J, Yao J, Li L (2019) A novel octree-based 3-D fully convolutional neural network for point cloud classification in road environment. IEEE Trans Geosci Remote Sens 57:7799–7818

Xiang B, Yao J, Lu X, Li L, Xie R, Li J (2018) Segmentation-based classification for 3D point clouds in the road environment. Int J Remote Sens 39:6182–6212. https://doi.org/10.1080/01431161.2018.1455235

Xu H, Cheng L, Li M, Chen Y, Zhong L (2015) Using octrees to detect changes to buildings and trees in the urban environment from airborne LiDAR Data. Remote Sens 7:9682–9704

Yadav M, Khan P, Singh AK, Lohani B (2021) An automatic hybrid method for ground filtering in mobile laser scanning data of various types of roadway environments. Autom Constr 126:103681

Yang B, Dong Z (2013) A shape-based segmentation method for mobile laser scanning point clouds. ISPRS J Photogramm Remote Sens 81:19–30. https://doi.org/10.1016/j.isprsjprs.2013.04.002

Yang B, Dong Z, Zhao G, Dai W (2015) Hierarchical extraction of urban objects from mobile laser scanning data. ISPRS J Photogramm Remote Sens 99:45–57

Yu Y, Li J, Guan H, Wang C (2015) Automated extraction of urban road facilities using mobile laser scanning data. IEEE Trans Intell Transp Syst 16:2167–2181

Zhang J, Duan M, Yan Q, Lin X (2014) Automatic vehicle extraction from airborne LiDAR data using an object-based point cloud analysis method. Remote Sens 6:8405–8423

Analyzing the Role of Geospatial Technologies and AI in Urban Infrastructure Planning and the Development of Smart Cities, Including Transportation Systems, Utilities, and Public Services

Vedprakash Maralapalle, Jayatheja Muktinutalapati, Bogireddy Chandra, Gangadhara Reddy Narala, and Rajiv Iyer

1 Introduction

It has been a more recent process human activities, which have greatly influenced the socio-economic and physical land divides of countries. The huge population from rural areas to urban areas in search of employment opportunities education facilities and health facilities. The UN estimates that 68% of the global population will be living in cities by 2050—an increase that will likely bring new strains to bear on urban systems, infrastructures, the environment, and communities (Openshaw 1995).

The process if urbanization implies that as the population of people in the urban centers increases, this calls for expansion of physical infrastructures such as transport systems, accommodation, electricity, water and sewerage and other public utilities

V. Maralapalle (✉)
L.S. Raheja School of Architecture, Mumbai, India
e-mail: civilved@gmail.com

J. Muktinutalapati
RICS School of Built Environment, Amity University Mumbai, Mumbai, India

B. Chandra
ALT University, Almaty, Kazakhstan

G. R. Narala
Department of Civil Engineering, Fiji National University, Suva, Fiji

R. Iyer
Department of CSE, Amity University Mumbai, Mumbai, India

M. O. Sarif and A. Sharifi (eds.), *Advanced Geospatial Intelligence and AI for Environmental Resilience and Sustainable Development*, Advances in Geographic Information Science, https://doi.org/10.1007/978-3-032-03714-5_4

and service providers. This process continues at a very fast pace and in most cases, it puts a lot of pressure on city planners to ensure that all the necessary facilities and amenities are created in response to this fast growing population such as transport, housing, utilities and the environment suffers the consequences thus including transport problems such as traffic congestion, lack of shelter among the population, overloading of the few available utilities and social problems such as pollution of the environment. New habitats experienced in several developing nations are characterized by urban development, and many people in such regions are living in slums without access to appropriate standard needs like clean drinking water, hygiene, and reliable light resources. Such issues call for innovative solutions in regard to the planning and management of infrastructural systems in cities (Benediktsson and Pesaresi 2019; Hall 2002).

Urbanization is compounded by climate change. Mega cities are among the most sensitive to the effects of climate change, such as flood, heat-waves, and storm. Coastal cities will be flooded due to rising sea levels, and water supply might be affected by emerging changes in rainfall patterns. Also, the developments arising from climate change such as the urban heat island effect; whereby urban centers tend to record higher temperatures compared to rural areas because of human endeavors and structures compounding the reality felt by the urban populace (Liu and Li 2022).

The first major issue is impact on sustainability of urbanization process. Another important topic is the implications of the model for sustainable development. Historical concepts of urban planning do not consider the sustainable, social, and, financial consequences of urban growth. For instance, a type of urbanization that is not facilitated by a good plan can result to depletion of ecosystems, deforestation and greenhouse emission. Poor transport systems and a bad system of waste disposal augment pollutions levels which ultimately degrades the living standards of the people (Harris and Batty 1993).

The emergence of Smart Cities provides the solution to these challenges with the use of technology in the development of sustainable efficient and habitable urban space. However, realizing these measures is occasionally inseparable from understanding spatial and temporal differentials, something that logical planning fails to capture. This is why technologies such as geospatial and artificial intelligence prove instrumental as they bring an endearing approach to the planning of cities that focus on data (ESRI 2021).

1.1 Importance of Geospatial Technologies and AI in Urban Planning

There is a special use of technologies that involve geographical location, including GIS, remote sensing and GPS in urban planning. These technologies allow one to gather, map and study spatial data and therefore planners get the feel of the urban infrastructure. For instance, GIS can be applied to develop land use planning, identify population distribution and changes, and study changes in the environment in the course of urbanization (Batty 2018; Zhang and Lin 2021; Sieber 2006).

Remote sensing has the capability of observing urban areas on a continual basis hence satellite imagery and aerial surveys. This capability is very useful especially for observing transition in the land use map, measuring the effect of development projects and identifying areas of environmental degradation. GPS technology additionally enhances these tools by giving geographical co-ordinates that are important in transport planning, disasters control, and infrastructural development (McHarg 1969).

The use of geospatial technologies brings in features not previously conceivable in the field of urban planning when coupled with AI. Machine learning and deep learning are subfields in AI that can analyze both spatial and non-spatial data and determine patterns and trends conducive to preparation of useful insights. For instance, AI applicant GIS systems can apply artificial intelligence techniques for traffic flow data to define better transportation networks, the future projection of cities to map out the utilization of land better, and lastly, they can estimate the susceptibility of structures to specific disasters (Bishop 2006). Employment of predictive models is another area where AI has made a major impact to urban planning. This is with the help of analyzing the data that has been gathered in the past and the data that is being gathered at the moment to predict the future outcomes that will help the planners to make right decisions. For instance, when it comes to evaluating proposed infrastructure solutions, AI can predict time loss due to traffic, pollution levels, and the occupancy of landscape, and therefore help to determine possible pathogenic results and remedial measures.

Geospatial technologies and AI also support democratic planning and management of urban areas. Such technologies being embraced fosters the creation of engagement tools whereby the stakeholders such as policy makers, developers and inhabitants-can physically embark on visualization of the proposed development projects. Figure 1 illustrates the integration of AI and geospatial technologies in urban planning for data-driven decision-making. This is good for the coordination and guarantee that the planning decisions are compatible with those of the populace (Guo et al. 2020). For smart cities, geospatial technologies and AI are core to build smart solutions improving the efficiency and intelligibility of Urban services. For example, the use of intelligent sensors placed in transport systems, it will be possible to quantify the traffics in real-time and adjust management to minimize traffic densities. Likewise, other resources including electricity and water can also be well distributed using GIS information so that the distribution is fair and wastage minimized.

Further, risk assessment and emergency response models rely on geospatial technologies and AI in providing greater disaster resilience. For instance, GIS is employed for showing regions affected by flooding, whereas AI is applied to estimate the probability of flooding based on climate and past experience information. Using such details, the government is in a position to erect framework and draw up disaster relief strategies that would reduce the effects of disasters on growing populations in urban areas (Lu and Zhang 2019). Therefore, geospatial technologies and AI can be

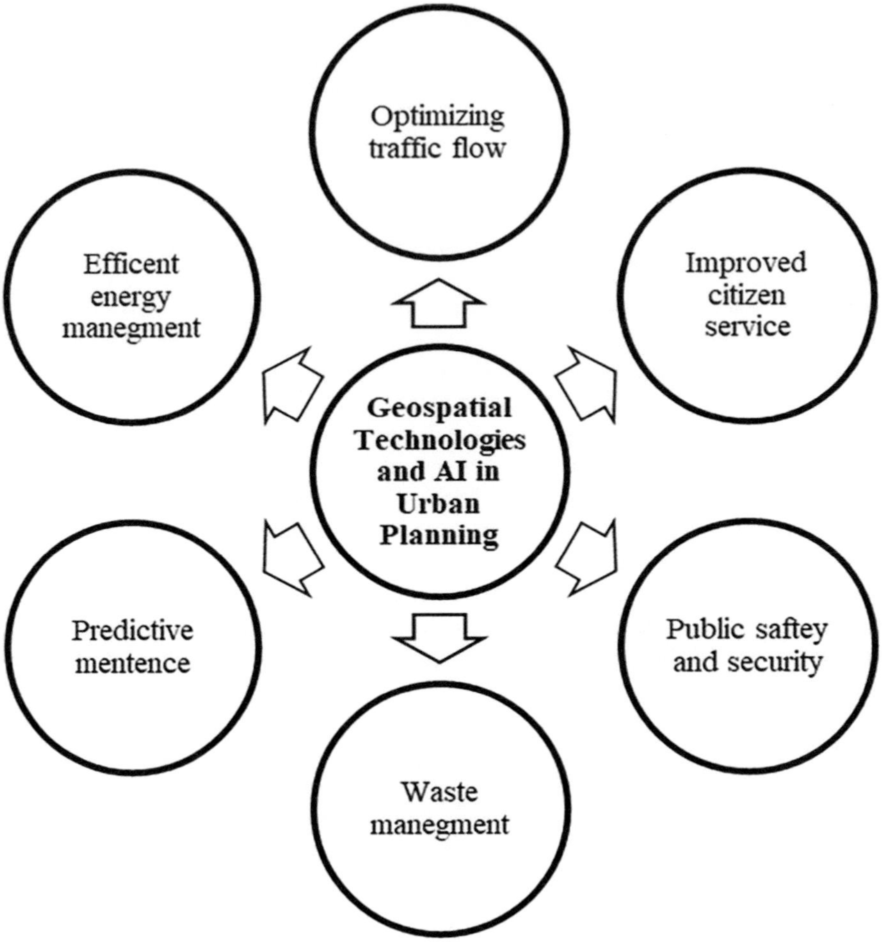

Fig. 1 Geospatial technologies and AI in urban planning

considered as revolutionizing urban planning as they present the cups to the problems caused by urbanization. It is in this regard that these technologies help planners to understand the space time relations as they plan for resilient sustainable and inclusive cities. In the future as the world experience a faster pace of urbanization the application of geospatial technologies and AI will be highly essential for the implementation of the smart cities, inhabitable, sustainable cities vision.

2 Geospatial Technologies in Urban Infrastructure

The process of developing infrastructure in the urban centers requires a complex strategy that embodies the system by which development and sustenance if infrastructural development is to take place. Geospatial information technologies are critical in meeting the increasing demand resulting from urbanization, shortages of resources, and environmental concerns. Some of the key geospatial technologies includes; Geographic Information System GIS, Remote Sensing and Spatial Database System. These technologies give the means and method of collecting, processing and mapping spatial data that is useful in managing the city's infrastructure. In this paper's following sections, its constituents and usage in mapping and analyses are discussed (UN-Habitat 2020; Cutter and Finch 2008).

2.1 Geographic Information Systems (GIS)

Far and away, GIS is instrumental in urban infrastructure planning and management tasks. It incorporates spatial data with attribute information for purposes of analysis, modelling and displaying the spatial relationships. By analyzing spatial data, GIS allows for the decision-maker to uncover more relationships and differences typical for urban environments (Cutter and Finch 2008). Key Features of GIS in Urban Infrastructure.

a. **Data Integration and Management**

   It operates information from satellite imagery, from aerial photographs to census data and field surveys. To urban planners this integration enhances options of viewing land use, zoning and information on demography.

b. **Spatial Analysis**

   Indeed, applying spatial algorithms, GIS conducts tasks as close proximity searches, buffer generation, and zone overlay. For example, when formation of public service provisions like hospital or school, GIS is used to identify an area where there is high population and physical access.

c. **Infrastructure monitoring**

   This makes GIS useful for monitoring current status of structures such as roads, utilities and the Whole transport system in urban areas. Using IoT sensors in relation to GIS systems increases the possibilities of tracking water supply pipes, electrical networks and traffic.

d. **Scenario Modeling and Simulation**

   Decision makers apply GIS to geo-reference different urban growth models and to evaluate the effects of undertaking certain developments. For instance, by using GIS one can simulate how the under construction metro line will impact on traffic flow and the surrounding land.

2.1.1 Applications of GIS in Urban Infrastructure

a. **Transportation Systems**: In the context of Intelligent Transportation System, GIS helps to address route finding and traffic congestion, development of the public transport system. Data feeding from the GPS devices on the ground is incorporated into the GIS to cone the status of the traffic and advise on probable routs.
b. **Utilities Management**: Public utility companies thus employ GIS in identifying and managing systems of water supply, sewerage, gas, and electricity. This enhances efficient maintenance and minimizes on vehicle breakdown circumstances.
c. **Urban Land Use Planning**: Environmental, social, and economic factors are used to place an overlay to demarcate areas meant for living, business, and industries.
d. **Disaster Management**: It also means GIS in a critical asset for identifying and marking the hazard areas, preparing the safe and efficient evacuation routes, and coordinating and controlling the general processes related to emergency contingencies like floods or earthquakes.

2.2 Remote Sensing and Spatial Databases

Remote sensing seeks to gather information about the earth surface by use of satellites and other airborne platforms with sensors. First, captured data helps identify information regarding land use and cover, vegetation, and urban sprawl. This information is kept and processed in spatial databases making it easier to be retrieved and analyzed. Figure 2 showcases the use of remote sensing technology for data collection and spatial analysis. Altogether, remote sensing and spatial databases are the technical foundations for the planning of infrastructure of urban areas (Smith et al. 2018).

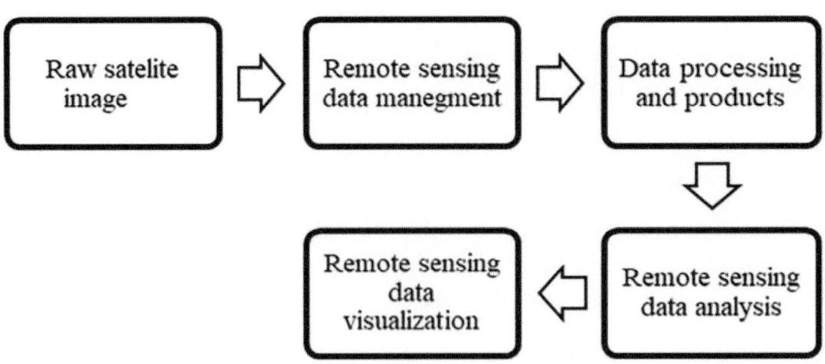

Fig. 2 Remote sensing technology

2.2.1 Remote Sensing Technologies in Urban Infrastructure

a. **Satellite Imagery**

   Some satellites include Landsat, Sentinel, and MODIS, in providing high-resolution images of urban environment. These images are useful in tracking proportional expansion of urban areas, dynamic changes in land use cover and even vegetation health and vigor.

b. **Aerial Surveys**

   Little autonomous drones and aircrafts along with cameras and LiDAR sensors offer precise data of terrain shape. This data is useful in applications that call for high accuracy, for example in planning a road network or in project management of construction projects.

c. **Thermal and Multispectral Imaging**

   Thermal imagery is used in measuring Urban Heat Islands and Multispectral imagery in estimating levels of pollution in water or soil.

2.2.2 Applications in Urban Infrastructure

a. **Environmental Monitoring**: Identification of air quality, deforestation, and water pollution in cities by remote sensing facilitates technological adaptations for various sustainable development.
b. **Urban Expansion Analysis**: Remote sensing technology is used on historical images to assess the city's development and forecast future development.
c. **Construction Planning**: Remote sensing data add substantive data to GIS thereby making it possible to plan integrated large structures like airports or highways accurately.
d. **Utility Infrastructure**: Surface models also involve the clients in the LiDAR and thermal imaging that can assist in the grounding of underground utilities and detect leakage or damage in pipeline systems.

2.3 Applications in Mapping and Analysis

Mapping and analysis, are the fundamental of urban infrastructure planning. Geography technology lays down the frame work of comprehensive maps that help in land usage, and disaster control among others (Kitchin and McArdle 2016; Talukdar et al. 2021; Simon 2016).

2.3.1 Mapping Urban Infrastructure

a. **Thematic Maps**

Thematic maps focus only on a particular characteristic of the city which may be the population density, transportation system, or quality of the environment. These maps help policy makers better determine where to direct infusions into infrastructure.

b. **3D Mapping and Visualization**

Current GIS applications and LiDAR create a three-dimensional representation of the built environment. These models play a major role in the depiction of heights of buildings, slopes of roads and drainage systems, which assist in formulation of sustainable urban design.

c. **Interactive Maps**

Social media applications such as the Google earth and Open street map provide for real time engagement of maps. These can be used by both planners and citizens for navigation, reporting, or as a tool on which to plot travel itineraries.

2.3.2 Analytical Applications in Urban Infrastructure

a. **Smart City Planning**

Spatial data contributes to artificial intelligence to enhance efficiency of resource utilization and forecast city dynamics. For example, geospatial analytics can point out where in the future traffic jams may appear and then address any problems.

b. **Sustainability Assessments**

GIS and remote sensing data enable determination of the degree of influence that the urban projects have on the environment. These assessments help the planners to reduce on the carbon footprints and improve on the spaces of green area.

c. **Crime Mapping and Prevention**

Spatial crime data helps the law enforcement agencies to determined areas that have most instances of crime and then be able to apply resources appropriately.

d. **Public Service Accessibility**

Figuratively the spatial analysis of data generates an understanding of the facility and accessibility of public amenities such as health facilities, school and parks and parity of access.Figuratively the spatial analysis of data generates an understanding of the facility and accessibility of public amenities such as health facilities, school and parks and parity of access.

e. **Emergency Response**

Geospatial technologies are part of mainly in cases of emergencies. Providing the geographic location of flood regions or following the actual occurrence of disasters make it easier and quicker to respond.

Technologies in GIS, remote sensing and spatial databases have variously impacted on the planning, implementation as well as the management of urban infrastructure. These technologies offer accurate information and efficient methods to deal with issues, which are unique in growing cities; for instance, scarcities, pollution, and inadequate infrastructure. When cities develop towards smarter and more sustainable forms of existence, then the position of geographical technologies will continue to be imperative with regard to designing sustainable and efficient urban landscapes (Hanski and Gaggiotti 2004; Shi and Tian 2010; Clarke 2003; Goodfellow et al. 2016).

3 Artificial Intelligence in Smart Cities

Smart City is a product of technology known as Artificial Intelligence (AI). AI by utilizing dependent on data and state-of-the-art computational resources enables efficient management of cities, improves public services and promotes sustainability. Figure 3 highlights the role of artificial intelligence in enhancing efficiency and sustainability in smart cities. In the following section, we discuss ML and predictive modeling, AI based decision support system, AI in transportation, utilities and public service (Craglia and Shanley 2015).

3.1 Machine Learning and Predictive Modeling in Smart Cities

It has become apparent that Machine Learning (ML), which is a component of Artificial Intelligence (AI), has helped redefine how urban issues can be solved, by providing tools to examine massive amounts of data, recognize entities within such data, and make predictions based on such data. In light of this, using predictive modeling as enhanced by ML, city planners can effectively predict likely bottlenecks in resource usage to improve planning and make cities sustainable. This section expands on the interaction of ML to smart cities, where it is used, the methods employed and the change it is likely to bring into smart city solutions.

Machine Learning has a significant part to contribute to the smarter cities since it can utilize data coming from various urban contexts. The conclusions arising therefrom assist to fine-tune processes, augment value delivery, and advance the standard of living for city dwellers (Craglia and Shanley 2015; Dodge et al. 2011).

a. **Data Analysis and Pattern Recognition**

   One of the primary functions of ML is processing extensive urban datasets to uncover meaningful patterns and trends.

i. Traffic Management: It uses the multiple convoy machine learning algorithms to read the traffic patterns derived from GPS devices, traffic sensors, and traffic cameras. These includes traffic signal timing, road infrastructure improvements and real-time services that support route guidance.

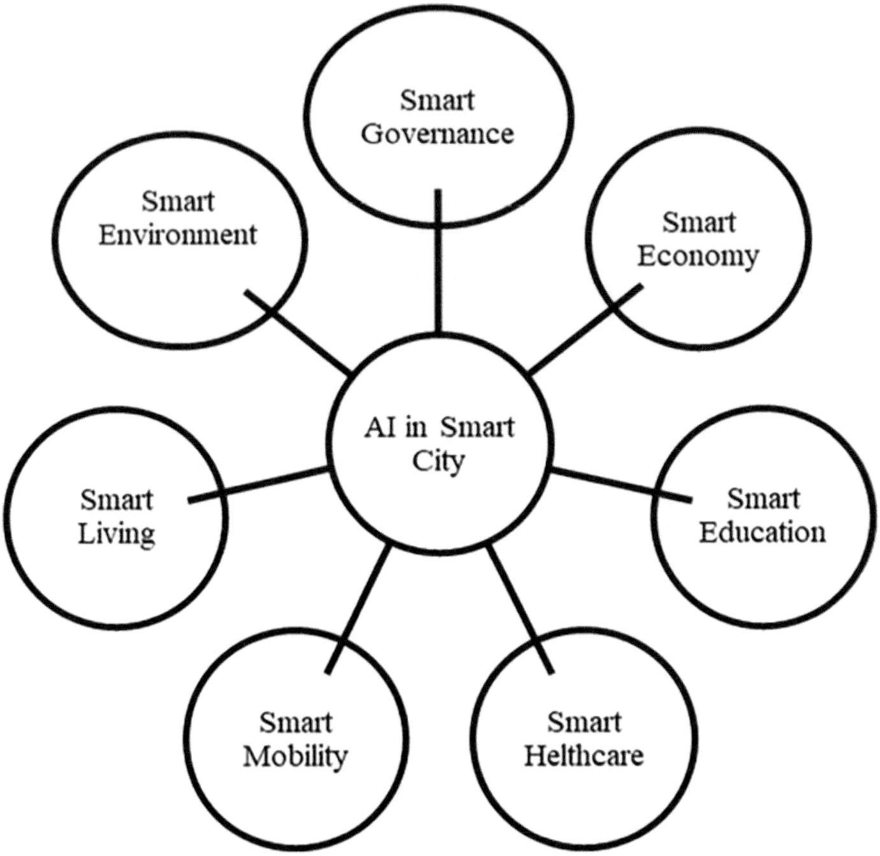

Fig. 3 Artificial intelligence in smart cities

ii. Energy Consumption: Others are applied in managing power consumption trends in residential, commercial, and industrial segments. These patterns assist the utilities in predicting how much demand would be in the future, how they can enforce energy conservation measures, and when there might be a failure of supply.
iii. Environmental Monitoring: By analyzing data coming from devices similar to air quality sensors, satellite images, or meteorological stations, ML models point to areas that are most likely to have high levels of pollutants, and the dispersion of contaminants. For example, energy meters installed in our homes gather information on the energy consumption patterns. Employing the e ML algorithms, the utility providers can then detect the inefficiencies, recommend to the users, and enhance the general energy use.

b. **Predictive Analytics**

Predictive analytics powered by ML models anticipates future urban scenarios, enabling proactive measures.

i. Traffic Congestion: Computing systems estimate traffic jams by using historical and current information and thus providing authorities with a tool for designing traffic control measures such as avoidance routes and dynamic tolls.

ii. Energy Demand: The power consumption is predicted by using techniques such as using climate and era or power usage frequency, so as to distribute power effectively.

iii. Pollution Levels: Climatic conditions, coupled with the indices of urban dynamics, are used by ML models for estimating the AQI and the 24 × 7 recommendations for action by authorities include banning operations in industries or providing more access to public transport. One main use case is the Singapore public transport to predict the demand during the rush hour through predictive modeling, thus using that, an embody of buses and trains is determined.

c. **Optimization of Resources**

Efficient resource allocation is a cornerstone of smart city development, and ML facilitates this by predicting demand and identifying inefficiencies.

i. Water Management: The supply of water is well managed by the processed data, which identifies leakage in supply networks consequently saving, and distributing the resources to the right place.

ii. Waste Management: The available predictive models on the volume of waste being produced and integration of efficient collection patterns lowers operating expenses and negative effects on the environment.

iii. Healthcare Services: In resource management, it assists in preparing for patient rate surge in urban health facilities, hence, proper staffing and equipment disposal. Not only do these applications raise the level of service delivery but they also make costly mistakes more efficient in terms of not wasting money or energy.

3.1.1 Key Techniques in Predictive Modeling

Machine Learning comprises of various approaches suitable for different problems that affect cities. It is understandably the case with every technique out there because every technique has some specificity that makes them suited for certain types of prediction tasks (Chen et al. 2021).

a. **Supervised Learning**

The supervised learning models work on learning data sets which are certified, identifying the relation between inputs and outputs. This technique is widely used for tasks requiring accurate predictions.

i. Energy Consumption Forecasting: The regression models determine the probable consumption of energy by arguably giving hourly, daily, and daily forecasts to the utilities.

ii. Crime Prediction: There are classification models of criminal occurrences; taking criminology data from the past, most lawyers use these models to foresee areas of high crime rates, which can then be addressed by police forces.
iii. Infrastructure Monitoring: Supervised models identify existence of structural abnormalities in bridge, road or any building by analyzing the data received from the sensors regarding their previously established standards. Traditional techniques of advanced analytics such as linear regression decision trees, and neural networks are typically used in such circumstances.

b. **Unsupervised Learning**

Unsupervised learning operates on unstructured data which, it seeks to categorize into various groups or categories. It is particularly useful in exploratory tasks.

i. Waste Management: Some clustering methods categories the urban areas according to the generative rates of waste so that the appropriate collection systems can be implemented.
ii. Traffic Density Analysis: Types of models: Unsaid clusters, regions with similar traffic flow patterns help in deciding on infrastructure deployment.
iii. Social Segmentation: Through a census and mobility approach to clustering, demographic spatial arrangements are distinguished aiding urban development polices. Among PCA and K-means clustering that is mentioned as unsupervised learning techniques, both are common in the field of urban.

c. **Reinforcement Learning**

Reinforcement learning (RL) involves training agents to make decisions in dynamic environments, where they learn from rewards and penalties.

i. Adaptive Traffic Control: RL algorithms modify the signal timings realizing the flow rates in real-time in a way that frees traffic clogs and time.
ii. Energy Management in Buildings: RL is applied in smart systems to maintain thermal comfort while minimizing energy consumption with regard to HVAC systems.
iii. Autonomous Vehicles: It beamed about the fact that RL helps autonomous vehicles avoid accident-prone areas in cities and handle them effectively. Other methods used in these systems are; Deep Q-Networks (DQNs) and policy gradient methods (Chen et al. 2021; Biljecki et al. 2016).

3.2 AI-Driven Decision Support Systems in Smart Cities

Artificial Intelligence (AI) becomes one of the most significant elements in modern governance environments and cities administration. Decision support systems (DSS) provide the administrative officers in the city immediate intelligence data, analytics by artificial intelligence, and visualization tools that enhance decision-making. Such systems are critically important for tackling multifaceted issues of urban environment, social and economic sustainability, as well as optimization of infrastructures and public services. For the purpose of this work, AI-enabled DSS shall be described as systems that can assimilate a vast quantity of data, originating from multiple sources, process than intelligently and provide meaningful insights derived from such data (He and Chen 2024). Here are the core features that make these systems invaluable for smart city development:

a. **Real-Time Monitoring**
   i. Data Integration: These systems aggregate data from Internet of Things (IoT) devices, satellite, and urban sensors. For example, traffic cameras and environmental sensors and utility meters are sending data in real time into the system.
   ii. Dynamic Alerts: Some examples of events are sudden surges in pollutants, event when traffic in a city is higher than expected, the system detects these and informs authorities.
   iii. Case Study: The management of the city of Barcelona employs a DSS connected to IoT appliances for the monitoring of traffic, water consumption, waste management, and other aspects to produce efficient services.

b. **Scenario Analysis**

   AI excels at simulating various scenarios to predict outcomes and assist in evaluating policy and infrastructure decisions.
   i. Policy Testing: DSS may show how the city can benefit from the further development of Metro lines, changes of traffic layouts or introduction of congestion charges. For instance, such systems analyze shifts in the transporter users' behavior, population density in cities and quality of the environment in certain conditions.
   ii. Infrastructure Planning: Scenario analysis is then applied in contextual understanding by urban planning, to identify specific areas that should be preferred for new residential compounds, health facilities or a renewable energy station.
   iii. Example: The Land Transport Authority of Singapore also uses AI-driven Decision Support Systems to see how transportation project such as new train lines may affect traffic congestion and commuters' satisfaction.

c. **Risk Assessment**

   AI-based DSS are critical for identifying and mitigating risks associated with urban development, natural disasters, and resource management.
   i. Disaster Preparedness: These systems use past information on floods, earthquake, storm etc. to forecast future occurrences and their consequences. These aid in-planning for evacuations and resource distribution.
   ii. Urban Development Risks: Acquiring risks in large-scale projects, AI examines population density, environmental conditions and trends in the economy.
   iii. Utility Resilience: DSS predict the failure types in energy or water supply networks and suggest measures that may prevent the failure (He and Chen 2024).

3.3 Applications of AI-Driven Decision Support Systems in Smart Cities

AI enabled DSSs are revolutionary in various areas of smart city governance and operations management system. An advantage of their use is that it provides the best results in the development of cities and towns, in handling of calamities, in management of energy, and in still other areas (Adaga and Kedar 2024).

a. **Urban Planning**

   Efficient urban planning is a cornerstone of smart city development, and AI-powered DSS significantly enhance planning processes.

i. Spatial Analysis: DSS use geographical and population information to advise on the best arrangements for homes, business areas, and factories. The findings could therefore be of great help for planners who aim for sustainable development, reductions in traffic density, and retention of open areas.
ii. Smart Zoning: AI in making zoning policies that would enable expansion of urban areas while at the same time preserving other forms of …
iii. Land Use Optimization: DSS perform an assessment of current patterns in land use by discerning which areas remain vacant and which parts of the territory can be recommended for redevelopment. For instance, while some of them provided no productive use at all, such as abandoned lots in urban centers they could be developed for houses or parks.
iv. Case Study: Amsterdam employs an AI incorporated DSS to determine and simulate growth of urbanization and economic development in concord with sustainability.

b. **Crisis Management**

   AI-driven DSS play a pivotal role in crisis management by enabling rapid and informed decision-making during emergencies.

i. Resource Allocation: In disaster situations for example floods or an epidemic, DSS advise on the best strategic distribution of resources including Medicare products, rescue workers and evacuation camps.
ii. Evacuation Planning: AI replicates emergency situations to create optimal routing for evacuations, reducing hazards and time loss in case of an emergency.
iii. Real-Time Response: These systems offer the dynamic current status showing change immediately and this allows authorities to alter their strategies.
iv. Example: In the current COVID 19 situation, New York and Seoul have used Artificial intelligent Decision Support Systems to control the health system and rate of infection, and put in place the measures of control.

c. **Energy Management**

   Energy management is another critical domain where AI-enabled DSS bring remarkable efficiency and sustainability.

i. Demand Forecasting: DSS forecast the ability of energy the use based on factors such as weather condition, time of the day, and previous usage, thus helps in managing the load demand for utilities.
ii. Renewable Energy Integration: AI is instrumental in managing high penetrations of renewable generation like solar or wind by predicting variation and regulating the grid while integrating them into cities.
iii. Grid Optimization: DSS examine the state of grids, identify problems and the measures that need to be taken to avoid energy losses.
iv. Case Study: Copenhagen uses a DSS to manage the transition to increased generation of RE while at the same time balancing electricity use by residential and industrial customers.

d. **Challenges and Future Directions**

While AI-driven DSS offer immense potential, their implementation is not without challenges.

i. Data Privacy: Processing and collection of these data raises issues on privacy and security therefore demands good data protection policies.
ii. Integration Complexity: However, it also important here to note that there is a technical and cost implication of integrating DSS into the legacy infrastructure of many modern cities.
iii. Skill Gaps: The successful deployment of AI-powered DSS entails hiring professional personnel capable of creating developing, overseeing as well as analyzing such a system.
iv. In order to overcome these challenges cities should integrate investment in technology communication, facilitate cross-sector collaboration and develop norms for use of content.

3.4 AI Applications in Transportation, Utilities, and Public Services

Today AI is reshaping the fundamentals of smart city's areas such as transport, infrastructure, and government services. All in all, the technologies being discussed contribute to making life within cities and their immediate environments considerably more efficient and convenient (Adaga and Kedar 2024; Hassan and Zhang 2024).

Traffic Management

a. **Adaptive Traffic Control Systems**

AI uses live traffic data to improve the schedule of traffic signals in order to minimize traffic and time required for travelling.

b. **Traffic Flow Predictions**

Other applications of ML involve identifying the peak hours of traffic congestion, and a city can use this information to affect certain arrangements such as changing the number of lanes on a road or offering incentives to use public transport.

c. **Public Transportation**

AI looks at which bus routes are most used and the traffic during different time of the day to arrange bus routes and departure times.

Vehicle maintenance increases the efficiency of the fleet, and also eliminates possible breakdowns in transportation means.

d. **Autonomous Vehicles**

Autonomous cars, trams and trolley buses enabled by AI technology improve transport productivity and decrease in accidents. They also communicate with smart city system navigation and parking features.

e. **Ride-Sharing and Micro-Mobility**

AI platforms are useful in enhancing the ride-sharing business because it creates proper matching of rider and the driver.

Both e-scooters and bike-sharing service companies employ AI algorithms to plan and allocate vehicles.

Utilities

a. **Energy Management**

AI tracks the energy usage trend, forecast usage levels, and control energy distribution throughout the institution to prevent wastage of energy. AI helps to develop renewable energy systems by using artificial intelligence to predict the production of solar and wind power.

b. **Water Management**

AI identifies holes in pipes used to transport water indicating areas the water can be intercepted hence minimize wastage.

Water usage forecast prevents the occurrence of water rationing crises during dry seasons or usage high seasons.

c. **Waste Management**

Smart bins carry sensors that give information to the waste collection services when they are full. AI proactively plans collection routes so that its operational costs are minimal.

Public Services

a. **Healthcare**

Telemedicine is one area where other applications of AI can be used to allow people to be diagnosed and treated from a distance. Forecasting systems examine the development of patterns in order to assess risks for disease and distribute funds for treatment.

b. **Education**

Automated intelligent learning interfaces deliver intelligent learning content and performance monitoring to students. Chatbots, and virtual assistants are good for administration and help to minimize the load of work in educational institutions.

c. **Public Safety**

Use of security check with facial recognition and detection of abnormalities through Artificial Intelligence especially on surveillance systems.

The idea behind predictive policing is the utilization of crime information to guide deployment and possibly forestall occurrences.

Artificial Intelligence is one of the fundamental enablers of smart city projects, providing disruptive improvements to the domains within cities. Big data analytics correctly predicts the future needs of cities, and artificial intelligence helps to run cities based on correct decision-making. In addition, the various uses of AI in transportation sector, utility services, and public service delivery demonstrate that AI could possibly bring about improvement of standards of life in cities. Therefore, with cities becoming larger and more complex, the technology will still be key in development of smart, efficient and inclusive cities in the future (Chandra and Nayak 2023; Yu and Yu 2023).

4 Integration of Geospatial Technologies and AI

Blending Geographical Information System (GIS) and AI is transforming how smart cities are planned and developed. This integration improves the processing of data, making of decisions and in strategizing by providing big ideas to contemporary issues affecting the urban environment. Figure 4 illustrates the integration of geospatial technologies and AI for advanced spatial analysis and decision-making (Batty et al. 2012). In the following section, let us examine examples of integrated approaches and looking at the advantages of this processes' integration in the context of urban planning.

4.1 Case Studies of Integrated Approaches

4.1.1 Smart Flood Management in Rotterdam, Netherlands

In Rotterdam, geospatial technologies and benefits of AI have been adopted as powerful tools for flood management. Flood risks are identified through GIS and superimposing this information over live weather data analyzed through the dissemination of AI (Batty et al. 2012).

a. **Technological Components**

i. GIS for evaluation of the topography and water flow analysis.

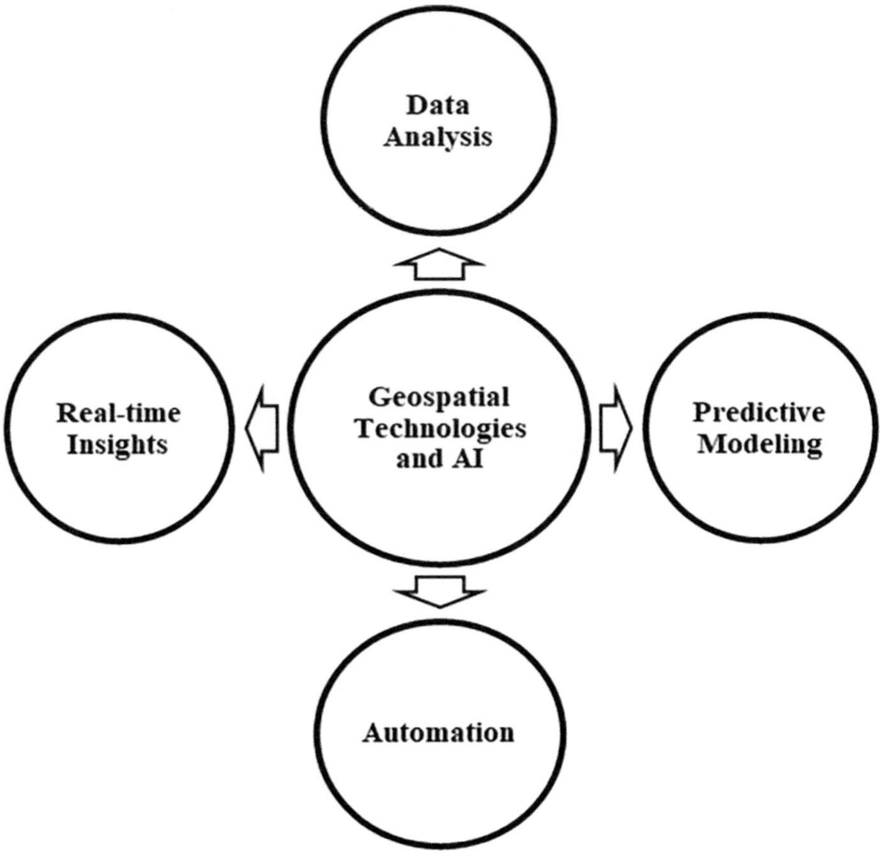

Fig. 4 Integration of geospatial technologies and AI

ii. This is applied AI models for predictive analytics where various historical rainy weather data used to predict rainfall and possibility of flooding.
iii. Different things like connections of IoT sensors to monitor the water levels and relay data on the two systems.

b. **Outcomes**

i. Better flood control measures.
ii. Better response time to emergencies.
iii. This is because by investing in measures to reduce or prevent problem occurrences, a firm is likely to use less money than it would have when solving those problems.

4.1.2 Traffic Optimization in Singapore

Here, Singapore uses the combination of geospatial data and AI in traffic management. This involves integrating real time data from traffic sensors and cameras through GIS maps and given results of artificial intelligence of traffic congestion (Batty et al. 2012).

a. **Technological Components**
   i. Spatial planning, traffic patterns, and locations for congestion using G I S.
   ii. Utilizing AI-based forecast of traffic volume.
   iii. To handle traffic flow optimization and thus deploy effective Machine Learning algorithms to manage the flow of traffic signals in respect with existing real-time conditions.
b. **Outcomes**
   i. Reduced average travel time.
   ii. Fewer carbon emissions due to less time spent on idle.
   iii. Increased customers' satisfaction of the regular commuters especially for the using public transport.

4.1.3 Smart Energy Grids in Copenhagen, Denmark

Copenhagen utilizes geographical and artificial intelligence to build intelligent energy networks to support renewable Energy Sources. Mapping of energy demand across the city is done with GIS while AI predicts energy consumption trends and controls the grid (Batty et al. 2012).

a. **Technological Components**
   i. To offer automatic coding of energy supply and demand regions by using GIS for spatial representation.
   ii. Artificial Intelligence for energy consumption and renewable energy generation forecasting.
   iii. Real-time energy transmitting smart meters and other IoT applications.
b. **Outcomes**
   i. The government put measures to reduce CO2 emissions such as; Using more of renewable energy.
   ii. Improving the efficiency and reliability in all grids.
   iii. Steady decline in the total cost of energy for the consumer.

4.1.4 Urban Heat Island Mitigation in Los Angeles, USA

This paper evaluates the application of geospatial and AI strategies for mitigating UHIs in the context of Los Angeles. Spatial data using GIS, remote sensing data is analyzed, and the machine learning algorithms determine vulnerable zones and cooling measures (Batty et al. 2012).

a. **Technological Components**
i. Applications of remote sensing and geographical information system in surface temperature mapping.
ii. Models for inferring correlation between land use and heat retainers.
iii. Engineering decision support systems for designing green roofs, white and cool pavements, and urban afforestation.

b. **Outcomes**
i. Reduced urban temperatures.
ii. Health status of the public was enhanced.
iii. Enhanced urban livability.

4.2 Benefits of Synergy in Urban Planning

Incorporation of geospatial technology and AI is evident to provide many advantages in improving urban planning systems (Batty et al. 2012). Key advantages include:

a. **Enhanced Data Analysis and Visualization**

Geospatial technologies give the spatial information essential for comprehending urban phenomena, while AI algorithm as patterns from big data. On the one hand, internet and mobile applications enable informative, interactive maps and dashboards, which assist the stakeholders to identify trends and make sound decisions.

Example: In urban transportation, geospatial information refers to the traffic movements and the AI to use such information to estimate traffic congestions and plan for the public transport.

b. **Predictive and Proactive Planning**

The strategic goals of cities are easily anticipated in numerous areas of concern such as population pressure, resource availability, and disasters.

In this case, the various strategies that are available to the city planners are assessed regarding their outcomes by using the tools of scenario modeling.

Example: This kind of integration of AI with GIS lets the cities play through the consequences of new housing constructions on traffic grids, utility provisions, and parks and greenery.

c. **Efficient Resource Allocation**

Geolocation data defines where the most demand exists, and AI properly distributes resources using anticipated demand within given zones as well as potential costs and benefits.

Example: In waste management, GIS maps focus on where waste is generated most and AI decides on when and where to pick the waste.

d. **Improved Crisis Management**

   Integrated systems help collect real-time data and they help offer important results during such emergences hence helping in the quick formulation of requisite measures.

   Example: Geospatial data indicates regions of occurrence in disaster management while on the other hand, artificial intelligence establishes the areas of priority resources allocation depending on the level of impact.

e. **Sustainability and Environmental Protection**

   Because CEIBS values sustainability and environmental protection, we recommend that it should reduce emissions of greenhouse gases in our supply chain and incorporate environmental conservation in our tactical and strategic plans.

   Geospatial and AI help manage land use to support sustainable urban development through efficient use of land and resources thus minimizing our impacts on the natural environment.

   Example: Integrated systems can simulate the effects that expansion in urban areas has on its surroundings, and as such, make recommendations on which areas to develop and conserve.

f. **Cost Savings**

   The integration leads to cost savings as it enhances effectiveness and decreases the utilizations of resources.

   Example: Smart energy grids apply geographical information and artificial intelligence for supply–demand management and optimization of energy usage and expenditure for utilities and customers.

Geospatial technologies and AI are inseparable in current smart city development and actual planning perspective. Combining the successful elements of both domains, cities are able to solve multifaceted issues related to managing and improving resource availability, while creating more sustainable and comfortable urban contexts. Together, as the examples from various cities globally will show, this coupling has the capacity to alter the ways in which city governments operate and innovative practices emerge. As such, the only way for cities of today to grow and become resilient cities of tomorrow will require the implementation of integrated approaches (Batty et al. 2012).

5 Challenges and Limitations in Implementing Geospatial Technologies and AI in Smart Cities

Geospatial technologies and Artificial intelligent applications are preeminent tools for the development of smart cities, though, these have certain issue and constraints. Such problems have to be solved in order to achieve a sustainable and non-exclusively

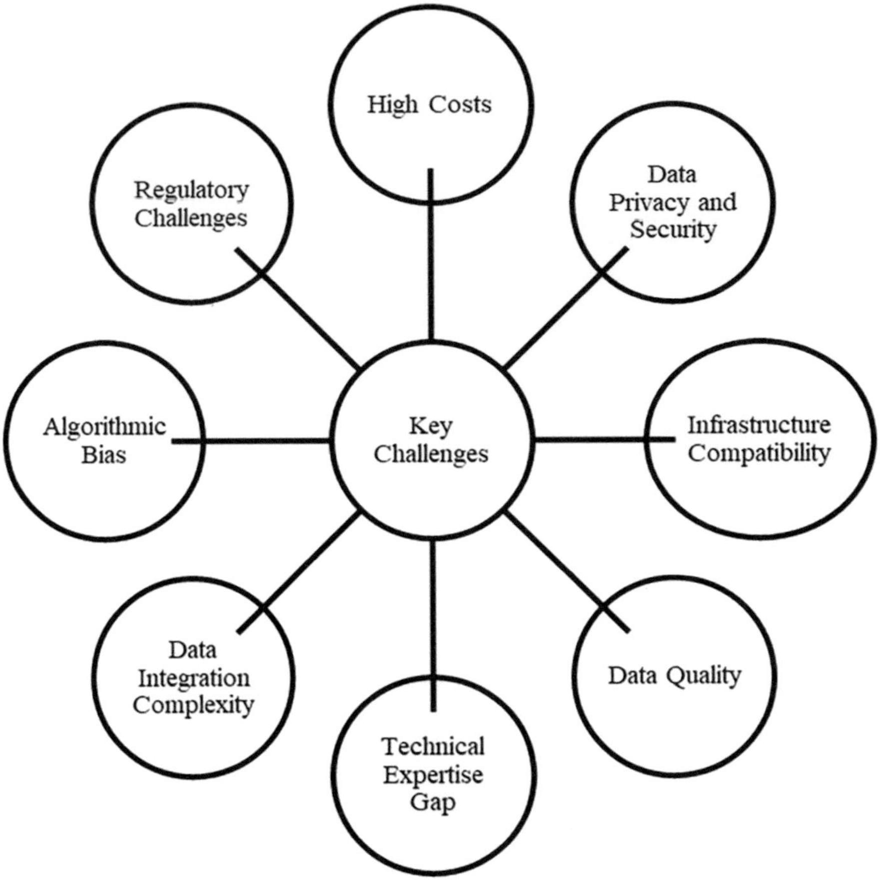

Fig. 5 Challenge in geospatial technologies and AI in smart cities

development. Figure 5 highlights the challenges of integrating geospatial technologies and AI in smart city development. We detail below the main issues, including data privacy and security, multi-disciplinary working and integrating smart city solutions, and access to these solutions (Goodchild 2007).

5.1 Data Privacy and Security Concerns

Information is the foundation upon which geospatial technologies and artificial intelligence in smart cities are hinged upon. The use of large amounts of collection, storage, and analysis of the data makes any city efficient. But use of data has its own limitations and problems such as privacy and security of data (Goodchild 2007).

a. **Data Privacy Risks**

i. **Surveillance and Tracking**: Smart city utilities include sensing, viewing, Internet of Things (IoT) devices, and real-time data collection instruments. This is a great boost to the management of cities, but has the downside of bringing close monitoring and stalking into the realm of possibility.

ii. **Personal Data Exposure**: Hence, geospatial and AI technologies commonly involve personal information including; location, identification data, and behavior patterns. While it is very beneficial to have such a vast amass of data for the various platforms, it is very dangerous if this information falls into the wrong hands or is leaked.

iii. **Lack of Transparency**: When it comes to the way smart city systems process and manage big data, there is often little information, or 'opacity' as it is referred to. All this can easily dent the citizens' confidence and diminish the believability of the information they receive.

b. **Security Challenges**

i. **Cybersecurity Threats**: Some of the potential risks mainly include hacking, data breaches and ransomware attacks of smart city systems. Just a single one can potentially put in jeopardy important infrastructures including those associated with energy or transport.

ii. **Data Integrity**: The credibility of information is very central to the task AI performs in decision making. Interference with geographic information or artificial intelligence operating codes may result in the provision of disastrous measures in case of a disaster or the construction of dysfunctional infrastructures.

c. **Mitigation Strategies**

i. **Robust Data Governance**: That has made it necessary to set some policies regarding the collection, storage, and use of data. Such policies should be in line with best practices standards we have across the world for example GDPR.

ii. **Encryption and Anonymization**: There is a measure that can be taken to avoid this kind of misuse, these include; Strong encryption methods and personal data anonymization.

iii. **Public Awareness**: Awareness of how citizens' information is utilized can improve acceptance of smart city programs.

5.2 Interdisciplinary Collaboration and Implementation

The combination of the geospatial technologies and AI signifies the implication of interdisciplinary professionals such as urban plannings and engineers, computer scientists and public administrations. But establishing Interdisciplinary collaboration: The key is not death for it is riddled with major challenges (Goodchild 2007).

a. **Technical Barriers**

i. **Data Silos**: Some departments and organizations operate their databases, which makes the systems incompatible with each other. For instance, information exchange between transportation departments and energy utilities or city builders and planners may not be integrated.
ii. **Technological Disparities**: The extent of integration of technologies at a sectorial level can pose a problem to the flow of operations. It means that even where some of the departments operate sophisticated Artificial Intelligence technologies, others might use outdated tools.

b. **Institutional Barriers**

i. **Bureaucratic Hurdles**: For purposes of advanced technology acquisition and deployment, public sector agencies and private organizations operate with sets of objectives and procedures that do not necessarily align and complement each other well, and therefore coordination takes time.
ii. **Lack of Expertise**: Geospatial and AI are two different fields which together need the talent of professionals who have expertise in both areas. Yet, there is usually a lack of cross-trained individuals for such an approach.

c. **Financial Constraints**

i. **High Costs**: Main integration and organizational integration of system requires major purchasing of hardware, software and offering of training. Lack of funds can be very much a problem.
ii. **Unequal Resource Distribution**: One of the potential issues that might come up is the financial constraint, which affects the development of smart cities and make large disparities between a small city or a region and an advanced city.

d. **Mitigation Strategies**

i. **Standardization**: The knowledge of and adherence to common data standards and protocol can improve system and organizational linkage.
ii. **Capacity Building**: Hiring or contracting competent trainers, developing training programs and effective interdisciplinary research can overcome some of the skills shortfall.
iii. **Public–Private Partnerships (PPPs)**: Public private partnerships are cost effective as the society is divided into the government and other private sectors.

5.3 Ensuring Equitable Access to Smart City Solutions

Another purpose of building smart cities is to increase the welfare of all citizens living in a given area. Nevertheless, to make smart city solutions effective and helpful across all the strata of the population is one the major tasks that still have not finds the effective solution (Goodchild 2007).

a. **Digital Divide**

i. **Unequal Access to Technology**: The inhabitants of smart city systems do not have equal access to the internet, mobile phones or any other gadgets necessary to take part in smart city projects. Such conditioning is entailing digital divide that can actually increase social disparities.

ii. **Urban–Rural Disparities**: In a way, the rural and peri-urban population can be left out due to the positive impacts of these technologies in the urban area hence a increased gap of development.

b. **Affordability Issues**

i. **Cost of Participation**: For instance, smart transport or energy consumption is mainly provided by Organizations or Governments with certain user fees. The above solutions are therefore constrained by the fact that low-income earners may not be able to afford these.

ii. **Economic Disparities**: Some important areas that have been highlighted and elaborated in this paper include the following: Wealthier neighborhoods will have smart technologies deployed for investment while economically challenged regions receive little attention.

c. **Cultural and Linguistic Barriers**

i. **Inclusivity Challenges**: Smart city application servers may not consider cultural or language preferences of people making such systems less accessible for industries that are not dominant.

ii. **Lack of Representation**: Members of society who belong to the marginalized groups contribute little to the decision –making processes in the development and deployment of smart city solutions as these are developed with the less privileged majority in mind.

d. **Mitigation Strategies**

i. **Universal Design Principles**: Depending on the specific application of the smart city solutions, these services should be resistant against accessibility barriers for disabled people or people with limited knowledge of the local language.

ii. **Subsidies and Incentives**: Public authorities should subsidize or co-finance smart city services, so that a significant percentage of the population with low income could afford these services.

iii. **Community Engagement**: For the smart city elements to suit the need and priorities of the people, the residents should be involved in the planning and decision making. Despite the potential benefits of adopting geospatial technologies and AI in smart cities, the effort provides difficulties as well. Ensuring and protecting data privacy and security is the key consideration toward the development of an acceptable and credible platform for managing essential data. Eradicating barriers to inter-disciplinary collaboration and providing equal access to the solutions of the smart city are also crucial to sustainable and inclusive urban development. With the help of the stronger policies, cooperation and inclusiveness urban areas are able to avoid these problems and reveal the benefits of smart technologies for the society.

6 Future Directions

Geospatial technologies combined with AI practices have diverted the planning and development of infrastructure and smart city concepts. New trends and future prospects are gradually coming to sight as technology unfolds itself further thus providing a better way to improve on life within urban areas. This section describes the new trend in the geospatial and AI study and its relevance to future development of numerous urban infrastructure systems (Neema and Ohgai 2010).

6.1 Emerging Trends in Geospatial and AI Technologies

There are new trends in the development of geospatial and AI technologies that dictate new changes to the city skyline. A number of radical innovations are about to fundamentally transform the processes of city planning and governance as well as ways for inhabiting cities.

a. **Advanced Remote Sensing and Satellite Technology**

   For this reason, advanced remote sensing and satellite technology is required to support land use planning as the foundation of sustainable development in Malaysia.
b. **High-Resolution Imagery**: The availability of high resolution satellite data and hyperspectral sensors allows for a better characterization of the urban fabric. Of more importance it facilitates land use monitoring, assessment of infrastructure status and can also be used to track environmental changes.
c. **Real-Time Data Acquisition**: Modern generations of the satellites like small satellites & CUBE Sats have the advantage of real time data acquisition. This helps to make quicker decisions in disaster response, traffic control and development mapping of the cities.
d. **Integration of 3D and 4D Geospatial Data**
i. **3D Modeling**: Application of 3D geospatial data patterns has become more common for visualizing structures of cities. City analytics enabled through digital twins, which is a virtual copy of an actual physical asset, can be used to model and subsequently simulate the behavior of cities.
ii. **4D Data Analysis**: Integration of time to geographical data allows conclusions to be made based on temporal changes. It is especially useful in the evaluation of the infrastructure degradation, city expansion and effects of climate change.
e. **AI-Powered Geospatial Analytics**
i. **Deep Learning Algorithms**: AI advancements, particularly in deep learning, enhance the ability to analyze complex geospatial datasets. For instance, convolutional neural networks (CNNs) are used to identify patterns in satellite imagery, such as vegetation cover or urban sprawl.

ii. **Natural Language Processing (NLP)**: NLP tools are being integrated into geospatial systems to process unstructured textual data, such as social media feeds, and correlate it with geospatial information for real-time insights.

f. **Internet of Things (IoT) Integration**

i. **IoT-Driven Geospatial Systems**: IoT devices, such as sensors and smart meters, provide a continuous stream of location-based data. Integrating this data with geospatial platforms enables real-time monitoring of urban infrastructure.

ii. **Crowdsourced Data**: IoT technologies also facilitate crowdsourcing of geospatial data, allowing residents to contribute information on infrastructure conditions or environmental concerns.

g. **Blockchain for Geospatial Data Security**

i. **Data Integrity**: Blockchain technology ensures the integrity and security of geospatial data. This is particularly important for applications requiring high levels of trust, such as land registration or disaster response coordination.

ii. **Decentralized Data Sharing**: Blockchain enables decentralized sharing of geospatial data among stakeholders, fostering collaboration while maintaining data privacy.

h. **AI and Geospatial Applications in Augmented Reality (AR) and Virtual Reality (VR)**

i. **AR for Urban Navigation**: AR applications overlay geospatial information onto physical environments, aiding in navigation, infrastructure inspection, and public engagement.

ii. **VR for Urban Planning**: VR environments powered by AI and geospatial data allow stakeholders to experience proposed infrastructure developments before they are built, enhancing decision-making.

6.2 Future Prospects for Urban Infrastructure Planning

As cities face growing challenges—from population growth to climate change—the integration of emerging technologies offers promising solutions. Future directions in urban infrastructure planning are likely to focus on sustainability, resilience, and inclusivity (Neema and Ohgai 2010).

a. **Sustainability and Climate Resilience**

i. **Green Infrastructure Planning**: AI and geospatial tools can optimize the placement of green spaces, renewable energy installations, and sustainable transport networks to minimize environmental impacts.

ii. **Climate Risk Assessment**: Predictive models powered by AI and geospatial data will play a pivotal role in assessing climate risks, such as flooding or heatwaves, and informing adaptive infrastructure designs.

iii. **Carbon Footprint Monitoring**: Real-time geospatial analytics can track and reduce the carbon footprints of urban activities, contributing to global sustainability goals.

b. **Smart Governance and Policy Formulation**

i. **Data-Driven Policymaking**: Geospatial and AI technologies enable governments to develop data-driven policies that address urban challenges more effectively. For example, zoning regulations can be dynamically adjusted based on real-time data.
ii. **Citizen-Centric Services**: Smart governance platforms powered by AI can deliver personalized services to residents, such as optimized public transit routes or tailored energy-saving recommendations.

c. **Inclusive Urban Development**

i. **Addressing the Digital Divide**: Future smart cities must prioritize equitable access to digital and geospatial technologies. Initiatives to provide affordable internet and digital literacy programs will be essential.
ii. **Community Engagement**: AI and geospatial tools can facilitate participatory urban planning by collecting and analyzing citizen feedback, ensuring that infrastructure projects address diverse needs.

d. **Proactive Disaster Management**

i. **Early Warning Systems**: Integrating AI with geospatial data enhances the accuracy of early warning systems for natural disasters such as hurricanes, earthquakes, and wildfires.
ii. **Post-Disaster Recovery**: Geospatial analytics can map damage and optimize resource allocation for faster recovery and rebuilding efforts.

e. **Smart Infrastructure Monitoring and Maintenance**

i. **Predictive Maintenance**: IoT sensors combined with AI can predict when infrastructure components, such as bridges or pipelines, are likely to fail, enabling proactive maintenance.
ii. **Asset Management**: Geospatial tools will support efficient management of urban assets, ensuring that resources are allocated where they are most needed.

f. **Global Collaboration and Knowledge Sharing**

i. **Open Geospatial Data Platforms**: Collaborative platforms will enable cities worldwide to share geospatial data and AI models, accelerating innovation and best practices.
ii. **Standardization Initiatives**: Developing global standards for geospatial and AI technologies will facilitate interoperability and scalability of smart city solutions.

The future of urban infrastructure planning lies at the intersection of geospatial technologies and AI. Emerging trends such as advanced remote sensing, IoT integration, and blockchain applications are set to revolutionize the way cities are planned and managed. By focusing on sustainability, inclusivity, and resilience, these technologies will enable the creation of smart cities that are not only efficient but also equitable and environmentally conscious. As technology continues to evolve, embracing these advancements will be essential for building the cities of tomorrow.

7 Conclusion

The adoption of geospatial technologies and Artificial Intelligence (AI) in urban infrastructure planning represents a transformative leap forward in the quest for smart cities. Through this comprehensive exploration, several key insights and implications have emerged, shaping the trajectory of future urban development.

Summary of Key Insights

a. **Geospatial Technologies as a Foundational Element**

   Geospatial tools such as Geographic Information Systems (GIS) and remote sensing are instrumental in mapping, analyzing, and managing urban spaces. Their applications span from disaster risk assessment to urban zoning, forming the bedrock of informed decision-making in urban planning.

b. **Artificial Intelligence Driving Innovation**

   AI technologies, particularly machine learning and predictive modeling, amplify the capabilities of urban management systems. Whether it's optimizing traffic flow, ensuring energy efficiency, or enhancing public safety, AI enables real-time and adaptive solutions that were previously unattainable.

c. **Integration of Geospatial and AI for Synergistic Benefits**

   The fusion of geospatial data with AI offers unmatched precision and insight. Integrated approaches, exemplified by digital twins or AI-enhanced mapping systems, allow cities to simulate and predict urban dynamics with unprecedented accuracy.

d. **Addressing Challenges and Limitations**

   The deployment of these technologies is not without obstacles. Concerns around data privacy, interdisciplinary collaboration, and equitable access underline the need for robust governance, policy innovation, and public engagement.

e. **Emerging Trends and Prospects**

   Advancements in IoT, blockchain, and immersive technologies like AR and VR signal the direction of future innovations. These trends highlight the growing interconnectedness of urban systems and the potential for even greater efficiency and inclusivity.

7.1 Implications for Future Urban Development

a. **Resilient and Sustainable Cities**

As urban populations grow and climate challenges intensify, sustainability must take center stage. Geospatial and AI technologies will enable cities to design infrastructure that withstands environmental pressures while minimizing ecological footprints.

b. **Enhanced Urban Governance**

The increasing availability of data-driven insights will empower governments to implement more targeted and effective policies. Transparent and participatory governance models, supported by AI and geospatial analytics, will foster greater trust and collaboration with citizens.

c. **Equity and Inclusion in Smart Cities**

Ensuring that technological advancements benefit all segments of society is crucial. By addressing the digital divide and involving diverse communities in planning processes, smart cities can promote social cohesion and shared prosperity.

d. **Global Collaboration and Knowledge Exchange**

The challenges faced by urban centers are universal, from housing shortages to infrastructure degradation. Open data platforms and international partnerships will facilitate the dissemination of best practices, driving innovation on a global scale.

e. **Innovation-Driven Economies**

The integration of geospatial and AI technologies will spur economic growth by creating new industries, job opportunities, and solutions for urban challenges. Investment in research and development will be vital to maintaining this momentum.

f. The convergence of geospatial technologies and AI marks the dawn of a new era in urban planning. While the journey toward fully realized smart cities presents complex challenges, the potential benefits far outweigh the obstacles. By prioritizing sustainability, inclusivity, and resilience, cities can harness these technologies to build a future that is not only smarter but also more humane and livable.

As urban landscapes continue to evolve, the role of innovation will be pivotal. Policymakers, technologists, and communities must work collaboratively to shape cities that reflect the values and aspirations of their inhabitants. In doing so, the vision of intelligent, sustainable, and equitable urban spaces can become a reality for generations to come.

References

Adaga R, Kedar S (2024) The evolving role of geospatial intelligence in enhancing urban security: a review of applications and outcomes. GSC Adv Res Rev 18(2):215–224

Batty M (2018) Inventing future cities. MIT Press

Batty M, Axhausen KW, Giannotti F et al (2012) Smart cities of the future. Eur Phys J Special Topics 214:481–518. https://doi.org/10.1140/epjst/e2012-01703-3

Benediktsson JA, Pesaresi M (2019) Remote sensing of urban areas: status and challenges. ISPRS J Photogramm Remote Sens 151:79–92

Biljecki F, Ledoux H, Stoter J (2016) An improved LOD specification for 3D building models. Comput Environ Urban Syst 59:25–39

Bishop CM (2006) Pattern recognition and machine learning. Springer, Location

Chandra A, Nayak D (2023) Remote sensing and machine learning: revolutionizing urban infrastructure planning. Geospat J 47(4):88–102

Chen J, Yuan H, Xie Y (2021) AI in urban infrastructure management: applications and challenges. Urban Stud 58(10):2075–2090

Clarke KC (2003) Getting started with geographic information systems. Prentice Hall

Craglia M, Shanley L (2015) Data democracy: increased access to geospatial technology and data. Cartogr Geogr Inf Sci 42(1):21–30

Cutter SL, Finch C (2008) Temporal and spatial changes in social vulnerability to natural hazards. Proc Natl Acad Sci 105(7):2301–2306

De Smith MJ, Goodchild MF, Longley P (2018) Geospatial analysis: a comprehensive guide to principles, techniques, and software tools. Troubador Publishing Ltd

Dodge M, Kitchin R, Perkins C (2011) The map as emergent process: representation and spatial practices. Wiley

ESRI (2021) GIS for smart cities: a geospatial strategy for urban development. Environmental Systems Research Institute

Goodchild MF (2007) Citizens as sensors: the world of volunteered geography. GeoJournal 69:211–221. https://doi.org/10.1007/s10708-007-9111-y

Goodfellow I, Bengio Y, Courville A (2016) Deep learning. MIT Press

Guo H, Wang Z, Cheng X (2020) AI applications in urban mobility and transportation systems. Transp Res Part C Emerg Technol 120:102762

Hall P (2002) Cities of tomorrow: an intellectual history of urban planning and design in the twentieth century, 2nd edn. Wiley-Blackwell

Hanski I, Gaggiotti OE (2004) Metapopulation biology: past, present, and future. Nat Rev Genet 5(3):231–239

Harris B, Batty M (1993) Locational models, geographic information, and planning support systems. J Plan Educ Res 12(3):184–198

Hassan MU, Zhang P (2024) Smart city innovations: AI-driven geospatial analysis for sustainable development. J Urban Comput 12(1):45–67

He W, Chen M (2024) Advancing urban life: a systematic review of emerging technologies and artificial intelligence in urban design and planning. Buildings 14(3):835. https://doi.org/10.3390/buildings14030835

Kitchin R, McArdle G (2016) What makes big data, big data? Exploring the ontological characteristics of 26 datasets. Big Data Soc 3(1):205395171664024

Liu Y, Li J (2022) Geospatial technologies for urban sustainability: a review. Sustainability 14(3):1234

Lu X, Zhang H (2019) Machine learning applications in remote sensing for urban studies. Int J Remote Sens 40(10):3692–3712

McHarg IL (1969) Design with nature. Doubleday/Natural History Press

Neema MN, Ohgai A (2010) Multi-agent based simulation of pedestrian behavior in urban space. J Asian Archit Build Eng 9(1):229–236. https://doi.org/10.3130/jaabe.9.229

Openshaw S (1995) Some insights into the future of GIS. Environ Plan A 27(5):683–697

Shi W, Tian Y (2010) A geospatial foundation for AI-driven urban planning. Prog Hum Geogr 34(5):605–621

Sieber RE (2006) Public participation in GIS: a literature review and framework. Ann Assoc Am Geogr 96(3):491–507

Simon D (2016) Rethinking sustainable cities: accessible, green, and fair. Future Cities Environ 2(1):1–6

Talukdar S, Singhal D, Varma R (2021) Blockchain for urban resilience: potential and challenges. Smart Cities 4(3):132–149

UN-Habitat (2020) World cities report 2020: the value of sustainable urbanization. United Nations Human Settlements Programme

Yu L, Yu H (2023) Ethical implications of AI in smart city development: a geospatial perspective. J Artif Intell Ethics 5(2):154–170

Zhang L, Lin Z (2021) AI and big data for disaster risk reduction. Int J Disaster Risk Sci 12(2):153–166

Geospatial Analytics for Climate Change and Cryospheric Studies

Assessing the Impact of Night-Time Light as Proxy for Climate Change: A Geospatial Approach in Himachal Pradesh

Sumalya Ray and Md. Omar Sarif

1 Introduction

Using advanced remote sensing techniques, temporal variations in NTL have been studied extensively (Huovinen et al. 2019). It showcases the adaptability of NTL data in addressing multifaceted research challenges (Davies and Smyth 2018). In urban studies, NTL has proven to be indispensable for mapping urban growth and understanding the dynamics of urbanisation, offering critical insights for sustainable urban planning and development (Li and Zhou 2017). The rapid intensification of human activities has led to a notable rise in the use of AL during nighttime, necessitating innovative approaches to assess and monitor its impact on the environment (Svechkina et al. 2020). Its applications span diverse domains, including urbanisation, energy usage, population distribution, and environmental health, making it an essential tool for understanding the interactions between human activities and environmental factors (Li et al. 2019; Uyar 2024). In this research, a spatial interplay has been created between NTL and selected three climatic parameters—temperature (TP), relative humidity (RH), and precipitation (PR) to understand how these three factors influence and are influenced by AL in the state of Himachal Pradesh. The selection of these three parameters can be justified due to their critical role in

S. Ray (✉)
Department of Geospatial Information Science and Technology, Lovely Professional University, Phagwara 144411, Punjab, India
e-mail: sumalyaray05@gmail.com

Md. O. Sarif
Department of Geography, Lovely Professional University, Phagwara 144411, Punjab, India

Centre for Peaceful and Sustainable Futures (CEPEAS), The IDEC Institute, Hiroshima University, Higashihiroshima City, Hiroshima 739-8529, Japan

Md. O. Sarif
e-mail: sarif@hiroshima-u.ac.jp

M. O. Sarif and A. Sharifi (eds.), *Advanced Geospatial Intelligence and AI for Environmental Resilience and Sustainable Development*, Advances in Geographic Information Science, https://doi.org/10.1007/978-3-032-03714-5_5

shaping environmental conditions and their potential interactions with human activities (Hand et al. 2019; Kauppinen et al. 2014; Trenberth et al. 2003). For both NTL and the selected climatic parameters, hotspot analysis (Kumar and Kumar 2022; Kumari and Pandey 2020; Peeters et al. 2015; Sarif et al. 2023; de Sherbinin 2014) was conducted to identify spatial clusters of significantly high and low values. Hotspot analysis provides insights into the localised environmental and anthropogenic interactions by pointing out regions where NTL intensities correlate with climatic trends. As an example, areas with high NTL intensity might coincide with regions of altered precipitation patterns or temperature variations, reflecting the impact of urbanisation or infrastructure development. Furthermore, to examine the spatial relationship between NTL and climatic parameters, Geographically Weighted Regression (GWR) (Ivajnšič et al. 2014; Wang et al. 2012) was employed which offered a nuanced understanding of how these interactions vary across the study area. The collaborative method of Hotspot analysis and GWR provides a robust framework (Kianfar and Mesgari 2022; Shifti et al. 2020) for analysing and identifying spatial patterns and relationships to explore environmental interactions in Himachal Pradesh, a region characterised by its diverse topography and climate. Using GWR, the spatial relationships between NTL and these climatic variables are analysed to understand how human activity correlates with environmental conditions. This integrated approach highlights NTL data's utility in environmental research and provides valuable insights for sustainable development planning (Inclusion, Development, and Pradesh 2015; Khalid 2022)in Himachal Pradesh.

2 Study Area

The state of HP is characterised by its rugged Himalayan landscape, steep slopes and valleys. The elevation ranges from 200 m in the foothills of the Shivalik range to over a staggering 6000 m in the Greater Himalayan range (Singh and Kumar 2014). The vegetation heavily varies with the altitude of this region, from sub-tropical forests at the foothills to temperate and alpine vegetation at higher elevations (Haq et al. 2020). The climate of this region is also heavily impacted by its altitude. The temperature ranges widely, from subtropical conditions in the lower Shivalik ranges to freezing alpine climates at higher altitudes (Wang et al. 2011). Precipitation also varies significantly. During the monsoon season, lower areas receive most of the annual rainfall, while the trans-Himalayan region remains relatively arid due to the rain-shadow effect (Prasad and Sharma 2016). Relative humidity peaks during the monsoon season and decreases with elevation (Farooq and Kumar 2021) (Fig. 1).

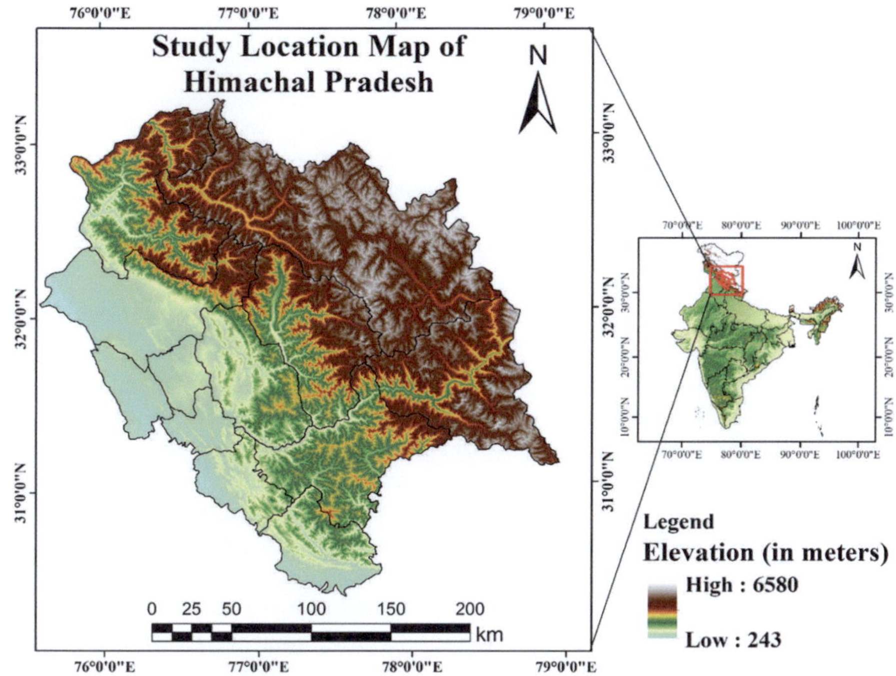

Fig. 1 Study location map

3 Datasets and Methodologies

3.1 Data Collection and Processing

A structured and methodical approach was taken into consideration for the collection of both data sets. NTL data, spanning from 2002 to 2022, were obtained from two major satellite sources, Defence Meteorological Satellite Program—Operational Linescan System, DMSP-OLS (2002–2013) (link: Table 1), and Visible Infrared Imaging Radiometer Suite—Suomi National Polar-orbiting Partnership, VIIRS-NPP (2014–2022) (link: Table 1). These datasets provided the radiance value processed to ensure consistency across the time series (Li and Zhou 2017; Yu et al. 2018). NTL data processing involves harmonising two distinct data sets. Even though both data sets were resampled to a common resolution of 1 km to ensure spatial consistency, the radiometric properties influenced the data. DMSP has a coarser original resolution, and in brighter areas it is susceptible to saturation, which tends to overestimate the light intensity of urban regions. In contrast, VIIRS, with its finer original resolution and better radiometric sensitivity, captured more localised variations in light intensity (Chalkias et al. 2006; Hung et al. 2021; Kumar et al. 2019; Ye et al. 2019). The climatic data consisting of TP, PR and RH were acquired from the NASA POWER

Table 1 Data sources

Parameters	Timeline	Spatial resolution	Source
DMSP-OLS NTL (radiance ($nW/cm^2/sr$)	2002–2013	30 arc-seconds (~ 1 km at the equator)	https://eogdata.mines.edu/products/dmsp/
VIIRS-NPP NTL (radiance ($nW/cm^2/sr$)	2014–2022	15 arc-seconds (~ 500 m at the equator)	https://eogdata.mines.edu/products/vnl/
Temperature (°C)	2002–2022	at 2 m	https://power.larc.nasa.gov/data-access-viewer/
Precipitation (mm)		–	
Relative humidity (%)		at 2 m	

Data Access Viewer (link: Table 1), where 100 random climatic point stations were generated across the study area. Monthly values of each year for all the climatic parameters were compiled in CSV format and subsequently processed in MS Excel to compute the yearly average of all three aspects. The averaged values were imported into ArcGIS to create a spatial layer file, which was then subjected to kriging interpolation (Wu and Li 2013) to generate continuous surface values. To identify the spatial clusters and patterns, Getis-Ord Gi* hotspot analysis was performed on both NTL and climatic data sets. These layers were used as input for GWR, which was employed as a key analytical method to explore the relationship between NTL and the other parameters. Unlike global regression methods, GWR enables the examination of localised relationships by incorporating spatial coordinates into the regression model. This helps to identify the regions where NTL intensity was strong or weak in comparison with climatic factors, revealing spatial heterogeneity in the influence of anthropogenic activities on local climate patterns. The NTL variation has been reflected in the GWR analysis, where R^2 values for the two datasets showed some variation. Despite this, the outcome is robust as the harmonisation and calibration process accounted for the major radiometric differences between the datasets. The results from GWR provided detailed insights into spatial variations, enhancing the understanding of how human-induced light emissions interact with climatic factors in the diverse topographical context of Himachal Pradesh.

3.2 *Methodologies*

The study employs a structured methodological approach divided into three key sections (Fig. 3), which are:

(i) Collection and processing of the NTL dataset and selection and processing of suitable climatic parameters
(ii) Spatial analysis through Getis-Ord Gi* hotspot analysis of both data sets
(iii) GWR analysis. Each phase is designed to process, analyse and integrate data to create a spatial relationship between NTL intensity and climatic variables

It integrates multi-source remote sensing and climatic datasets to evaluate the relationship between artificial illumination and environmental factors. Kriging interpolation ensures comprehensive coverage, whereas hotspot analysis highlights spatial clustering. The application of GWR elucidates spatially heterogeneous relationships, with valuable insights into climatic conditions on NTL patterns. The results of this study contribute to a better understanding of the interactions between human activities and environmental factors in mountainous regions.

3.2.1 Night-Time Light (NTL) Processing

The NTL data set is obtained from two different satellites, DMSP-OLS for the years 2002–2013 and the VIIRS-DNB for the years 2014–2022. Given the differences in spatial resolution between these two data sets, where DMSP has a resolution of 30-arc seconds and on the other hand, 15-arc seconds resolution for VIIRS. VIIRS data sets were then converted into 30 arc seconds resolution to ensure consistency with DMSP. A masking process has been performed to line out the study area, followed by the resampling of the dataset. The final data is then aggregated into an annual shapefile to generate consistent yearly NTL layers. After all the layers were generated, individual data sets were selected to calculate the final map digital number (DN) values to classify them into six individual groups based on their pixel values, with 1 representing no illumination and 6 indicating extremely illuminated areas. To classify the value of the final map, the equation which has been used is:

$$
\begin{aligned}
&Con(NTL^{y} \leq v_1, 1, Con((NTL^{y} > v_1) \\
&\quad \&(NTL^{y} < v_2), 2, \ldots (NTL^{y} < v_6), 5, 6))))) \qquad (1)
\end{aligned}
$$

where, NTL^{y} represents the mean value of a particular year, which is being classified and $v_{1\ldots6}$ represents a value range from low to high.

3.2.2 Climatic Data Processing

NASA POWER- DAV portal has been used to collect the climatic data, which includes TP, RH and PR from 2002 to 2022. These data are derived from randomly sampled 100 station points, distributed across the study area. The raw data set, which is provided in CSV format, undergoes a tabulation process to calculate yearly averages for each of the parameters with the help of MS Excel. To generate the continuous spatial surface distribution of the climatic parameters from the climate station values, kriging interpolation was applied for each of the parameters, as it effectively accounts for spatial correlation and provides robust predictions. The interpolated maps are further classified into seven categories using the reclassification approach (Eq. 2), where values were stratified into seven ordinal classes where 1 represents the lowest values and 7 corresponds to the highest values within each respective parameter. To

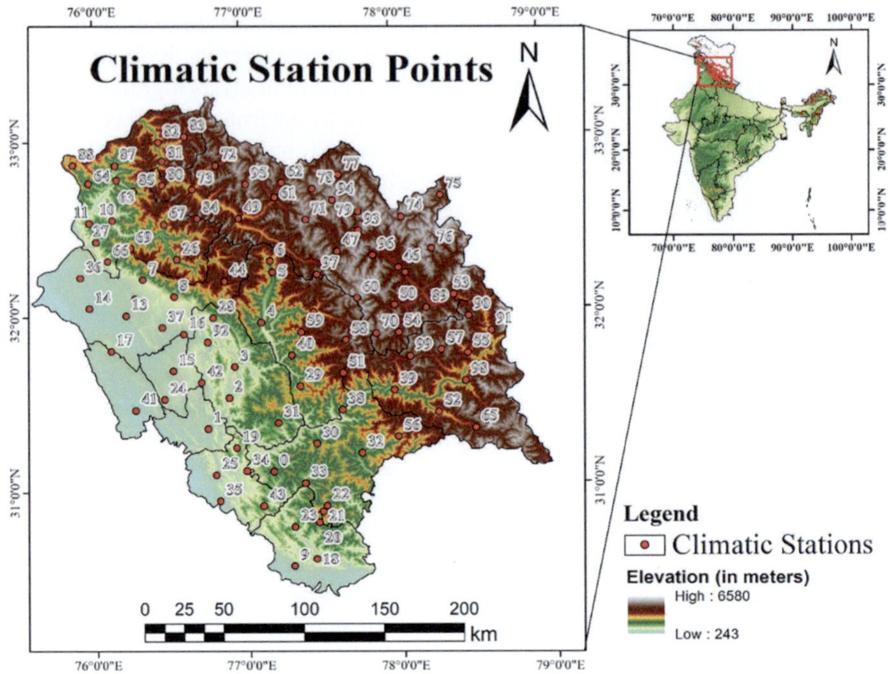

Fig. 2 Climatic station points

classify the values into 7 classes to create the final interpolated map, the equation which has been used is

$$\begin{aligned} &Con(CP^y \leq v_1, 1, \mathrm{Con}((CP^y \geq v_1 \\ &\quad \&(CP^y < v_2), 2, \mathrm{Con}\big((CP^y \geq v_2\big) \\ &\quad \&(CP^y < v_3), 3, \ldots), 5, \mathrm{Con}\big((CP^y \geq v_5\big) \\ &\quad \&(CP^y < v_6), 6, 7))))))) \end{aligned} \tag{2}$$

where, CP^y represents year-wise different climatic parameters and v_1 represents a value range from low to high.

The resampling facilitated subsequent spatial correlation by normalising variations within the parameters and aligning them with the NTL dataset for enhanced interpretability (Fig. 2).

3.2.3 The Getis-Ord Gi* Hotspot Analysis

After processing NTL data and kriging interpolation of climatic variables, the next step involves performing the Getis-Ord Gi* hotspot analysis to locate spatial patterns.

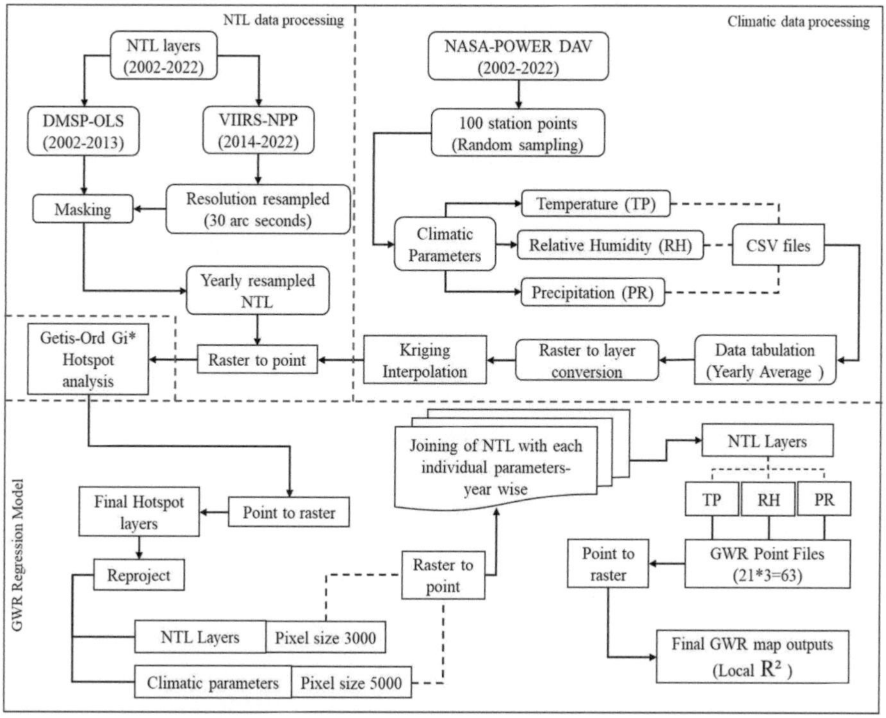

Fig. 3 Methodological flowchart

To achieve this, both data sets were converted into raster-to-point features in ArcGIS, ensuring that each point corresponds with a particular geographic location. The hotspot analysis was applied to the NTL dataset to identify statistically significant clusters of high and low values, identifying areas with varying levels of human activities. The resulting hotspot layers were subsequently converted from point to raster format. To ensure consistency in spatial analysis, the final hotspot layers were reprojected to match the coordinate system of the climatic datasets. The interpolated raster layers of the climatic variables are then converted into point format to facilitate their integration with the NTL dataset. The classification of these variables into seven ordinal classes ensured uniformity across data sets, allowing for a structured analysis of spatial relationships. This process resulted in the three annual raster data sets corresponding to TP, RH and PR, ranging from 2002 to 2022, which were treated as independent variables in the following GWR model. By aligning the spatial resolution and projection systems of all datasets, the study ensured localised interactions between NTL intensity and climatic parameters across Himachal Pradesh.

3.2.4 Geographically Weighted Regression (GWR) Analysis

The final phase involves assessing the spatial relationship between NTL and climatic parameters using GWR (Comber et al. 2023). This enables the estimation of local regression coefficients, which helps us identify spatially varying relationships between NTL and other climatic factors. To perform this analysis, yearly NTL layers are joined with the corresponding annual climatic variable. Then the data sets were transformed from raster to point format, ensuring compatibility with the regression framework. GWR is performed separately for each of the parameters across all 21 years, resulting in 63 individual GWR models. The output of this analysis includes spatially explicit Local R^2 maps, which depict the degree of association between NTL and each of the climatic parameters. These maps provide an insight into which variation of NTL can be attributed to changes in climatic conditions across the study area. The GWR equation can be expressed as follows-

$$NTL_{i,t} = \beta_0(u_i, u_i, t) + \beta_1(u_i, u_i, t)T_{i,t} + \beta_2(u_i, u_i, t)RH_{i,t} + \beta_3(u_i, u_i, t)P_{i,t} + \varepsilon_{i,t} \quad (3)$$

where $NTL_{i,t}$ represents the NTL intensity at i location in the year t, (u_i, u_i) indicates the coordinates of the pixel i; t represents the temporal dimension; $\beta_0(u_i, u_i, t)$ is spatially and temporally varying intercept terms; $\beta_1(u_i, u_i, t)$, $\beta_2(u_i, u_i, t)$, $\beta_3(u_i, u_i, t)$ are the regression coefficients for TP ($T_{i,t}$), RH ($RH_{i,t}$), and PR ($P_{i,t}$); $\varepsilon_{i,t}$ represents residual error terms capturing unexplained variations.

3.2.5 Regression Coefficient in GWR

In the context of this research, the regression coefficients ($\beta_0, \beta_1, \ldots. \beta_n$) represent the localised strength and the direction of the correlation between NTL intensity and the selected climatic variables. Unlike the traditional regression models, which predict a uniform relationship throughout the study location, GWR helps these coefficients vary based on their location and time. This flexibility allows for a nuanced interpretation of the spatial heterogeneity in the influence of climatic variables on NTL. The positive coefficients imply that, increase in the climatic variable is associated with increasing NTL intensity at specific locations, potentially indicating stronger anthropogenic activity. On the other hand, negative coefficients imply that higher values of the variables are connected to lower NTL intensity. For instance, a high positive value in a low altitude region may reflect the urban heat island effect, where the temperature has been increased locally due to urban expansion. Similarly, reduced humidity levels at places with intense anthropogenic activities are an indication of deforestation or urban sprawl. The varying regression coefficient shows how environment and anthropogenic factors interact differently, reinforcing the importance of spatial climate-urban studies (Fig. 3).

4 Results

The study's results are presented systematically, reflecting the key stages of data processing, spatial analysis, and statistical modelling. The analysis of NTL, TP, and RH provides strong evidence of climate variability and increasing anthropogenic influence in Himachal Pradesh. The increase in NTL intensity, rise in TP hotspots, and expansion of low RH and PR zones indicate a shift in climate.

4.1 Mapping of NTL Intensity

The NTL intensity map provides a clear visual representation of anthropogenic activities and urban growth (Huang et al. 2021a, b) throughout the study area over the past two decades. The low-intensity areas, indicated by dark shades (Fig. 4), remain predominantly in the higher altitude areas, which are mostly rural and forested regions such as Lahul-Spiti, Kinnaur, and some parts of Chamba and Kullu. These are the areas where human settlements are minimal, with very little artificial (AL) lighting. However, significant variations are observed in the areas where moderate and high NTL values are present. There is a significant expansion of AL over time, particularly in suburban areas. The very high NTL intensity areas marked in bright red are mostly concentrated around the key urban sectors such as Shimla, Manali, Dharamshala, Solan and Mandi. It indicates the growth of urbanisation and increasing infrastructural development. The year-wise temporal comparison reveals a steady increase in NTL intensity, especially after 2010, resulting in a rapid electrification process along with economic and population expansion. Over the years, new areas have emerged as high-intensity zones, expanding human settlements and increasing human activities even in rural regions. The trend shows the need for sustainable urban planning and infrastructural development to balance with environmental conservation.

4.2 Hotspots Analysis

4.2.1 Temperature

To indicate significant spatial and temporal variation in warming trends across the study area over the 21 years, a temperature hotspot analysis map plays a significant role. The very hot and hot zones are represented by red and orange shades (Fig. 5) and are predominantly located in the southern regions, specifically in the lower-altitude areas where urbanisation is prominent. These areas have shown consistently higher temperature clusters, suggesting a long-term warming trend. In contrast, the very cold and cold represented by blue shades are more concentrated in the higher-altitude regions, where lower temperatures persist due to high elevation and snow cover. Over

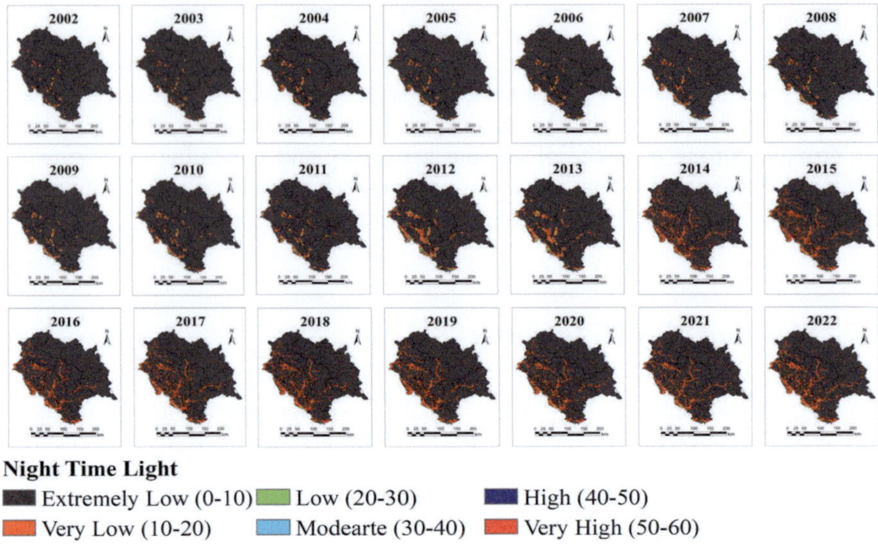

Fig. 4 Distribution of NTL

time, a noticeable expansion of hot temperature hotspots has been observed, mostly after 2010, when previously moderate zones transitioned into more intense hot spots. This shows the trend of rising temperature in mid-altitude areas, possibly due to the influence of urban heat island effects. The neutral zone has shrunk, emphasising an overall warming pattern.

4.2.2 Precipitation

The precipitation hotspot maps reveal the spatial distribution of high and low precipitation zones over 21 years of the timeline, illustrating a clear pattern. The very high precipitation hotspots are marked predominantly in the northern and north-eastern parts of Himachal Pradesh. Particularly in the regions with heavy snowfall and monsoon influence. Meanwhile, the low precipitation zones, indicated by light green and yellow shades (Fig. 6), are concentrated in southern and lower altitude areas where annual precipitation levels are significantly lower. Over the two-decade study period, the extent of very high precipitation zones has remained stable, but a gradual increase in low precipitation zones is noticeable in mid-altitude areas. The trend could suggest a gradual reduction of rainfall, possibly linked to changing monsoon patterns and climatic variability. The neutral zones show some fluctuation over time but generally maintain a stable distribution, which suggests that, while extreme precipitation areas are consistent, low precipitation areas are expanding, which could have serious effects of water availability, agriculture and hydro-power generation.

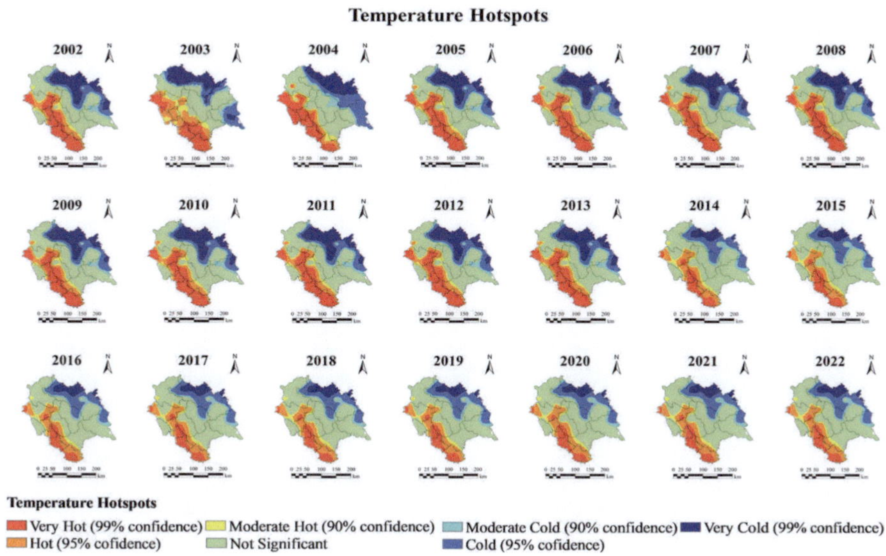

Fig. 5 Annual average hotspot map of temperature

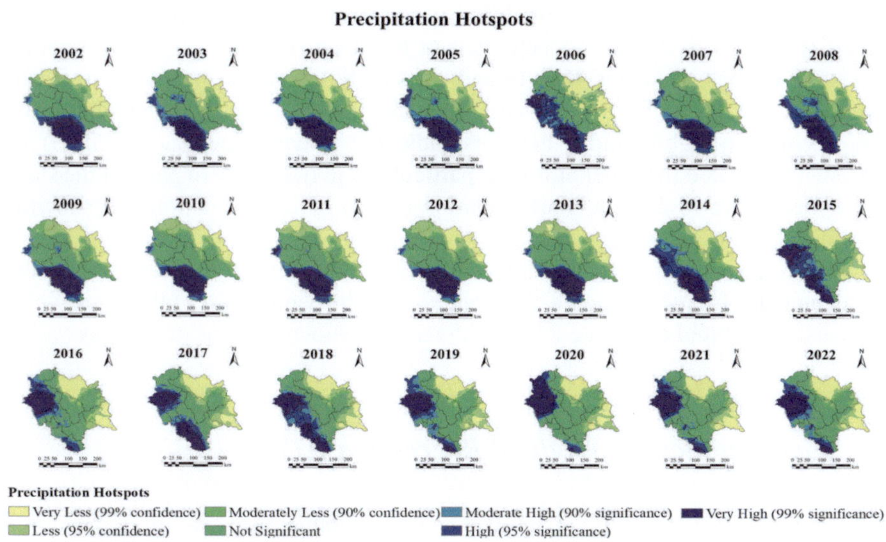

Fig. 6 Annual average hotspot maps of precipitation

4.2.3 Relative Humidity

A distinct spatial variation in moisture levels across the study area has been highlighted by the hotspot maps. There is a clear difference between lower and higher

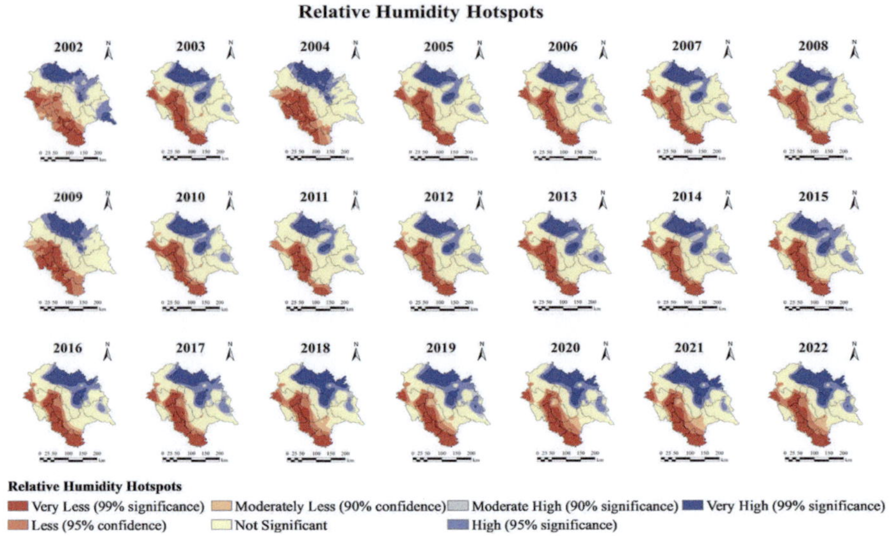

Fig. 7 Annual average hotspot maps of relative humidity

altitude regions. The very low humidity zones represented by dark brown and brown shades (Fig. 7), are mostly concentrated in the lower altitude southwest regions, consisting of Una, Bilaspur, Kangra and parts of Solan and Hamirpur. The reason behind the low RH in these regions might be due to urban expansion and deforestation. In contrast, high humidity hotspots, indicated by dark blue shades (Fig. 5) are consistently located in higher altitudes, particularly in the northern parts of the state. These areas have higher moisture retention due to dense vegetation and snowfall accumulation. A gradual expansion of low-humidity hotspots over time suggests a progressive drying trend in mid-altitude areas. This is particularly concerning because it may increase frequent drought-like conditions. The central part of the state, which previously exhibited moderate RH, is increasingly showing signs of dryness.

4.3 *Geographically Weighted Regression (GWR)*

Table 2 represents the results of the GWR analysis conducted to examine the relationship between NTL and the other three climatic variables across Himachal Pradesh from 2002 to 2022. It (Table 2) includes R^2 values, adjusted R^2 values, and local R^2 values (maximum and mean) for each year, providing insights into how these relationships vary over space and time.

Table 2 R^2, R^2 adjusted, and local R^2 values based on GWR between NTL and climatic parameters

Year	Variables											
	NTL versus TEMP				NTL versus RH				NTL versus PR			
	R^2	R^2 Adjusted	Local R^2		R^2	R^2 Adjusted	Local R^2		R^2	R^2 Adjusted	Local R^2	
			Max	Mean			Max	Mean			Max	Mean
2002	0.818	0.706	0.688	0.017	0.81	0.694	0.672	0.016	0.81	0.695	0.671	0.016
2003	0.811	0.698	0.401	0.018	0.801	0.686	0.371	0.017	0.81	0.686	0.369	0.017
2004	0.875	0.795	0.5593	0.018	0.874	0.792	0.566	0.018	0.874	0.792	0.549	0.019
2005	0.88	0.804	0.7804	0.02	0.876	0.795	0.762	0.018	0.876	0.796	0.761	0.018
2006	0.874	0.792	0.702	0.019	0.868	0.783	0.632	0.018	0.868	0.783	0.636	0.018
2007	0.885	0.809	0.645	0.021	0.879	0.801	0.638	0.02	0.879	0.801	0.628	0.02
2008	0.873	0.792	0.644	0.02	0.867	0.783	0.479	0.019	0.867	0.783	0.483	0.0198
2009	0.831	0.724	0.656	0.055	0.825	0.714	0.638	0.02	0.825	0.714	0.652	0.054
2010	0.875	0.795	0.717	0.02	0.87	0.788	0.685	0.019	0.87	0.788	0.687	0.0193
2011	0.88	0.8	0.576	0.023	0.875	0.793	0.556	0.021	0.875	0.793	0.556	0.021
2012	0.841	0.743	0.887	0.054	0.833	0.732	0.859	0.052	0.833	0.732	0.857	0.052
2013	0.827	0.722	0.728	0.051	0.819	0.711	0.701	0.05	0.819	0.712	0.706	0.05
2014	0.385	0.244	0.327	0.014	0.358	0.226	0.325	0.0121	0.364	0.228	0.33	0.012
2015	0.366	0.224	0.27	0.014	0.333	0.204	0.268	0.012	0.335	0.205	0.264	0.012
2016	0.432	0.264	0.434	0.015	0.398	0.24	0.41	0.013	0.401	0.242	0.413	0.013
2017	0.409	0.256	0.266	0.014	0.382	0.235	0.264	0.013	0.384	0.236	0.259	0.013
2018	0.439	0.281	0.405	0.017	0.415	0.261	0.451	0.015	0.417	0.262	0.451	0.015
2019	0.51	0.347	0.563	0.02	0.484	0.326	0.486	0.018	0.486	0.326	0.489	0.0186
2020	0.508	0.344	0.344	0.0188	0.477	0.321	0.322	0.016	0.482	0.323	0.327	0.017

(continued)

Table 2 (continued)

Year	Variables											
	NTL versus TEMP				NTL versus RH				NTL versus PR			
	R^2	R^2 Adjusted	Local R^2		R^2	R^2 Adjusted	Local R^2		R^2	R^2 Adjusted	Local R^2	
			Max	Mean			Max	Mean			Max	Mean
2021	0.48	0.334	0.264	0.2648	0.461	0.318	0.246	0.015	0.461	0.318	0.24	0.015
2022	0.58	0.344	0.385	0.0194	0.467	0.325	0.451	0.0185	0.471	0.326	0.351	0.017

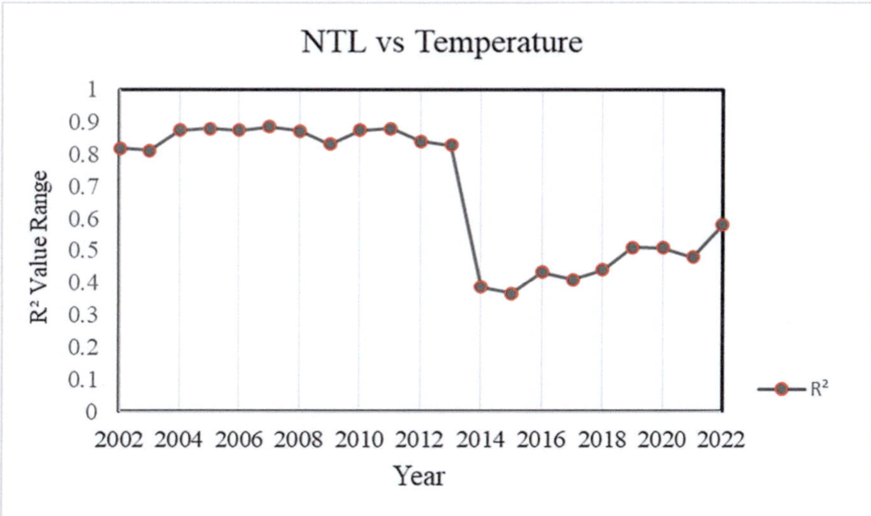

Fig. 8 R^2 trend between NTL and temperature

4.3.1 Relationship Between NTL and Temperature

The R^2 value for NTL and TP indicates a strong positive correlation before 2013, with values consistently above 0.80, peaking at 0.88 in 2011 (Figs. 8 and 9). The adjusted R^2 values also remain high, indicating the robustness of the regression model. The local R^2 max values, which represent the highest spatial correlation in specific locations, range between 0.64 and 0.78, indicating spatial heterogeneity in the relationship. This aligns with the expectation that increased urbanisation, as reflected in NTL, contributes to higher temperatures. However, there is a sharp decline observed post-2013, where the R^2 value dropped to 0.385 in 2014 and fluctuated at lower values thereafter. This indicates a weaker statistical correlation. The major reason behind this shift can be attributed to the transition from DMSP-OLS data sets to VIIRS-NPP datasets, as VIIRS provides higher accuracy and avoids overestimation in comparison to DMSP. Despite this change, localised influences remain evident as local R^2 max values reach 0.385 in 2022, proving the continued spatial relevance of NTL in influencing temperature patterns.

4.3.2 Relationship Between NTL and Precipitation

The R^2 values for NTL and PR follow a similar trend with a stronger correlation before 2013, where values exceeding 0.87 are observed in many years (Figs. 10 and 11). The highest correlation is recorded in 2005 and 2007, where it reaches 0.876 and 0.879, respectively. This indicates that there is a significant influence of anthropogenic activities. The adjusted R^2 values are consistently high, confirming

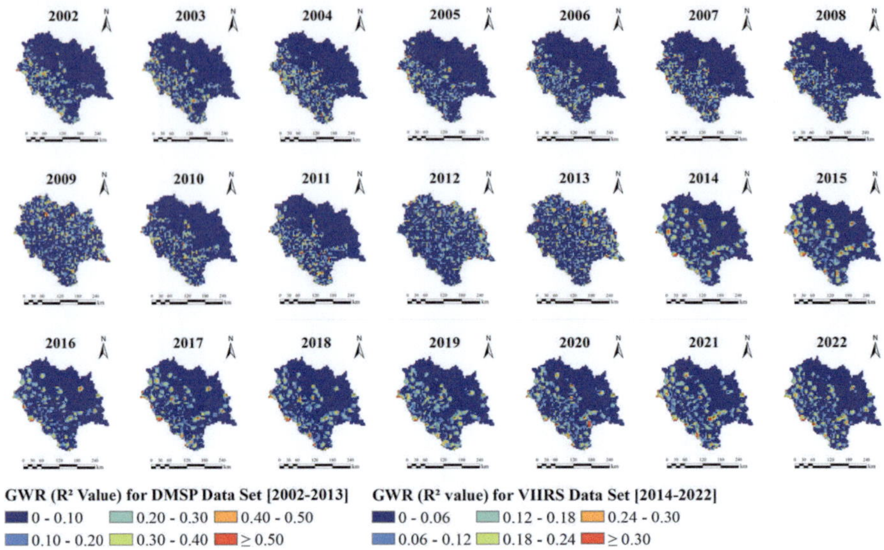

Fig. 9 GWR between NTL and temperature

the reliability of the regression model. However, after the transition of datasets from DMSP to VIIRS in 2013, R^2 values declined to 0.335, with subsequent years showing a lower correlation. Nonetheless, localised influences remain evident as local R^2 max values of 0.451 in 2018, reinforcing the idea.

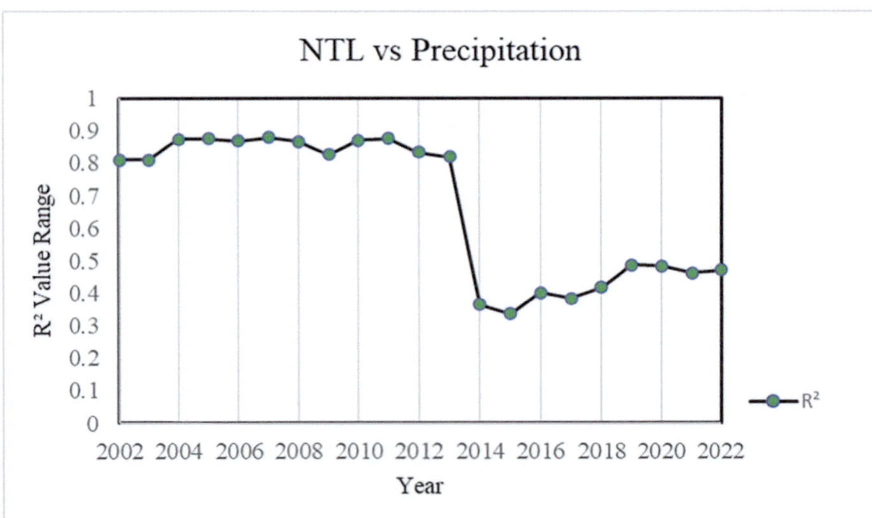

Fig. 10 R^2 trend between NTL and precipitation

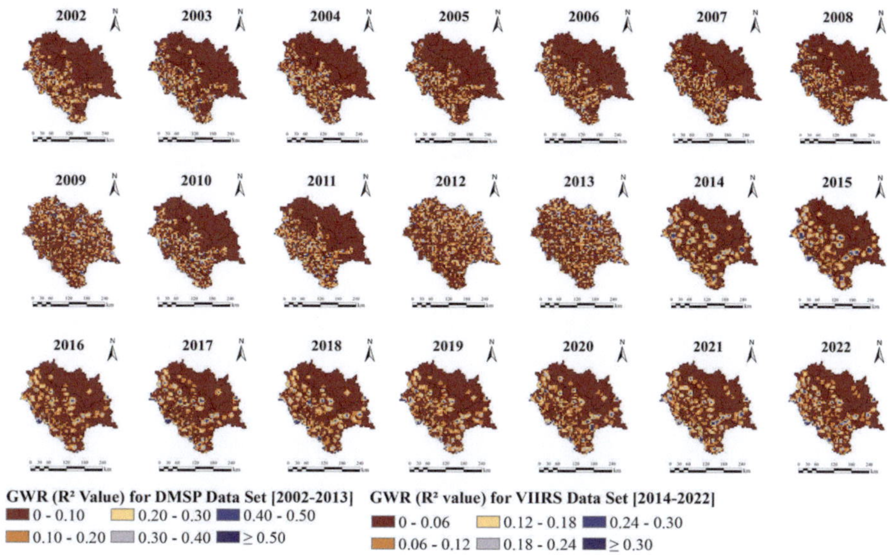

Fig. 11 GWR between NTL and relative humidity

4.3.3 Relationship Between NTL and Relative Humidity

Similarly, the R^2 values for NTL and RH exhibit a strong correlation before 2013 (Figs. 12 and 13). This indicates that urbanisation significantly influenced local humidity levels, likely through reduced vegetation and altered evaporation rates. Before 2013, the correlation peaked as much as 0.879 in 2007 and maintained values above 0.80 in most years. After 2013, the R^2 values experienced a substantial drop due to the shift from DMSP to VIIRS data. Despite the decrease, spatial variation persists, indicated by moderate R^2 max values.

4.3.4 Temporal Variation in NTL-Climate Relationship (Pre and Post 2013)

The GWR analysis revealed a prominent shift in the correlation between NTL and climatic factors when comparing the period of pre- and post-2013. Before 2013, the DMSP-OLS satellite data were used; the regression model shows a consistently strong association between NTL and other climatic variables. The higher R^2 values indicated a robust and spatially significant correlation, which suggested that urbanisation played a major role in influencing local climate. However, there is a prominent drop in correlation strength after 2013, corresponding with the shift of satellite data from DMSP to VIIRS-NPP, due to the discontinuation of DMSP data, and it is more advanced and recent. VIIRS provides an improved radiometric accuracy and eliminates saturation issues. The consequent lower R^2 values do not imply a diminished

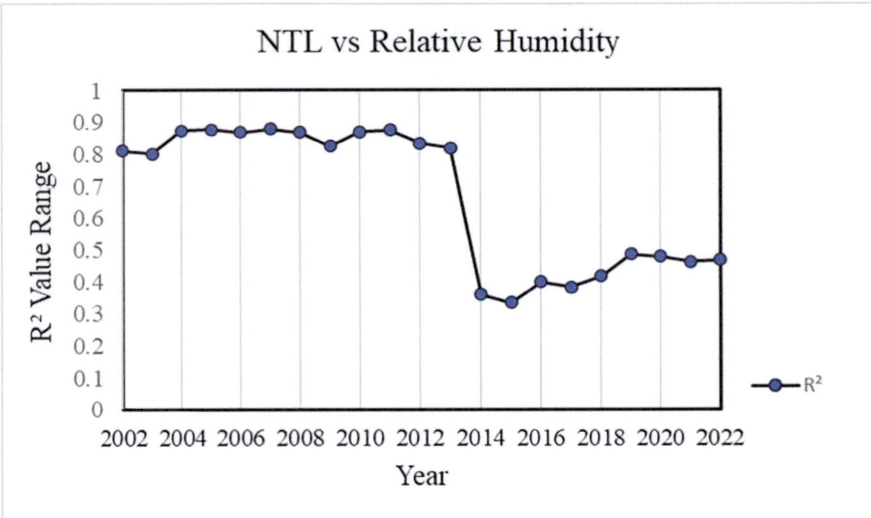

Fig. 12 R^2 trend between NTL and relative humidity

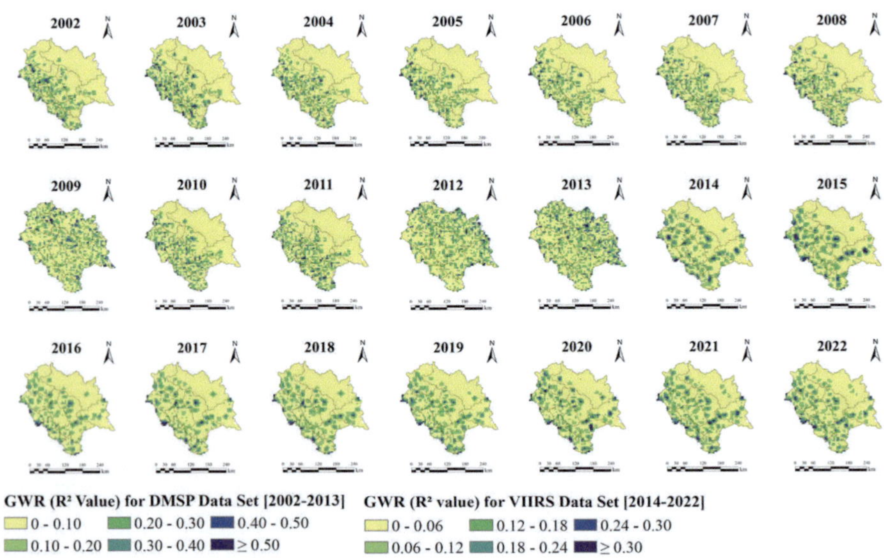

Fig. 13 GWR between NTL and relative humidity

ecological interaction. Despite the decline, the local R^2 values remained significant, reflecting persistent spatial heterogeneity and confirming that anthropogenic activities continued to influence climatic variables in different regions of Himachal Pradesh. By grouping the analysis, the study highlights both the consistency of urban climate linkage in the early period and the importance of methodological considerations in long-term geospatial studies, especially when integrating a multi-sensor database.

5 Discussion

The GWR analysis reveals a strong correlation between NTL intensity and selected climatic parameters till 2013, followed by a notable decline due to improved accuracy in NTL measurements with VIIRS datasets. Despite the shift, localised effects remain significant, emphasising the importance of spatially explicit investigation when analysing the interaction between human activities and different climatic variables. This study provides the importance of data source variations in long-term analysis. The significant shift in the correlation creates an abrupt change that corresponds to the transition from DMSP to VIIRS. The VIIRS data set is known to provide more accurate and calibrated NTL measurements compared to DMSP, which tends to have saturation issues caused by its lower resolution. Consequently, the lower R^2 values post-2013 do not indicate a weaker relation between NTL and other variables but rather a more precise representation of NTL intensities with reduced overestimation. Despite the decline in overall R^2 values, the local R^2 values still indicate spatial variation in correlation strength, depicting stronger associations even in the VIIRS datasets. Overall, the GWR results in a strong association between NTL and climatic parameters, particularly before 2013, when the DMSP datasets were in use. The post-2013 decline in R^2 values is a primary result of the transition to the VIIRS set, which provides lower NTL values, leading to a reduction in the correlation. Despite this, localised relationships remain significant, suggesting that human influences continue to change climatic patterns in different parts of Himachal Pradesh.

5.1 Key Findings

In this research work, there is a detailed geospatial assessment of the relationship between NTL intensity, which is taken as a factor representing human interference and its impact on the climate, where TP, RH, and PR have been taken into consideration. The key findings from this research work are discussed below,

There has been a noticeable progression and intensification of NTL over the period (2002–2022). High-intensity clusters have emerged around the already existing urban areas such as Shimla, Mandi, Dharamshala, and Manali. There is a progression of high-illumination zones which have been expanded over time, indicating the intense growth of urban centres and electrification.

In the southern and central zones, temperature hotspots have increased significantly, resulting in mid-altitude places such as Solan, Bilaspur, and some parts of

Kangra to fall under persistent hot zones. There is a significant shift in warming trends, suggesting the intensification of the heat island effect.

There has been an increase in low precipitation zones in the central Himachal Pradesh, but the high precipitation zones in the northeast remain stable. The expansion of low precipitable areas is linked to the changing patterns of monsoon, potentially influenced by deforestation and land cover change patterns due to human activities.

The central areas, which were once under the influence of moderate RH, have shifted to low RH zones, which increase the susceptibility to drought. The reason behind this is mostly due to the reduction of evapotranspiration from the loss of natural vegetation around areas such as Una, Bilaspur, and Kangra.

The GWR analysis before 2013 (DMSP-OLS) shows a strong spatial correlation between NTL and climatic variables, where R^2 values are often exceeding 0.85, and local R^2 max values reaching 0.78, indicating strong interrelation. Post 2013 (VIIRS-NPP), the R^2 values dropped to 0.44 due to the change in data source, where the decline in correlation does not imply weakening of correlation, but a shift in calibrated data standards because VIIRS tends to capture subtle changes in illumination more precisely.

There has been a pattern of drying, warming, and urban spread throughout the study region; the suburban areas are facing climate stress, with visible changes in climatic conditions. These key findings indicate the validation of the hypothesis, where the increasing values of NTL are a proxy for understanding environmental and climatic variability in an ecologically fragile, mountainous region such as the state of Himachal Pradesh.

5.2 Ecological Impacts

The strong correlation between NTL intensity and climatic parameters indicates that the study location is experiencing an increased artificial illumination, which has a noticeable impact on the local climate. Some of the ecological consequences are the intensity of the urban heat island effect and local warming, where there is a consistently high NTL intensity, especially in urban centres, where the artificial infrastructure retains heat, disturbing the local thermal balance. It has a significant impact on nocturnal species, which rely on natural light cycles for their behaviours. Shift in NTL patterns may cause the natural rhythm of life, causing disturbances in behavioural patterns, reproductive cycle, and migration (Bennie et al. 2016; Brüning et al. 2015; Gaston et al. 2013, 2015; van Geffen et al. 2015; Hatori et al. 2017; Raap et al. 2015; Rodríguez et al. 2015; Svechkina et al. 2020). With the expansion of NTL intensity, there is a growing fragmentation of natural habitat. The spatial decline of RH in high NTL regions indicates less evapotranspiration due to deforestation, which makes the ecosystem prone to drought, especially in transitional mid-altitude regions. The correlation between NTL and PR and its decline in specific zones could be indicative of land-use shifts due to infrastructural development, forest clearance, etc., which can directly affect the hydrological cycle by reducing rainfall, affecting ground water and surface water availability. Even though the higher elevation areas are not much affected by NTL intensity, increasing overall development and small pockets can destabilise the fragile ecosystem. The ecological impact due to the effect of anthropogenic activities needs an urgent sustainable development planning in Himachal Pradesh, balancing development with the preservation of the environment amid ongoing climatic changes.

5.3 Policy Relevance

The findings of this research work have a significant implication on climate-resilient urban planning and environmental policy, particularly in ecologically sensitive regions such as Himachal Pradesh. Policy makers should leverage insights from this research to prioritise region-specific interventions, such as regulating land-use transformation, promoting sustainable development planning (Cattivelli 2021; Dannevig et al. 2012; NITI Ayog 2021; Raj 2014; Regmi et al. 2016; Simonson et al. 2021). The spatial variability of the regression values suggests that uniform frameworks might not be effective, which requires a localised planning. To ensure sustainable development, urban expansion must be controlled by environmental thresholds with proper planning to mitigate the heat island effect, along with reducing ALP. Integrating RS tools and geospatial modelling into environmental assessment can support and foster long-term climate adaptation planning at the local and regional level.

6 Conclusion

This study provides valuable insight into the evolving interaction between urbanisation, which has been depicted by NTL, and climate dynamics in the state of Himachal Pradesh. GWR has been used to assess the localised impact of NTL on TP, RH and PR, revealing a correlation between them. Even though a significant shift has been noticed due to the change in datasets, it does not indicate an actual weakening of the relationship but rather highlights the challenges of integrating multi-sensor datasets for spatio-temporal analysis. The higher sensitivity of VIIRS-NPP to low-intensity light sources may have introduced subtle variations in NTL trends that DMSP-OLS did not capture. the finding shows the importance of maintaining data continuity in long-term climatic studies, as methodological shifts in satellite data can influence statistical interpretations. Despite the challenges, the research provides a comprehensive understanding of urbanisation's impact on climate, especially in a mountainous region like Himachal Pradesh, where topographical and meteorological factors bring further complexity. It also shows the importance of hybrid datasets, data calibration techniques for multi-sensor datasets and the importance of ground observations to enhance the robustness in further analysis. Policymakers and urban planners must consider the insights to develop climate-resilient urban growth, focusing on sustainability by integrating remote sensing, geospatial techniques and climate modelling for future understanding, ultimately contributing to the sustainable development of environmentally sensitive regions such as Himachal Pradesh.

References

Bennie J, Davies TW, Cruse D, Gaston KJ (2016) Ecological effects of artificial light at night on wild plants. J Ecol 104(3):611–620. https://doi.org/10.1111/1365-2745.12551

Brüning A, Hölker F, Franke S, Preuer T, Kloas W (2015) Spotlight on fish: light pollution affects circadian rhythms of European perch but does not cause stress. Sci Total Environ 511:516–522. https://doi.org/10.1016/j.scitotenv.2014.12.094

Cattivelli V (2021) Climate adaptation strategies and associated governance structures in mountain areas. The case of the alpine regions. Sustainability 13(5):1–25. https://doi.org/10.3390/su13052810

Chalkias C, Petrakis M, Psiloglou B, Lianou M (2006) Modelling of light pollution in suburban areas using remotely sensed imagery and GIS. J Environ Manage 79(1):57–63. https://doi.org/10.1016/j.jenvman.2005.05.015

Comber A, Brunsdon C, Charlton M, Dong G, Harris R, Lu B, Lü Y, Murakami D, Nakaya T, Wang Y, Harris P (2023) A route map for successful applications of geographically weighted regression. Geogr Anal 55(1):155–178. https://doi.org/10.1111/gean.12316

Dannevig H, Rauken T, Hovelsrud G (2012) Implementing adaptation to climate change at the local level. Local Environ 17(6–7):597–611. https://doi.org/10.1080/13549839.2012.678317

Davies TW, Smyth T (2018) Why artificial light at night should be a focus for global change research in the 21st century. Glob Change Biol 24(3):872–882. https://doi.org/10.1111/gcb.13927

de Sherbinin A (2014) Climate change hotspots mapping: what have we learned? Clim Change 123(1):23–37. https://doi.org/10.1007/s10584-013-0900-7

Farooq Z, Kumar R (2021) Spatial and temporal trend analysis of relative humidity in the himalayan region: a case study. Arab J Geosci 14(21). https://doi.org/10.1007/s12517-021-08624-9

Gaston KJ, Bennie J, Davies TW, Hopkins J (2013) The ecological impacts of nighttime light pollution: a mechanistic appraisal. Biol Rev 88(4):912–927. https://doi.org/10.1111/brv.12036

Gaston KJ, Visser ME, Hölker F (2015) The biological impacts of artificial light at night: the research challenge. Philos Trans R Soc B Biol Sci 370(1667). https://doi.org/10.1098/rstb.2014.0133

Hand R, Keenlyside NS, Omrani NE, Bader J, Greatbatch RJ (2019) The role of local sea surface temperature pattern changes in shaping climate change in the North Atlantic Sector. Clim Dyn 52(1–2):417–438. https://doi.org/10.1007/s00382-018-4151-1

Haq MA, Baral P, Yaragal S, Rahaman G (2020) Assessment of trends of land surface vegetation distribution, snow cover and temperature over entire Himachal Pradesh using MODIS datasets. Nat Resour Model 33(2):1–26. https://doi.org/10.1111/nrm.12262

Hatori M, Gronfier C, Van Gelder RN, Bernstein PS, Carreras J, Panda S, Marks F, Sliney D, Hunt CE, Hirota T, Furukawa T, Tsubota K (2017) Global rise of potential health hazards caused by blue light-induced circadian disruption in modern aging societies. NPJ Aging Mech Dis 3(1):5–7. https://doi.org/10.1038/s41514-017-0010-2

Huang C, Zhuang Q, Meng X, Guo H, Han J (2021a) An improved nightlight threshold method for revealing the spatiotemporal dynamics and driving forces of urban expansion in China. J Environ Manage 289(April 2019):112574. https://doi.org/10.1016/j.jenvman.2021.112574

Hung LW, Anderson SJ, Pipkin A, Fristrup K (2021b) Changes in night sky brightness after a countywide LED retrofit. J Environ Manage 292(October 2020):112776. https://doi.org/10.1016/j.jenvman.2021.112776

Huovinen P, Ramírez J, Caputo L, Gómez I (2019) Mapping of spatial and temporal variation of water characteristics through satellite remote sensing in Lake Panguipulli, Chile. Sci Total Environ 679:196–208. https://doi.org/10.1016/j.scitotenv.2019.04.367

Inclusion, Social, Sustainable Development, and Himachal Pradesh (2015) SCALING social inclusion and sustainable development

Ivajnšič D, Kaligarič M, Žiberna I (2014) Geographically weighted regression of the urban heat Island of a small city. Appl Geogr 53:341–353. https://doi.org/10.1016/j.apgeog.2014.07.001

Kauppinen J, Heinonen J, Malmi P (2014) Influence of relative humidity and clouds on the global mean surface temperature. Energy Environ 25(2):389–399. https://doi.org/10.1260/0958-305X.25.2.389

Khalid AM (2022) Developmental success and the sustainability challenge in a Mountain Region: case of Himachal Pradesh in India. Int J Reg Dev 9(1):47. https://doi.org/10.5296/ijrd.v9i1.19518

Kianfar N, Mesgari MS (2022) GIS-based spatio-temporal analysis and modeling of COVID-19 incidence rates in Europe. Spatial Spatio Temp Epidemiol 41(June 2021):100498. https://doi.org/10.1016/j.sste.2022.100498

Kumar S, Kumar A (2022) Hotspot and trend analysis of forest fires and its relation to climatic factors in the Western Himalayas. Nat Hazards 114(3):3529–3544. https://doi.org/10.1007/s11069-022-05530-5

Kumar P, Rehman S, Sajjad H, Tripathy BR, Rani M, Singh S (2019) Analyzing trend in artificial light pollution pattern in India using NTL sensor's data. Urban Clim 27:272–283. https://doi.org/10.1016/j.uclim.2018.12.005

Kumari B, Pandey AC (2020) MODIS based forest fire hotspot analysis and its relationship with climatic variables. Spat Inf Res 28(1):87–99. https://doi.org/10.1007/s41324-019-00275-z

Li X, Zhou Y (2017) Urban mapping using DMSP/OLS stable night-time light: a review. Int J Remote Sens 38(21):6030–6046. https://doi.org/10.1080/01431161.2016.1274451

Li L, Tao Y, Zhao L, Zhan Y, Zheng F, Zhang Y, Mumtaz F, Wang C (2019) Characteristics and trend analysis of the relationship between land surface temperature and nighttime light intensity levels over China. Infrared Phys Technol 97(January):381–390. https://doi.org/10.1016/j.infrared.2019.01.018

NITI Ayog (2021) Himachal Pradesh state action plan on climate change. https://www.nitiforstates.gov.in/policy-viewer?id=GSSNAD000287

Peeters A, Zude M, Käthner J, Ünlü M, Kanber R, Hetzroni A, Gebbers R, Ben-Gal A (2015) Getis-Ord's hot- and cold-spot statistics as a basis for multivariate spatial clustering of orchard tree data. Comput Electron Agric 111:140–150. https://doi.org/10.1016/j.compag.2014.12.011

Prasad R, Sharma A (2016) Spatial distribution of annual rainfall in Himachal Pradesh and its implication in farming. Himachal J Agric Res 42(2):115–125

Raap T, Pinxten R, Eens M (2015) Light pollution disrupts sleep in free-living animals. Sci Rep 5:1–8. https://doi.org/10.1038/srep13557

Raj J (2014) Uttarakhand action plan on climate change: transforming crisis into opportunity. https://moef.gov.in/uploads/2017/08/Uttarakhand-SAPCC.pdf

Regmi BR, Star C, Filho WL (2016) An overview of the opportunities and challenges of promoting climate change adaptation at the local level: a case study from a community adaptation planning in Nepal. Clim Change 138(3–4):537–550. https://doi.org/10.1007/s10584-016-1765-3

Rodríguez A, García D, Rodríguez B, Cardona E, Parpal L, Pons P (2015) Artificial lights and seabirds: is light pollution a threat for the threatened balearic petrels? J Ornithol 156(4):893–902. https://doi.org/10.1007/s10336-015-1232-3

Sarif MO, Gupta RD, Murayama Y (2023) Assessing local climate change by spatiotemporal seasonal LST and six land indices, and their interrelationships with SUHI and hot–spot dynamics: a case study of Prayagraj City, India (1987–2018). Remote Sens 15(1):179. https://doi.org/10.3390/rs15010179

Shifti DM, Chojenta C, Holliday EG, Loxton D (2020) Application of geographically weighted regression analysis to assess predictors of short birth interval hot spots in Ethiopia. PLoS ONE 15(5):1–18. https://doi.org/10.1371/journal.pone.0233790

Simonson WD, Miller E, Jones A, García-Rangel S, Thornton H, McOwen C (2021) Enhancing climate change resilience of ecological restoration—a framework for action. Perspect Ecol Conserv 19(3):300–310. https://doi.org/10.1016/j.pecon.2021.05.002

Singh RB, Kumar P (2014) Geographic and socio-economic realities of Himachal Pradesh, Northwestern Himalaya. In: Livelihood security in Northwestern Himalaya, pp 11–26. https://doi.org/10.1007/978-4-431-54868-3_2

Svechkina A, Portnov BA, Trop T (2020) The impact of artificial light at night on human and ecosystem health: a systematic literature review. Landsc Ecol 35(8):1725–1742. https://doi.org/10.1007/s10980-020-01053-1

Trenberth KE, Dai A, Rasmussen RM, Parsons DB (2003) The changing character of precipitation. Bull Am Meteorol Soc 84(9):1205–1217+1161. https://doi.org/10.1175/BAMS-84-9-1205

Uyar N (2024) Monitoring Turkey's nighttime light and environmental change. Environ Monit Assess 196(9). https://doi.org/10.1007/s10661-024-13032-z

van Geffen KG, van Eck E, de Boer RA, van Grunsven RHA, Salis L, Berendse F, Veenendaal EM (2015) Artificial light at night inhibits mating in a geometrid moth. Insect Conserv Div 8(3):282–287. https://doi.org/10.1111/icad.12116

Wang K, Sun J, Cheng G, Jiang H (2011) Effect of altitude and latitude on surface air temperature across the Qinghai-Tibet Plateau. J Mt Sci 8(6):808–816. https://doi.org/10.1007/s11629-011-1090-2

Wang C, Zhang J, Yan X (2012) The use of geographically weighted regression for the relationship among extreme climate indices in China. Math Probl Eng 2012:369539. https://doi.org/10.1155/2012/369539

Wu T, Li Y (2013) Spatial interpolation of temperature in the United States using residual Kriging. Appl Geogr 44:112–120. https://doi.org/10.1016/j.apgeog.2013.07.012

Ye WEI, Zuoqi C, Chunliang XIU, Bailang YU, Hongxing LIU (2019) Siting of dark sky reserves in China based on multi-source spatial data and multiple criteria evaluation method 29(6):949–961

Yu B, Tang M, Wu Q, Yang C, Deng S, Shi K, Peng C, Wu J, Chen Z (2018) Urban built-up area extraction from log-transformed NPP-VIIRS nighttime light composite data. IEEE Geosci Remote Sens Lett 15(8):1279–1283. https://doi.org/10.1109/LGRS.2018.2830797

Decadal Study on Supraglacial Debris Cover of Kaer Glacier, Eastern Himalayas

Sivaranjani Sivalingam and Saravanan Subbarayan

1 Introduction

On a global scale, 4–7% of the debris is found in the supraglacial layer of the complete frozen surface (Brown and Brock 2023; McCarthy et al. 2022; Scherler et al. 2018). In high-relief orogenic regions like the Himalayas, glaciers are often found in complex, dynamic topographies, where an unremitting layer of rock fragments covers the lower portion of the zone of ablation (Racoviteanu et al. 2021). This debris consists of materials generated by glacier abrasion or excavation, rock fragments, and sediment deposited on the glacier surface. These fragments result from gelifluction, ice or snow slide, and gravity-driven sliding, which cause rock and debris to implode on the glacier (Kirkbride et al. 2023; Wu et al. 2023). Supraglacial debris principally accumulates in the ablation zone of glaciers, often displaying spectral features similar to those of the adjoining rock environment in satellite images. This similarity complicates the task of distinguishing debris-covered areas from bare rock (Casey and Benn 2011; Alifu et al. 2016). The application of computerized practices to map glaciers coated with debris can enhance the mapping progression by minimizing the influence of physical visual analysis and reducing the particular biases that often arise (Chu et al. 2022). However, this debris complicates the delineation of the glacier morphological outlines.

Several techniques, such as threshold segmentation, object-based classification, synthetic aperture radar interferometry, and machine learning approaches, have been used to map pristine ice/snow and supraglacial debris (Dammann et al. 2019). These techniques effectively remove clean snow and ice, but they still have issues understanding supraglacial debris. Mass balance research, glacial lake outburst flood

S. Sivaranjani · S. Saravanan (✉)
Department of Civil Engineering, National Institute of Technology, Tiruchirappalli, Tiruchirappalli, India
e-mail: ssaravanan@nitt.edu

M. O. Sarif and A. Sharifi (eds.), *Advanced Geospatial Intelligence and AI for Environmental Resilience and Sustainable Development*, Advances in Geographic Information Science, https://doi.org/10.1007/978-3-032-03714-5_6

prediction, and glacier dynamics monitoring are just a few of the glaciological applications that depend on accurate and detailed data on the dimensions and geographical range of debris-covered glaciers (Sivalingam et al. 2023; Majeed et al. 2021). The response of a glacier to climate change depends on several aspects, including lake morphology (Milner et al. 2017), ice flow velocity (Sivalingam et al. 2022), and glacier melt rates (Hu et al. 2023; Khadka et al. 2023), which are greatly influenced by supraglacial debris. It is crucial to comprehend how debris cover contributes to the generation of glacial meltwater to forecast regional water supplies and the rise in sea levels worldwide (Herreid and Pellicciotti 2020). As has been extensively reported in the literature, loss of material from the debris of the glaciers is specifically connected with the establishment of subglacial and englacial lakes (Racoviteanu et al. 2022; Dubey and Goyal 2020). The complete effect of these lakes on the potential for local hazards is yet unknown, though, especially in light of probable future climate scenarios (Schmidt et al. 2020; Allen et al. 2022). In tandem with the glacier equilibrium line, the higher bound of the debris gradually resettles up-glacier as glaciers undergo thinning and mass loss (Mölg et al. 2019; Scherler and Egholm 2020). The gradual exposure of englacial debris and surface ice ablation are the direct causes of this shift, which thickens the debris as the glacier recedes.

The total impression of the debris covered on the glacier surface and its behavior is still poorly understood, despite significant development of our comprehension of the dynamics of glacier surfaces and their significance. The precise mechanisms by which debris cover influences glacier dynamics, such as ice flow, melt rates, and the glacier's long-term stability, are still being studied, despite the growing recognition that the presence of thick debris layers can slow down the reciprocation of glaciers to climatic changes (Glaciation 1998). The relationship between debris cover and glacier behavior will become increasingly crucial as glaciers lose mass and their debris-covered areas grow. This phenomenon significantly affects long-term ice mass loss, hazard potential, and regional hydrological impacts. The relationship above is significant since there is mounting evidence that glaciers with heavy debris layers may respond to climate change more slowly, which might postpone or obscure the effects of higher temperatures on glacier dynamics. Recognizing the role of glaciers buried under debris is crucial for enhancing climate forecasts and hazard assessments in glacierized regions, given the broader consequences for water supplies, sea-level rise, and glacial hazards.

Multiple studies have observed the difficulty of accurately determining glaciers' actual termini due to extensive debris cover (Hussain et al. 2024). Researchers can find that the stationary ice areas are close to the front of nearly all debris-covered glaciers, particularly those with a substantial debris coating. When it comes to climate change, these glaciers often react more slowly than clean-ice glaciers, which are more directly affected by variations in precipitation and temperature (Pratap et al. 2015). Monitoring and forecasting glacier behavior is made more difficult by the dense debris, which slows down the melting process and makes it more difficult to spot changes in the glacier's terminus. Supraglacial lakes, which frequently arise in low-altitude areas of debris-covered glaciers, are made easier by dense debris cover, which also obscures the glacier terminal. These lakes could eventually merge to become much bigger lakes, but they usually begin as small, isolated bodies of water (Zeller et al. 2024; Haritashya et al. 2018). These lakes' dynamics are particularly concerning because their water can evaporate quickly, sometimes in minutes, and generate disastrous outcomes, such as abrupt glacial lake outbursts (GLOFs) (Sahu et al. 2023). Such lakes have the potential to collapse, causing enormous flood surges that might seriously endanger infrastructure and towns downstream and disturb regional ecosystems. The reaction of a glacier to climate forcing, especially in ablation rates, is greatly influenced by the thickness of the debris layer (Dubey et al. 2024; Zhang et al. 2023).

Depending on its thickness, debris can accelerate or slow down the melting of ice. The underlying ice may receive more heat from thinner debris layers, improving ablation and hastening glacier melt. Thicker debris layers, on the other hand, may serve as an insulating barrier, slowing the ablation rate and delaying the melting of glaciers (Nicholson et al. 2018; Patel et al. 2021). Recognizing the long-term response of glaciers to climate change requires understanding the correlation between debris thickness and the melting process of glaciers. A solid understanding of debris cover's impact on glacier behavior is necessary for precise estimates of glacial mass loss, the effects that come with it, and the complexity it introduces to the environment. This mass loss entails determining how glaciers could react to specific future climatic scenarios and evaluating glacier dangers like GLOFs. More research is required to understand better the mechanisms by which debris influences glacier dynamics and to create more precise models for forecasting future glacier behavior under changing climatic conditions, as debris-covered glaciers remain a protruding facet of glacierized regions, especially in high-relief mountain ranges. Combining this impact with the dynamic topography of glacier surfaces results in highly diverse ablation and the creation of surface lakes and ponds, typical characteristics of glaciers covered in receding debris (Bush and Bishop 2017; Rowan et al. 2021). However, little is known about the mechanisms and underlying causes influencing the supraglacial ponds' spatial distribution. Therefore, in light of climate change, water security, and continuous glacier changes, mapping and precisely evaluating supraglacial debris cover is crucial to comprehending these glaciers' impending growth and sustainability.

Recent developments in remote sensing methods offer potential tools to enhance our comprehension of debris-covered glacier surfaces and their erraticism, particularly in regions with limited access. The present study is one example that examines the variations in the supraglacial debris cover of a transboundary glacier from 2014 to 2023 using Landsat-8 data about a threshold on the NDSI.

2 Recent Developments

Remote sensing technology is a scalable and economical method of collecting data over broad areas, allowing researchers to gain full knowledge of glacier dynamics and debris distribution. This technology has significantly improved our ability to monitor the supraglacial debris cover in the Himalayas, as standard field-based approaches for mapping debris cover are typically logistically complex and time-consuming, especially in remote places.

Using high-resolution satellite images from platforms like Landsat, Sentinel, and World View is one of the most critical developments in this sector (Yang et al. 2024). The precise images offered by these satellites are examined using various techniques to distinguish between snow, ice, and debris. Other spectral indices, such as the NDSI, have shown promise in detecting and measuring debris cover (Man et al. 2014). Researchers can reliably characterize surface features by applying precise thresholds to these indices, which improves debris mapping precision.

Machine learning approaches are also practical tools for mapping supraglacial debris. Random Forest and Support Vector Machines are algorithms trained on labeled datasets to automatically classify debris cover in new images. These methods improve mapping accuracy and are more efficient at processing massive datasets than older techniques (Sharda and Srivastava 2024; Alifu et al. 2020). Furthermore, integrating multi-sensor data, which includes optical, thermal, and radar information, provides a more comprehensive view of glacier surfaces, allowing for better differentiation of debris types and conditions.

High-resolution mapping using Unmanned Aerial Vehicles (UAVs) is another significant advancement. UAVs can obtain high spatial resolution aerial images at lower altitudes (Gök et al. 2023). This method has proven very beneficial in localized research, enabling thorough analysis of debris morphology and its influence on the melt processes of the glacier.

Modern remote sensing tools have provided vital intuitions into the dynamics of supraglacial debris cover in the Himalayan region. Researchers have discovered debris extent, thickness, and distribution patterns, correlating these changes to climate change indicators like temperature rises and glacial retreat. Furthermore, tracking supraglacial debris over time has improved understanding of its role in modifying glacier melt rates, which has significant consequences for downstream water resources.

3 Study Area and Its Rationale

Kaer Glacier (35° 54′ 59″N and 74° 21′ 47″ E) is a transboundary glacier with Global Land Ice Measurements from Space (GLIMS) Id "G088097E27918N" (Fig. 1) positioned at the cradle of the Pumqu Stream on the northern slope of the central Himalayas, surpassing Sikkim's South Lhonak Glacier (Sharma et al. 2024). This glacier (RGI2000-v7.0-G-15-07982) encompasses 30.484 km^2 and has a span of 8.7 km. Debris covers an area of 1.1 km^2 from the glacier's snout to its ablation zone. This glacier was once known as Longbasaba Glacier, and its tongue has numerous serac clusters and supraglacial lakes. The altitude of the glacier ranges from 5,481 to 6,245 meters above mean sea level.

Ice avalanches creating crevasses in the glacier front created Proglacial Lake near the parent glacier. The glacier front retreated 40 m year between 1977 and

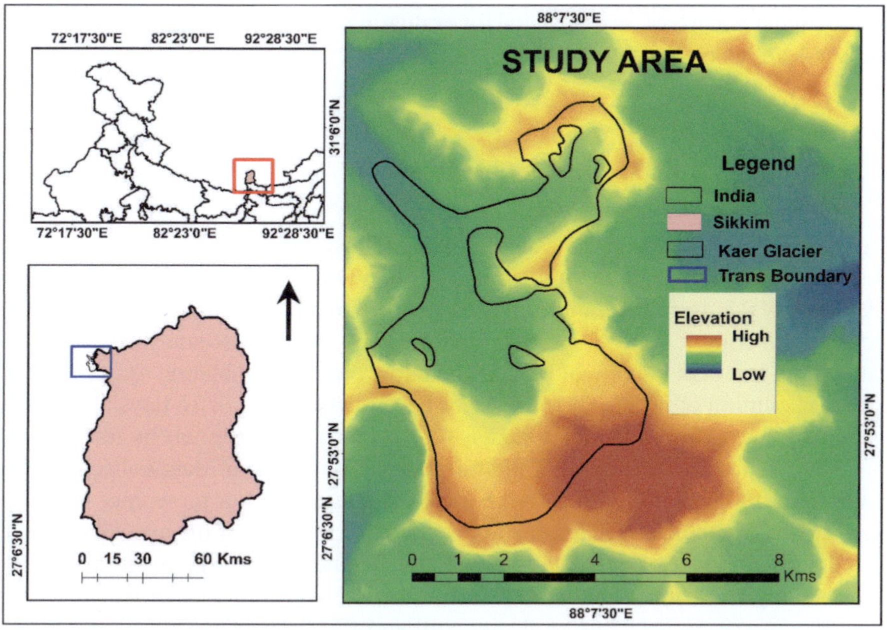

Fig. 1 Study area-Kaer Glacier overlaid on the Copernicus GLO-30 Digital Elevation Model (COP30) representing the Transboundary region, Western part of Sikkim, Himalayas

2009, which caused glacial lakes to expand. According to the literature, Kaer Lake is at high risk of outbursts (Rowan et al. 2021; Xin- et al. 2008). Furthermore, from 2005 to 2009, there was an increasing retreat of the glacier terminal (63.8 m/year), as well as a rapid increase of the Kaer Lake area of 0.040 km^2/year (Wei et al. 2020; Sharma et al. 2022). Recent research has highlighted significant changes in the Longbasaba Glacier, which has experienced a notable retreat accompanied by frequent ice avalanches. These avalanches release large volumes of ice into the Proglacial Lake, positioning them as a primary cause for the breach of moraine dams and ensuing GLOF events (Wei et al. 2021).

Additionally, the lowering of the glacier surface has been identified as a critical factor contributing to the overall mass loss associated with glacier shrinkage. This work underscores the significance of comprehending the interaction between glacier-related dynamics and risks to inform future management and mitigation strategies for communities affected by these phenomena. The investigation of debris cover is crucial, as it directly influences the melting processes of glaciers. In particular, a thinner debris layer accelerates the pace of melting, whereas a more extensive cover has the reverse effect. The ablation zone region impacted by debris is mainly determined by the amount of supraglacial debris cover (SDC). Coastal areas are significantly affected by increasing sea levels brought on by glacier melting (Silvano and Milva 2023). Although the overall number of glaciers may grow or shrink, researchers have difficulties examining the cover of supraglacial material. This study mainly imparts that SDC cover is vital for evaluating its influence on glacier dynamics and sea-level changes.

4 Materials and Methods

This chapter elaborates on the satellite data employed in the current study and the methodological flow adopted for estimating the NDSI. Multispectral scenes of path 139 and row 041 from the Landsat-8 satellite were obtained from the Earth Explorer website, covering a decadal study period of 2014–2023. This analysis prioritized data with a minimal canopy of clouds and snow during the ablation season, with a spatial resolution of 30 m. The data series has a revisit cycle of 16 days. The green and shortwave infrared (SWIR) bands were utilized for this analysis to calculate the NDSI ratios (Pratibha and Kulkarni 2018). The epoch of data collection was determined by availability during the ablation season, with a preference for July, August, and September (Dhanush and Priya 2023). Because of the unavailability of data due to cloud cover, the data for June and October are considered in this study. The Randolph Glacier Inventory (RGI) was used to define the glacier boundaries in this study. An upgraded genre of glacier outlines of RGI-Version 7.0 was used in this work for all the years by maintaining a stable glacier area throughout to calculate the percentage of increase in the SDC area. The datasets utilized for the investigation are detailed in Table 1.

Figure 2 illustrates the method used to estimate the NDSI using the Band 3 and 5 of Landsat-8 imageries.

4.1 Pre-processing of the Raw Data

The time series images obtained from the Earth Explorer platform were initially used to calibrate digital numbers (DN). As this study employs multisensory data,

Table 1 Summary of the satellite images utilized for the SDC area estimation

Data	Landsat-8 OLI	
Path and row	139 and 041	
Band used	Band 3 (green), Band 5 (SWIR)	
Spatial resolution	30 m	
Acquisition dates	2014-10-22	2019-06-14
	2015-10-09	2020-10-06
	2016-10-27	2021-06-03
	2017-10-30	2022-08-25
	2018-10-17	2023-06-09

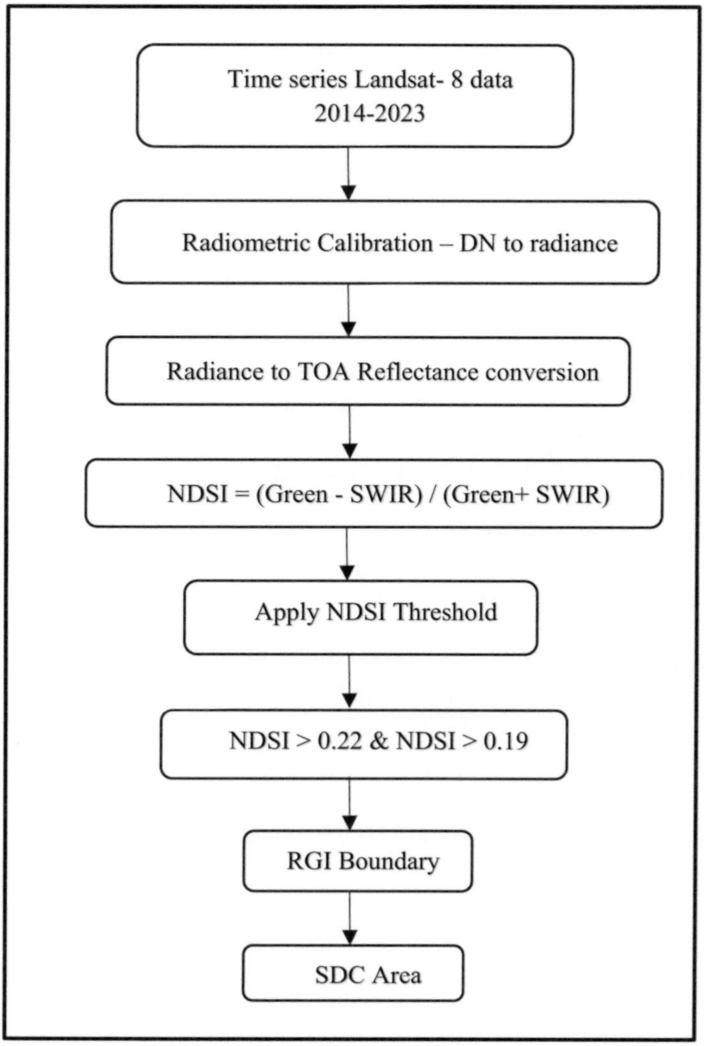

Fig. 2 The methodology adopted to estimate SDC area

radiometric errors may occur due to variations in sensor features, scene illumination, distance separating the earth and sun, atmospheric disorders, and other parameters. The DNs were transformed into a more evocative radiometric measure through a process of radiometric correction to address these challenges (Kääb et al. 2016). This calibration involved transforming DN values into spectral radiance using instrument calibration parameters, which were then converted to planetary top-of-atmosphere (TOA) reflectance. The correction procedures adhered to the guidelines at https://www.usgs.gov/landsat-missions/using-usgs-landsat-level-1-data-product for the Operational Land Imager (OLI) sensor. This process was carried out using a semi-automatic plugin on the open-source software platform Quantum GIS. As a result, 10 Green and 10 SWIR bands, free of radiometric errors, were obtained for the NDSI calculation of the study area.

4.2 Post-processing of the Data

The NDSI band ratio approach was used post-processing to evaluate the glacier's Snow-Covered Area (SCA) referring to the extent of land surface covered by snow at a given time (Xie et al. 2020). The NDSI is a well-established method for snow cover mapping that effectively distinguishes snow from clouds and can identify snow cover even in shadowed mountainous regions. Snow and ice exhibit maximum reflectance in visible bands and very minimum reflectance in shortwave infrared (SWIR) bands, thus allowing for effective band rationing with applicable thresholds to map snow cover (Riggs et al. 2015). Moreover, using the NDSI has provided more apparent mapping paralleled to other techniques. This method is unaffected by topographic variations and can detect snow even in areas of mountains covered by shades. The NDSI is calculated using the formula (1).

$$\text{NDSI} = \frac{Green\,Refelectance\,(Band\;3) - SWIR\,Refelectance\,(Band\;5)}{Green\,Refelectance\,(Band\;3) + SWIR\,Refelectance\,(\text{Band}\;5)} \tag{1}$$

4.3 Mapping of Supraglacial Debris Distribution

Due to the challenges and logistical constraints of field-based studies, we employ remote sensing methods to gather information over extensive spatial levels, distinguish topographies such as debris in glaciated areas, and detect the changes in the cryosphere over time. Various thresholds were applied to the NDSI ratios calculated from the pre-processed TOA datasets of Landsat-8 (Bhardwaj et al. 2015). For the years 2015, 2019, 2021, and 2023, an NDSI threshold of 0.19 was used, while for the years 2014, 2016, 2017, 2018, 2020, and 2022, a threshold of 0.22 was adopted by our analysis to discriminate debris pixels from snow pixels. By applying these thresholds to the NDSI band ratios, differentiation between the areas covered by snow and ice and those covered by debris has been identified (Fig. 3). Using two NDSI thresholds (0.19 and 0.22) in the SDC analysis of Landsat 8 data ensures consistent classification across years by accounting for sensor variability, atmospheric effects, and seasonal changes (Paul and Kääb 2005). A lower threshold (0.19) reduces overestimation in years with residual snow, while a higher one (0.22) minimizes false positives in more apparent conditions (Racoviteanu et al. 2008). Debris thickness and grain size variations also influence NDSI values, requiring adaptive thresholding. Optimizing thresholds improves classification accuracy, considering regional and

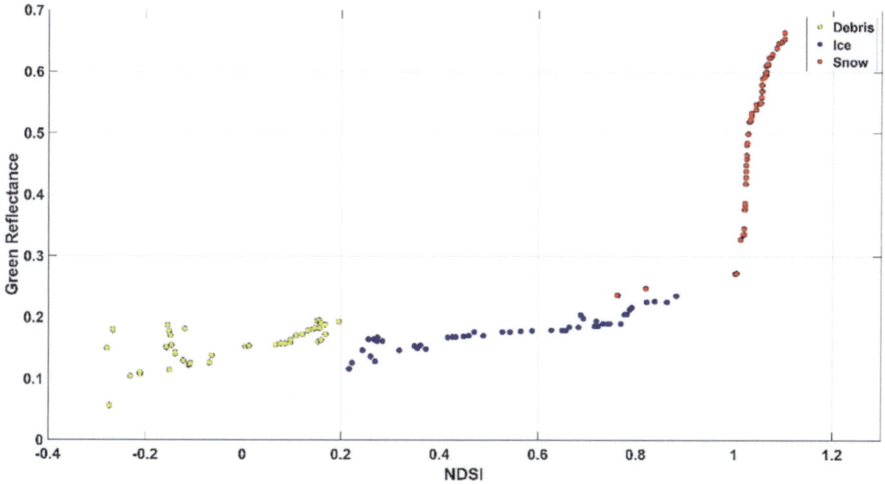

Fig. 3 Threshold adopted over NDSI to distinguish the ice pixels from Debris pixels

temporal differences (Shukla et al. 2010). The post-processed records are compared to the RGI V7.0 polygons to represent the glacier boundary.

5 Results and Discussion

5.1 Decadal Scale Analysis of SDC

The investigation focused on the Kaer Glacier, examining decadal changes to estimate the Supraglacial Debris Cover (SDC) area from 2014 to 2023. According to the RGI 7.0 glacier boundary, the glacier covers an area of 30.484 square kilometers, which served as the basis for the entire analysis. Figure 3 displays the SDC maps of the Kaer Glacier for the periods 2014–2018 and 2019–2023, respectively. The estimated SDC area calculated using NDSI band ratio techniques for the decadal study is presented in Table 2 and Fig. 4. It is clear that the NDSI effectively distinguishes between ice-covered and debris-covered areas, resulting in accurate glacier mapping. Table 2 summarizes the pixel counts derived from the NDSI images and the corresponding Supraglacial Debris Cover areas in square kilometers for each year from 2014 to 2023. The data indicates notable fluctuations in the SDC area over the years, highlighting changes in debris cover on the Kaer Glacier. Based on Table 2, in this analysis, the subsequent deductions can be drawn regarding the SDC of Kaer Glacier:

The pixel counts derived from the NDSI images indicate a marked variability in debris coverage across the decade, with the SDC area showing both increases and decreases in specific years. In 2014, the SDC area was estimated at 0.2394 km^2, based on a pixel count 266 (Fig. 5a). This relatively low value suggests that debris coverage

Table 2 SDC area of Kaer Glacier

Year	Pixel count	Area_SDC (km^2)	Percentage of SDC area
2014	266	0.2394	0.84
2015	985	0.8865	3.13
2016	795	0.7155	2.52
2017	1701	1.5309	5.40
2018	1052	0.9468	3.34
2019	1322	1.1898	4.19
2020	1889	1.7001	5.99
2021	1265	1.1385	4.01
2022	1212	1.0908	3.85
2023	647	0.5823	2.05

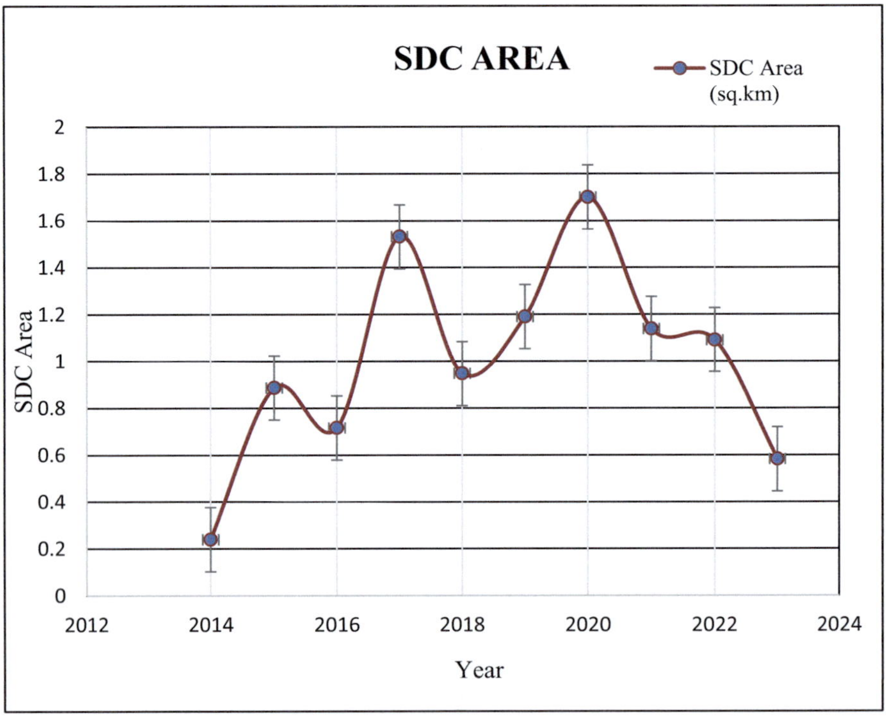

Fig. 4 Decadal SDC area of Kaer Glacier from 2014 to 2023

was minimal during this period. The sparse debris may indicate a combination of factors, including climatic conditions that favored a higher snow cover and limited exposure to the underlying glacier surface. In 2015, there was a marked increase in SDC, with the area expanding to 0.8865 km^2 and the pixel count rising to 985 (Fig. 5b), suggesting an accumulation of debris, possibly driven by enhanced melting conditions that exposed more glacial surfaces. The growth in debris cover may be attributed to a combination of warmer temperatures and changes in precipitation patterns that facilitated the transport and deposition of debris materials onto the glacier. In 2016, the SDC area slightly decreased to 0.7155 km^2, with a pixel count of 795 (Fig. 5c). This decline may reflect variability in weather patterns, possibly indicating a colder year or changes in snow accumulation that could have covered previously exposed debris. The decrease emphasizes how debris cover is dynamic and susceptible to transient climate changes.

It is clear from the observed pattern between 2014 and 2016 how sensitive supraglacial debris cover is to environmental factors. While the reduction in 2016 reflects a return to more stable conditions, where debris may have been lost owing to natural processes or partially hidden by snow, the rise in debris cover in 2015 shows a reaction to favorable melting conditions. Notably, the SDC area expanded significantly in 2017, reaching 1.5309 km^2, corresponding with the highest pixel count of 1701 (Fig. 5d). This peak suggests a notable accumulation of debris, potentially influenced by climatic conditions that favored increased melting and the exposure of underlying debris fields. The subsequent years, particularly 2019 and 2020,

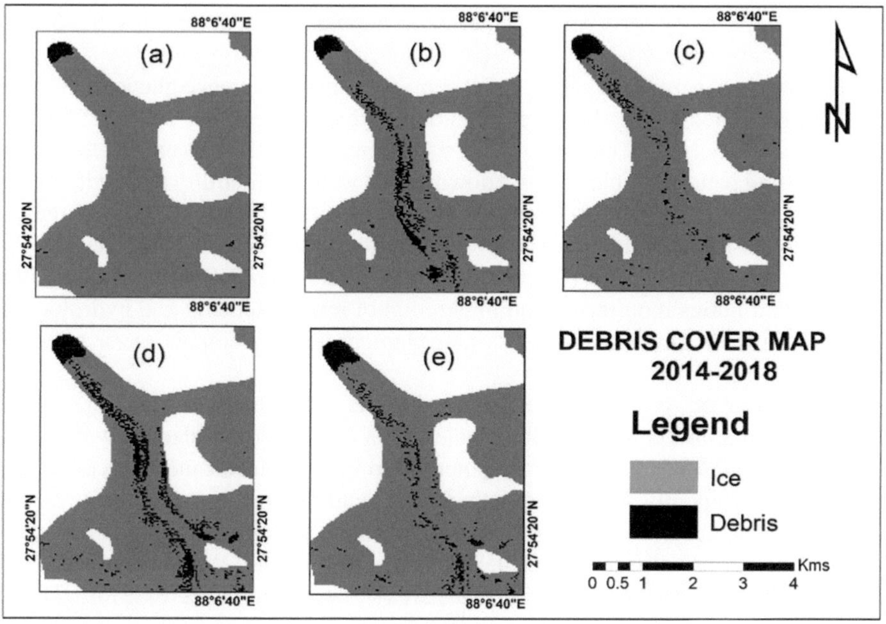

Fig. 5 SDC map of Kaer Glacier **a** 2014, **b** 2015, **c** 2016, **d** 2017, **e** 2018

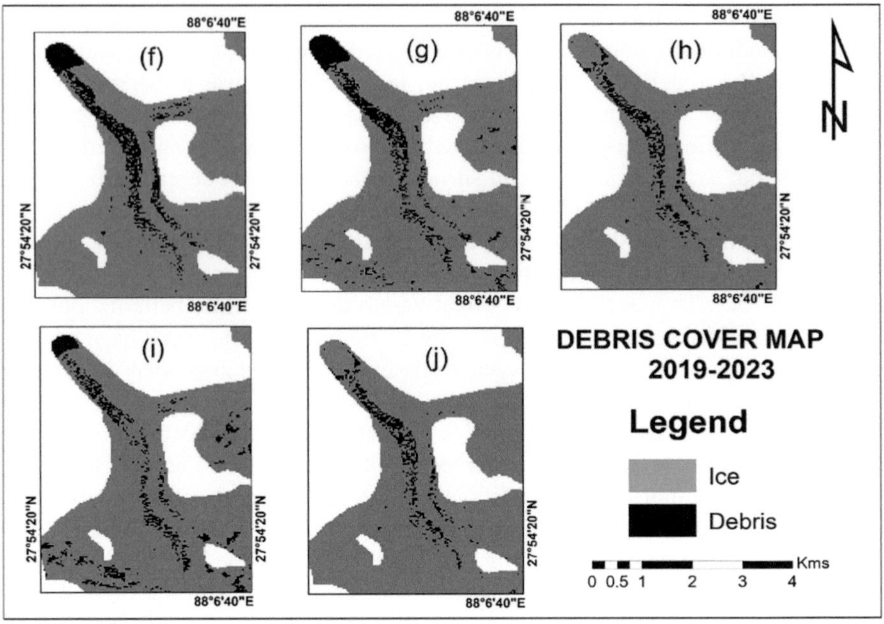

Fig. 6 SDC map of Kaer Glacier **f** 2019, **g** 2020, **h** 2021, **i** 2022, **j** 2023

maintained relatively high SDC areas of 1.1898 km^2 and 1.7001 km^2, respectively, indicating persistent debris coverage across larger glacier sections (Figs. 5e, 6f and g).

The observed increase in supraglacial debris cover can influence glacier melt rates by altering surface energy balance, potentially affecting regional water security in the Himalayas (Scherler et al. 2011). Debris-covered glaciers often exhibit heterogeneous melting patterns, which impact downstream water availability and hydropower potential (Miles et al. 2021). Moreover, the expansion of debris-covered areas is closely linked to the formation and evolution of proglacial and supraglacial lakes, increasing the risk of GLOFs (Richardson and Reynolds 2000). This underscores the need for continuous monitoring and integration of remote sensing and hydrological modeling in the region's disaster risk management and sustainable water resource planning.

The distribution of debris is not confined to the snout region; instead, it extends into the lower ablation zones of the glacier. This finding emphasizes how dynamic the supraglacial debris is, having the potential to impact glacier melt processes and hydrology prominently. The fact that debris is found in these areas indicates that as the glacier recedes, debris may become more visible, which might affect melt rates by changing surface albedo and insulating the ice underneath. With values of 1.1385 km^2 and 1.0908 km^2, respectively, the SDC area showed a slight decrease in 2021 and 2022 (Fig. 6h and i). Due to shifting climatic conditions, such as milder temperatures or different precipitation patterns, which may have influenced the accumulation of debris, this trend may indicate the stability of debris cover or an immediate decrease.

However, in 2023, the SDC area dropped significantly to 0.5823 km^2, corresponding to a decrease in pixel count to 647 (Fig. 6j). This decline may indicate a substantial reduction in supraglacial debris cover, possibly due to increased melting leading to the redistribution or removal of debris materials. The decrease in SDC could have implications for future glacier dynamics and melt rates, highlighting the need for continued monitoring.

Overall, the analysis demonstrates that supraglacial debris cover on the Kaer Glacier is a complex and dynamic feature influenced by climatic conditions and glacier dynamics. The observed trends suggest a need for further research into the interactions between debris cover and glacier melt processes and the long-term implications for water resources in downstream areas dependent on glacial runoff. Estimations conducted in August are preferred over those in July or September. This preference arises because, in August, the entire zone of ablation is visible due to intensified melting throughout this month of each hydrological year (HY). In contrast, September can bring unexpected snowfall, leaving parts of the ablation area covered in snow. The lowest SDC was recorded in 2014. These variations may be attributed to a higher presence of solid snow or significantly reduced melting, reflecting the glacier's conditions during that year. Nevertheless, the SDC area increased between 2015–2017 and 2019–2022, suggesting a recession of the glaciers.

5.2 *Temperature and Precipitation Analysis with SDC Changes*

Temperature and precipitation significantly influence supraglacial debris cover (SDC) as they both impact glacier melting, debris supply, and transport mechanisms. The climate information utilized in this research (Fig. 7) was gathered from the Climate Lhasa website and provided by the weather station 555,910 (ZULS) at Lhasa Gonggar Airport, which is situated at latitude 29.66, longitude 91.13, and an elevation of 3650 m. The given dataset demonstrates notable changes in these climate characteristics over time, consistent with the observed shifts in the scope of SDC.

5.2.1 Temperature Trends and Their Impact on SDC

Warming Trend (2014–2023)

Over the period, temperatures exhibit an increasing trend, peaking at 10.8 °C in 2023. Temperature fluctuations strongly influence glacier melt, impacting the exposure and redistribution of debris. Higher temperatures accelerate glacier ice loss, increasing the debris-covered area due to the melting of underlying ice.

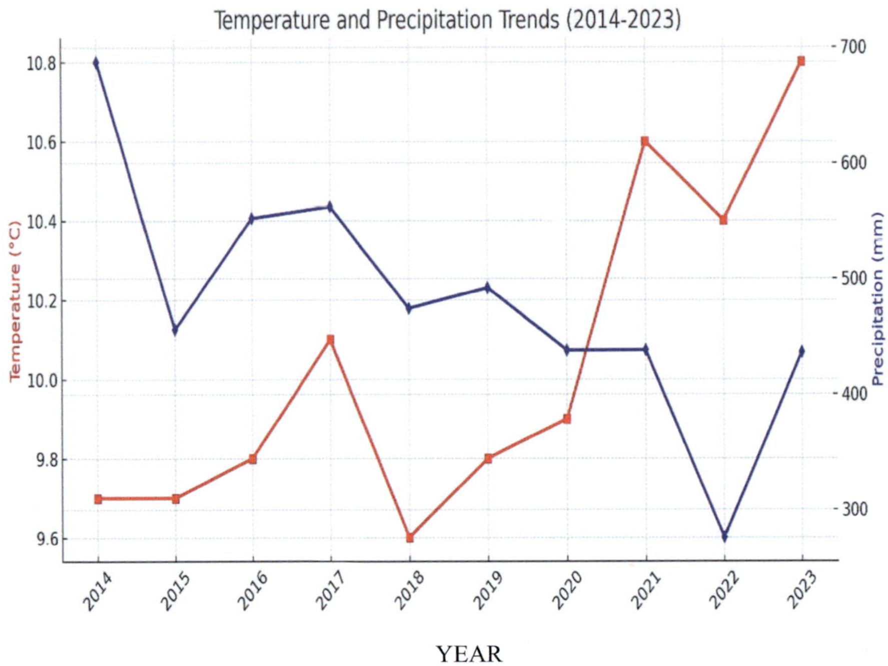

Fig. 7 Average annual temperature and precipitation trends (2014–2023)

Periods of High Temperature and SDC Expansion (2015–2020)

A gradual increase in temperature from 9.7 °C in 2014 to 9.9 °C in 2020 corresponds with a growth in SDC (from 0.2394 km^2 in 2014 to 1.7001 km^2 in 2020). This increase in temperature suggests that glacier thinning and ice melt contributed to debris expansion on the glacier surface. Melting exposes more englacial and subglacial debris, increasing supraglacial debris cover.

Post-2020 Peak and SDC Decline (2021–2023)

After 2020, temperatures increased to 10.6 °C in 2021 and 10.8 °C in 2023, yet SDC decreased (from 1.7001 km^2 in 2020 to 0.5823 km^2 in 2023). This trend suggests that excessive melting reduced debris retention, leading to possible debris flushing into proglacial streams and melt ponds. The thinning glacier surface may have collapsed in some regions, leading to a localized loss of debris cover.

5.2.2 Precipitation Trends and Their Impact on SDC

Declining Precipitation Trend (2014–2023)

Precipitation decreased from 687.32 mm in 2014 to 436 mm in 2023, with a significant drop in recent years (2021–2022). Reduced precipitation can lead to lower debris input from surrounding slopes, affecting the availability of new material for supraglacial deposition.

Periods of Moderate Precipitation and SDC Expansion (2015–2020)

During 2015–2020, precipitation values remained moderate (455–562 mm), which coincides with an increase in the SDC area. Moderate snowfall helps preserve glacier mass, but when combined with warming, it can accelerate melt, revealing more debris. Additionally, rainfall contributes to debris transport from lateral moraines onto the glacier surface.

5.2.3 Post-2020 Drying and SDC Reduction (2021–2023)

The sharp decline in precipitation (especially 275.84 mm in 2022) correlates with a decrease in the SDC area. Less precipitation reduces the supply of new debris from rockfalls and limits debris redistribution by meltwater, reducing debris on the glacier surface. It is clear that precipitation played a role in debris expansion during earlier years, but its recent decline has led to lower debris accumulation and potential erosion of existing debris.

6 Study Limitations Resolution Constraints

Although the NDSI works well for distinguishing between snow and debris, the satellite images' geographic resolution may influence how well it performs. Lower-resolution data might not adequately capture more minor debris characteristics, which could result in an underestimation or incorrect debris cover classification.

6.1 Atmospheric Interference

Atmospheric conditions can significantly influence the reflectance numbers utilized in the NDSI computations. Accurately identifying snow and debris is more difficult because of the noise introduced into the data by cloud cover, haze, and fluctuating illumination. Techniques for atmospheric correction can lessen these impacts, but they are not perfect.

6.2 Seasonal Variability

Seasonal fluctuations in glacier surface conditions affect debris cover mapping. Snow levels vary throughout the year, and NDSI is sensitive to their presence. If not adequately considered, seasonal fluctuations may result in inaccurate estimates of debris cover.

6.3 Threshold Selection

The efficacy of the NDSI technique is dependent on establishing proper thresholds for discriminating between snow and debris. These criteria are subjective and may necessitate considerable calibration for various environmental conditions, compromising the results' reliability.

6.4 Data Availability

The study's timeframe is liable to the accessibility of high-quality satellite imagery. Data gaps caused by satellite scheduling, maintenance, or lousy weather circumstances may impede thorough analysis throughout the desired decadal timeframe.

7 Uncertainty Estimation

Since the SDC area is derived from pixel counts in Landsat 8 imagery (30 m resolution), there is an inherent uncertainty due to potential pixel misclassification. The estimation uses a buffer approach by adding and subtracting one pixel to account for possible misclassification errors. The uncertainty range remains constant at 0.0018 km^2 for all years, accounting for a $\pm$ 1 pixel variation. This method provides a simple but effective estimate of uncertainty due to pixel misclassification in remote sensing data. Other possible causes of uncertainty involve the misclassification of pixels due to spectral confusion (e.g., debris-covered ice vs. surrounding terrain); shadow effects in satellite images may lead to missing or extra pixels in debris-covered glacier regions. In addition, the mixed pixels at the boundaries of debris-covered areas can cause overestimation or underestimation, and seasonal variations & sensor errors in Landsat 8 imagery may contribute to pixel misclassification.

8 Future Implications

8.1 Integration of Multi-source Data

Future studies should combine data from many sources, such as optical, thermal, and radar images. Combining these datasets improves the accuracy of debris cover mapping and provides an improved understanding of glacier movements.

8.2 Advanced Machine Learning Techniques

Machine learning techniques can improve classification precision and expedite the mapping procedure. Future studies should employ supervised machine learning techniques (Random Forest, SVM, and XGBoost) and deep learning models (CNNs, U-Net) to enhance debris-ice distinction through optical-SAR fusion and terrain-based analysis learning to enhance debris recognition and measure changes over time.

8.3 Long-Term Monitoring

Establishing a long-term monitoring strategy based on NDSI and other remote sensing techniques will be critical to understanding the trends and implications of changes in supraglacial debris cover. Regular assessments will help discover small changes and enhance predictive models for glacier behavior.

8.4 Incorporating Field Data

Ground truth data collection using field surveys can help to evaluate remote sensing results and improve debris-mapping reliability. Future research should prioritize collecting in-situ data to better understand supraglacial processes and give context for remote sensing assessments.

8.5 Climate Change Impact Studies

Understanding the role of supraglacial debris in glacier melt and hydrogeology is crucial, especially given climate change predictions. Future studies should investigate how changes in debris cover affect glacier response and downstream water resources, using hydrological modeling to estimate potential implications.

9 Conclusion

A simple computational method and publicly available data are used in this study to understand the decadal fluctuations (2014–2023) in the supraglacial debris cover of the transboundary glacier "Kaer." The NDSI was calculated, and the Landsat-8 dataset was analyzed using the Quantum Geographic Information System (QGIS) software. Our analysis established two threshold values statistically to distinguish the debris from the snow and ice in the NDSI. This study found that the SDC area increased by over 45% throughout the study, which was associated with either a prolonged negative mass balance or sustained mass loss. The rise in SDC may be due to inefficient debris transport on the ablation-dominated terrain.

Furthermore, if temperature influences the rates at which debris is transported to glaciers, the magnitude of debris cover may shift dramatically as the climate changes. Paleoclimatic reconstructions based on historical glacial extents should account for the quantity of SDC and its consequence on glacial mass stabilities in steep locations. Debris cover is predicted to have a more significant effect on mass balances as glaciers retreat. Even though they are situated in almost identical elevation ranges, the "clean," "debris," and "debris with lakes" glaciers all show differences in the area they lost owing to the climate crisis. Compared to "clean" glaciers, most transboundary glaciers near Sikkim are debris-covered, have moraine-obstructed lakes at their snout region, and have lost a greater area. These obstructed lakes were formed at the glacier terminal due to the progressive development and coalescence of supraglacial lakes, even if some debris-covered glaciers displayed stable fronts. According to this research, debris cover on glaciers may stimulate the formation of glacial lakes. As a result, glaciers connected with lakes with debris are retreating more than clean glaciers.

Acknowledgements The authors are grateful to the Randolph Glacier Inventory (RGI 7.0) Consortium, 2023, for supplying the glacier outlines (https://nsidc.org/data/nsidc-0770/versions/7) and the United States Geological Survey (USGS) for providing the COP 30 DEM (https://earthexplorer.usgs.gov/).

Conflict of Interest Disclosure The authors disclosed no possible conflicts of interest.

AI Disclosure AI is not employed in this research work.

References

Alifu H, Vuillaume JF, Johnson BA, Hirabayashi Y (2020) Machine-learning classification of debris-covered glaciers using a combination of Sentinel-1/-2 (SAR/optical), Landsat 8 (thermal) and digital elevation data. Geomorphology 369:107365. https://doi.org/10.1016/j.geomorph.2020.107365

Alifu H, Johnson BA, Tateishi R (2016) Delineation of debris-covered glaciers based on a combination of geomorphometric parameters and a TIR/NIR/SWIR Band Ratio. IEEE J Sel Top Appl Earth Obs Remote Sens 9:781–792. https://doi.org/10.1109/JSTARS.2015.2500906

Allen SK, Sattar A, King O, Zhang G, Bhattacharya A, Yao T, Bolch T (2022) Glacial lake outburst flood hazard under current and future conditions: worst-case scenarios in a transboundary Himalayan basin. Nat Hazards Earth Syst Sci 22:3765–3785. https://doi.org/10.5194/nhess-22-3765-2022

Bhardwaj A, Joshi PK, Snehmani SL, Singh MK, Singh S, Kumar R (2015) Applicability of Landsat 8 data for characterizing glacier facies and supraglacial debris. Int J Appl Earth Obs Geoinf 38:51–64. https://doi.org/10.1016/j.jag.2014.12.011

Brown GL, Brock BW (2023) Carbon flux in supraglacial debris over two ablation seasons at Miage Glacier, Mont Blanc massif. European Alps Front Earth Sci 11:1–17. https://doi.org/10.3389/feart.2023.1200779

Bush ABG, Bishop MP (2017) Glaciers, topography, and climate. Elsevier Inc. https://doi.org/10.1016/B978-0-12-809665-9.00158-0

Casey KA, Benn DI (2011) Optical remote sensing of debris covered glaciers characterization of glacier debris cover via in situ and optical remote sensing methods: a case study in the Khumbu Himalaya, Nepal Optical remote sensing of debris covered glaciers. TCD 5:499–564. https://doi.org/10.5194/tcd-5-499-2011

Chu X, Yao X, Duan H, Chen C, Li J, Pang W (2022) Glacier extraction based on high-spatial-resolution remote-sensing images using a deep-learning approach with attention mechanism. Cryosphere 16:4273–4289. https://doi.org/10.5194/tc-16-4273-2022

Dammann DO, Eriksson LEB, Nghiem SV, Pettit EC, Kurtz NT, Sonntag JG, Busche TE, Meyer FJ, Mahoney AR (2019) Iceberg topography and volume classification using TanDEM-X interferometry. Cryosphere 13:1861–1875. https://doi.org/10.5194/tc-13-1861-2019

Dhanush S, Priya MG (2023) Supraglacial debris cover for Ny-Ålesund using Sentinel-2 data, pp 391–400

Dubey S, Goyal MK (2020) Glacial lake outburst flood hazard, downstream impact, and risk over the Indian Himalayas. Water Resour Res 56:1–21. https://doi.org/10.1029/2019WR026533

Dubey S, Sattar A, Gupta V, Goyal MK, Haritashya UK, Kargel JS (2024) Transboundary hazard and downstream impact of glacial lakes in Hindu-Kush Karakoram Himalayas. Sci Total Environ 914:169758. https://doi.org/10.1016/j.scitotenv.2023.169758

Glaciation G (1998) Glaciers & glaciation. Choice Rev Online 35, 35-6240–35-6240. https://doi.org/10.5860/choice.35-6240

Gök DT, Scherler D, Anderson LS (2023) High-resolution debris-cover mapping using UAV-derived thermal imagery: Limits and opportunities. Cryosphere 17:1165–1184. https://doi.org/10.5194/tc-17-1165-2023

Haritashya UK, Kargel JS, Shugar DH, Leonard GJ, Strattman K, Watson CS, Shean D, Harrison S, Mandli KT, Regmi D (2018) Evolution and controls of large glacial lakes in the Nepal Himalaya. Remote Sens 10:1–31. https://doi.org/10.3390/rs10050798

Herreid S, Pellicciotti F (2020) The state of rock debris covering Earth's glaciers. Nat Geosci 13:621–627. https://doi.org/10.1038/s41561-020-0615-0

Hu M, Zhou G, Lv X, Zhou L, Wang X, He X, Tian Z (2023) Warming has accelerated the melting of glaciers on the Tibetan Plateau, but the debris-covered glaciers are rapidly expanding. Remote Sens 15. https://doi.org/10.3390/rs15010132

Hussain N, Rehman HU, Iqbal M, Zafar MA, Sarwar MK, Tariq MAUR (2024) Assessing the supraglacier extent of Karakoram glaciers located in Hunza River Basin, Pakistan. AQUA Water Infrastr Ecosyst Soc 73:1884–1898. https://doi.org/10.2166/aqua.2024.182

Kääb A, Winsvold SH, Altena B, Nuth C, Nagler T, Wuite J (2016) Glacier remote sensing using sentinel-2. Part I: radiometric and geometric performance, and application to ice velocity. Remote Sens 8. https://doi.org/10.3390/rs8070598

Khadka N, Chen X, Sharma S, Shrestha B (2023) Climate change and its impacts on glaciers and glacial lakes in Nepal Himalayas. Reg Environ Chang 23:1–14. https://doi.org/10.1007/s10113-023-02142-y

Kirkbride MP, Sherriff SC, Rowan AV, Egholm DL, Quincey DJ, Miles E, Hubbard B, Miles K (2023) Provenance and transport of supraglacial debris revealed by variations in debris geochemistry on Khumbu Glacier, Nepal Himalaya. Earth Surf Process Landforms 48:2737–2753. https://doi.org/10.1002/esp.5657

Majeed U, Rashid I, Najar NA, Gul N (2021) Spatiotemporal dynamics and geodetic mass changes of glaciers with varying debris cover in the Pangong Region of Trans-Himalayan Ladakh, India Between 1990 and 2019. Front Earth Sci 9:1–16. https://doi.org/10.3389/feart.2021.748107

Man QX, Guo HD, Liu G, Dong PL (2014) Comparison of different methods for monitoring glacier changes observed by Landsat images. IOP Conf Ser Earth Environ Sci 17:2–7. https://doi.org/10.1088/1755-1315/17/1/012127

McCarthy M, Miles E, Kneib M, Buri P, Fugger S, Pellicciotti F (2022) Supraglacial debris thickness and supply rate in High-Mountain Asia. Commun Earth Environ 3. https://doi.org/10.1038/s43247-022-00588-2

Miles E, McCarthy M, Dehecq A, Kneib M, Fugger S, Pellicciotti F (2021) Health and sustainability of glaciers in High Mountain Asia. Nat Commun 12. https://doi.org/10.1038/s41467-021-23073-4

Milner AM, Khamis K, Battin TJ, Brittain JE, Barrand NE, Füreder L, Cauvy-Fraunié S, Gíslason GM, Jacobsen D, Hannah DM, Hodson AJ, Hood E, Lencioni V, Ólafsson JS, Robinson CT, Tranter M, Brown LE (2017) Glacier shrinkage driving global changes in downstream systems. Proc Natl Acad Sci U S A 114:9770–9778. https://doi.org/10.1073/pnas.1619807114

Mölg N, Bolch T, Walter A, Vieli A (2019) The role of debris cover in the evolution of Zmuttgletscher, Switzerland, since the end of the Little Ice Age. Role debris Cover Evol. Zmuttgletscher, Switzerland, since end Little Ice Age, pp 1–34

Nicholson LI, McCarthy M, Pritchard HD, Willis I (2018) Supraglacial debris thickness variability: impact on ablation and relation to terrain properties. Cryosphere 12:3719–3734. https://doi.org/10.5194/tc-12-3719-2018

Patel LK, Sharma P, Singh A, Oulkar S, Pratap B, Thamban M (2021) Influence of supraglacial debris thickness on thermal resistance of the glaciers of Chandra basin, Western Himalaya. Front Earth Sci 9:1–15. https://doi.org/10.3389/feart.2021.706312

Paul F, Kääb A (2005) Perspectives on the production of a glacier inventory from multispectral satellite data in Arctic Canada: Cumberland Peninsula, Baffin Island. Ann Glaciol 42:59–66. https://doi.org/10.3189/172756405781813087

Pratap B, Dobhal DP, Mehta M, Bhambri R (2015) Influence of debris cover and altitude on glacier surface melting: a case study on Dokriani Glacier, central Himalaya, India. Ann Glaciol 56:9–16. https://doi.org/10.3189/2015AoG70A971

Pratibha S, Kulkarni A. V.: Decadal change in supraglacial debris cover in Baspa basin, Western Himalaya. Curr Sci 114:792–799. https://doi.org/10.18520/cs/v114/i04/792-799

Racoviteanu AE, Nicholson L, Glasser NF (2021) Surface composition of debris-covered glaciers across the Himalaya using linear spectral unmixing of Landsat 8 OLI imagery. Cryosphere 15:4557–4588. https://doi.org/10.5194/tc-15-4557-2021

Racoviteanu AE, Williams MW, Barry RG (2008) Optical remote sensing of glacier characteristics: a review with focus on the Himalaya, pp 3355–3383. https://doi.org/10.3390/s8053355

Racoviteanu AE, Nicholson L, Glasser NF, Miles E, Harrison S, Reynolds JM (2022) Debris-covered glacier systems and associated glacial lake outburst flood hazards: challenges and prospects. J Geol Soc Lond 179. https://doi.org/10.1144/jgs2021-084

Richardson SD, Reynolds JM (2000) An overview of glacial hazards in the Himalayas. Quat Int 65–66:31–47. https://doi.org/10.1016/S1040-6182(99)00035-X

Riggs G, Hall D, Román MO (2015) VIIRS_snow_cover_ATBD_2015

Rowan AV, Egholm DL, Quincey DJ, Hubbard B, King O, Miles ES, Miles KE, Hornsey J (2021) The role of differential ablation and dynamic detachment in driving accelerating mass loss from a debris-covered Himalayan Glacier. J Geophys Res Earth Surf 126:1–20. https://doi.org/10.1029/2020JF005761

Sahu R, Ramsankaran R, Bhambri R, Verma P, Chand P (2023) Evolution of supraglacial lakes from 1990 to 2020 in the Himalaya-Karakoram region using cloud-based Google Earth engine platform. J Indian Soc Remote Sens 51:2379–2390. https://doi.org/10.1007/s12524-023-01773-2

Scherler D, Egholm DL (2020) Production and transport of supraglacial debris: insights from cosmogenic 10Be and numerical Modeling, Chhota Shigri Glacier, Indian Himalaya. J Geophys Res Earth Surf 125:1–26. https://doi.org/10.1029/2020JF005586

Scherler D, Bookhagen B, Strecker MR (2011) Spatially variable response of Himalayan glaciers to climate change affected by debris cover. Nat Geosci 4:156–159. https://doi.org/10.1038/ngeo1068

Scherler D, Wulf H, Gorelick N (2018) Global assessment of supraglacial debris-cover extents. Geophys Res Lett 45:11798–11805. https://doi.org/10.1029/2018GL080158

Schmidt S, Nüsser M, Baghel R, Dame J (2020) Cryosphere hazards in Ladakh: the 2014 Gya glacial lake outburst flood and its implications for risk assessment. Nat Hazards 104:2071–2095. https://doi.org/10.1007/s11069-020-04262-8

Sharda S, Srivastava M (2024) Mapping of debris-covered glaciers using object-based machine learning technique. J Indian Soc Remote Sens 52:399–411. https://doi.org/10.1007/s12524-024-01832-2

Sharma A, Gupta M, Sharma N (2024) Glacier facies characterisation in transboundary West Sikkim Himalaya from TerraSAR-X; GLCM based classification approach. J Spat Sci 69:43–59. https://doi.org/10.1080/14498596.2022.2164085

Sharma A, Gupta M, Sharma N, Gupta S (2022) Line of sight glacier velocity estimation of transboundary glaciers in Eastern Himalayas using high resolution TerraSAR-X data, pp 1–6

Shukla A, Arora MK, Gupta RP (2010) Synergistic approach for mapping debris-covered glaciers using optical-thermal remote sensing data with inputs from geomorphometric parameters. Remote Sens Environ 114:1378–1387. https://doi.org/10.1016/j.rse.2010.01.015

Silvano F, Milva P (2023) Sea level rise impacts in coastal areas and possible mitigation engineering approaches. Ann Civ Environ Eng 7:030–036. https://doi.org/10.29328/journal.acee.1001053

Sivalingam S, Murugesan GP, Dhulipala K, Kulkarni AV, Pandit A (2022) Essential study of Karakoram glacier velocity products extracted using various techniques. Geocarto Int 37:7381–7397. https://doi.org/10.1080/10106049.2021.1974954

Sivalingam S, Murugesan GP, Kulkarni AV, Dhulipala K, Devaraj S (2023) Study of dynamics in surface ice flow rate of glaciers in Hunza basin, Karakoram. Environ Sci Pollut Res 30:62782–62802. https://doi.org/10.1007/s11356-023-26441-3

Wei J, Liu S, Wang X, Zhang Y, Jiang Z, Wu K, Zhang Z, Zhang T (2021) Longbasaba glacier recession and contribution to its proglacial lake volume between 1988 and 2018. J Glaciol 67:473–484. https://doi.org/10.1017/jog.2020.119

Wei J, Liu S, Zhang T, Wang X, Zhang Y, Jiang Z, Wu K, Zhang Z (2020) Terminal motions of Longbasaba Glacier and their mass contributions to proglacial lake volume during 1988–2018. Cryosph Discuss 1–27

Wu Y, Zhang W, Zhang S, Mao Z, Chen A, Li Z, Zhang Q, Guo Z, Jiang X, Long Y (2023) Emerging signal of englacial debris on one clean surface glacier based on high spatial resolution remote sensing data in Northeastern Tibetan Plateau. Remote Sens 15. https://doi.org/10.3390/rs15153899.

Xie F, Liu S, Wu K, Zhu Y, Gao Y, Qi M, Duan S, Saifullah M, Tahir AA (2020) Upward expansion of supra-glacial debris cover in the Hunza Valley, Karakoram, During 1990∼2019. Front Earth Sci 8:1–18. https://doi.org/10.3389/feart.2020.00308

Xin- W, Liu S, Guo W, Xu J (2008) Assessment and simulation of glacier lake outburst floods for longbasaba and Pida lakes, China. Mt Res Dev 28:310–317. https://doi.org/10.1659/mrd.0894

Yang X, Xie F, Liu S, Zhu Y, Fan J, Zhao H, Fu Y, Duan Y, Fu R, Guo S (2024) Mapping debris-covered glaciers using high-resolution imagery (GF-2) and deep learning algorithms. Remote Sens 16. https://doi.org/10.3390/rs16122062

Zeller L, McGrath D, McCoy SW, Jacquet J (2024) Seasonal to decadal dynamics of supraglacial lakes on debris-covered glaciers in the Khumbu region, Nepal. Cryosphere 18:525–541. https://doi.org/10.5194/tc-18-525-2024

Zhang T, Wang W, An B, Wei L (2023) Enhanced glacial lake activity threatens numerous communities and infrastructure in the Third Pole. Nat Commun 14. https://doi.org/10.1038/s41467-023-44123-z

Assessing the Impact of Climate Change on Glaciers Through Region Wise Mass Balance and AI Tools, Contributing to Understanding of Environmental Changes in Upper Indus Basin and Kashmir Himalayas

Suhail Ahmad Dar, Md. Omar Sarif, and Sumalya Ray

1 Introduction

In the twenty-first century, climate change is just one of many pressing challenges. The increase in the Earth's surface temperature is not the sole concern there is also a decline in precipitation, which disrupts the water cycle. This has significant consequences on the availability of water for drinking, agriculture, hydropower, industrial use, and domestic purposes, further highlighting the urgency of the situation (Rashid et al. 2017). Climate change involves the long-term alteration of temperatures, weather patterns, and atmospheric conditions, largely driven by human activities (Rasul and Molden 2019). A key factor is the burning of fossil fuels such as coal, oil, and gas, which releases greenhouse gases like carbon dioxide (CO_2) and methane (CH_4) into the atmosphere (Dar and Sarif 2024). These gases trap heat, leading to a rise in the Earth's overall temperature leading to the melting of glacier which store 70 percent of fresh water having implications on the livelihood of millions of people

S. A. Dar (✉) · Md. O. Sarif
Department of Geography, Lovely Professional University, Phagwara, Punjab, India
e-mail: suhaild7@gmail.com

Md. O. Sarif
e-mail: sarif@hiroshima-u.ac.jp

Md. O. Sarif
Center for Peaceful and Sustainable Futures (CEPEAS), The IDEC Institute, Hiroshima University, 1-5-1 Kagamiyama, Higashihiroshima City, Hiroshima, Japan

S. Ray
Department of Geospatial Information Science and Technology, Lovely Professional University, Phagwara, Punjab, India

M. O. Sarif and A. Sharifi (eds.), *Advanced Geospatial Intelligence and AI for Environmental Resilience and Sustainable Development*, Advances in Geographic Information Science, https://doi.org/10.1007/978-3-032-03714-5_7

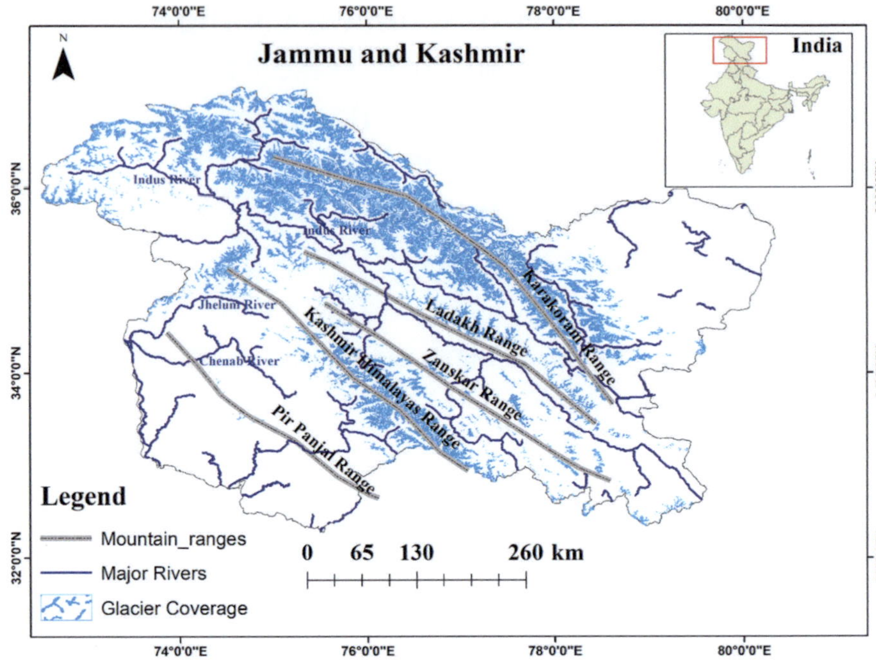

Fig. 1 Study area of Upper Indus Basin and Kashmir Himalayas

around the world (Zhang et al. 2019). Due to high temperature, there is more energy demands for cooling purposes, which puts more pressure on carbon-based fuels and hydroelectricity (Dar and Sarif 2025). The climate change had led to global recession of the glaciers both continental and alpine glaciers, which ultimately leads in global increase in sea level around the world (Marazi and Romshoo 2018). The role of geospatial technology in the studying the glaciers is immense because glaciers are found in the harsh environments of the earth where the field studies are tough (Mandal et al. 2024a). The current study focuses on the geodetic mass balance region wise in North Western Himalayas using multi satellite data. We have divided the Kashmir Himalayas into six regions as shown in Fig. 1 to understand the mass balance of all glaciers founded in the respective regions and same kind of approaches are applied by different researchers thought out the world (Mandal et al. 2024a; Zheng 2000; Berthier et al. 2007; Azam et al. 2018; Brun et al. 2017; Srinivasalu et al. 2024).

2 Study Area

The Kashmir Himalayas is the northwest part of Himalayas and falls under the greater Himalayas having average elevation of 1850 m above the sea level whereas the minimum elevation is 205 m and the maximum elevation is 8564 m based on

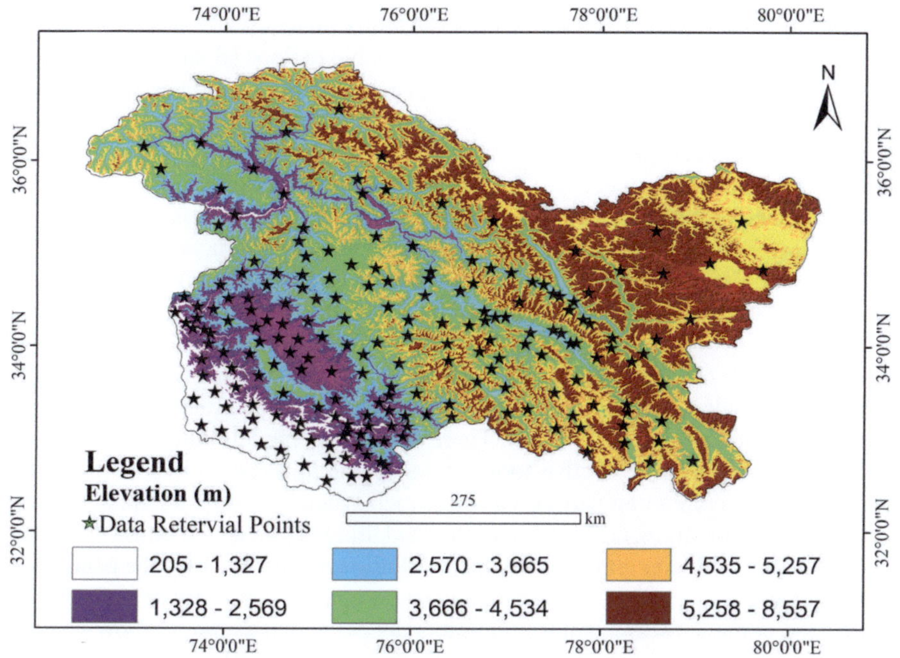

Fig. 2 Data collection points with elevation in meters

satellite data of SRTM 1 Arc-Second Global digital elevation (Murtaza and Romshoo 2017) (Fig. 2). The Kashmir Himalayas have several high mountain ranges in the world such as Karakorum, Ladakh, Zanskar and Pir Panjal Range. And the one of the biggest glaciers are found in this region outside polar region such as Siachen, Baifo, Shafat, Drangdrung, Machoi, Harmukh, Parkhachik, Kolohai, Hoksar, and Thajwas (Ahmed et al. 2024). The region is having plentiful of waters resources and there can be found more than 10,000 glaciers in the western Himalayan region. The worlds famous Valley of Kashmir is located in this region between Pir-Panjal and Zanskar range and also the cold desert of Ladakh is located between Zanskar range and Karakorum range, most of the glaciers are founded above 4000 m above the sea level in this region (Azam et al. 2016).

2.1 *Role of Geospatial Technology in Monitoring of Glaciers*

Geospatial technology is playing a vital role in monitoring of glaciers throughout the world. Offering advanced tools and techniques to assess the glacier dynamics to understand the long-term changes in response to the climatic variability (Ahmed 2022). The integration of remote sensing and GIS and GPS enables the researchers to track and analyse glacier at multiple spatial and temporal scales which helps us understand the glacier dynamics over vast and inaccessible regions (Bisht 2015). Remote sensing technology like satellite imagers especially (InSAR) and aerial surveys is invaluable in glacier monitoring. It allows researchers acquisition of high- resolution, multi-spectral images that can detect surface changes, mass balance, glacier thickness and Snow cover (Joya et al. 2021). Techniques such as Interferometric Synthetic Aperture Radar (InSAR) and optical stereo imagery enable precise measurements of glacier elevation changes and ice flow velocities, which are crucial for calculating glacier mass balance. These techniques are implemented by using python-based algorithms, which make the pre-processing of data very efficient and large datasets can be handled easily. The geodetic method, for example, utilizes satellite-derived digital elevation models (DEMs) to estimate glacier volume changes, making it possible to monitor thousands of glaciers simultaneously and assess large-scale trends in glacial mass loss or gain (Azam et al. 2018). There are various models which are AI based such as GIS is a highly effective tool for the spatial analysis and visualization of glacier data. By combining various datasets, such as temperature, precipitation, and topographic information, GIS facilitates the modelling of glacier dynamics and helps identify trends in glacier retreat or advance. Additionally, it allows for the generation of interpolated maps and models that capture the spatial variability of both glaciological and climatic factors, contributing to a deeper understanding of how glaciers respond to climate change on both regional and global scales (Romshoo et al. 2015). There are lot of physical challenges in the glaciology which are levied due to Remote Sensing and GIS. Glacier are dynamic entities and reacts differently under same climatic conditions (Romshoo et al. 2022a). With the help of geospatial technology, the glaciers are studied in terms of mass balance, Depth analysis, Glacial area loss, Snout retreat, Surface velocity, Aerosol contamination, Overdeepning, Snow cover monitoring, Snow and Glacier Melt Runoff Models (Li et al. 2024; Romshoo et al. 2024).

2.2 *Role of AI in Glaciology*

The use of AI in glacier inventory and mapping presents a promising opportunity, providing a transformative method to enhance both the efficiency and precision of glacier monitoring efforts (Nurakynov et al. 2024). AI applications offer another promising avenue in the field of snow and ice discrimination, providing an innovative solution to enhance the accuracy and efficiency of remote sensing analysis. By training extensively on diverse datasets, including satellite imagery and ground-based observations, AI models can rapidly learn to identify the subtle spectral and textural differences unique to snow and ice, surpassing the limitations of traditional manual interpretation or spectral analysis methods (Haq et al. 2019). This capability not only accelerates the processing of large remote sensing datasets but also enables precise quantification of snow and ice extent, which is crucial for climate research, hydrologic modelling, and environmental monitoring efforts. AI holds significant potential to revolutionize the efficiency and accuracy of calculations in ice dynamics modelling, offering another promising application in this research area (Li et al. 2024). By utilizing large volumes of observational data, satellite imagery, and remote sensing datasets, AI-driven models can capture the complex interactions involving ice flow, mass balance, and calving dynamics. This capability not only accelerates the modelling process but also provides researchers with deeper insights into the various factors influencing glacier dynamics and their responses to environmental changes (Jin et al. 2022). Artificial Intelligence (AI) is playing an increasingly vital role in glaciology by providing advanced methods to process and interpret large, complex datasets related to glaciers and the cryosphere. A major application of AI in this field is the automated interpretation of satellite imagery to identify glacier boundaries, snow coverage, debris, crevasses, and glacial lakes, even in difficult terrains and regions with frequent cloud cover (Tanniru et al. 2025). AI also supports time-series analysis, which allows scientists to monitor glacier retreat, changes in ice volume, and calving events more quickly and precisely. Moreover, AI techniques are being used to estimate ice thickness and subglacial topography by integrating existing datasets with digital elevation models (DEMs) and surface flow data. Machine learning is also employed to model glacier mass balance and forecast glacier responses to changing climate conditions, deepening our understanding of glacier-climate dynamics (Jin et al. 2022). In hazard assessment, AI helps predict events such as glacial lake outburst floods (GLOFs) and ice avalanches. Various AI tools and frameworks assist these efforts, including deep learning models like U-Net and Convolutional Neural Networks (CNNs) for image segmentation, and traditional algorithms like Random Forests and Support Vector Machines (SVMs) for classification (Nurakynov et al. 2024). Long Short-Term Memory (LSTM) models are useful for analysing time-series data. Additionally, platforms like Google Earth Engine (GEE), equipped with AI capabilities, enable extensive glacier monitoring at scale. These technological advances are significantly improving the efficiency, accuracy, and scope of cryospheric research (Kumar et al. 2023). The integration of Artificial Intelligence (AI) with geospatial technologies has significantly enhanced the study and monitoring of glacier behaviour in glaciology. While these innovations have expanded research capabilities, they also come with notable challenges and limitations that need to be recognized. A major obstacle is the availability and reliability of data. Many glaciated regions, particularly in high-altitude or polar environments, lack consistent and high-resolution satellite coverage. Since AI systems depend on extensive and accurate datasets for effective training and validation, any gaps or errors in remote sensing data can lead to misleading or skewed results. Furthermore, the variability in spatial and temporal resolutions across different data sources complicates the tracking of long-term glacier changes. Another issue is the limited adaptability of AI models. Algorithms developed for specific glaciers or regions may not transfer well to other areas with different environmental or topographic conditions. This limits the general use of AI models and calls for localized calibration and validation to ensure accurate outcomes. The complexity and opacity of AI models, especially deep learning networks, also pose significant concerns. These models often lack interpretability, making it difficult to trace or explain how results are generated. In glaciology, where understanding the underlying physical processes is essential, this lack of clarity can be a major barrier to trust and scientific insight. Additionally, AI-geospatial integration demands high computational capacity. The processing of detailed satellite imagery and the training of sophisticated models require powerful computing systems and technical know-how—resources that may not be readily available to all institutions, especially in less technologically advanced regions. Another persistent challenge is validating AI predictions. Reliable in-situ data is critical for verifying the accuracy of AI-generated results, yet such data is often minimal or completely absent in many remote glacier environments. This limitation undermines confidence in the outcomes derived from AI models. Beyond technical constraints, the use of AI in environmental science also raises important ethical and policy-related questions. Concerns over data ownership, access rights, and the fair distribution of technological advantages are becoming increasingly relevant. Ensuring equitable access to both data and analytical tools is essential for inclusive and responsible research practices. In summary, while AI and geospatial tools present powerful opportunities for advancing glaciological research, their full potential can only be realized by addressing existing limitations through collaborative, interdisciplinary efforts, enhanced data systems, and transparent model development.

2.3 Glacial Mass Balance

The glacial mass balance is an important parameter of glacier studies by which we can assess the current situation of the glacier (Cogley 2009). Glacier mass balance is a crucial variable that needs to be monitored in order to understand the response of the climate system to climate change (Romshoo et al. 2023). It is a direct and sensitive response to climate change, making it an essential component for monitoring the climate system (Singh et al. 2018). There are several techniques available for measuring glacier mass balance, including glaciological, hydrological, geodetic, and remote sensing methods (Gaddam et al. 2024). Among these methods, the glaciological approach is considered the most accurate for measuring glacier mass balance. However, it is also the most time-consuming and expensive as it requires physically traversing the glacier surface in challenging terrain and harsh weather conditions (Cogley 2009). As a result, there has been less research done on glacial mass balance in the high alpine glaciers of Himalayas. The glaciological method for measuring glacier Mass balance involves taking direct measurements on the glacier surface. This approach involves monitoring the difference in mass between two dates during a specific hydrological year at multiple locations on the glacier surface. The glacial mass balance at a specific point is determined by multiplying the difference in surface density with the resultant mass. This is done by taking measurements of exposed stakes of two different dates and measuring snow depth over a known stratigraphic surface, such as past summer and winter surfaces. Variations in glacier mass can be determined by excavating snow pits in the accumulation area down to a reference surface to measure snow density (Romshoo et al. 2022b). In estimating glacier mass balance using the geodetic method involves tracking changes in surface elevation over time, by using two DEMs of different time period providing insight into glacier mass balance and its climate-driven changes. This method relies on remote sensing technologies, such as satellite imagery, aerial photography, and LiDAR (Light Detection and Ranging), to monitor variations in glacier elevation and surface topography (Kumar et al. 2017). Recent techniques have been developed with the help of technology advancements the instruments like GPR (Ground Penetrating Radar), Terrestrial Laser Scanner and UAV mapping had made estimation of glacial mass balance easy by using these equipment's.

3 Methodology and Data Used

For estimation of mass balance, we have using geodetic method of mass balance by using SRTM DEM of 30 m and Copernicus DEM of 30 m for years 2000 and 2012 (Klug et al. 2018). These datasets have been used by many glaciologists around the world to estimate the geodetic mass balance (Kumar et al. 2017; Gardelle et al. 2013; Vincent et al. 2021). This method is widely used method in the field of glaciology for estimation of mass balance of the glaciers worldwide (Azam et al. 2016; Romshoo et al. 2023; Yu et al. 2024; Mandal et al. 2002). SRTM (Shuttle Radar Topography Mission) data plays a crucial role in geodetic mass balance assessments by providing a foundational digital elevation model (DEM) from the year 2000 (Nuth 2011). This dataset serves as a reference point for comparing glacier surface elevations with those from more recent satellite or aerial observations. By measuring the differences in elevation over time, scientists can estimate changes in glacier volume, which are then converted into mass balance figures. This method is especially valuable in remote and high-altitude regions like the Himalayas, where direct field measurements are challenging to obtain and having high vertical accuracy (Orlandi et al. 2019). The Copernicus Digital Elevation Model (DEM), produced through the European Union's Copernicus Programme, has become a vital tool in geodetic mass balance research aimed at tracking glacier dynamics. Offering high-resolution and enhanced vertical precision, the Copernicus DEM delivers dependable elevation data necessary for identifying changes in glacier surface heights over time (Santillan 2023). When analysed alongside earlier datasets such as the SRTM, it facilitates the computation of elevation variations, which are then used to derive glacier volume and mass changes. Due to its global reach and uniform quality, it is particularly beneficial for studying glacier mass balance in remote and topographically complex areas like the Himalayas (Fischer 2011).

Table 1 Data used

Data	Resolution	Date of acquisition	Source
SRTM DEM	30 m	11–21, Feb 2000	NASA Earth Explorer
Copernicus DEM (GLO 30)	30 m	01/01/2011–07/01/2015	European Space Agency
Glacier outlines	NA	Sep 2023	Randolph Glacier Inventory 7.0
Climatic data	NA	2002–2015	NASA Power Project

3.1 Meteorological Data

The climatic data was retrieved from the NASA Power Project because of remote locations of the glaciers the ground based meteorological data was not present we selected the 200 distinct point with help of google earth pro tagging the 200 points for whole study area. Interpolation of climatic data was done using IDW method in ARG GIS 10.8 for temperature and Precipitation to understand the effect of climatic change on mass balance of glaciers (Source: https://power.larc.nasa.gov/).

3.2 Dem Data Acquisition

The SRTM data downloaded from NASA Earth Explorer and Copernicus was downloaded from European Space Agency, and Glacier outlines were download from Randolph Glacier inventory 7.0 (Table 1).

3.3 Pre-Processing and Co-registration Process

The Given data sets were reprojected using project raster in ARG GIS 10.8 and the datum was det to WGS 1984 UTM 43N (Nuth 2011). For Resampling both data sets were set to 30 m (Tricht 2023). After reprojection and resampling, the co-registration process was done by using Berthiers code using ENVI IDL (Berthier et al. 2016). This process typically involves comparing the SRTM DEM and Copernicus DEM to identify the horizontal (X and Y) shifts (Kumar et al. 2017). Sometimes, even if the DEMs are horizontally aligned, there may be vertical biases due to different acquisition methods or processing techniques (Mandal et al. 2024b). We calculated the mean bias by computing the mean elevation difference over stable terrain and used for uncertainty estimation (Dussaillant et al. 2018).

3.4 Elevation Change

The fundamental measurement in the geodetic method is the change in surface elevation of the glacier between two time periods and is estimated using Raster calculator by subtracting the earlier DEM (2000) from the more recent one (2012). Here elevation changes in surface reflects the thickness change in glacier the formula is mentioned below (Abdullah et al. 2020).

$$\Delta H = h_1 - h_2 \tag{1}$$

Here h_1 is the surface elevation extracted from COP 30 Dem, h_2 is the surface elevation extracted from SRTM 30 m Dem and ΔH Change in elevation (positive indicates thickening, negative indicates thinning) (Mukherjee et al. 2018).

3.5 Volume Change

The volume is calculated by multiplying the area into elevation change of particular cell and then the volume change is estimated by sum the all volume in attribute and the below formula is used

$$\Delta V = \sum_{I=1}^{n} \Delta h_i . A_i \tag{2}$$

Here Δh_i is Elevation change at grid cell i and A_i is the Area at grid cell I and n is total number of grid cells. By this we can estimate the change in volume of glacier (Zheng 2000).

3.6 Mass Change (ΔM)

For estimation of Mass Change the Volume is multiplied with the density of the ice which typically ranges between (850–900 kg/m^3). The below equation is used (Fischer 2011)

$$\Delta M = \rho \cdot \Delta \mathrm{V} \tag{3}$$

Here ρ is the density of the Ice which was taken as 850 kg/m^3 and ΔV is the total change in the volume and ΔM indicates change in the mass of the glacier which is denoted in Giga-tons (Cogley 2009).

3.7 Geodetic Mass Balance

Mass balance represent the gain are the loss of the glacial ice mass per unit area at glacier over a certain period of time and it was estimated by using Eq. 4 given below (Piermattei et al. 2024).

$$B = \frac{\Delta M}{A \cdot \Delta t * \rho Water} \tag{4}$$

Here ΔM is the change in mass of the glacier and A is the area of the glacier and Δt is the time period and ρ water is density of water and B represent the mass balance in mwea^{-1} (meter water equivalent per year) (Garg 2021).

3.8 Uncertainty and Error Propagation

For uncertainty with additional term to account for the radar penetration error (σ penetration) assuming that all the errors are uncorrelated and random. The uncertainty of glacier-wide specific elevation change (Δh) is computed. While calculating geodetic mass balance, uncertainties in both elevation change (Δh) and density (ρ) need to be considered for evaluation (Brun et al. 2017). The uncertainty in elevation change (Δh) depends on the resolution and accuracy of the DEMs used (Mukul et al. 2015). If the DEMs come from different sources, the inherent resolution and precision of the instruments (e.g., satellite, airborne, or LiDAR) will affect the uncertainty. SRTM has a vertical accuracy of ± 5 to 6 m (Orlandi et al. 2019), while as Copernicus has vertical accuracy ± 2 to 5 m (Santillan 2023). Uncertainties in elevation change rates are assessed by analysing elevation changes in off-glacier areas. In this study, elevation changes were adjusted to remove outliers using the widely adopted Normalized Median Absolute Deviation (NMAD) method, which has been frequently utilized in various glaciological studies. as mentioned in Eqs. 5 and 6 given below (Piermattei et al. 2024).

$$\Delta h_{Uncertainty} = \sqrt{(\Delta h_{SRTM})^2 + (\Delta h_{Copernicus})^2 + (\sigma_{penetration})^2} \tag{5}$$

where Δh SRTM: vertical accuracy of the SRTM DEM, Δh Copernicus is vertical accuracy of the Copernicus DEM. σ penetration is the radar penetration of SRTM taken ± 1.5 to 2 (Nuth 2011). The combined vertical uncertainty was calculated to be ± 8 m. The calculation is performed using the following formula given below (Mandal et al. 2020)

$$\delta_{\Delta M} = |\Delta_M| \cdot \sqrt{\left(\frac{\delta_{\Delta h}}{\Delta h}\right)^2 + \left(\frac{\delta_A}{A}\right)^2 + \left(\frac{\delta_\rho}{\rho}\right)^2} \tag{6}$$

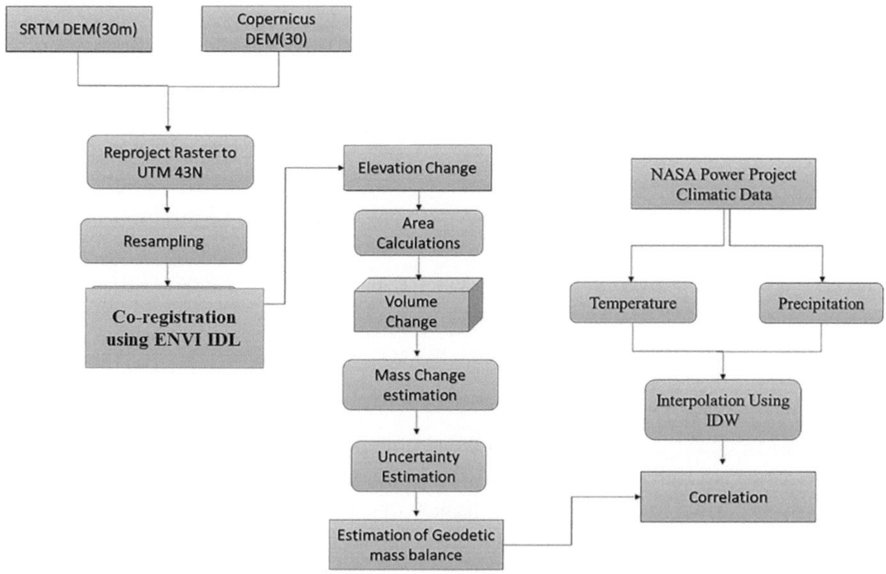

Fig. 3 Flow chart for glacial mass balance estimation

In estimating the uncertainty of mass change (δ Δ_M), factors such as glacier elevation change uncertainty ($\delta_{\Delta h}$), glacier area uncertainty (δ_A), and the error in the density assumption (δρ) are all taken into account, where ρ is the density of ice, δ_ρ ice is the uncertainty in density of ice which is taken as ± 60 kg/m^3 (Huss 2013; Florentine et al. 2023). For uncertainty in Elevation Change (δΔh) was taken as using Eq. 5 (Srinivasalu et al. 2024) (Fig. 3).

4 Results and Discussion

4.1 Mass Balance and Mass Loss

Mass balance is an important parameter by which we can access the health of the glacier and can interpret is the glacier losing or gaining its mass. For this purpose, we selected the geodetic mass balance because it can cover thousands of glaciers once at a time so the coverage it much larger in this method. The results indicated that the glacier present in the Western Himalayas mainly exhibits the negative mass balance except the Karakorum range. The Table 2 provides a comparative analysis of glacier mass loss and mass balance across six major mountain ranges in the western Himalayas, covering the time period from 2000 to 2012. The study indicated that Pir-Panjal Ladakh, Shamsbari, Greater Himalayas Zanskar exhibits negative mass balance with − 1.47 ± 0.12 mwea^{-1}, − 0.70 ± 0.06 mwea^{-1}, − 0.60 ± 0.06, −

Table 2 Glaciers mass balance, elevation change and mass change in gigatons

Range	Mean elevation (m)	Mean slope (°)	Area (km^2)	Elevation change (ma^{-1})	Mass change (Gta^{-1})	Mass balance ($mwea^{-1}$)
Pir Panjal	4218	20.68	40	– 1.6	– 0.05 ± 0.005	– 1.47 ± 0.12
Shamsbari	4700	22.27	479	– 0.72	– 0.28 ± 0.03	– 0.60 ± 0.06
Greater Himalayas	4900	22.58	660	– 0.58	– 0.32 ± 0.03	– 0.50 ± 0.04
Zanskar	5032	23.59	1918	– 0.48	– 0.75 ± 0.06	– 0.40 ± 0.04
Ladakh	5200	24.43	1592	– 0.82	– 1.09 ± 0.08	– 0.70 ± 0.06
Karakorum	5259	31.62	16,493	+ 0.16	+ 2.19 ± 0.09	+ 0.14 ± 00.1

0.50 ± 0.04 $mwea^{-1}$, – 0.40 ± 0.04 $mwea^{-1}$, respectively. The highest annual mass loss of recorded in Ladakh region, followed by Zanskar region, followed by Kashmir Himalayas as shown in (Table 2). The highest mass loss is observed in Ladakh range and Zanskar range which means the melting is higher in this region. While as the Karakorum region exhibits a positive mass balance of + 0.14 ± 00.1 $mwea^{-1}$ with mass gain of + 2.19 ± 0.0.09 Gt per year. The melting is observed all over the western Himalayas except the Karakorum range and this can be attributed to its mean elevation being highest and the slope being highest among all which is 31.62 degrees, rest of the regions have mean slope less than 25 degrees and mean elevation less than 5200 m above the sea level.

4.2 *Elevation Changes in the Western Himalayas*

The maximum elevation changes were observed in Pir-Panjal, Ladakh range with—mean annual elevation change of – 1.6 ma^{-1}, – 0.82 ma^{-1} respectively followed by Shamswari region with – 0.72 ma^{-1} followed by Greater Himalayas with – 0.58 ma^{-1}. And the Karakorum region is gaining the elevation of + 0.16 ma^{-1} also known as Karakorum anomaly. These are some serious decreases in the elevation of the glaciers which means the glaciers are losing mass at very rates. Same rates are being observed over the western Himalayas as estimated by following researchers (Azam et al. 2016; Romshoo et al. 2022a; Mandal et al. 2024b; Dong et al. 2013). The Fig. 4 represents the cumulative elevation changes of the glacier in the region over 12 years. The highest cumulative elevation changes were recorded in Pir-Panjal range because this range is receiving the highest snowfall in winters and also heavy rates of recession are being recorded in this region (Rashid et al. 2022a, 2022b). The elevation changes in Fig. 4 represents that most of the changes in glacier are happening in ablation zone, where the melting is occurring and in accumulation zone of glacier the elevation change is showing positive indicating the accumulation in the mass in the glaciers.

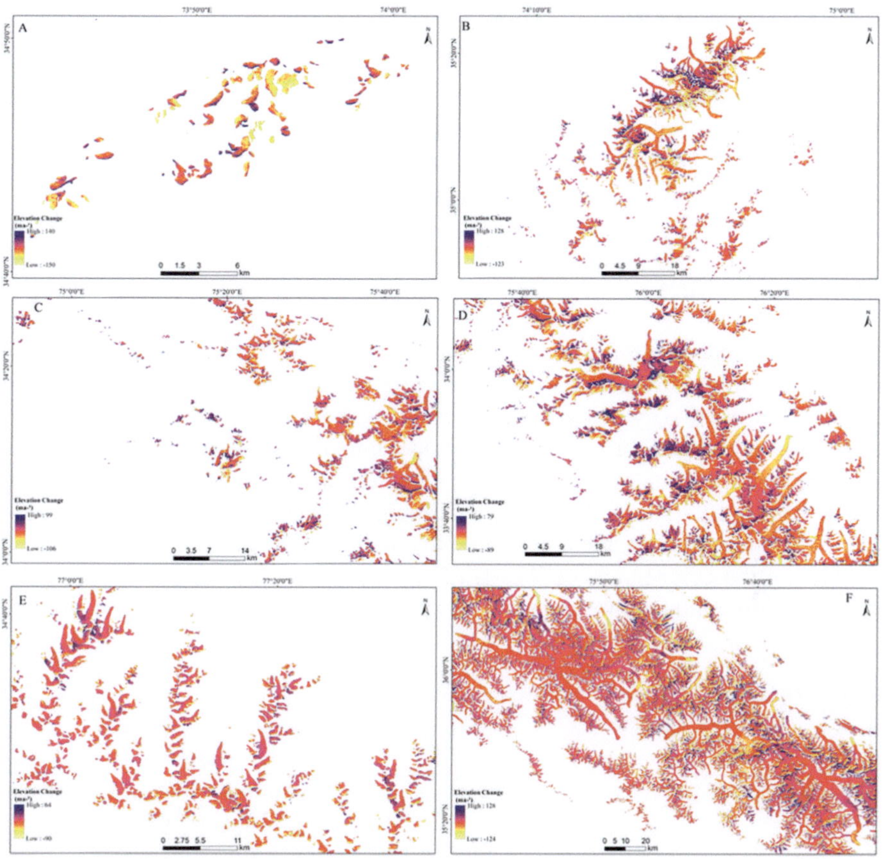

Fig. 4 Elevation changes in glaciers across the six regions (**a** is Pir-Panjal, **b** is Shamasbari, **c** is Greater Himalayas, **d** is Zanskar, **e** is Ladakh and **f** is Karakorum range)

4.3 Comparative Analysis of the Mass Balance Region Wise

The results from our study indicated that the same recession rates were also found by other researchers in the north western Himalayan region and around the world (Mandal et al. 2024a; Zheng 2000; Gardelle et al. 2013).

4.4 Analysis of Climatic Data

The meteorological was analysed from 2002 to 2015 throughout the western Himalayas by collecting the data from 200 points using google earth. The interpolation of data was done using IDW technique in ARG GIS 10.8 on temperature and precipitation and the precipitation maps were drawn for 14 years.

4.4.1 Temperature

The average annual temperature was recorded to be 3.90 °C for 14 years from 2002 to 2015. The highest was observed in 2002 followed by 2007 and 2004 as shown in Table 3. There were six years where temperature was above average and 8 years where it was below average. From interpolated temperature maps based on the geographical areas the western Himalayas have diverse climate, here the annual mean temperature ranges from − 12.70 to 26 °C as shown in figure the overall trend of the temperature for 14 years seems to be stable with little fluctuations. The highest temperatures are recorded in the outer plains of Western Himalayas flowed by Kashmir valley and the Surrounding mountains ranges. while as the lowest temperatures are recorded in

Table 3 Glacial mass balance region wise and mass budget

Region	Time period	Mass balance ($mwea^{-1}$)	Mass budget (GTa^{-1})	Authors
Hindu Kush	2000–2011	− 0.12 ± 0.16	− 0.7 ± 1.0	Gardelle et al. (2013)
Karakorum	2000–2011	+ 0.10 ± 0.16	+ 1.9 ± 3.1	Gardelle et al. (2013)
Pamir	2000–2011	+ 0.14 ± 0.14	+ 1.3 ± 1.3	Gardelle et al. (2013)
Ladakh	2000–2021	− 0.37 ± 0.07	–	Mandal et al. (2024a)
Zanskar	2000–2011	− 0.99 ± 0.43	− 2.34 ± 0.84	Abdullah et al. (2020)
West Nepal	2000–2011	− 0.32 ± 0.14	− 2.2 ± 1.0	Gardelle et al. (2013)
Bhutan	2000–2011	− 0.22 ± 0.13	− 0.9 ± 0.5	Gardelle et al. (2013)
South Asia west	2000–2023	− 0.17 ± 0.05	− 5.4 ± 1.6	Zheng (2000)
South Asia east	2000–2023	− 0.52 ± 0.08	− 7.3 ± 1.2	Zheng (2000)
Central Asia	2000–2023	− 0.22 ± 0.06	− 10.4 ± 2.7	Zheng (2000)
Western Canada and USA	2000–2023	− 0.68 ± 0.06	− 9.0 ± 0.9	Zheng (2000)
Russian Arctic	2000–2023	− 0.32 ± 0.04	− 16.1 ± 2.0	Zheng (2000)

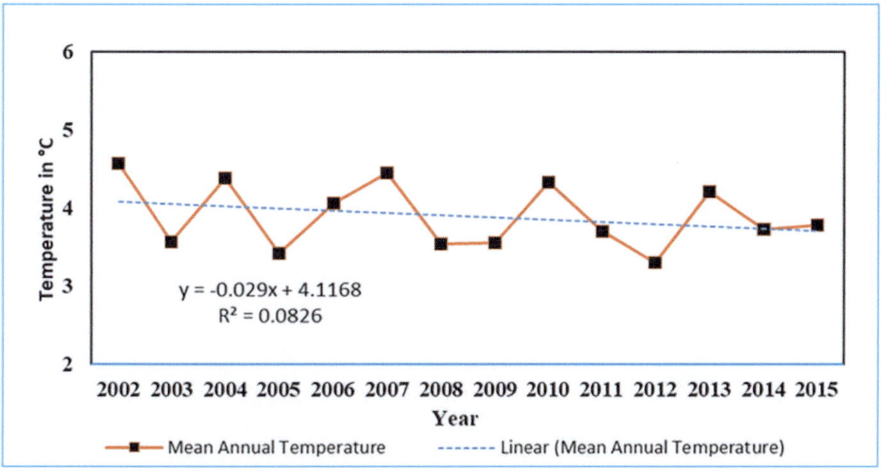

Fig. 5 Annual temperature from 2002 to 2015 of western Himalayas

Karakorum range followed by Zanskar and Ladakh range as shown in interpolated temperature maps (Fig. 5)

4.4.2 Precipitation

The average annual precipitation for 14 years was 446.53 mm per year. And there were 7 years where precipitation was below average and 7 years where precipitation was above average which shows the uncertainty in the precipitation pattern. The highest precipitation was observed in 2015 followed by 2006 and 2003, while lowest precipitation years were 2002 followed by 2004 and 2009 as shown in Fig. 6. The region is very diverse if we talk of range of Precipitation in 14 years it is 8.41–1205 mm. The overall trend of precipitation was observed to be statistically significant with 15 mm per year increase of precipitation as shown in figure. The interpolated maps showed that the maximum precipitation happens in the Pir-Panjal range followed by Kashmir Valley and Greater Himalayas of Kashmir and the lowest precipitation falls in Karakorum range followed by Ladakh and Zanskar region as shown in (Figs. 7, 8 and Table 4).

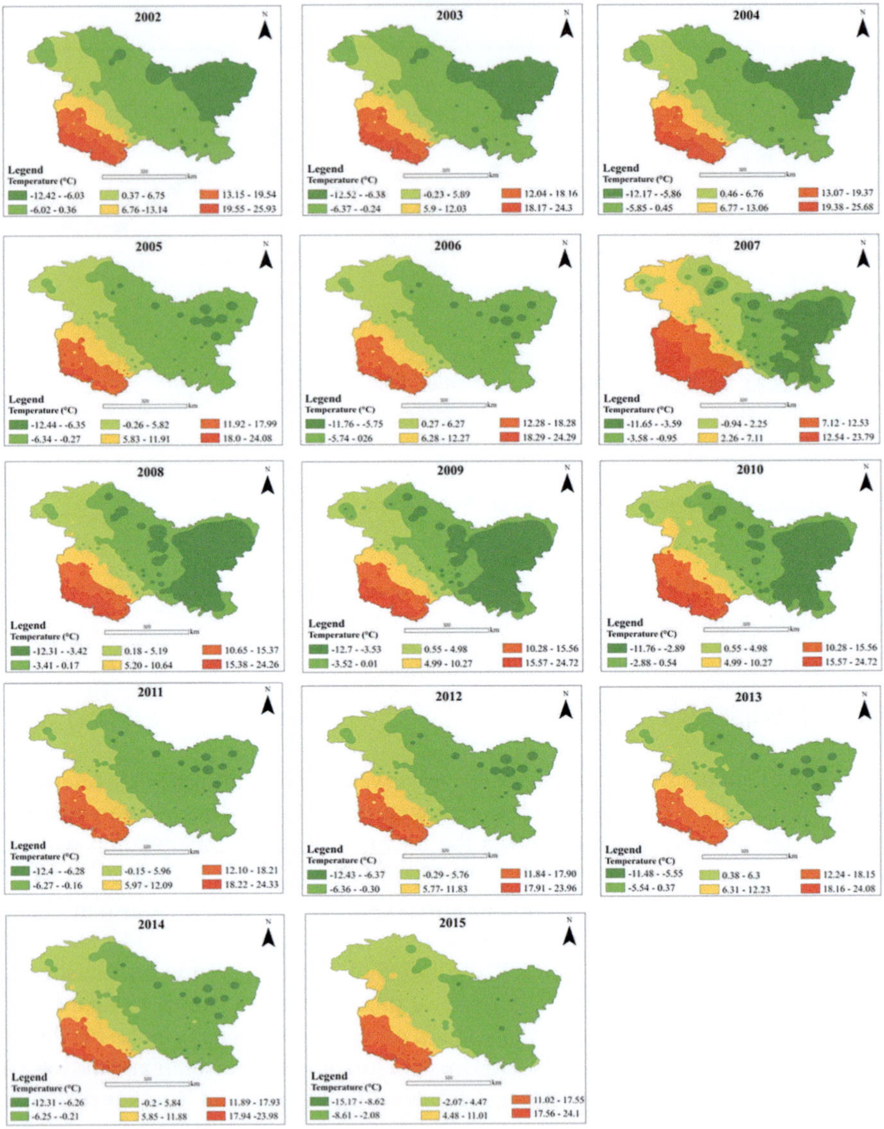

Fig. 6 Interpolated temperature maps from 2002 to 2015 of Western Himalayas

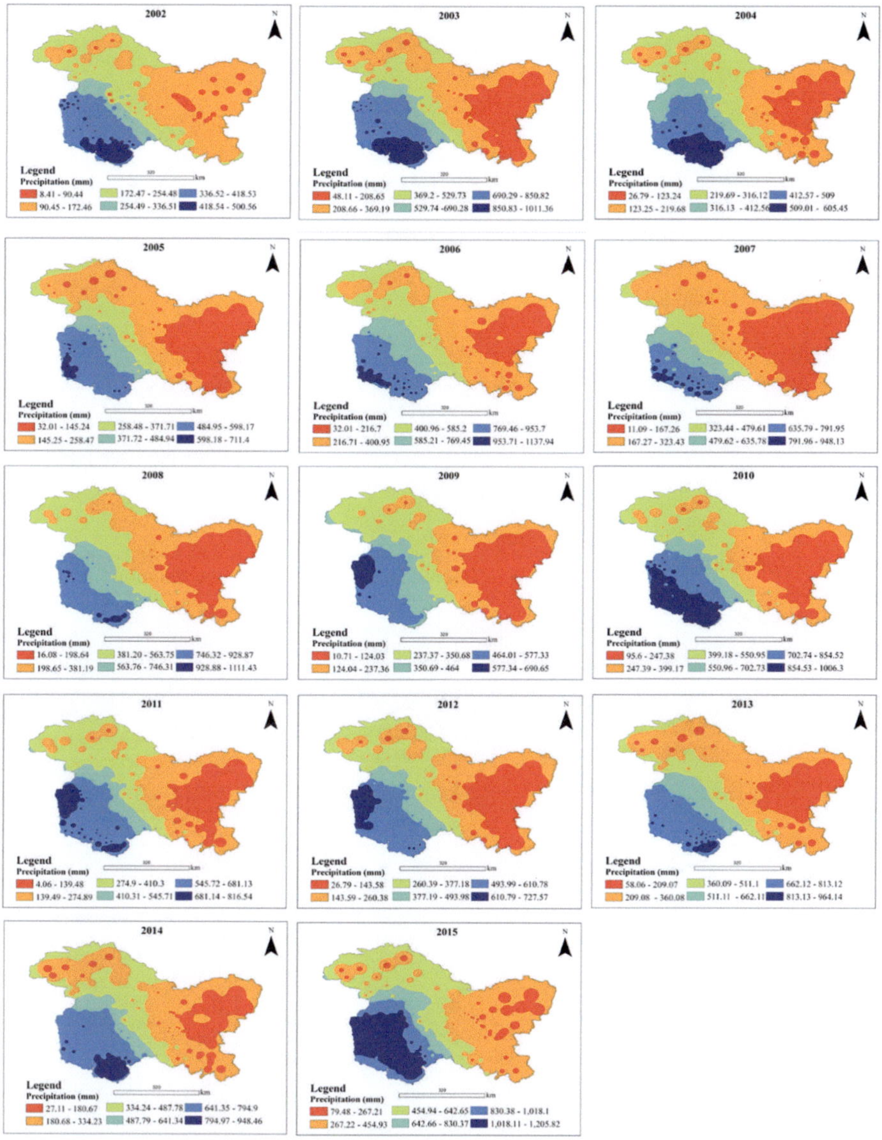

Fig. 7 Interpolated precipitation maps from 2002 to 2015

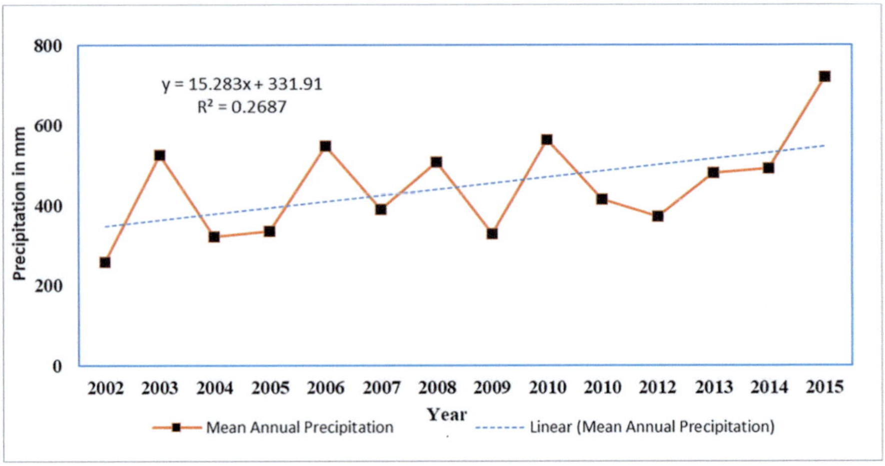

Fig. 8 Mean annual precipitation from 2002 to 2015

Table 4 Representing annual mean temperature and precipitation data from interpolated 200 points

Year	Temperature (°C)	Precipitation (mm)
2002	4.57	258.95
2003	3.57	525.29
2004	4.38	321.89
2005	3.42	335.08
2006	4.06	547.25
2007	4.45	389.05
2008	3.54	506.70
2009	3.55	328.56
2010	4.32	563.41
2011	3.70	414.87
2012	3.30	370.62
2013	4.21	480.33
2014	3.72	491.06
2015	3.78	718.37
AVG	3.90	446.53

5 Discussion

The observed trends in glacier mass balance, temperature, and precipitation across the Western Himalayas reveal significant regional variations that reflect the complex interplay of climate change and topography. The dominant pattern of negative mass balance in the Ladakh, Zanskar, Pir-Panjal, Shamsbari, and Greater Himalayan ranges underscores the substantial ice loss occurring in these regions, which is closely tied to rising temperatures. The negative mass balance in these areas, with the most pronounced mass loss in the Ladakh and Zanskar ranges, suggests heightened glacier melting as reported by earlier studies (Mir and Majeed 2018; Shukla and Qadir 2016; Rashid and Majeed 2018) likely driven by their lower elevations and less steep slopes compared to the Karakoram range. The Karakoram anomaly, where glaciers exhibit a positive mass balance, can be attributed to its high mean elevation and steeper slopes, which likely provide insulation from the warming temperatures observed across the region. The analysis of temperature trends between 2002 and 2015 indicates that while the overall temperature trend has remained stable, there are considerable fluctuations, ranging between 3.30 and 4.57 °C. This stability in temperature, despite the ongoing glacier mass loss, suggests that even small variations in temperature, combined with other factors such as topography and localized climate patterns, can have substantial effects on glacial dynamics. Furthermore, the variation in precipitation patterns, reveal an average annual precipitation of 446.53 mm ranging between 258.95 and 718 mm from 2002 to 2015 which shows that there is lot of uncertainty in the precipitation pattern, highlights the complexity of climate change impacts in this region. The highest precipitation observed in the Pir-Panjal range, contrasted with the drier Karakoram, Ladakh, and Zanskar ranges, demonstrates that precipitation is not uniformly distributed as showed in (Fig. 7), contributing further to the spatial variations in glacier behaviour. Overall, the results emphasize the vulnerability of glaciers in the Western Himalayas to climate change, particularly in the lower and less steep regions where glacier retreat is most pronounced. The Karakoram range's resilience to warming underscores the importance of localized geographic factors in moderating climate change impacts. This time period has shown that the precipitation has increased over the Northwest Himalayan region, but the glaciers are still having the negative mass balance it means there are multiple factors involved in the glacier melting of the glacier and one of the causes is clearly known that is increasing in the aerosol content in atmosphere and at glacier surfaces and second factor is change in the form of precipitation and seasonality. To understand the glacier dynamics, we need more ground based meteorological data which is scare in this region. This study highlights the need for continued monitoring of both temperature and precipitation trends in the region, as future changes in these parameters could have profound impacts on water resources, ecosystems, and communities dependent on glacial meltwater.

5.1 *Implications of Glacier Mass Loss on Downstream Communities*

The mass loss in glaciers has significant impact on downstream communities, affecting the agriculture, water security, and livelihoods, initially, there will be increased meltwater in the river flow but as glaciers continue to shrink and loss mass, the flow will eventually drop during the seasons where water is most needed leading to water scarcity. This threatens the irrigation, hydropower, drinking water supplies. This retreat will lead to expansion in glacier lakes. Moreover, the formation of new unstable glacier lakes raises the risk of Glacier Lake Outburst Floods (GLOFs), which can cause the catastrophic damage to infrastructure and settlements. The changes in river regimes disrupt the aquatic ecosystems threatening the biodiversity, the water sharing agreement by neighbouring countries will be affected by creating dispute when water levels will change. Glacier retreat can lead to social challenges such as the displacement of communities, erosion of cultural heritage, and the disruption of long-standing water management practices, especially among Indigenous and highland populations. These impacts collectively increase community vulnerability, highlighting the urgent need for adaptation measures and strategies to reduce disaster risks.

5.2 *Possible Drivers in Glacier Retreat*

The mass loss of glaciers in the north-western Himalayas is interlinked linked to large scale climatic systems such as the mid-latitude Westerlies, the El Niño–Southern Oscillation (ENSO), and Western Disturbances (WDs), which are embedded in the Westerlies, are the primary source of winter precipitation mainly snowfall for this region (Shafiq et al. 2024). These disturbances originate from the Mediterranean and travel across Central Asia, delivering crucial snow to glacier accumulation zones. However, climate change appears to be modifying the intensity, frequency, and trajectory of WDs, potentially leading to reduced or erratic snowfall (Hunt et al. 2024). The frequency of Western Disturbances has remained relatively steady over recent decades, with no significant long-term increase or decrease observed in most studies. For example, a study in Jammu & Kashmir from 1980 to 2019 found a total of 600 WD cases, The analysis showed significant decreasing trends of WD frequencies ($p < 0.01$) during November and March with slope values − 0.61and − 2.75 respectively and increasing trends were observed during February with slope value 2.79 ($p < 0.01$) and January with slope value 0.41 but were non-significant (Ahmed et al. 2022). ENSO plays a significant role in modulating both the Indian Monsoon and Western disturbance. During El Niño events, the Indian Summer Monsoon tends to weaken, leading to reduced summer precipitation. This can lower accumulation on glaciers, especially at higher elevations where snowfall may still occur during summer months. El Niño exerts a significant influence on Western Disturbances primarily by altering atmospheric circulation, particularly the subtropical westerly jet. This modulates the intensity, positioning, and moisture transport of Western Disturbances, which in turn impacts winter precipitation over the Himalayas and surrounding regions (Dad et al. 2021). Understanding this interaction is key for improving weather forecasts and managing water resources in affected area (Qiu et al. 2022). The Karakoram region has experienced an approximately 10% increase in precipitation intensity linked to Western Disturbances (WDs) in recent decades. Notably, WDs contribute around 65% of the total seasonal snowfall in the area, highlighting their significant influence on the region's unique mass-balance anomaly (Javed et al. 2022). There is growing scientific evidence that aerosol deposition contributes to glacier melting; however, the spatiotemporal distribution of aerosols across the Indian Himalayas remains poorly understood a study conducted by Rashid et al. (2022a) Indicated that The annual variation in aerosol levels for each parameter indicates a rising trend from 1980 to 2020 across all mountain ranges within the study area, with the highest concentrations observed in the Pir-Panjal range and the lowest in the Zanskar range. This upward trend in aerosol presence may be attributed to the global rise in anthropogenic activities, particularly the combustion of fossil fuels and biomass (Kang et al. 2019). This ultimately leads to decrease in albedo of the glacier surface, which ultimately leads to the rapid melting of the glaciers (Kang et al. 2020).

6 Conclusion

This study provides critical insights into the changing climate dynamic of the North Western Himalayas by examining mass balance, temperature, and precipitation trends across various regions. The results indicate that Pir-Panjal, Ladakh, Shamswari, Greater Himalayas and, Zanskar, exhibit negative mass balances, with the most substantial glacier annual mass losses recorded in the Ladakh (-1.09 ± 0.08 GT) and Zanskar (-0.75 ± 0.06 GT) ranges due to presence of vast glacial area. while as Karakoram range shows a positive mass balance of $+0.14 \pm 0.01$ mwea^{-1}, with an annual mass gain of $+2.19 \pm 0.09$ Gt, marking a clear distinction from the other regions, often referred to as the "Karakoram anomaly. This indicated the different regions have different behaviour towards climate change. The other aspects are also playing in the role in glacier mass los which can be the non-climatic parameters such as physiography, slope, elevation, aspect. The annual average temperature for 14 years from 2002 to 2015 recorded to be 3.90 °C, which sharp deviations from the mean which means there are uncertainty in the temperature, which can lead to the melting of the glaciers. Precipitation patterns over the same period reveal an average annual precipitation of 446.53 mm ranging between 258.95 and 718 mm from 2002 to 2015 which shows that there is lot of uncertainty in the precipitation pattern. The region experienced seven years of below-average precipitation and seven years of above-average precipitation, indicating a high degree of variability. The interpolated maps showed with the maximum precipitation occurring in the Pir-Panjal range, followed by the Kashmir Valley and Greater Himalayas followed by Zanskar,

Ladakh and Karakorum Range. These findings emphasize the complex and diverse climate of the Western Himalayas, highlighting both rapid glacier retreat in certain regions and increasing precipitation trends. The contrasting trends, particularly the mass gain in Karakoram and mass loss in other ranges, underscore the region's sensitivity to global climate change. The ongoing monitoring is crucial for understanding these changes and their potential impacts on local water resources and ecosystems. The implementation of sustainable development goal 13 is very much needed in this region because being the eco-sensitive zone and policy making and implementation at community level is needed to combat the climate change and its adverse effects.

References

Abdullah T, Romshoo SA, Rashid I (2020) The satellite observed glacier mass changes over the Upper Indus Basin during 2000–2012. Sci Rep 10:1–9

Ahmed M, Das B, Lotus S, Ali M (2022) A study on frequency of western disturbances and precipitation trends over Jammu & Kashmir, India: 1980–2019. Mausam 73:283–294

Ahmed R et al (2022) Glacial Lake outburst flood hazard and risk assessment of Gangabal Lake in the Upper Jhelum Basin of Kashmir Himalaya using geospatial technology and hydrodynamic modeling. Remote Sens 14

Ahmed R et al (2024) Assessing climate trends in the Northwestern Himalayas: a comprehensive analysis of high-resolution gridded and observed datasets. Geomatics Nat Hazards Risk 15

Azam MF et al (2016) Meteorological conditions, seasonal and annual mass balances of Chhota Shigri Glacier, western Himalaya, India. Ann Glaciol 57:328–338

Azam MF et al (2018) Review of the status and mass changes of Himalayan-Karakoram glaciers. J Glaciol 64:61–74

Berthier E et al (2007) Remote sensing estimates of glacier mass balances in the Himachal Pradesh (Western Himalaya, India). Remote Sens Environ 108:327–338

Berthier E, Cabot V, Vincent C, Six D (2016) Decadal region-wide and glacier-wide mass balances derived from multi-temporal aster satellite digital elevation models. Validation over the mont-blanc area. Front Earth Sci 4

Bisht SM (2015) Study of Gangotri glacier dynamics using remote sensing and geospatial tools, vol 88. M.Tech. Thesis

Brun F, Berthier E, Wagnon P, Kääb A, Treichler D (2017) A spatially resolved estimate of High Mountain Asia glacier mass balances from 2000 to 2016. Nat Geosci 10:668–673

Cogley JG (2009) Geodetic and direct mass-balance measurements: comparison and joint analysis. Ann Glaciol 50:96–100

Dad JM, Muslim M, Rashid I, Reshi ZA (2021) Time series analysis of climate variability and trends in Kashmir Himalaya. Ecol Indic 126:107690

Dar SA, Sarif MO (2024) Status on snow cover area, mass balance, glacial area loss, surface velocities and applications of snowmelt runoff model over Kashmir Himalayas and Upper Indus Basin: a review. Polar Sci 101075. https://doi.org/10.1016/j.polar.2024.101075

Dar SA, Sarif MO (2025) Modelling of snow cover area in relation with climatic variability over the Sind basin of Kashmir Himalayas (2002–2022). Phys Chem Earth 138:103843

Dong P, Wang C, Ding J (2013) Estimating glacier volume loss using remotely sensed images, digital elevation data, and GIS modelling. Int J Remote Sens 34:8881–8892

Dussaillant I, Berthier E, Brun F (2018) Geodetic mass balance of the northern patagonian icefield from 2000 to 2012 using two independent methods. Front Earth Sci 6:1–13

Fischer A (2011) Comparison of direct and geodetic mass balances on a multi-annual time scale. Cryosphere 5:107–124

Florentine C, Sass L, McNeil C, Baker E, O'Neel S (2023) How to handle glacier area change in geodetic mass balance. J Glaciol 1–7. https://doi.org/10.1017/jog.2023.86

Gaddam VK et al (2024) Mass Balance Assessment using the "AARTI" approach in the Baspa Basin, Indian Himalayas. Remote Sens Earth Syst Sci. https://doi.org/10.1007/s41976-024-00133-z

Gardelle J, Berthier E, Arnaud Y, Kääb A (2013) Region-wide glacier mass balances over the Pamir-Karakoram-Himalaya during 1999– 2011. Cryosphere 7:1263–1286

Garg S et al (2021) Revisiting the 24 year (1994–2018) record of glacier mass budget in the Suru sub-basin, western Himalaya: overall response and controlling factors. Sci Total Environ 800

Haq MA, Ghosh A, Rahaman G, Baral P (2019) Artificial neural network-based modeling of snow properties using field data and hyperspectral imagery. Nat Resour Model 32:1–36

Hunt KMR et al (2024) Western disturbances and climate variability: a review of recent developments. Egusph [preprint] 43–112

Huss M (2013) Density assumptions for converting geodetic glacier volume change to mass change. Cryosphere 7:877–887

Javed A et al (2022) Does the recent revival of western disturbances govern the Karakoram Anomaly? J Clim 35:4383–4402

Jin D et al (2022) An improvement of snow/cloud discrimination from machine learning using geostationary satellite data. Int J Digit Earth 15:2355–2375

Joya E, Bromand MT, Murtaza KO, Dar RA (2021) Current glacier status and ELA changes since the Late Pleistocene in the Hindu Kush Mountains of Afghanistan. J Asian Earth Sci 219

Kang S et al (2019) Linking atmospheric pollution to cryospheric change in the Third Pole region: current progress and future prospects. Natl Sci Rev 6:796–809

Kang S, Zhang Y, Qian Y, Wang H (2020) A review of black carbon in snow and ice and its impact on the cryosphere. Earth-Science Rev 210:103346

Klug C et al (2018) Geodetic reanalysis of annual glaciological mass balances (2001–2011) of Hintereisferner, Austria. Cryosphere 12:833–849

Kumar GV, Kulkarni AV, Gupta AK, Sharma P (2017) Mass balance estimation using geodetic method for glaciers in Baspa basin, Western Himalaya. Curr Sci 113:486–492

Kumar M, Gupta T, Baghel V (2024) A comprehensive analysis of the 2023 Himalayan disasters in Himachal Pradesh: multi-sectoral impacts and strategic mitigation approaches. E3S Web Conf 596

Li W, Hsu C, Tedesco M (2024) Advancing Arctic sea ice remote sensing with AI and deep learning : now and future. EGUsphere 1–36

Mandal A et al (2020) Understanding the interrelationships among mass balance, meteorology, discharge and surface velocity on Chhota Shigri Glacier over 2002–2019 using in situ measurements. J Glaciol 66:727–741

Mandal A et al (2024b) Glacier mass balance and its climatic and nonclimatic drivers in the Ladakh region during 2000–2021 from remote sensing data. J Glaciol. https://doi.org/10.1017/jog.2024.19

Mandal A et al (2024) Glacier mass balance and its climatic and nonclimatic drivers in the Ladakh region during 2000–2021 from remote sensing data. J Glaciol 70:e19

Mandal A et al (2002) Understanding the interrelationships among mass balance , meteorology , discharge and surface velocity on Chhota Shigri Glacier over 2002–2019 using in situ measurements 259:727

Marazi A, Romshoo SA (2018) Streamflow response to shrinking glaciers under changing climate in the Lidder Valley, Kashmir Himalayas. J Mt Sci 15:1241–1253

Mir RA, Majeed Z (2018) Frontal recession of Parkachik Glacier between 1971–2015, Zanskar Himalaya using remote sensing and field data. Geocarto Int 33:163–177

Mukherjee K, Bhattacharya A, Pieczonka T, Ghosh S, Bolch T (2018) Glacier mass budget and climate reanalysis data indicate a climatic shift around 2000 in Lahaul-Spiti, western Himalaya. Clim Change 148:219–233

Mukul M, Srivastava V, Mukul M (2015) Analysis of the accuracy of Shuttle Radar Topography Mission (SRTM) height models using International Global Navigation Satellite System Service (IGS) Network. J Earth Syst Sci 124:1343–1357

Murtaza KO, Romshoo SA (2017) Recent glacier changes in the Kashmir Alpine Himalayas, India. Geocarto Int 32:188–205

Nurakynov S, Merekeyev A, Baygurin Z, Sydyk N, Akhmetov B (2024) Application of Artificial Intelligence in Glacier Studies: a state-of-the-art review. Water 16

Nuth C, Kääb (2011) Co-registration and bias corrections of satellite elevation data sets for quantifying glacier thickness change. Cryosphere 5:271–290

Orlandi AGGG et al (2019) Vertical accuracy assessment of the processed SRTM data for the Brazilian territory. Bol Ciencias Geod 25:1–14

Piermattei L et al (2024) Observing glacier elevation changes from spaceborne optical and radar sensors—an inter-comparison experiment using ASTER and TanDEM-X data. Cryosphere 18:3195–3230

Qiu T, Huang W, Wright JS, Yang Z, Wang B (2022) Impacts of western disturbances on wintertime precipitation over the Southeastern Tibetan Plateau. J Geophys Res Atmos 127:1–17

Rashid I, Majeed U (2018) Recent recession and potential future lake formation on Drang Drung glacier, Zanskar Himalaya, as assessed with earth observation data and glacier modelling. Environ Earth Sci 77:1–13

Rashid I, Romshoo SA, Abdullah T (2017) The recent deglaciation of Kolahoi valley in Kashmir Himalaya, India in response to the changing climate. J Asian Earth Sci 138:38–50

Rashid I et al (2022a) Aerosol variability and glacial chemistry over the western Himalayas. Environ Chem 19:312–327

Rashid I, Najar NA, Majeed U, Rasool W (2022b) Ascertaining glacier dynamics and geodetic mass changes in the Pangong Region of Trans-Himalayan Ladakh using remote sensing data. Data Br 42:108176

Rasul G, Molden D (2019) The global social and economic consequences of mountain cryospheric change. Front Environ Sci 7

Romshoo SA, Murtaza KO, Abdullah T (2022b) Towards understanding various influences on mass balance of the Hoksar Glacier in the Upper Indus Basin using observations. Sci Rep 12:1–14

Romshoo SA, Abdullah T, Murtaza KO, Bhat MH (2023) Direct, geodetic and simulated mass balance studies of the Kolahoi Glacier in the Kashmir Himalaya, India. J Hydrol 617:129019

Romshoo SA, Abdullah T, Ameen U, Bhat MH (2024) Glacier thickness and volume estimation in the upper Indus basin using modelling and ground penetrating radar measurements. Ann Glaciol. https://doi.org/10.1017/aog.2024.2

Romshoo SA, Abdullah T, Rashid I, Bahuguna IM (2022) Explaining the differential response of glaciers across different mountain ranges in the north-western Himalaya, India. Cold Reg Sci Technol 196

Romshoo SA et al (2015) Implications of shrinking cryosphere under changing climate on the streamflows in the lidder catchment in the Upper Indus Basin, India. Arctic Antarct Alp Res 47:627–644

Santillan JRRR (2023) Vertical accuracy evaluation of the forest and buildings removed Copernicus Dem (Fabdem) over the Philippines. Int Arch Photogramm Remote Sens Spat Inf Sci ISPRS Arch 48:311–318

Shafiq M et al (2024) Assessment of climate variability and trends in different physiographic zones of North Western Himalayas. Theor Appl Climatol 155:5557–5576

Shukla A, Qadir J (2016) Differential response of glaciers with varying debris cover extent: evidence from changing glacier parameters. Int J Remote Sens 37:2453–2479

Singh S, Kumar R, Dimri AP (2018) Mass balance status of Indian Himalayan glaciers: a brief review. Front Environ Sci 6

Srinivasalu P, Kulkarni AV, Srinivas VV, Satheesh SK (2024) Catchment scale volume and mass changes for glaciers in the Parvati basin of Western Himalaya during 2000–2015. Environ Earth Sci 83:1–19

Tanniru S, Singh DK, Singh KK, Ramsankaran R (2025) Exploring machine learning ' s potential for estimating high resolution daily snow depth in Western Himalaya Using. https://doi.org/10.1029/2024EA003849

Van Tricht L et al (2023) Monitoring the annual geodetic mass balance of Bordu and Sary-Tor glaciers using UAV data. Ann Glaciol 1–10. https://doi.org/10.1017/aog.2023.71

Vincent C et al (2021) Geodetic point surface mass balances: a new approach to determine point surface mass balances on glaciers from remote sensing measurements. Cryosphere 15:1259–1276

Yu F et al (2024) Historical reconstruction of glacier mass balance and its contribution to water resources in the Sawir Mountains from 2000 to 2020. Sci Total Environ 944:173703

Zhang F, Thapa S, Immerzeel W, Zhang H, Lutz A (2019) Water availability on the Third Pole: a review. Water Secur 7

Zheng W et al (2025) Community estimate of global glacier mass changes from 2000 to 2023. Nature

Climate Change Impacts on Various Ecosystems: A Geospatial and AI Perspective

Mallika Bhuyan, Preeti Sardar, C. Jeganathan, and Girish S. Pujar

1 Introduction

Natural climatic variations occur across various time scales, but the rapid warming observed in recent decades poses a significant threat to ecosystem stability. This accelerated climate change impacts species individually and disrupts their interactions with other organisms and habitats, leading to alterations in ecosystem structures, functions, and the essential services they provide to humanity. As a pervasive global challenge, climate change jeopardizes biodiversity and ecosystem health, underscoring the importance of understanding ecological responses to anticipate and adapt to these changes effectively (Diaz et al. 2019).

Climate change encompasses a complex interplay of stressors, including rising temperatures, elevated atmospheric CO^2 levels, and an increasing frequency of extreme events such as severe droughts, all of which are impacting ecosystems at various scales (Weiskopf et al. 2020). A study by Leemans and Eickhout (2004)

M. Bhuyan · P. Sardar · C. Jeganathan (✉)
Department of Remote Sensing and Geoinformatics, Birla Institute of Technology, Mesra, Ranchi, Jharkhand, India
e-mail: jeganathanc@bitmesra.ac.in

M. Bhuyan
e-mail: mallikabhuyan21@gmail.com

P. Sardar
e-mail: psardar1994@gmail.com

C. Jeganathan
Sarala Birla University, Ranchi, Jharkhand, India

G. S. Pujar
Earth Observation and Disaster Management Program Office, ISRO Headquarters, Department of Space, Bengaluru, Karnataka, India
e-mail: pujar@isro.gov.in

M. O. Sarif and A. Sharifi (eds.), *Advanced Geospatial Intelligence and AI for Environmental Resilience and Sustainable Development*, Advances in Geographic Information Science, https://doi.org/10.1007/978-3-032-03714-5_8

highlights that even modest global temperature increases, ranging from 1 to 2 °C, can significantly disrupt species, ecosystems, and landscapes, pushing their adaptive capacity to its limits. This rapid pace of change is expected to exacerbate biodiversity loss and reduce the availability of essential ecosystem services, amplifying the vulnerability of natural systems. Consequently, combating climate change has been prioritized as one of the 17 Sustainable Development Goals (SDG 13) outlined by the United Nations (UN) Agenda 2030 (UN 2019). Therefore, effective monitoring and assessment of these changes are crucial to developing adaptive strategies and sustainable policies.

Geospatial techniques, including Geographic Information Systems (GIS), remote sensing, and spatial modeling, have emerged as indispensable tools in climate change research. Remote sensing provides critical data on variables such as vegetation, temperature, and land-use changes, while GIS integrates this information to assess vulnerabilities, predict risks, and model future scenarios such as sea-level rise and extreme weather events (Myneni et al. 1995; Janke 2010; IPCC 2021). These techniques are also vital for tracking greenhouse gas emissions, evaluating carbon stocks, and designing climate-resilient strategies, including renewable energy planning and disaster risk reduction (Janke 2010; Goetz et al. 2009). By delivering high-resolution, actionable insights, geospatial tools enhance global efforts in climate adaptation and mitigation.

Similarly, artificial intelligence (AI) offers transformative potential in tackling the complex challenges posed by climate change. AI supports efforts to identify areas vulnerable to climate-related hazards, predict disasters such as floods, wildfires, and landslides, and develop targeted strategies to protect ecosystems and communities (Feng et al. 2015). Advanced AI-powered models also aid policymakers in anticipating climate impacts and designing effective adaptation measures, including early warning systems to enhance disaster preparedness. However, ethical considerations, such as biases in training data, must be addressed to prevent unintended consequences and ensure equitable solutions. By integrating ethical frameworks, AI can play a transformative role in developing inclusive and effective climate adaptation strategies (Jain et al. 2023).

Geospatial Artificial Intelligence (GeoAI) is another emerging technique that combines advancements in spatial science with cutting-edge artificial intelligence (AI) techniques, including machine learning, deep learning, and high-performance computing, to extract valuable insights from spatial big data (Janowicz et al. 2020; Mai et al. 2025). Geospatial AI offers invaluable tools in combating climate change through diverse applications. It leverages AI models to analyze satellite imagery and weather data, enabling the monitoring of deforestation, forecasting of floods and wildfires, and assessment of ecosystem health (Atmaja et al. 2024). These capabilities are essential for developing effective mitigation strategies and establishing early warning systems to address climate-related challenges. Further, GeoAI aids in climate adaptation by projecting the effects of rising sea levels, extreme weather, and other climate-related impacts on vulnerable regions, empowering communities and policymakers to devise robust, region-specific adaptation plans (Feng et al. 2015).

Its ability to process and analyze vast datasets in real-time makes this technique an indispensable tool in addressing the challenges posed by a rapidly changing climate.

Recognizing the urgency of addressing the impacts of climate change, this chapter focuses on two critical ecosystems that play indispensable roles in maintaining ecological balance and supporting biodiversity: forest ecosystems and coastal ecosystems. Additionally, the role of geospatial science and AI in addressing such impacts on these ecosystems is explored, highlighting their potential to enhance monitoring, prediction, and the development of effective mitigation and adaptation strategies.

2 Forest Ecosystems

2.1 Use of Geospatial Techniques and AI in Monitoring Forest Changes Due to Climate Change

Forests are vital for maintaining the Earth's water, energy, and carbon balance while providing essential ecosystem services. They play a key role in mitigating climate change by regulating atmospheric gases, sequestering carbon, preserving water cycles, stabilizing soils, and supporting biodiversity. Forests also help mitigate flood risks by absorbing excess water, enhancing soil stability on steep slopes, and moderating local climates.

However, climate change, primarily driven by human activities, is profoundly altering forest ecosystems and their functions on a global scale. Rising temperatures, shifting precipitation patterns, elevated atmospheric CO^2 and the increasing frequency of extreme weather events are posing significant threats to biodiversity and forest stability. These changes manifest in various ways, including shifts in phenological events, more frequent and severe forest fires, changes in species composition, and fluctuations in carbon storage. Recognizing and understanding these vulnerabilities is crucial for developing adaptive strategies that can mitigate adverse impacts and bolster the resilience of forest ecosystems. The integration of geospatial techniques and AI has emerged as a robust approach for assessing and predicting forest dynamics under changing climatic conditions. Some of the areas where geospatial techniques and AI (including Machine learning) have come into use to evaluate changes in forests due to climate change have been discussed below.

Vegetation Dynamics and Phenological Shifts. Phenological shifts, including altered timing of leaf-out, flowering, and fruiting, are among the most prominent effects of climate change on ecosystems. These changes disrupt the synchrony between species interactions, such as pollination and seed dispersal, leading to cascading ecological effects (Pan et al. 1979; Zhang and Steiner 2022). Recent studies highlight that shifts in phenological events are advancing at unprecedented rates, particularly in temperate regions, and are strongly correlated with warming temperatures (Richardson et al. 2013; Cleland et al. 2007). Such disruptions can jeopardize ecosystem services, including food production and biodiversity conservation (Aravind et al. 2013; Ramaswami et al. 2019).

The combined use of geospatial technologies and AI facilitates the detection of phenological shifts, such as changes in flowering and leafing cycles, induced by climate variability. AI-powered models analyze long-term satellite data to identify patterns and predict future phenological trends. A study conducted by Xin et al. (2020) revealed that machine learning methods, including neural networks and random forests, outperform rule-based methods in retrieving vegetation phenology from satellite data. These methods provide more accurate start-of-season (SOS) values when compared to field observations, highlighting their utility in understanding satellite-based land surface phenology. Another study conducted by Lamjiak et al. (2021) utilizing Self-Organizing Maps (SOMs) and Long Short-Term Memory (LSTM) models, highlights that leaf phenology in Thailand's dry dipterocarp forests exhibits species-specific responses to climate variability and El Nino, with rainfall timing and amount significantly influencing lag times in phenological patterns. Many other researchers (Aravind et al. 2013; Ramaswami et al. 2019; Xin et al. 2020) have applied machine learning and artificial intelligence in studying forest dynamics and phenological changes due to climate change.

Forest Fires. Climate change has significantly increased the frequency, intensity, and duration of forest fires, which are exacerbated by prolonged droughts, higher temperatures, and changing precipitation patterns (Flannigan et al. 2009; Abatzoglou and Williams 2016). Recent assessments reveal that wildfire seasons are lengthening globally, with areas like the Amazon and boreal forests experiencing unprecedented fire activity (Richardson et al. 2022; Jones et al. 1979). Forest fires not only result in immediate biodiversity loss but also release substantial amounts of carbon dioxide, contributing to a positive feedback loop that accelerates climate change. With the advent of advanced machine learning and deep learning models, it has become effectively possible to predict the forest fire likelihood and risk levels (Achu et al. 2021) as well as vulnerable zones (Tien Bui et al. 2019), which can help establish cost-effective forest fire prediction and early warning systems (Akila et al. 2023; Bera et al. 2022).

Carbon Stock Variability. Forests act as vital carbon sinks, but their capacity to sequester carbon is increasingly threatened by climate-induced stressors. Variations in carbon stocks, particularly aboveground biomass density (AGBD), are strongly influenced by changes in temperature, precipitation, and disturbances such as logging and fires (Pan et al. 1979; Baccini et al. 1979). Recent satellite-based studies show a significant decline in AGBD in tropical forests, attributed to deforestation and climate anomalies (Harris et al. 2021). Such changes underscore the urgent need to protect and restore forest ecosystems to mitigate climate change impacts. Many researchers worldwide have used machine learning (Cheng et al. 2024; Vasudeva et al. 2021; Liu et al. 2024) and deep learning (Zhang et al. 2023) approaches for carbon stock estimation, which can, in turn, help in mitigating and addressing climate change impacts on the forests. In India, (Ahirwal et al. 2021) used various machine learning models like AdaBoost, Bagging, Random Forest and XGBoost to estimate the carbon stock of the Indian Himalayan region.

Changes in Forest Structure. Climate change-induced alterations in forest structure, such as reductions in canopy height, tree density, and AGBD, have profound implications for ecosystem health and resilience. Advanced remote sensing techniques, including LiDAR (Kacic et al. 2023) and high-resolution satellite imagery (Zhang et al. 2020; Wang et al. 2021), reveal that forest structural changes are accelerating, especially in vulnerable regions like the Amazon and Southeast Asia (Lang et al. 2023; Tanase et al. 2024). These structural shifts impact habitat quality, water cycling, and carbon storage, emphasizing the need for adaptive forest management strategies. The growing availability of Unmanned Aerial Vehicles (UAVs), along with advanced AI techniques of data analysis, has revolutionized the monitoring of forest structural changes (Neuville et al. 2021; Zhang et al. 2021). In this context, (Yadav et al. 2021) successfully mapped the vertical structure and biomass of Indian tropical forest plantations using TanDEM-X PolInSAR data, integrating physical scattering models with advanced machine learning to achieve precise measurements.

2.2 *Case Study of Wall-to-Wall AGBD Prediction Using GEDI Data and Machine Learning*

Introduction. Accurately estimating AGBD is critical for understanding the global carbon cycle, managing forest resources, and mitigating climate change impacts. Forests, as major carbon sinks, play a pivotal role in reducing atmospheric CO_2 levels and regulating climate dynamics (Pan et al. 1979). However, robust and spatially comprehensive AGBD estimations remain a challenge due to the heterogeneity of forest structures and limited ground-based observations, particularly in tropical and subtropical regions. In recent years, advances in remote sensing technologies and machine learning algorithms have provided innovative pathways for addressing these challenges.

Global Ecosystem Dynamics Investigation (GEDI) data, which includes precise lidar-derived measurements of forest structure, has revolutionized the ability to estimate AGBD at high spatial resolution (Dubayah 2020; Duncanson 2020). GEDI's waveform lidar technology captures three-dimensional forest structural information, enabling the direct estimation of biophysical attributes such as canopy height, foliage height profiles, and canopy cover, which are strongly correlated with biomass. However, GEDI's sampling is footprint-based, providing sparse data points across landscapes, necessitating the integration of machine learning models to upscale these measurements to wall-to-wall coverage.

Machine learning techniques have emerged as powerful tools for leveraging remote sensing data for AGBD prediction. These algorithms, including Random Forests, Gradient Boosting Machines, and Deep Learning models, can effectively handle the non-linear relationships between input features and target variables, even in the presence of complex interactions (Lang et al. 2022; Wang et al. 2021). By combining GEDI lidar metrics with other spatially contiguous datasets such as optical imagery (e.g., Sentinel-2) and synthetic aperture radar (SAR) data (e.g., Sentinel-1), these models can predict AGBD across continuous landscapes, addressing gaps in spatial coverage and improving the resolution of biomass maps (Potapov et al. 2021; Tamiminia et al. 2024).

This study focuses on developing a wall-to-wall AGBD prediction model by integrating GEDI lidar data with auxiliary remote sensing datasets such as high-resolution LISS IV (5 m), Sentinel-1 and PALSAR data using advanced machine learning algorithms. The research aims to (i) evaluate the potential of GEDI lidar metrics in predicting AGBD at regional scales, (ii) explore the contribution of multi-spectral and SAR data in improving model performance, and (iii) generate high-resolution AGBD maps for a diverse forested landscape. This approach not only improves the spatial resolution of biomass estimates but also provides valuable insights into forest dynamics and carbon storage potential.

Study Area. The Kaziranga National Park, situated in the Nagaon and Golaghat districts of Assam, and located between latitudes 26° 30′ N to 26° 45′ N and longitudes 93° 08′ E to 93° 36′ E, encompasses an area of about 429.93 km^2 (Fig. 1). Spanning the southern floodplains of the Brahmaputra River, the park features an east-to-west slope and is bordered by the foothills and snow-clad peaks of the eastern Himalayas (Goswami et al. 2019). Its terrain consists of grasslands, wetlands, riverine lakes, and elevated woodlands, which provide critical habitats for wildlife, especially during the annual Brahmaputra floods.

As a key region within the Eastern Himalayan biodiversity hotspot, Kaziranga hosts a wide range of rare and endangered species, including the one-horned rhinoceros, tigers, elephants, swamp deer, wild buffalo, and wild boars. The park's vegetation, comprising semi-evergreen forests, mixed deciduous woodlands, and grasslands, is sustained by periodic flooding, which enriches soil nutrients and supports biodiversity. Recognized as a UNESCO World Heritage Site in 1985, the park exemplifies ecological richness and conservation success (Borah et al. 2018).

Despite its ecological significance, Kaziranga faces mounting challenges from intensifying floods and human activities such as poaching, encroachment, and habitat loss. Monitoring AGBD is critical for understanding forest productivity, carbon storage, and ecosystem resilience, offering vital information to inform conservation and management strategies in this globally significant landscape.

Methodology. Wall-to-wall mapping of Forest Aboveground Biomass Density in Kaziranga National Park was conducted using variables from multisource remote sensing datasets for the year 2022.

Datasets Used. The datasets used for predicting AGBD in Kaziranga National Park include freely available high-resolution LISS IV optical imagery (5.8 m), GEDI lidar data (25 m), Sentinel-1 C-band SAR imagery (10 m), SRTM DEM (30 m), and ALOS

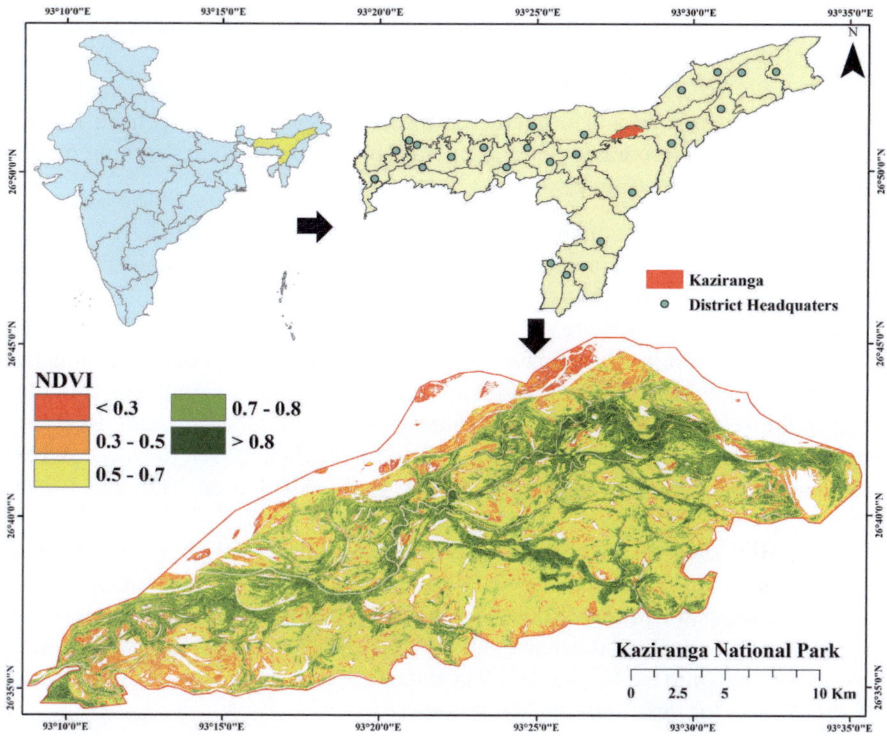

Fig. 1 Location of Kaziranga National Park showing the variation of NDVI in the region

PALSAR (25 m) data. Each dataset is subjected to pre-processing prior to inclusion in the models.

Linear Imaging Self Scanning Sensor (LISS) IV data downloaded from the Bhoonidhi portal of NRSC are originally available in DN values, which are then converted to spectral radiance and surface reflectance using the MODTRAN-based QUAC module within the ENVI software package (Saini et al. 2016). The Surface reflectance (SR) bands are then used to derive the vegetation indices to be used as input parameters in the models.

This study utilized the GEDI Level 4A (L4A) Version 2 dataset to estimate AGBD in (Mg/ha). Only high-quality data were retained in the analysis by implementing quality filtering to remove degraded and low-quality shots. The initial data processing and preparation were conducted using the Google Earth Engine platform.

Sentinel-1 data was downloaded from the Copernicus Data Hub and pre-processed in the SNAP platform using the Sentinel 1 Toolbox. Backscatter metrics (VV and VH) are calibrated, Speckle Filtered, terrain corrected and resampled to obtain corrected backscatter metrics along with other radar indices.

PALSAR yearly global mosaic data in 25 m resolution has been used in this study. The raw HH/HV polarization data was converted into gamma nought values in decibels (dB), which were then used to calculate various radar matrices (Bhuyan et al. 2024).

Further, SRTM DEM data has been used to obtain elevation, slope, and aspect metrics for the study area. The predicted forest canopy height map from Bhuyan et al. (2024) was incorporated as an input variable in the machine-learning models employed in this study (Fig. 2).

Methods. A total of 38 predictor variables were considered for this study (Table 1). All the variables were resampled to 5 m to standardize the data into one resolution. The predicted AGBD map was thus generated in 5 m.

Including redundant features in machine learning models can lead to increased errors and longer computation times. To address this, Recursive Feature Elimination (RFE) was applied in this study to identify and remove less significant variables, ensuring the selection of the most relevant inputs for the models. Using RFE with Random Forest, 11 key variables were identified for the final model: Elevation, CHM, Slope, Sen_Sum, Sen_VV, GRVI, B3, Palsar_Ratio, SR, Palsar_NDI, and Palsar_Sum.

The AGBD data were filtered to remove the values above 150 Mg/ha as outliers. A fishnet grid matching the size of the GEDI pixels was created to extract average values from all the predictor rasters. Rows containing NA values were carefully removed from all the predictor variables and subsequently, 1498 rows of observations were retained for input into the ML models. The data were then randomly split into training and testing sets for input into the ML models.

Two different types of machine learning models, viz., Random Forest and Extreme Gradient Boosting, were trained and tested for the prediction of AGBD in this study for the year 2022 using the CARET package in RStudio (Kuhn 2023). Random

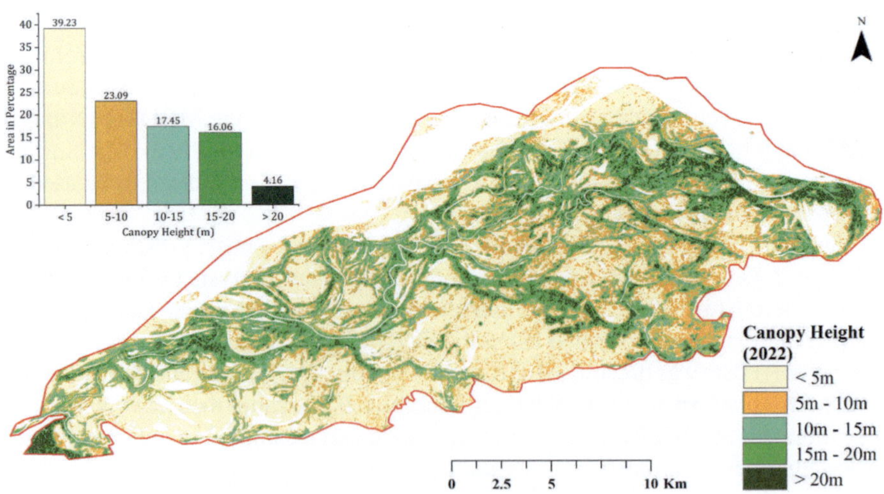

Fig. 2 Forest Canopy height map of KNP in 2022 (*Source* Bhuyan et al. 2024)

Table 1 Predictor variables considered for mapping AGBD in KNP

Data	Predictor variables
LISS IV	Difference Vegetation Index (DVI), Green Difference Vegetation Index (GDVI), Global Environmental Monitoring Index (GEMI), Green Ratio vegetation Index (GRVI), Green Normalized Difference Vegetation Index (GNDVI), Modified Chlorophyll Absorption ration Index-Improved (MCARI-I), Infrared Percentage Vegetation Index (IPVI), Modified Non-Linear Index (MNLI), Modified Simple Ratio (MSR), Non-Linear Index (NLI), Modified Triangular Vegetation Index (MTVI), Modified triangular vegetation Index-Improved (MTVI-I), Normalized Difference Vegetation Index (NDVI), Renormalized Difference Vegetation Index (RDVI), Optimized Soil Adjusted Vegetation Index (OSAVI), Red Green Ratio Index (RGRI), Simple Ratio (SR), Sum Green Index (SGI), Soil Adjusted Vegetation Index (SAVI), Transformed Difference Vegetation Index (TDVI)
SENTINEL—1 SAR	VV, VH, VV + VH, VV/VH, Normalized Difference Index (NDI), Modified Radar Vegetation Index (mRVI), Radar Vegetation Index (RVI)
PALSAR	HV, HH, HV + HH (Palsar_SUM), HV/HH (Palsar_Ratio)
SRTM	Slope, Elevation, Aspect
GEDI	Forest Canopy Height Map (CHM)

Forests are robust ensemble models suitable for classification and regression, effectively managing high-dimensional data with parameters like mtry and ntree. XGBoost, a powerful gradient-boosting framework, stands out for its efficiency, scalability, and ability to reduce overfitting, often delivering superior performance in both speed and accuracy.

R-squared (R^2) and Root Mean Square Error (RMSE) were used to assess the performance of the models. The hyperparameters of both models went through a tenfold cross-validation to find the parameters that performed the best. Thus, the combination of hyperparameters that yielded the best results was chosen as the optimal configuration for the models to predict AGBD. The best model was then chosen to perform the wall-to-wall AGBD prediction of Kaziranga National Park.

Results and Discussion

AGBD Prediction and Model Accuracy. Among the two machine learning models tested it was observed that Random forest gave the best results with an RMSE of 27.21 Mg/ha, achieving an R^2 of 0.52 with the other variables (Table 2). In contrast, the XGB model produced an RMSE of 28.04 Mg/ha and an R^2 of 0.49. As such, the RF model was used for the final AGBD predictions in the region (Fig. 3).

The random forest based variable importance analysis revealed canopy height to be the most important variable in mapping AGBD, followed by the elevation of the area (Fig. 4).

Spatial Patterns of AGBD in Kaziranga. The AGBD of Kaziranga National Park, as predicted using the RF model, has been categorized into six classes: < 20 Mg/ha, 20–40 Mg/ha, 40–60 Mg/ha, 60–80 Mg/ha, 80–100 Mg/ha, and > 100 Mg/ha.

Table 2 Tuned hyperparameters and performance metrics of models used for AGBD prediction (The performance of the best model is highlighted in bold)

Tuned hyperparameters		RMSE	R^2	MAE
RF	mtry = 3, ntree = 800	**27.21**	**0.52**	19.73
XGB	nrounds = 500, eta = 0.01, max depth = 2, gamma = 0.05, col sample = 1, min child = 3, sub-sample = 0.5	28.04	0.49	20.33

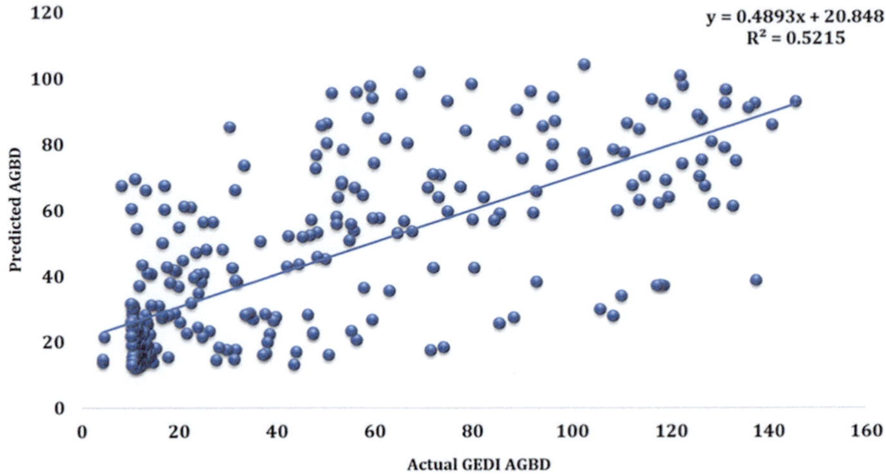

Fig. 3 Actual versus predicted GEDI AGBD in KNP

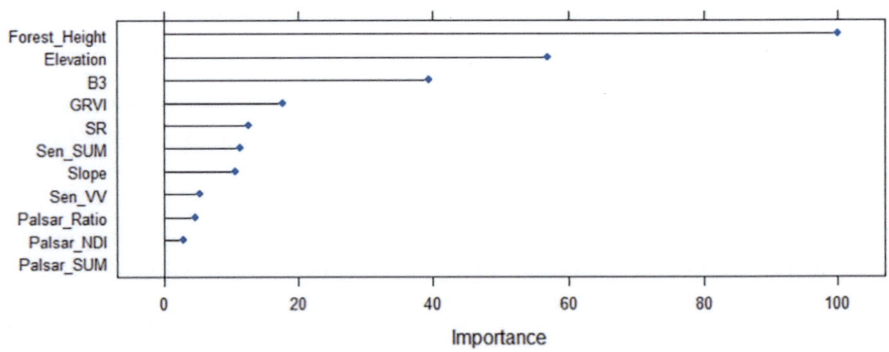

Fig. 4 Variable Importance of predictors in AGBD mapping after RFE

The results indicate that a significant portion of the region (57.4%) has an AGBD of less than 40 Mg/ha, primarily due to KNP's grassland dominance, characterized by canopy heights of less than 5 m, which limits woody biomass in the area (Fig. 5).

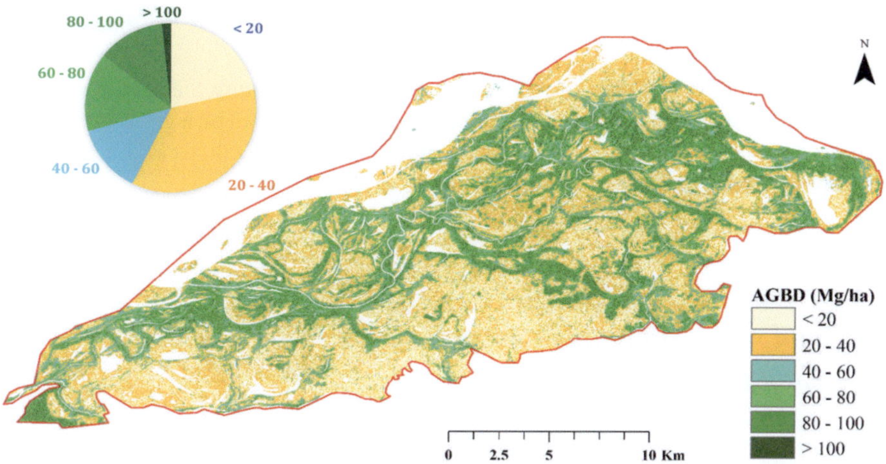

Fig. 5 Predicted AGBD map of KNP in 2022

Approximately 28.19% of the park exhibits moderate AGBD values between 40 and 80 Mg/ha. Areas with higher woody biomass (> 80 Mg/ha) are limited to 14.4% of the park, with only 1.76% of the area showing biomass levels exceeding 100 Mg/ha (Table 3). These high biomass zones are mostly found in regions where canopy heights surpass 15 m. Gupta and Sharma (2022) reported similar findings in their analysis of AGBD in a mixed tropical forest in India.

Conclusion. This study presents the continuous mapping of AGBD using various machine-learning approaches. The integration of LISS IV data with GEDI and SAR datasets demonstrated its effectiveness in predicting AGBD (Bhandari et al. 2023). The long-term availability of LISS IV data offers significant potential for such predictions.

Among the models tested, the Random Forest algorithm outperformed XGBoost in AGBD prediction and was consequently used for the final estimation. Canopy height emerged as the most critical predictor, followed closely by elevation. The analysis revealed that a majority of KNP has biomass levels below 40 Mg/ha, largely due to the dominance of grasslands. Areas with woody biomass exceeding 60 Mg/

Table 3 Area in percentage of AGBD classes in KNP

Classes	Area (in km^2)	Percentage
< 20	73.97	21.43
20–40	124.19	35.97
40–60	45.61	13.21
60–80	51.73	14.98
80–100	43.64	12.64

ha were confined to a small portion of the park mostly along the riverbanks of the region.

This continuous mapping of AGBD is crucial for estimating the region's carbon sequestration potential. Additionally, such kinds of studies can be a great aid in planning for mitigating forest fire risks and formulating effective conservation strategies to prevent further degradation of forest cover.

2.3 Limitations and Future Scope of Geospatial Techniques and AI in Monitoring Forest Changes Due to Climate Change

Limitations. The use of geospatial techniques and AI for monitoring forest changes, though quite transformative, has several associated challenges. A major limitation lies in data quality and availability. Accurate monitoring relies heavily on high-resolution satellite imagery and consistent temporal datasets, which are often unavailable for regions with persistent cloud cover, such as tropical forests of North East India (Shrestha et al. 2019). Moreover, inconsistencies in data acquisition across sensors complicate long-term monitoring efforts.

Another significant limitation is the computational complexity associated with processing large-scale spatial datasets and deploying advanced AI models. Deep learning algorithms, in particular, require substantial computational resources, which are often difficult to access. Additionally, these models, while powerful, are often perceived as "black boxes" owing to their lack of transparent processing, making it difficult to interpret their outputs and apply them confidently in decision-making processes (Hassija et al. 2024).

Another major challenge is acquiring field data, which is essential for training and validating AI models. Remote and inaccessible forest regions often lack adequate ground truth data, limiting the accuracy and applicability of the models. Also, conducting field studies is often labour and cost-intensive, which becomes difficult at times. Furthermore, AI models trained on specific ecosystems may fail to generalize to different forest types, where variations in climatic conditions and species composition demand tailored approaches. Ethical concerns also arise in geospatial monitoring, particularly when forest regions intersect with indigenous lands, raising issues of data privacy and consent.

Future Scope. Despite these limitations, the application of geospatial techniques and AI in monitoring forest changes is a rapidly growing field of study. Advances in satellite technologies, such as high-resolution multispectral and hyperspectral imaging, and the increasing availability of open-access data from agencies like NASA, ESA, and ISRO are enhancing data availability and quality. The integration of emerging sensors, such as NASA's GEDI, NASA-ISRO NISAR, ECOSTRESS, and ESA's upcoming BIOMASS satellite, holds the potential for improving the estimation of aboveground biomass and many other critical indicators of forest dynamics, health and carbon stock.

The development of explainable AI (XAI) techniques represents a significant step forward, addressing the issue of model interpretability and building trust in AI-driven forest monitoring (Hassija et al. 2024). Hybrid models that combine AI with physical modelling approaches offer the potential for more accurate predictions of forest dynamics under changing climatic conditions. Similarly, the integration of geospatial data with socioeconomic and ecological datasets can enable holistic monitoring, linking vegetation changes with human activities and biodiversity trends.

Real-time monitoring and early warning systems are another area of growth. AI-driven platforms have the capacity to enhance the prediction of disturbances such as forest fires, droughts, and deforestation, allowing for proactive interventions. UAVs equipped with AI-based sensors can supplement satellite data, providing detailed insights into hard-to-reach areas (Zhang et al. 2021; Han et al. 2019).

AI models can also be adapted to support conservation policies by predicting long-term trends, such as carbon stock variability and species migration patterns, under different climate scenarios. Enhanced global collaboration between governments, research institutions, and industries is critical to accelerate the development of geospatial-AI tools. Promoting open-source software and data-sharing platforms will democratize access to these advanced techniques, particularly in resource-constrained regions.

Finally, incorporating indigenous and local knowledge systems into geospatial-AI frameworks can provide a more nuanced understanding of forest changes, fostering sustainable forest management practices. Coupling AI models with long-term climate projections can further refine predictions of future impacts on forests, aiding in the development of adaptive management strategies (Rashid et al. 2015; Mishra et al. 2014; Kundu et al. 2018).

Therefore, by addressing current limitations and leveraging emerging technologies, geospatial techniques and AI can play a transformative role in mitigating the impacts of climate change on forests, ensuring their sustainable management and conservation.

3 Coastal Ecosystems

3.1 Use of Geospatial Techniques and AI in Monitoring Changes in Coastal Ecosystems Due to Climate Change

Coastal ecosystems, including mangroves, coral reefs, and salt marshes, are highly sensitive to climate change impacts such as sea-level rise, increased storm intensity, and temperature fluctuations. These ecosystems play a vital role in biodiversity conservation, carbon sequestration, and providing ecosystem services to coastal communities. Monitoring these changes is crucial for sustainable management and mitigation strategies. Geospatial techniques and AI offer innovative tools for high-resolution spatial and temporal data analysis to track these changes effectively (Politi et al. 2019).

While these ecosystems are typically stable, geospatial studies using remote sensing, time series data, and AI have provided valuable insights into their change rates and impacts (Marzouk and Azab 2024). However, current methodologies face limitations and uncertainties, highlighting the need for ongoing monitoring and improved strategies to enhance data accuracy. These techniques are significant for scientific policy decisions, especially as many coastal regions experience rapid alterations that threaten ecosystems. Advanced technologies like remote sensing and AI are essential for understanding coastal ecosystem dynamics and addressing mounting pressures from climate change, which alters hydrology and biodiversity. Although changes often occur gradually, they can surpass critical thresholds, leading to dramatic shifts and highlighting the need for continuous monitoring (Nicholls et al. 2007).

Monitoring Coastal Dynamics and Sea-Level Rise. Coastal ecosystems are undergoing rapid changes due to natural and human-induced factors, with tropical and temperate regions facing significant disruptions. Understanding the interactions between coastal dynamics and climate change is critical, as these areas are at the forefront of climate impacts. Coastal dynamics, including water flows, sediment transport, and habitat changes, are heavily influenced by sea-level rise (SLR). Between 1901 and 2010, global sea levels rose by 0.19 m, with projections indicating a rise of 0.26–0.98 m by 2100 due to thermal expansion and melting ice sheets (IPCC, Climate Change 2014). Coastal habitats, particularly intertidal zones, are at high risk, with up to 150 million people globally projected to be affected by flooding by 2050, doubling after 2100 without intervention (Dasgupta et al. 2007). The ongoing challenges to coastal ecosystems necessitate a focus on understanding SLR and its interactions with climate change. Effective monitoring and experimental networks are paramount for addressing these challenges and should be prioritized to enhance our understanding of coastal dynamics in the broader Earth system (Nichols 2019; Benveniste 2019).

Remote sensing and GIS tools are essential for tracking changes in coastal ecosystems. Satellite imagery, such as Landsat, Sentinel, and MODIS, allows for monitoring shoreline changes, vegetation cover, and sediment dynamics. For example, geospatial analyses have been used to assess mangrove loss due to rising sea levels and human encroachment (Hamilton and Casey 2016). High-resolution LiDAR data combined with GIS models can map elevation changes and predict areas at risk from SLR, providing critical information for coastal zone management (Cooper et al. 2013; Gesch 2018).

AI enhances these analyses by automating feature extraction and change detection. Machine learning models like convolutional neural networks (CNNs) have been employed to analyze satellite imagery for identifying shoreline erosion, accretion, and vegetation health, reducing manual effort and increasing accuracy (Scardino et al. 2022; Giang et al. 2023).

Assessing Habitat Loss and Ecosystem Degradation. Coastal ecosystems, including wetlands, mangroves, corals, and seagrasses, are among the most threatened habitats due to anthropogenic and natural pressures. Understanding these impacts is critical for conservation planning. While traditional field surveys provide detailed habitat assessments, they are resource-intensive and spatially limited. Remote sensing offers comprehensive views of coastal ecosystems and their changes over time (Tuia et al. 2022). Anthropogenic pressures often drive degradation, though natural processes also contribute. Habitat loss significantly impacts biodiversity and ecosystem services, leading to economic losses. Restoration efforts since the 1990s have shown varying success, with recovery often falling short of natural habitat conditions due to socio-economic and environmental factors (Giri et al. 2011).

Hyperspectral remote sensing combined with AI algorithms can identify and classify coastal habitats based on spectral signatures. For example, deep learning techniques have detected coral bleaching events from satellite data, revealing the spatial extent and severity of damage (Hedley et al. 2016). Similarly, machine learning models monitor seagrass cover changes, critical for carbon storage and marine biodiversity (Katija et al. 2021). Mangroves, often called "blue carbon ecosystems," play a significant role in mitigating climate change. Synthetic aperture radar (SAR) is particularly effective for mapping mangroves in regions with frequent cloud cover. When integrated with AI, these tools enable accurate estimations of aboveground biomass, carbon stocks, and mangrove extent, supporting conservation and restoration initiatives (Fatoyinbo et al. 2008).

Predicting Extreme Weather Impacts. Coastal ecosystems are highly vulnerable to extreme weather events like hurricanes and cyclones, raising concerns about their long-term resilience. Advanced modeling techniques, including AI and machine learning, are essential for forecasting and managing these impacts. Rapid response models guide disaster management efforts, addressing the effects of hurricanes, cyclones, and storm surges on coastal habitats and infrastructure. For instance, predictive models have been used to assess the impacts of events like the Deepwater Horizon Oil Spill and Hurricane Sandy (Record et al. 2013; Sobel et al. 2021).

For example, recurrent neural networks (RNNs) and ensemble models have been used to predict storm surge heights and their potential impact on coastal vegetation by integrating datasets such as historical storm tracks, sea surface temperatures, and wind speeds (Wei and Nguyen 2022). Reliable predictive models tailored for specific regions are essential to protect habitats and communities. These models enhance disaster management and foster adaptive practices for mitigating climate impacts (Hinkel et al. 2014).

3.2 *Case Study of the Application of Geospatial Technologies and AI in Monitoring the Health of Mangrove Forests and Automatic Shoreline Mapping in Sundarbans*

Introduction. Mangroves as one of the most critical "blue carbon ecosystems," are some of the highly productive forests found in tropical and subtropical coastal regions. These ecosystems play a vital role in biodiversity conservation, coastal protection, and carbon sequestration while acting as natural barriers against storms and coastal erosion. However, mangroves face significant anthropogenic pressures, including deforestation, habitat fragmentation, and climate-induced stressors such as sea-level rise and increasing frequency of extreme weather events (Chakraborty 2017; Alongi 2022). These threats have accelerated the degradation of mangroves, disrupting biodiversity, ecosystem services, and the livelihoods of dependent coastal communities.

Globally, the urgency of mangrove conservation has gained momentum, as reflected in the targets set by the Global Biodiversity Framework (GBF) to protect marine and coastal ecosystems (Singh et al. 2022). Despite these conservation efforts, mangroves remain vulnerable to climate change impacts such as habitat inundation, salinity intrusion, and coastal erosion caused by declining river discharges and modifications in land use. The Sundarbans, the world's largest contiguous mangrove forest, exemplifies these vulnerabilities, facing risks of extensive inundation due to rising sea levels and increasing habitat degradation.

While mangroves have some adaptive mechanisms to cope with environmental changes, these responses may not suffice in the face of rapid climatic shifts. Understanding and addressing the threats to mangrove ecosystems require advanced tools and techniques for monitoring and management. In this context, AI and ML have emerged as transformative technologies, enabling efficient analysis of spatial, spectral, and temporal data to assess ecosystem health and changes over time (Ferraris et al. 2023).

The integration of remote sensing data with AI/ML techniques has significantly advanced research on habitat health and density. Remote sensing allows for the quantification of habitat changes driven by climate factors, such as rising temperatures, land cover alterations, rainfall variability, and sea-level rise (Gargaran et al. 2024). New indices, such as the Enhanced Vegetation Index (EVI), have been developed to monitor vegetation health, density, and stress levels. EVI has been particularly effective in assessing the health and density of mangrove forests, offering a nuanced understanding of their response to climatic and anthropogenic stressors.

In addition to vegetation analysis, AI/ML techniques have been successfully applied to other aspects of mangrove monitoring. For instance, Support Vector Machines (SVMs), a widely used ML algorithm, have proven effective in shoreline mapping, a critical component of understanding and mitigating the impacts of coastal erosion on mangroves. The combination of remote sensing data, vegetation indices, and ML algorithms enables more accurate and comprehensive assessments, supporting conservation and restoration efforts.

This study focuses on leveraging ML techniques to monitor the health and density of mangrove forests using EVI, and automatic shoreline change detection using SVM. These advanced methodologies provide critical insights into mangrove ecosystems, facilitating better management and conservation strategies.

Methodology

Study Area. The Sundarbans Mangrove Forest, the world's largest tidal halophytic forest, spans 10,000 km^2 across India and Bangladesh at the confluence of the Ganges, Brahmaputra, and Meghna rivers (Fig. 6). In India, it covers parts of North and South 24 Parganas districts in West Bengal, constituting 40% of the forest area, with the rest in Bangladesh (Dutta et al. 2022). The region features tidal waterways, creeks, mudflats, and 106 islands, 54 of which are inhabited, and 52 are forested (Chakraborty 2017).

Located between 21°31′ N to 22°30′ N latitude and 88°10′ E to 89°51′ E longitude, the Sundarbans experience a tropical climate with high humidity (70–88%) (Sahana

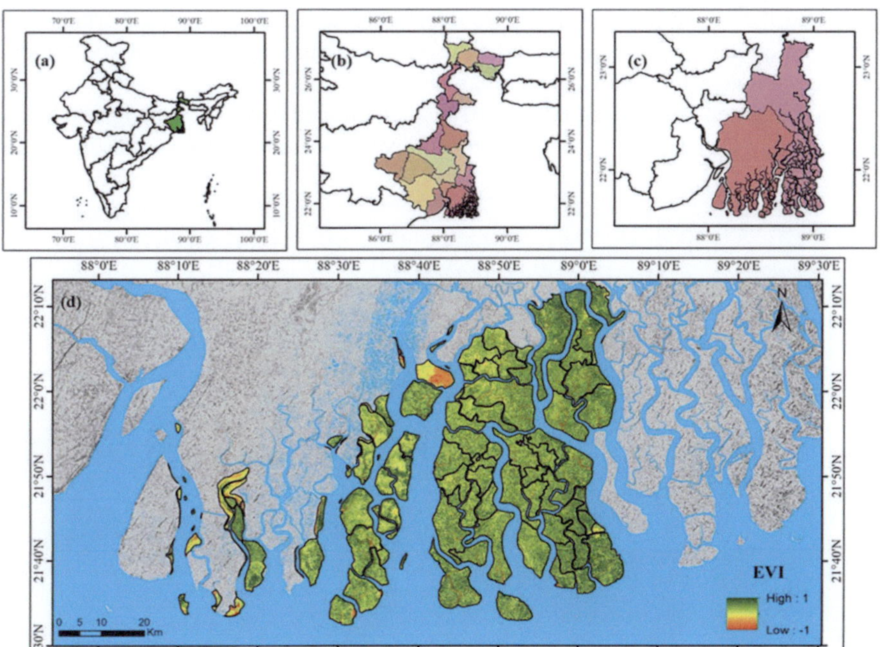

Fig. 6 Spatial representation of the Sundarbans Reserve Forest and its major islands. **a** Location of West Bengal within the geographical context of India. **b** Administrative boundaries of districts within West Bengal. **c** The Sundarbans region extending across South and North 24 Parganas. **d** EVI map showing vegetation distribution within the Indian Sundarbans, where areas of high EVI (dense vegetation) are shown in green, and areas of low EVI (sparse vegetation) are depicted in red

et al. 2022). Winters are mild (21–31 °C), summers are hot, and the monsoon season (June–September) brings 80% of the annual rainfall, averaging 1920 mm. Tidal fluctuations range from 3.5 to 5 m, particularly in July and August, influenced by limited freshwater flow from major rivers.

Home to rich biodiversity, the Sundarbans supports diverse mangrove species and wildlife, including the Bengal tiger. The northern areas are dominated by Avicennia species, while the southern regions feature Rhizophora, Bruguiera, and Heritiera fomes (Dutta et al. 2022). The ecosystem faces significant threats from climate change, sea-level rise, extreme weather events, and human activity. As a UNESCO World Heritage Site (1987) and Ramsar site (2019), conservation efforts focus on ecosystem protection and sustainable livelihoods through ecotourism and habitat conservation (Neogi et al. 2017).

Methods. This study employs a structured workflow to assess mangrove vegetation health, density, and shoreline dynamics using Landsat data (2001–2020). The methodology integrates remote sensing indices, temporal analysis, and machine learning for precise monitoring and classification (Fig. 7).

Landsat satellite imagery (2001–2020) was used as the primary dataset which was assessed through the Google Earth Engine (GEE) platform. Cloud masking was performed to remove atmospheric interferences. Temporal median composites were then created to fill gaps and smooth the data, ensuring consistent and reliable time-series imagery for subsequent analysis.

The Enhanced Vegetation Index (EVI) was calculated to monitor mangrove health and density over time. The long term mean EVI was computed for each time point

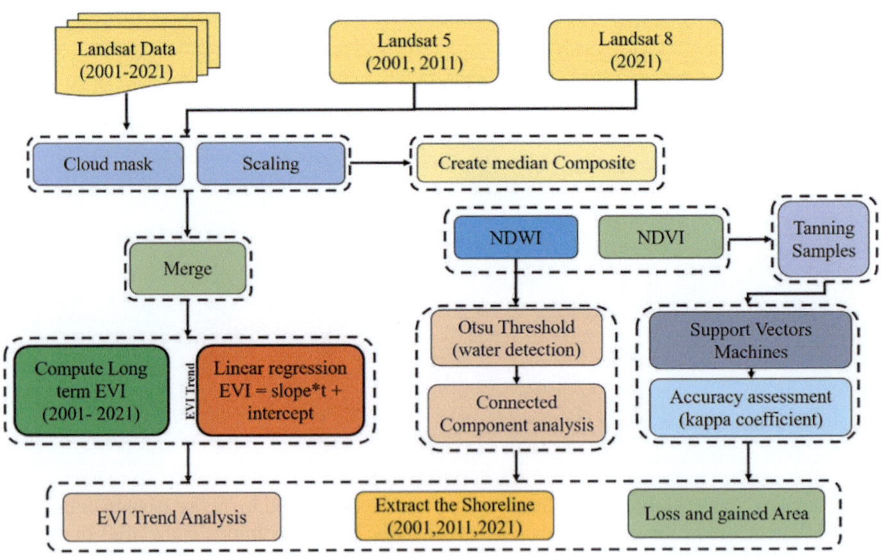

Fig. 7 Methodology for shoreline detection

to identify trends in mangrove vegetation condition. These trends provided insights into long-term health dynamics and responses to environmental changes.

Shoreline changes were mapped using the Normalized Difference Water Index (NDWI). A supervised machine learning method, Support Vector Machine (SVM), was employed to classify land cover, including mangrove vegetation. In this investigation, three distinct classes were identified: water, land, and other types, which include mudflats and fallow lands. For each class, a total of 200 training samples were collected, with 70% allocated for training purposes and the remaining 30% reserved for testing. To enhance the classification results, Otsu's Dynamic Thresholding technique was implemented to automatically determine the threshold values for identifying water pixels. Following this, inland water pixels were eliminated, and the resulting image was transformed into a polygon format. Ultimately, this polygon underwent a manual normalization process, leading to the extraction of the coastline. This approach quantified shoreline changes, highlighting coastal erosion and its impact on mangrove ecosystems.

The classified results and vegetation trends were combined with shoreline maps to provide a comprehensive assessment of mangrove dynamics. This integration linked shoreline changes with vegetation conditions, emphasizing the interconnected effects of environmental and climatic drivers.

Results and Discussion

Vegetation Trends and Mangrove Health Analysis. The analysis of vegetation dynamics in the Sundarbans using EVI offers valuable insights into the health, density, and resilience of the mangrove forests (Fig. 8). This visual representation helps in understanding both the spatial and temporal trends of vegetation over a 20-year period, highlighting the effects of environmental stressors and signs of ecological recovery. The temporal trend analysis reveals both positive and negative shifts in vegetation health across the region. Quantitatively, about 3228.85 km^2 showed an increasing positive EVI trend, indicating improved vegetation health. However, around 1143.81 km^2 experienced a decline, primarily due to extreme climatic events like cyclones and rising sea levels.

A significant drop in EVI values is observed during major cyclonic events such as Aila (2009) and Amphan (2020). High-intensity cyclones often lead to tree uprooting, canopy damage, and increased salinity from storm surges, all of which contribute to sharp declines in EVI values. These impacts are clearly captured in the negative trend zones on the graph. The blue-shaded areas in the temporal trend graph correspond to these declines, highlighting periods of stress and degradation. Conversely, the green-highlighted areas represent recovery phases, where natural regrowth, along with local and government-led restoration efforts, has contributed to vegetation resurgence. Previous studies have shown similar trends, noting that post-cyclone natural regeneration plays a critical role in ecosystem recovery. Mangroves, particularly species like Avicennia marina and Rhizophora mucronate, are known for their remarkable capacity for natural regeneration, especially in saline and dynamic environments (Alongi 2022; Neogi et al. 2017).

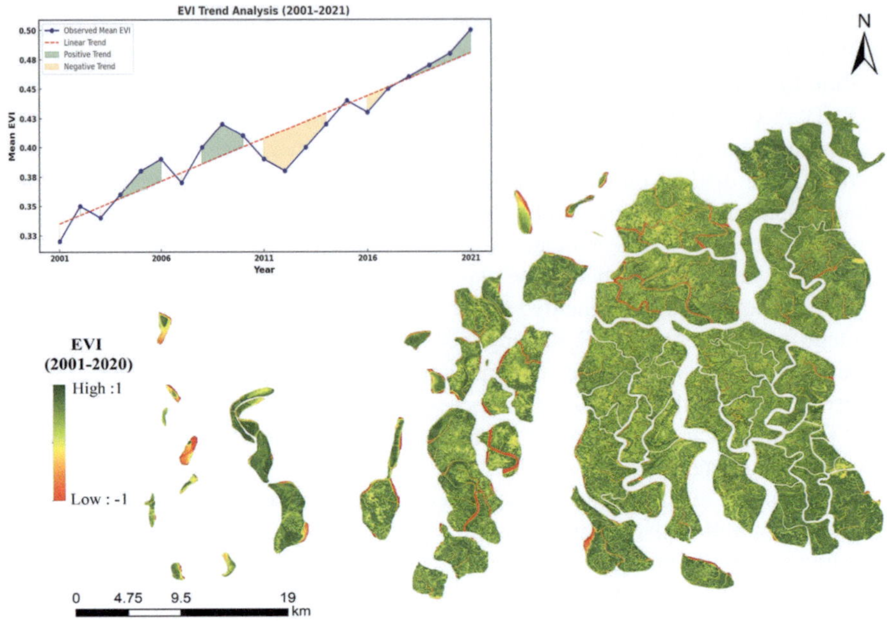

Fig. 8 Spatial and graphical representation of long term mean of EVI and EVI trend analysis

Classification and Shoreline Dynamics. Support Vector Machine (SVM) classification was conducted for the years 2001, 2011, and 2021 to assess decadal changes in shoreline configuration. The land cover was classified into three main categories: water bodies, vegetation, and other land types (Fig. 9). Classification accuracy ranged from 80 to 98%, reflecting a reliable assessment of land cover changes. A key finding from the results is the gradual reduction in the deltaic landmass due to coastal erosion, with the Bangaduni Island region serving as a prominent example of intense shoreline retreat. Over the past two decades, the island has experienced significant erosion along its seaward edge. Studies by Chakraborty (2017) and Singh et al. (2022) confirm similar trends of coastal retreat in other deltaic regions impacted by climate change.

Using the Normalized Difference Water Index (NDWI) as a reference, shoreline changes were analyzed across different time frames (Fig. 10). The results indicate that between 2001 and 2011, approximately 29.46 km^2 of land was gained, whereas 54.94 km^2 was lost to erosion or submergence. A similar trend was observed from 2011 to 2021, with a land gain of 29.14 km^2 and a loss of 50.94 km^2. These findings suggest a net loss of coastal land, reinforcing concerns about the long-term sustainability of the Sundarbans ecosystem. The impact of sea-level rise and human interventions on mangrove habitats has been extensively documented in previous research (Sahana et al. 2022; Das et al. 2021).

The spatial analysis of shoreline retreat in the Sundarbans reveals significant landward shifts across multiple regions. A visual overlay of historical shoreline positions from 2001 (green), 2011 (yellow), and 2021 (black) highlights the progressive erosion

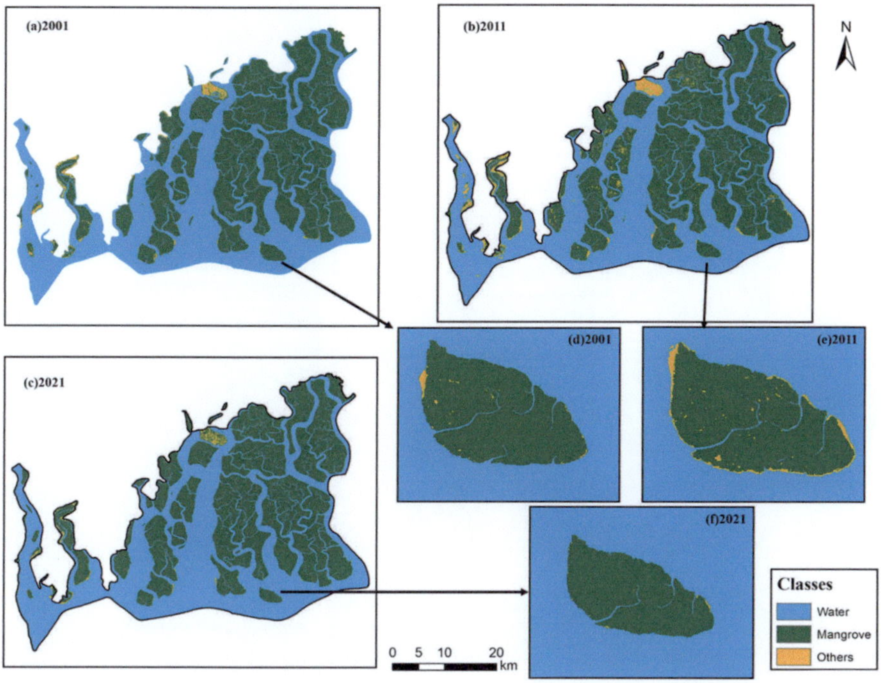

Fig. 9 Classification maps of the study area for **a** 2001, **b** 2011, and **c** 2021, showing land cover categorized into water bodies, vegetation, and other land types. The Ban-gaduni Island region is highlighted in **d** 2001, **e** 2011, and **f** 2021 to emphasize the shoreline retreat and erosion over the two decades

of coastal areas (Fig. 10). The false-color composite (FCC) image, where red represents mangrove forests, shows a noticeable reduction in mangrove cover over time. The sequential retreat of the shoreline underscores the ongoing loss of land, primarily driven by climate change-induced sea level rise. This process is further exacerbated by increasing storm surges and tidal amplification, accelerating coastal erosion and threatening the stability of this critical ecosystem.

Repeated cyclonic disturbances, notably Cyclones Aila (2009), Fani (2019), and Amphan (2020), have significantly intensified erosional forces acting on these fragile deltaic margins. The high-intensity storm events accelerate sediment displacement and hinder natural accretion, leading to the submergence or disappearance of low-lying zones. These vulnerable areas often coincide with regions historically prone to tidal fluctuations, where mangrove vegetation is either sparse or degraded, further reducing the natural buffering capacity against coastal erosion. These observations align with global studies on mangrove vulnerability to climate change, highlighting the increasing threat posed by extreme weather events to these critical coastal ecosystems (Marzouk and Azab 2024; Record et al. 2013).

Implications and Conservation Strategies

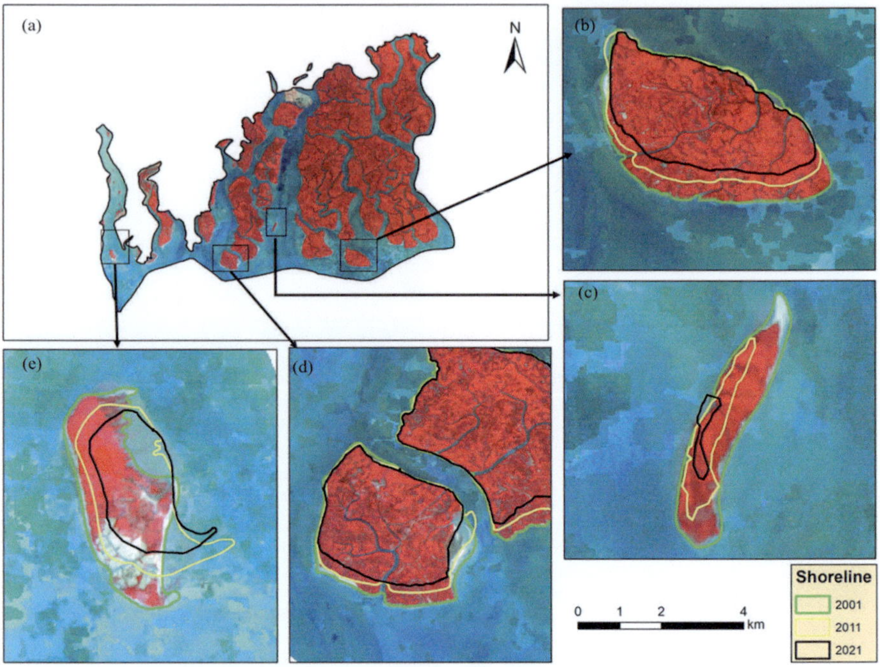

Fig. 10 Spatial representation of shoreline change: **a** base map of 2001; **b** shoreline change in Bangaduni Island; **c** shoreline change in Halliday Island; **d** shoreline change in Bulchery Island; **e** shoreline change in Jambudwip

The findings of this study underscore the vulnerability of the Sundarbans mangrove ecosystem to climate change and anthropogenic pressures. Despite periods of vegetation recovery, the overall trend indicates that coastal erosion and habitat loss remain pressing concerns. Conservation efforts should focus on enhancing the resilience of vulnerable coastal zones through afforestation programs, mangrove restoration initiatives, and sustainable land-use policies.

Previous studies have emphasized the effectiveness of community-based conservation and ecosystem-based adaptation strategies in mitigating coastal vulnerability (Politi et al. 2019; Benveniste 2019). Considering the observed shoreline retreat and vegetation loss, proactive measures such as the implementation of bio-shields, community-based conservation programs, and improved coastal zoning regulations are essential. The study also highlights the importance of continuous monitoring using geospatial techniques and AI-driven models to track ecosystem changes and inform policy decisions.

Conclusion. The analysis reveals that while some areas of the Sundarbans have shown resilience and recovery, the overall trend reveals significant environmental challenges, particularly in terms of shoreline erosion and mangrove degradation. The integration of geospatial techniques, remote sensing, and AI-based models provides a robust framework for monitoring these changes and devising mitigation strategies. Future research should focus on incorporating real-time monitoring systems, integrating socio-economic factors into conservation planning, and exploring adaptive management approaches to safeguard the Sundarbans ecosystem against future climate threats.

3.3 Limitations and Future Scope of the Use of AI/ML in Monitoring Coastal Ecosystems

The integration of geospatial techniques, particularly satellite remote sensing and AI-driven analytics, has significantly advanced the monitoring of coastal ecosystems. These technologies facilitate the detection of habitat changes, shoreline dynamics, and environmental stressors such as seawater intrusion, coastal erosion, and sediment deposition. However, several challenges persist, limiting their effectiveness. The availability of high-resolution satellite and drone imagery remains a constraint, particularly for remote and localized coastal regions, where cloud cover and atmospheric disturbances can affect data quality (Zhu et al. 2012). Additionally, AI models are highly sensitive to training data biases, which may result in misclassifications, especially in complex coastal environments with rapidly changing conditions (Maxwell et al. 2018). The intricate interactions between land, ocean, and atmosphere in deltaic regions further complicate monitoring, making it difficult to achieve accurate delineation of coastal features and track real-time changes in sediment dynamics and tidal fluctuations. The challenge of real-time coastal monitoring is also compounded by the limited temporal resolution of existing remote sensing datasets, restricting the capacity to detect sudden environmental shifts such as extreme storm events or rapid coastal degradation (Politi et al. 2019).

Looking forward, the evolution of AI, big data analytics, and Internet of Things (IoT) integration presents promising avenues for enhancing coastal ecosystem assessments (Matyukira and Mhangara 2024). The upcoming NASA-ISRO Synthetic Aperture Radar (NISAR) mission, along with the continued operation of Copernicus Sentinel-1 and Sentinel-2 satellites, will provide higher temporal and spatial resolution imagery, enabling more precise and frequent coastal monitoring (Cavanaugh et al. 2025). Additionally, the fusion of multi-sensor remote sensing data—including optical, synthetic aperture radar (SAR), and LiDAR—can improve the accuracy of coastal vulnerability assessments, particularly in mangrove ecosystems and low-lying deltaic zones prone to inundation (Alotaibi and Nassif 2024; Masria 2024). AI-driven predictive models, particularly deep learning frameworks, hold the potential to enhance forecasting of coastal land cover changes and ecosystem resilience to climate-driven disturbances (Blais and Akhloufi 2025). Moreover, the integration of drone-based remote sensing and real-time monitoring networks can offer localized, high-resolution insights into shoreline retreat, vegetation health, and sediment transport patterns, facilitating adaptive coastal management strategies (Cavanaugh et al.

2025; Fatoyinbo and Simard 2013). As AI techniques continue to evolve, addressing data biases, enhancing transfer learning capabilities across different coastal regions, and improving the interpretability of AI models will be critical for ensuring robust, scalable, and actionable insights for coastal conservation and climate resilience efforts.

4 Conclusion

The integration of geospatial techniques with AI/ML algorithms marks a paradigm shift in how we monitor, analyze, and address the impacts of climate change on ecosystems worldwide. This chapter has highlighted the immense potential of these advanced technologies to provide timely and accurate insights into ecosystem dynamics, offering tools for both researchers and policymakers to understand and mitigate the effects of a rapidly changing climate. While challenges such as data limitations, computational demands, and the need for interdisciplinary collaboration persist, the progress in remote sensing, machine learning, and predictive modelling underscores the critical role these innovations play in sustainable ecosystem management. Together, geospatial science and AI lay the foundation for future advancements in ecosystem monitoring, ensuring more informed decision-making in the face of unprecedented climate variability.

This chapter highlights the monitoring of forest ecosystems through the integration of high-resolution satellite imagery, LiDAR data, and AI algorithms which have advanced our ability to track canopy dynamics, carbon sequestration, and phenological shifts, unveiling critical insights into resilience and regeneration. Similarly, in coastal ecosystems, these technologies have proven indispensable for monitoring shoreline erosion, mangrove degradation, and coral reef health, offering strategies for mitigating the impacts of sea-level rise and extreme weather events. By bridging the gap between cutting-edge technology and ecological science, this chapter has demonstrated how geospatial techniques and AI not only enhance our understanding of climate-induced ecosystem changes but can also provide actionable solutions for conserving and restoring both forested and coastal landscapes. As we move forward, the convergence of these tools holds the promise of addressing global environmental challenges with greater precision and efficacy, fostering a future where ecosystems are better equipped to adapt to the evolving climate.

Acknowledgements The authors thank NASA for providing the freely available GEDI and SRTM data. NRSC is also acknowledged for providing the high resolution LISS IV data freely. Furthermore, the authors are grateful to ESA and JAXA for granting access to the SENTINEL-1 SAR and PALSAR data respectively. The first author would like to thank the INSPIRE program (Inspire Fellowship Registration number – IF190591) of the Department of Science and Technology, India, for providing the required funding for conducting this research through the Sanction Order number (DST/INSPIRE Fellowship/2019/IF190591). The authors are also immensely grateful for the support provided by ISRO-RESPOND for carrying out this research work.

Conflicts of Interest On behalf of all authors, the corresponding author states that there is no conflict of interest whatsoever.

AI Disclosure In the process of writing this article, the authors used the following AI technologies and writing tools: Grammarly (Free version) was used to correct errors in spelling and grammar, and ChatGPT (Open AI) was used to rewrite some sentences of the manuscript to enhance readability.

References

Abatzoglou JT, Williams AP (2016) Impact of anthropogenic climate change on wildfire across western US forests. Proc Natl Acad Sci 113(42):11770–11775. https://doi.org/10.1073/pnas.1607171113

Achu AL, Thomas J, Aju CD, Gopinath G, Kumar S, Reghunath R (2021) Machine-learning modelling of fire susceptibility in a forest-agriculture mosaic landscape of southern India. Ecol Inform 64:101348. https://doi.org/10.1016/j.ecoinf.2021.101348

Ahirwal J, Nath A, Brahma B, Deb S, Sahoo UK, Nath AJ (2021) Patterns and driving factors of biomass carbon and soil organic carbon stock in the Indian Himalayan region. Sci Total Environ 770:145292. https://doi.org/10.1016/j.scitotenv.2021.145292

Akila T, Menaka M, Vasanth P, Jancee BV (2023) Forecasting naturally occurring forest fires using AI and machine learning in wireless sensor networks. In: 2023 5th international conference on inventive research in computing applications (ICIRCA). IEEE, pp 498–502. https://doi.org/10.1109/ICIRCA57980.2023.10220860

Alongi DM (2022) Climate change and mangroves. In: Mangroves: biodiversity, livelihoods and conservation. Springer Nature Singapore, Singapore, pp 175–198. https://doi.org/10.1007/978-981-19-0519-3_8

Alotaibi E, Nassif N (2024) Artificial intelligence in environmental monitoring: in-depth analysis. Discover Artif Intell 4(1):84. https://doi.org/10.1007/s44163-024-00198-1

Aravind NA, Ganeshaiah KN, Shaanker RU (2013) Indian monsoons shape dispersal phenology of plants. Biol Lett 9(6). https://doi.org/10.1098/rsbl.2013.0675

Atmaja T, Setiawati MD, Kurisu K, Fukushi K (2024) Advancing coastal flood risk prediction utilizing a GeoAI approach by considering mangroves as an Eco-DRR strategy. Hydrology 11(12):198. https://doi.org/10.3390/hydrology11120198

Baccini A, Walker W, Carvalho L, Farina M, Sulla-Menashe D, Houghton RA (2017) Tropical forests are a net carbon source based on aboveground measurements of gain and loss. Science 358(6360):230–234. https://doi.org/10.1126/science.aam5962

Benveniste J et al (2019) Requirements for a coastal hazards observing system. Front Mar Sci 6. https://doi.org/10.3389/fmars.2019.00348

Bera B, Shit PK, Sengupta N, Saha S, Bhattacharjee S (2022) Forest fire susceptibility prediction using machine learning models with resampling algorithms, Northern part of Eastern Ghat Mountain range (India). Geocarto Int 37(26):11756–11781. https://doi.org/10.1080/10106049.2022.2060323

Bhandari K, Srinet R, Nandy S (2023) Forest height and aboveground biomass mapping by synergistic use of GEDI and sentinel data using random forest algorithm in the Indian Himalayan Region. J Indian Soc Remote Sens. https://doi.org/10.1007/s12524-023-01792-z

Bhuyan M, Jeganathan C, Pujar GS (2024) Enhancing forest canopy height mapping in Kaziranga National Park, Assam, by Integrating LISS IV and SAR data with GEDI LiDAR data Using Machine Learning. In: ISPRS annals of the photogrammetry, remote sensing and spatial information sciences. Copernicus Publications, pp 39–44. https://doi.org/10.5194/isprs-annals-X-3-2024-39-2024

Blais M-A, Akhloufi MA (2025) Advances in remote sensing and deep learning in coastal boundary extraction for erosion monitoring. Geomatics 5(1):9. https://doi.org/10.3390/geomatics5010009

Borah SB, Sivasankar T, Ramya MNS, Raju PLN (2018) Flood inundation mapping and monitoring in Kaziranga National Park, Assam using Sentinel-1 SAR data. Environ Monit Assess 190(9). https://doi.org/10.1007/s10661-018-6893-y

Cavanaugh KC, Bell TW, Aerni KE, Byrnes JEK, McCammon S, Smith MM (2025) New technologies for monitoring coastal ecosystem dynamics. Ann Rev Mar Sci 17(1):409–433. https://doi.org/10.1146/annurev-marine-040523-020221

Chakraborty SK (2017) Mangrove ecosystem of Sundarbans, India. Biodiv Ecol Threats Conserv. https://www.researchgate.net/publication/285637880

Cheng F, Ou G, Wang M, Liu C (2024) Remote sensing estimation of forest carbon stock based on machine learning algorithms. Forests 15(4):681. https://doi.org/10.3390/f15040681

Cleland EE, Chuine I, Menzel A, Mooney HA, Schwartz MD (2007) Shifting plant phenology in response to global change. Trends Ecol Evol 22(7):357–365. https://doi.org/10.1016/j.tree.2007.04.003

Cooper HM, Fletcher CH, Chen Q, Barbee MM (2013) Sea-level rise vulnerability mapping for adaptation decisions using LiDAR DEMs, vol 37, no 6, pp 745–766. https://doi.org/10.1177/0309133313496835

Das K, Sarkar S, Mukherjee A, Das P, Pathak A (2021) Observing tidal and storm generated wave height impact on groundwater levels in a tropical delta (the Sundarbans). J Hydrol 603:126813. https://doi.org/10.1016/j.jhydrol.2021.126813

Dasgupta S, Laplante B, Meisner C, Wheeler D, Yan J (2007) The impact of sea level rise on developing countries: a comparative analysis. In: World Bank policy research working paper. http://econ.worldbank.org

Diaz S et al (2019) Summary for policymakers of the global assessment report on biodiversity and ecosystem services of the Intergovernmental Science-Policy Platform on Biodiversity and Ecosystem Services. Secretariat of the Intergovernmental Science-Policy Platform on Biodiversity and Ecosystem Services, Bonn, Germany. www.ipbes.net

Dubayah R et al (2020) The global ecosystem dynamics investigation: high-resolution laser ranging of the Earth's forests and topography. Sci Remote Sens 1. https://doi.org/10.1016/j.srs.2020.100002

Duncanson L et al (2020) Biomass estimation from simulated GEDI, ICESat-2 and NISAR across environmental gradients in Sonoma County, California. Remote Sens Environ 242. https://doi.org/10.1016/j.rse.2020.111779

Dutta D, Kumar T, Jayaram C, Jain P (2022) Mangrove health analysis using multi-temporal Sentinel-2 multi-spectral instrument (MSI) data. In: Mangroves: biodiversity, livelihoods and conservation. Springer Nature Singapore, Singapore, pp 71–106. https://doi.org/10.1007/978-981-19-0519-3_4

Fatoyinbo TE, Simard M, Washington-Allen RA, Shugart HH (2008) Landscape-scale extent, height, biomass, and carbon estimation of Mozambique's mangrove forests with Landsat ETM+ and Shuttle Radar Topography Mission elevation data. J Geophys Res Biogeosci 113(G2). https://doi.org/10.1029/2007JG000551

Fatoyinbo TE, Simard M (2013) Height and biomass of mangroves in Africa from ICESat/GLAS and SRTM. Int J Remote Sens 34(2):668–681. https://doi.org/10.1080/01431161.2012.712224

Feng Q, Liu J, Gong J (2015) Urban flood mapping based on unmanned aerial vehicle remote sensing and random forest classifier—a case of Yuyao, China. Water 7(4):1437–1455. https://doi.org/10.3390/w7041437

Ferraris S, Meo R, Pinardi S, Salis M, Sartor G (2023) Machine learning as a strategic tool for helping cocoa farmers in Côte D'Ivoire. Sensors 23(17):7632. https://doi.org/10.3390/s23177632

Flannigan M, Stocks B, Turetsky M, Wotton M (2009) Impacts of climate change on fire activity and fire management in the circumboreal forest. Glob Chang Biol 15(3):549–560. https://doi.org/10.1111/j.1365-2486.2008.01660.x

Gargaran JPS et al (2024) Mangrove ecosystem in Asia: review and synthesis of ecosystem services and economic valuation methods. Environ Exp Biol 22(2):59–70. https://doi.org/10.22364/eeb.22.06

Gesch DB (2018) Best practices for elevation-based assessments of sea-level rise and coastal flooding exposure. Front Earth Sci 6. https://doi.org/10.3389/feart.2018.00230

Giang TL et al (2023) Coastal landscape classification using convolutional neural network and remote sensing data in Vietnam. J Environ Manage 335:117537. https://doi.org/10.1016/j.jenvman.2023.117537

Giri C et al (2011) Status and distribution of mangrove forests of the world using earth observation satellite data. Glob Ecol Biogeogr 20(1):154–159. https://doi.org/10.1111/j.1466-8238.2010.00584.x

Goetz SJ et al (2009) Mapping and monitoring carbon stocks with satellite observations: a comparison of methods. Carbon Balance Manag 4(1):2. https://doi.org/10.1186/1750-0680-4-2

Goswami VR, Yadava MK, Vasudev D, Prasad PK, Sharma P, Jathanna D (2019) Towards a reliable assessment of Asian elephant population parameters: the application of photographic spatial capture–recapture sampling in a priority floodplain ecosystem. Sci Rep 9(1). https://doi.org/10.1038/s41598-019-44795-y

Gupta R, Sharma LK (2022) Aboveground biomass prediction by fusing gedi footprints with optical and SAR data using the random forest in the mixed tropical forest, India. In: International geoscience and remote sensing symposium (IGARSS). Institute of Electrical and Electronics Engineers Inc., pp 5460–5463. https://doi.org/10.1109/IGARSS46834.2022.9883443

Hamilton SE, Casey D (2016) Creation of a high spatio-temporal resolution global database of continuous mangrove forest cover for the 21st century (CGMFC-21). Glob Ecol Biogeogr 25(6):729–738. https://doi.org/10.1111/GEB.12449

Han L et al (2019) Modeling maize above-ground biomass based on machine learning approaches using UAV remote-sensing data. Plant Methods 15(1). https://doi.org/10.1186/s13007-019-0394-z

Harris NL et al (2021) Global maps of twenty-first century forest carbon fluxes. Nat Clim Chang 11(3):234–240. https://doi.org/10.1038/s41558-020-00976-6

Hassija V et al (2024) Interpreting black-box models: a review on explainable artificial intelligence. Cognit Comput 16(1):45–74. https://doi.org/10.1007/s12559-023-10179-8

Hedley J et al (2016) Remote sensing of coral reefs for monitoring and management: a review. Remote Sens 8(2):118. https://doi.org/10.3390/rs8020118

Hinkel J et al (2014) Coastal flood damage and adaptation costs under 21st century sea-level rise. Proc Natl Acad Sci 111(9):3292–3297. https://doi.org/10.1073/pnas.1222469111

IPCC (2021) Technical summary. In: Climate change 2021: the physical science basis. Contribution of working group I to the sixth assessment report of the intergovernmental panel on climate change. Cambridge University Press, pp 35–144. https://doi.org/10.1017/9781009157896.002

IPCC, Climate Change 2014 (2014) Synthesis report. Contribution of working groups I, II and III to the fifth assessment report of the intergovernmental panel on climate change. IPCC, Geneva

Jain H, Dhupper R, Shrivastava A, Kumar D, Kumari M (2023) AI-enabled strategies for climate change adaptation: protecting communities, infrastructure, and businesses from the impacts of climate change. Comput Urban Sci 3(1):25. https://doi.org/10.1007/s43762-023-00100-2

Janke JR (2010) Multicriteria GIS modeling of wind and solar farms in Colorado. Renew Energy 35(10):2228–2234. https://doi.org/10.1016/j.renene.2010.03.014

Janowicz K, Gao S, McKenzie G, Hu Y, Bhaduri B (2020) GeoAI: spatially explicit artificial intelligence techniques for geographic knowledge discovery and beyond. Int J Geogr Inf Sci 34(4):625–636. https://doi.org/10.1080/13658816.2019.1684500

Jones et al (1979) Global rise in forest fire emissions linked to climate change in the extratropics. Science 386(6719). https://doi.org/10.1126/science.adl5889

Kacic P, Thonfeld F, Gessner U, Kuenzer C (2023) Forest structure characterization in Germany: novel products and analysis based on GEDI, Sentinel-1 and Sentinel-2 Data. Remote Sens 15(8):1969. https://doi.org/10.3390/rs15081969

Katija K et al (2021) Visual tracking of deepwater animals using machine learning-controlled robotic underwater vehicles. In 2021 IEEE winter conference on applications of computer vision (WACV), IEEE, pp 859–868. https://doi.org/10.1109/WACV48630.2021.00090
Kuhn M (2023) Classification and regression training [R package caret version 6.0-94]. Comprehensive R archive network (CRAN): 6.0-94. https://CRAN.R-project.org/package=caret. Accessed 10 Jun 2023
Kundu S, Mondal A, Khare D, Hain C, Lakshmi V (2018) Projecting climate and land use change impacts on actual evapotranspiration for the Narmada river basin in Central India in the future. Remote Sens 10(4). https://doi.org/10.3390/rs10040578
Lamjiak T, Kaewthongrach R, Sirinaovakul B, Hanpattanakit P, Chithaisong A, Polvichai J (2021) Characterizing and forecasting the responses of tropical forest leaf phenology to El Nino by machine learning algorithms. PLoS ONE 16(8):e0255962. https://doi.org/10.1371/journal.pone.0255962
Lang N, Jetz W, Schindler K, Wegner JD (2023) A high-resolution canopy height model of the Earth. Nat Ecol Evol 7(11):1778–1789. https://doi.org/10.1038/s41559-023-02206-6
Lang N, Kalischek N, Armston J, Schindler K, Dubayah R, Wegner JD (2022) Global canopy height regression and uncertainty estimation from GEDI LIDAR waveforms with deep ensembles. Remote Sens Environ 268. https://doi.org/10.1016/j.rse.2021.112760
Leemans R, Eickhout B (2004) Another reason for concern: regional and global impacts on ecosystems for different levels of climate change. Glob Environ Chang 14(3):219–228. https://doi.org/10.1016/j.gloenvcha.2004.04.009
Liu X, Wang R, Shi W, Wang X, Yang Y (2024) Research on estimation model of carbon stock based on airborne LiDAR and feature screening. Sustainability 16(10):4133. https://doi.org/10.3390/su16104133
Mai G et al (2025) Towards the next generation of geospatial artificial intelligence. Int J Appl Earth Obs Geoinf 136:104368. https://doi.org/10.1016/j.jag.2025.104368
Marzouk M, Azab S (2024) Modeling climate change adaptation for sustainable coastal zones using GIS and AHP. Environ Monit Assess 196(2):147. https://doi.org/10.1007/s10661-023-12287-2
Masria A (2024) Bridging coastal challenges: the role of remote sensing and future research. Reg Stud Mar Sci 73:103502. https://doi.org/10.1016/j.rsma.2024.103502
Matyukira C, Mhangara P (2024) Advances in vegetation mapping through remote sensing and machine learning techniques: a scientometric review. Eur J Remote Sens 57(1). https://doi.org/10.1080/22797254.2024.2422330
Maxwell AE, Warner TA, Fang F (2018) Implementation of machine-learning classification in remote sensing: an applied review. Int J Remote Sens 39(9):2784–2817. https://doi.org/10.1080/01431161.2018.1433343
Mishra V, Shah R, Thrasher B (2014) Soil moisture droughts under the retrospective and projected climate in India. J Hydrometeorol 15(6):2267–2292. https://doi.org/10.1175/JHM-D-13-0177.1
Myneni RB et al (1995) Optical remote sensing of vegetation: modeling, caveats, and algorithms. Remote Sens Environ 51(1):169–188. https://doi.org/10.1016/0034-4257(94)00073-V
Neogi SB et al (2017) Sundarban mangroves: diversity, ecosystem services and climate change impacts. Asian J Med Biol Res 2(4):488–507. https://doi.org/10.3329/ajmbr.v2i4.30988
Neuville R, Bates JS, Jonard F (2021) Estimating forest structure from UAV-mounted LiDAR point cloud using machine learning. Remote Sens 13(3):352. https://doi.org/10.3390/rs13030352
Nicholls RJ et al (2007) Coastal systems and low-lying areas. In: Parry ML, Canziani OF, Palutikof JP, van der Linden PJ, Hanson CE (eds) Climate change 2007: impacts, adaptation and vulnerability. Contribution of working group II to the fourth assessment report of the intergovernmental panel on climate change. Cambridge University Press, Cambridge
Nichols CR et al (2019) Collaborative science to enhance coastal resilience and adaptation. Front Mar Sci 6. https://doi.org/10.3389/fmars.2019.00404
Pan Y et al (2011) A large and persistent carbon sink in the world's forests. Science 333(6045):988–993. https://doi.org/10.1126/science.1201609

Politi E, Paterson SK, Scarrott R, Tuohy E, O'Mahony C, Cámaro-García WCA (2019) Earth observation applications for coastal sustainability: potential and challenges for implementation. Anthropocene Coasts 2(1):306–329. https://doi.org/10.1139/anc-2018-0015

Potapov P et al (2021) Mapping global forest canopy height through integration of GEDI and Landsat data. Remote Sens Environ 253:112165. https://doi.org/10.1016/j.rse.2020.112165

Ramaswami G, Datta A, Reddy A, Quader S (2019) Tracking phenology in the tropics and in India: the impacts of climate change. In: The implications of climate change for Indian biodiversity: an overview, pp 45–59

Rashid I et al (2015) Projected climate change impacts on vegetation distribution over Kashmir Himalayas. Clim Chang 132(4):601–613. https://doi.org/10.1007/S10584-015-1456-5

Record S, Charney ND, Zakaria RM, Ellison AM (2013) Projecting global mangrove species and community distributions under climate change. Ecosphere 4(3):1–23. https://doi.org/10.1890/ES12-00296.1

Richardson AD, Keenan TF, Migliavacca M, Ryu Y, Sonnentag O, Toomey M (2013) Climate change, phenology, and phenological control of vegetation feedbacks to the climate system. Agric for Meteorol 169:156–173. https://doi.org/10.1016/J.AGRFORMET.2012.09.012

Richardson D et al (2022) Global increase in wildfire potential from compound fire weather and drought. NPJ Clim Atmos Sci 5(1):23. https://doi.org/10.1038/s41612-022-00248-4

Sahana M, Areendran G, Sajjad H (2022) Assessment of suitable habitat of mangrove species for prioritizing restoration in coastal ecosystem of Sundarban Biosphere Reserve, India. Sci Rep 12(1):20997. https://doi.org/10.1038/s41598-022-24953-5

Saini V, Tiwari RK, Gupta RP (2016) Comparison of FLAASH and QUAC atmospheric correction methods for Resourcesat-2 LISS-IV data. In: Earth observing missions and sensors: development, implementation, and characterization IV, SPIE, p 98811V. https://doi.org/10.1117/12.2228097

Scardino G, Scicchitano G, Chirivì M, Costa PJM, Luparelli A, Mastronuzzi G (2022) Convolutional neural network and optical flow for the assessment of wave and tide parameters from video analysis (LEUCOTEA): an innovative tool for coastal monitoring. Remote Sens 14(13):2994. https://doi.org/10.3390/rs14132994

Shrestha DP, Saepuloh A, van der Meer F (2019) Land cover classification in the tropics, solving the problem of cloud covered areas using topographic parameters. Int J Appl Earth Obs Geoinf 77:84–93. https://doi.org/10.1016/j.jag.2018.12.010

Singh M, Schwendenmann L, Wang G, Adame MF, Mandlate LJC (2022) Changes in mangrove carbon stocks and exposure to sea level rise (SLR) under future climate scenarios. Sustainability 14(7):3873. https://doi.org/10.3390/su14073873

Sobel AH et al (2021) Tropical cyclone frequency. Earths Future 9(12). https://doi.org/10.1029/2021EF002275

Tamiminia H, Salehi B, Mahdianpari M, Goulden T (2024) State-wide forest canopy height and aboveground biomass map for New York with 10 m resolution, integrating GEDI, Sentinel-1, and Sentinel-2 data. Ecol Inform 79:102404. https://doi.org/10.1016/j.ecoinf.2023.102404

Tanase MA et al (2024) Long-term annual estimation of forest above ground biomass, canopy cover, and height from airborne and spaceborne sensors synergies in the Iberian Peninsula. Environ Res 259:119432. https://doi.org/10.1016/j.envres.2024.119432

Tien Bui D, Hoang N-D, Samui P (2019) Spatial pattern analysis and prediction of forest fire using new machine learning approach of multivariate adaptive regression splines and differential flower pollination optimization: a case study at Lao Cai province (Viet Nam). J Environ Manage 237:476–487. https://doi.org/10.1016/j.jenvman.2019.01.108

Tuia D et al (2022) Perspectives in machine learning for wildlife conservation. Nat Commun 13(1):792. https://doi.org/10.1038/s41467-022-27980-y

UN (2019) The sustainable development goals report. New York

Vasudeva V, Nandy S, Padalia H, Srinet R, Chauhan P (2021) Mapping spatial variability of foliar nitrogen and carbon in Indian tropical moist deciduous sal (*Shorea robusta*) forest using machine learning algorithms and Sentinel-2 data. Int J Remote Sens 42(3):1139–1159. https://doi.org/10.1080/01431161.2020.1823043

Wang Y, Zhang X, Guo Z (2021) Estimation of tree height and aboveground biomass of coniferous forests in North China using stereo ZY-3, multispectral Sentinel-2, and DEM data. Ecol Indic 126:107645. https://doi.org/10.1016/j.ecolind.2021.107645

Wang H, Seaborn T, Wang Z, Caudill CC, Link TE (2021) Modeling tree canopy height using machine learning over mixed vegetation landscapes. Int J Appl Earth Obs Geoinf 101:102353. https://doi.org/10.1016/j.jag.2021.102353

Wei Z, Nguyen HC (2022) Storm surge forecast using an encoder–decoder recurrent neural network model. J Mar Sci Eng 10(12):1980. https://doi.org/10.3390/jmse10121980

Weiskopf SR et al (2020) Climate change effects on biodiversity, ecosystems, ecosystem services, and natural resource management in the United States. Sci Total Environ 733(137782). https://doi.org/10.1016/j.scitotenv.2020.137782

Xin Q, Li J, Li Z, Li Y, Zhou X (2020) Evaluations and comparisons of rule-based and machine-learning-based methods to retrieve satellite-based vegetation phenology using MODIS and USA National Phenology Network data. Int J Appl Earth Obs Geoinf 93:102189. https://doi.org/10.1016/j.jag.2020.102189

Yadav S, Padalia H, Sinha SK, Srinet R, Chauhan P (2021) Above-ground biomass estimation of Indian tropical forests using X band Pol-InSAR and random forest. Remote Sens Appl 21:100462. https://doi.org/10.1016/j.rsase.2020.100462

Zhang Y, Ma J, Liang S, Li X, Li M (2020) An evaluation of eight machine learning regression algorithms for forest aboveground biomass estimation from multiple satellite data products. Remote Sens 12(24):4015. https://doi.org/10.3390/rs12244015

Zhang H, Bauters M, Boeckx P, Van Oost K (2021) Mapping canopy heights in dense tropical forests using low-cost UAV-derived photogrammetric point clouds and machine learning approaches. Remote Sens 13(18):3777. https://doi.org/10.3390/rs13183777

Zhang X, Jia W, Sun Y, Wang F, Miu Y (2023) Simulation of spatial and temporal distribution of forest carbon stocks in long time series—based on remote sensing and deep learning. Forests 14(3):483. https://doi.org/10.3390/f14030483

Zhang Y, Steiner AL (2022) Projected climate-driven changes in pollen emission season length and magnitude over the continental United States. Nat Commun 13(1). https://doi.org/10.1038/s41467-022-28764-0

Zhu X, Liu D, Chen J (2012) A new geostatistical approach for filling gaps in Landsat ETM+ SLC-off images. Remote Sens Environ 124:49–60. https://doi.org/10.1016/j.rse.2012.04.019

Geospatial and AI Solutions for Agriculture and Forest Management

Early Identification and Pathogenesis of Leaf Diseases in *Magnifera Indica* Using Transfer Learning

Arshad Husain and Subhojit Bhowmick

1 Introduction

Horticulture is considered one of the main source of the economy in India. Like individuals, plants also experience the ill effects of infections which influence the normal growth of a plant. Diseases can be in any part of the plant including leaves, flowers, friuts and stems. Because of the complexity and the large number of crops and cultivated plants, the number of diseases is likewise large. Subsequently, a pathologist may often fail to analyze a disease precisely. The exact and timely finding of plant infections shields crops from quantitative and qualitative loss. Most cultivators have an absence of information about the effective identification of plant diseases. The recognizable plant illness by the naked eye is tedious, requires persistent checking, and is less precise. At Present, many works are ongoing in the field of technology, which can be utilized successfully in the field of crop health observation, the identification of sicknesses in plants, and so on. This type of framework gives solid, exact outcomes and decreases the time, cost, and labor for maintaining and guaranteeing quality real-time applications. In the field of farming, there are many opportunities for researchers to apply AI techniques from different perspectives, like the identifcation of plants, early detection of dieases, pesticides, nutrient requirements, and so on. In this work, it is considered certain specific diseases which occur on the leaves of the plant, such as Anthracnose, Bacterial Canker, Cutting Weevil, Die Back, Gall Midge, Powdery Mildew, and Sooty Mold. The DenseNet201, NasNet-Mobile, MobileNet, and VGG19 models have been used for the detection of diseases. The work is begun by first performing a literature review of existing approaches in this domain of research and gaps in the same are consequently identified. Later, the final output layers of the chosen models are modified to predict eight classes from

A. Husain (✉) · S. Bhowmick
Bennett University, Greater Noida, Uttar Pradesh, India
e-mail: arshad.husain18@gmail.com

M. O. Sarif and A. Sharifi (eds.), *Advanced Geospatial Intelligence and AI for Environmental Resilience and Sustainable Development*, Advances in Geographic Information Science, https://doi.org/10.1007/978-3-032-03714-5_9

the dataset. After testing on the dataset, 8×8 confusion matrices are created for the models. Finally, various performance-measuring techniques are used to determine the performance of the models.

A pre-trained Convolutional Neural Network (CNN) model called AlexNet was proposed, and the model was trained on 7222 grape leaf images and 1266 mango leaf images. The accuracy achieved in this study for identifying grape and mango leaf diseases was 99.03% and 89% respectively (Rao et al. 2021). In another study, a CNN based model was proposed which used FrCNet (Saleem et al. 2021a). The proposed model segmented the images with an accuracy of 99.2%. A segmentation technique based on the vein pattern of the leaf was also proposed (Saleem et al. 2021b). In the identification step, a cubic support vector machine is used. The proposed model gives an accuracy of 95.5%. A machine learning based system was proposed that predicts leaf diseases with an accuracy of 69% (Gining et al. 2021). A multilayer CNN architecture was proposed which classified mango leaves infected with Anthracnose. The accuracy achieved by this model was 97% (Singh et al. 2019). An Artificial Neural Network (ANN) technique was proposed that identified early diseases in plant leaves with small blobs of the diseases, which could be detected with high-resolution images (Pham et al. 2020). In addition, a Neural Network Ensemble (NNE) method for mango leaf disease prediction was proposed, which can classified the diseases with an average accuracy of 80% (Mia et al. 2020). A modified version of the VGGNet model, called as V2IncepNet which combines the best features of both VGGNet and the Inception module, was introduced (Venkatesh and Hegde 2020). The model proposed by the author detected Anthracnose disease on mango leaves with an accuracy of 92%.

2 Data Acquisition and Methodology

This section explains the parts of the proposed work: data acquisition, description of models, training, and testing of the dataset.

2.1 Dataset Acquisition

The dataset used in this study is the MangoLeafBD dataset, which is available on Mendeley and Kaggle. The dataset consists of 4000 images in total. Out of these, 1800 are distinct leaf images, while the remaining images were prepared by rotating and zooming. These images belong to eight classes namely, Anthracnose, Bacterial Canker, Cutting Weevil, Die Back, Gall Midge, Powdery Mildew, Sooty Mould, and Healthy Leaf images (Fig. 1).

Fig. 1 Disease in mango leaf from left to right, Anthracnose, Bacterial Canker, Cutting Weevil, Die Back, Gall Midge, Healthy, Powdery Mildew, Sooty Mould

2.2 Model Description

In this proposed work, the traditional way of constructing CNNs was replaced, and four well-known models were used for training the dataset. VGG19, DenseNet201, NASNetMobile, and MobileNet were employed. These models had been trained on datasets with 1000 classes. The output layer was modified to include 8 neurons, the dataset used in this study consisted of only 8 classes. Pre-trained models were utilized as a starting point for image identification in the transfer learning technique. These models were created for a task but were reused as starting point for different tasks. Such models were pre trained and weights had already assigned to each layer. In transfer learning, the output layers of the state-of-the-art models were generally modified according to requirements. In this study, the dataset consisted of eight classes. All of the models considered had 1000 neurons in output layer. This means that they had been trained on the dataset with 1000 classes. As the dataset used here consisted of only 8 classes, the output layer was modified such that it contains 8 neurons. VGG19 is a CNN with 19 layers. This model had initially been trained on more than a million images. These images were classified into 1000 classes. This model consisted of 16 convolutional layers, 3 Fully Connected Layers, 5 MaxPool Layers, and 1 SoftMax Layer. The input for VGG19 was 224 × 224 RGB images. DenseNet201 was Dense CNN with 201 layers. Each of these layer was connected to the next layer in a feed forward fashion. DenseNet201 had been trained to classify images into 1000 different classes. The input for DenseNet201 was a 224 × 224 RGB image. MobileNetV2 was a lightweight model developed to run on mobile devices. This model was 53 layers deep. This model took 224 × 224 RGB images as input. This model had been trained to classify images into 1000 different classes. For this study, the output layer of the model was modified to only eight neurons belonging to the eight classes in dataset.

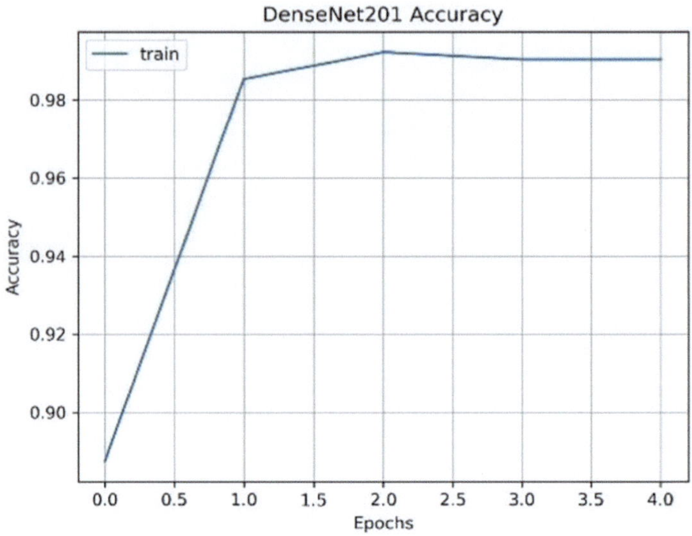

Fig. 2 DenseNet201 accuracy on each epoch

2.3 Training and Testing

The results of four algorithms, VGG19, DenseNet201, NASNetMobile, and MobileNet were studied. These algorithms were trained on 4000 images classified into 8 classes. The models were trained based on 5 epochs, meaning the algorithms iterated through and learned from the dataset 5 times. After training, the models were tested on the test dataset, and the accuracies achieved were recorded. The original dataset of 4000 images was divided into a ratio of 4:1 for training set and testing respectively. The training set included 3200 images, and the test set consisted of 800 images. An 8 × 8 confusion matrix was created for each model (Figs. 2, 3 and 4).

3 Result and Discussion

For the analysis of the algorithms studied, seven performance metrics were utilized. Accuracy, Precision, Sensitivity, Specificity, FNR, and FPR were considered for the performance metrics. The equations of the respective metrics were as follows.

$$\text{Accuracy} = ((\text{TP} + \text{TN})/(\text{TP} + \text{FP} + \text{FN} + \text{TN})) \times 100\% \quad (1)$$

$$\text{Precision} = (\text{TP}/(\text{TP} + \text{FP})) \times 100\% \quad (2)$$

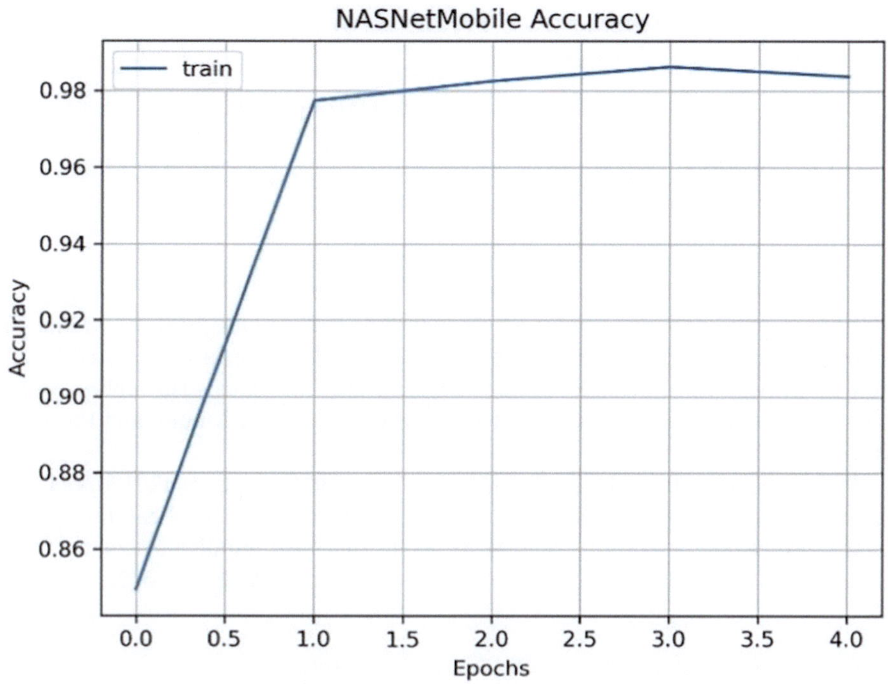

Fig. 3 NASNetMobile accuracy on each epoch

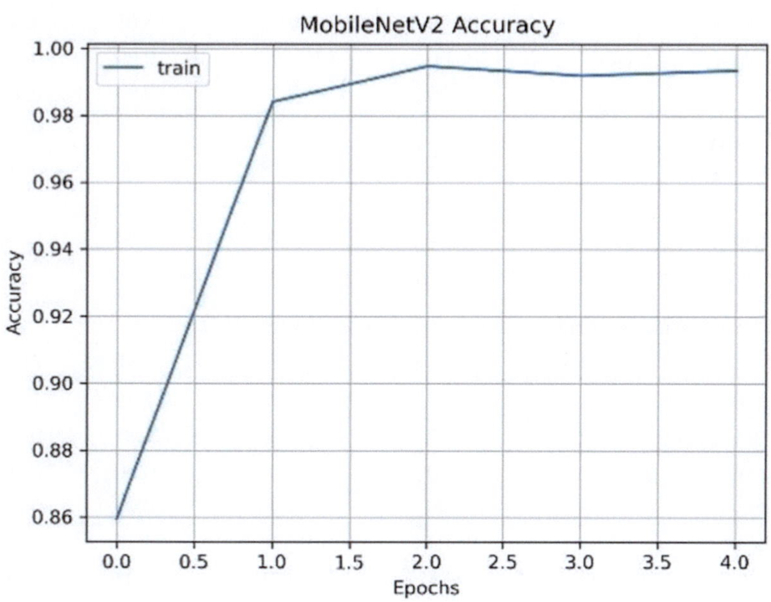

Fig. 4 MobileNetV2 accuracy on each epoch

$$\text{Sensitivity} = (\text{TP}/(\text{TP} + \text{FN})) \times 100\% \tag{3}$$

$$\text{Specificity} = (\text{TN}/(\text{FP} + \text{TN})) \times 100\% \tag{4}$$

$$\text{FNR} = (\text{FN}/(\text{TP} + \text{FN})) \times 100\% \tag{5}$$

$$\text{FPR} = (\text{FP}/(\text{FP} + \text{TN})) \times 100\% \tag{6}$$

Table 1 exhibits class wise performance matrices for the VGG19 Model. It achieved an overall accuracy of 99.02% on the mango leaf disease test dataset. Precision, Sensitivity, Specificity, FNR, and FPR were also calculated. The highest precision and sensitivity were achieved by Cutting Weevil, Sooty Mould, and Powdery Mildew.

Table 2 exhibits class wise performance matrices for the DenseNet201 Model. We can see that it achieved an overall accuracy of 97.00% on the mango leaf disease test dataset. Precision, Sensitivity, Specificity, FNR, and FPR are also calculated. The highest precision and sensitivity were achieved by Cutting Weevil and Die Back.

Table 1 Performance evaluation matrix for each class (VGG19)

Class	Precision (%)	Sensitivity (%)	Specificity (%)	FNR (%)	FPR (%)
Anthracnose	97.41	96.58	99.42	2.59	0.58
Bacterial Canker	96.97	100.00	100.00	3.03	0.00
Cutting Weevil	100.00	100.00	100.00	0.00	0.00
Die Back	100.00	96.70	99.58	0.00	0.42
Gall Midge	98.04	99.71	99.71	4.76	0.29
Healthy	94.29	99.14	99.14	0.00	0.86
Powdery Mildew	100.00	100.00	100.00	0.00	0.00
Sooty Mould	100.00	100.00	100.00	4.40	0.00

Table 2 Performance evaluation matrix for each class (DenseNet201)

Class	Precision (%)	Sensitivity (%)	Specificity (%)	FNR (%)	FPR (%)
Anthracnose	97.41	94.17	98.98	2.59	1.02
Bacterial Canker	98.99	98.99	99.86	1.01	0.14
Cutting Weevil	100.00	100.00	100.00	0.00	0.00
Die Back	100.00	100.00	100.00	0.00	0.00
Gall Midge	95.24	95.24	99.28	4.76	0.72
Healthy	97.98	97.98	99.71	2.02	0.29
Powdery Mildew	94.00	94.00	99.14	6.00	0.86
Sooty Mould	95.60	95.60	99.44	4.40	0.56

Table 3 Performance evaluation matrix for each class (NASNetMobile)

Class	Precision (%)	Sensitivity (%)	Specificity (%)	FNR (%)	FPR (%)
Anthracnose	94.83	94.02	98.98	5.17	1.02
Bacterial Canker	97.98	95.10	99.29	2.02	0.71
Cutting Weevil	100.00	100.00	100.00	0.00	0.00
Die Back	97.62	97.62	99.72	6.82	0.28
Gall Midge	91.35	98.71	98.71	9.52	1.29
Healthy	94.23	99.14	99.14	1.01	0.86
Powdery Mildew	96.00	96.00	99.43	4.00	0.57
Sooty Mould	95.70	99.44	99.44	2.20	0.56

Table 4 Performance evaluation matrix for each class (MobileNetV2)

Class	Precision (%)	Sensitivity (%)	Specificity (%)	FNR (%)	FPR (%)
Anthracnose	97.41	100.00	100.00	2.59	0.00
Bacterial Canker	100.00	100.00	100.00	0.00	0.00
Cutting Weevil	100.00	100.00	100.00	0.00	0.00
Die Back	98.86	94.57	99.30	1.14	0.70
Gall Midge	97.14	94.44	99.14	2.86	0.86
Healthy	97.98	97.98	99.71	2.02	0.29
Powdery Mildew	94.00	100.00	100.00	6.00	0.00
Sooty Mould	95.60	93.55	99.15	4.40	0.85

Table 3 exhibits the class wise performance matrices of NASNetMobile model. The model achieved an overall accuracy of 97.50% on the test data of infected mango leaf images. Precision and Sensitivity values were highest for Cutting Weevil disease.

In Table 4 we record the class wise performance matrices for the MobileNet model. This model was able to achieve highest accuracy among all models. Overall accuracy recorded by this model is 97.60%. Precision and Sensitivity values were maximum for Bacterial Canker and Cutting Weevil.

4 Conclusion

Mangifera Indica, widely known as king of fruits (Mango) contributes major portion of Indian food economics. It also plays vital role in occupation of various farmers in India and other countries specially during summers. Therefore, to improve the crop yield is important aspects in their economics. The leaves of these plants are vulnerable and depends upon various variables.

In present study, the transfer learning approaches were successfully applied to identify and cluster the diseases in mango leaves. Four pretrained CNN models, namely VGG19, DenseNet201, NASNetMobile, and MobileNet models were utilized on the ManoLeafBD dataset consisting of 4000 images categorized into seven unique and distinct classes of mango leaf diseases and one healthy class. Among the models, MobileNetV2 achieved the highest overall accuracy. The Highest overall accuracy of 99.02% was recorded by VGG19 model. In future, the study may focus on predicting more disease with this CNN technique and improve the accuracy for predicting the diseases.

References

Gining RAJM, Othman MA, Nasir NFM, Zakaria Z, Yusof MZM (2021) harumanis mango leaf disease recognition system using image processing technique. Indones J Electr Eng Comput Sci 23(1):378–386. https://doi.org/10.11591/ijeecs.v23.i1.pp378-386

Mia MdR, Roy S, Das SK, Rahman MdA (2020) Mango leaf disease recognition using neural network and support vector machine. Iran J Comput Sci 3(3):185–193. https://doi.org/10.1007/s42044-020-00057-z

Pham TN, Tran LV, Dao SVT (2020) Early disease classification of mango leaves using feed-forward neural network and hybrid metaheuristic feature selection. IEEE Access 8:189960–189973. https://doi.org/10.1109/ACCESS.2020.3031914

Rao US, Varalakshmi P, Jayanth NK (2021) Deep learning precision farming: grapes and mango leaf disease detection by transfer learning. Glob Transit Proc 2(2):535–544. https://doi.org/10.1016/j.gltp.2021.08.002

Saleem R, Shah JH, Sharif M, Ansari GJ (2021) Mango leaf disease identification using fully resolution convolutional network. Comput Mater Contin 69(3):3581–3601. https://doi.org/10.32604/cmc.2021.017700

Saleem R, Shah JH, Sharif M, Yasmin M, Yong HS, Cha J (2021) Mango leaf disease recognition and classification using novel segmentation and vein pattern technique. Appl Sci 11(24):11901. https://doi.org/10.3390/app112411901

Singh UP, Chouhan SS, Jain S, Jain S (2019) Multilayer convolution neural network for the classification of mango leaves infected by anthracnose disease. IEEE Access 7:43721–43729. https://doi.org/10.1109/ACCESS.2019.2907383

Venkatesh NY, Hegde SU (2020) Transfer learning-based convolutional neural network model for classification of mango leaves infected by Anthracnose. In: IEEE International conference for innovation in technology (INOCON), pp 1–7. https://doi.org/10.1109/INOCON50539.2020.9298269

Examining the Role of Geospatial Data and AI in Conducting Comprehensive Forestry Inventories and Mapping to Support Sustainable Forest Management

Manohar Yadav and Mukesh Kumar Verma

1 Introduction

Forestry inventories and mapping are essential for realizing the goals of Sustainable Forest Management (SFM), since they supply vital data required for the monitoring, assessment, and management of forest ecosystems. By methodically gathering and examining data regarding forest composition, structure, and dynamics, these tools facilitate informed decision-making to reconcile environmental, economic, and social priorities (Wolfslehner and Seidl 2010). Forest inventories provide quantitative data on tree species distribution, biomass, carbon stocks, and other ecological characteristics, which are crucial for assessing forest health and biodiversity (Knoke et al. 2021). Geospatial technology and data facilitate mapping, which generates precise representations of forest landscapes, so supporting land-use planning, conservation, and sustainable resource management (Sharma et al. 2024).

The amalgamation of sophisticated geospatial data with Artificial Intelligence (AI) has transformed the domain, facilitating more accurate and efficient forestry inventories and mapping. Technologies including Light Detection and Ranging (LiDAR), hyperspectral photography, and unmanned aerial vehicles (UAVs) yield

M. Yadav · M. K. Verma (✉)
Geographic information system (GIS) Cell, Motilal Nehru National Institute of Technology Allahabad, Prayagraj, Uttar Pradesh 211004, India
e-mail: mukeshverma02feb@gmail.com

M. Yadav
e-mail: ssmyadav@mnnit.ac.in

Present Address:
M. K. Verma
AIT-CSE, Chandigarh University, Mohali, Punjab 140413, India

M. O. Sarif and A. Sharifi (eds.), *Advanced Geospatial Intelligence and AI for Environmental Resilience and Sustainable Development*, Advances in Geographic Information Science, https://doi.org/10.1007/978-3-032-03714-5_10

Fig. 1 Challenges in traditional forestry inventory methods

high-resolution data, and AI-driven models enhance data processing and interpretation (Ali et al. 2024). These developments enable adaptive forest management, aiding projects such as carbon accounting, biodiversity conservation, and sustainable timber harvesting (Gardner 2010). Furthermore, they enable local communities by promoting participatory forest management and equitable resource allocation, consistent with the fundamental tenets of sustainable forest management (Ngome and Yeom 2024). This chapter examines the revolutionary capabilities of geospatial data and AI in executing thorough forestry inventories and mapping, highlighting its essential function in safeguarding forest resources for present and future generations. Figure 1 delineates the constraints of conventional forestry inventory techniques, highlighting their labor-intensive characteristics and restricted spatial reach.

Conventional forestry inventory techniques encounter numerous obstacles that impede their efficacy in delivering precise and comprehensive data for forest management. A key restriction is the time-consuming and labor-intensive nature of terrestrial data collection. Field surveys necessitate comprehensive manual measurements, which can be resource-demanding and susceptible to human error, particularly in thick or isolated forested regions (Luo et al. 2024). These conventional methods frequently yield restricted spatial coverage and infrequent data collection, complicating the ability to monitor the dynamic changes within forest ecosystems (Coppin et al. 2002). Figure 2 illustrates the challenges of data gathering, particularly the difficulty in predicting structural characteristics inside intricate forest ecosystems.

Moreover, conventional techniques face challenges in precisely predicting structural factors such as biomass and carbon stocks, which are essential for evaluating ecosystem services and their role in climate change mitigation (Sahle et al. 2018). Intricate forest features, including multi-tiered canopies, exacerbate data collection

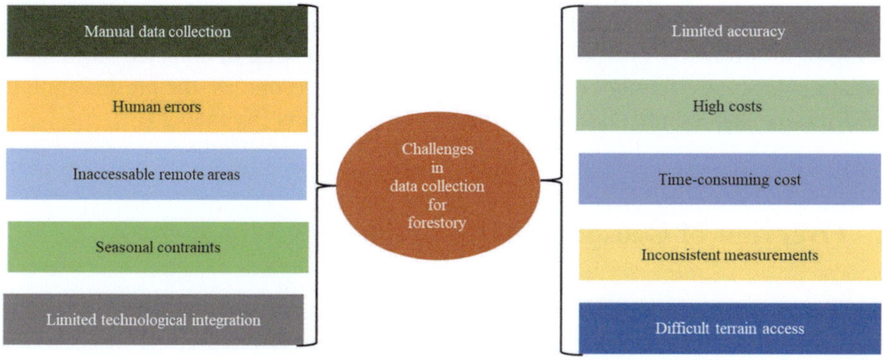

Fig. 2 Challenges in data collection for forestry

and processing challenges, resulting in mistakes in forest inventories (Yin and Wang 2016). Furthermore, the absence of real-time data capture constrains the capacity to properly monitor deforestation, forest degradation, or disease outbreaks.

A notable problem is the dependence on generalized allometric equations for predicting forest attributes, which may overlook species-specific or site-specific differences, hence impacting the accuracy of the inventory (Paul et al. 2013). Moreover, conventional methods are less adept at integrating varied datasets, such as remote sensing and socio-economic data, which are increasingly essential for comprehensive forest management strategies. Confronting these difficulties necessitates the implementation of sophisticated technology such as geospatial data and artificial intelligence to improve efficiency, precision, and scalability in forestry inventory methodologies.

This chapter seeks to examine the amalgamation of geospatial data and Artificial Intelligence (AI) for executing thorough forestry inventories and mapping, which are critical elements of Sustainable Forest Management (SFM). The swift progress in geospatial technologies, including LiDAR, hyperspectral imaging, and Unmanned Aerial Vehicles (UAVs), along with AI-driven analytics, has transformed forest data collection and analysis (Ali et al. 2024). This chapter aims to demonstrate how these tools provide more precise, economical, and scalable evaluations of forest characteristics, including biomass, species diversity, and carbon stocks. Additionally, it examines the capacity of AI to automate data processing, identify trends, and deliver predictive insights to improve decision-making in forest management (Buchelt et al. 2024).

This chapter also aims to critically assess the advantages and drawbacks of these advanced technologies in facilitating sustainable forest management (SFM). Although geospatial data and AI provide substantial benefits, like enhanced accuracy, diminished fieldwork, and real-time monitoring capabilities, their implementation is accompanied by obstacles. The factors encompass the substantial expense of equipment, the intricacy of data processing, and the requirement for technical proficiency (Coleman et al. 2016). This chapter seeks to provide a balanced perspective to assist academics, policymakers, and practitioners in the effective integration of new technologies into forestry operations. This chapter's ideas seek to enhance sustainable forest management through the optimal use of forest resources, the preservation of biodiversity, and the alleviation of climate change effects (Keenan 2015).

2 Overview of Geospatial Data in Forestry Management

Geospatial data is essential in forestry management, facilitating effective monitoring, analysis, and decision-making. These data sets offer essential insights into forest cover, biomass, biodiversity, and general forest health, which are critical for sustainable forest management methods. The progression of geospatial technology has markedly enhanced forestry applications, enabling more precise and prompt evaluations of forest resources. Geospatial data in forestry is often obtained from three primary sources: satellite images, aerial data, and ground-based technologies, each providing distinct benefits for forest monitoring and management. Figure 3 delineates the contributions of geospatial technologies such as LiDAR, UAVs, and satellite imaging to sustainable forest management.

Satellite-based remote sensing offers extensive data and is extensively utilized for land cover mapping, forest change monitoring, and forest inventory assessments. Optical imagery from platforms such as Landsat and Sentinel-2 provides significant

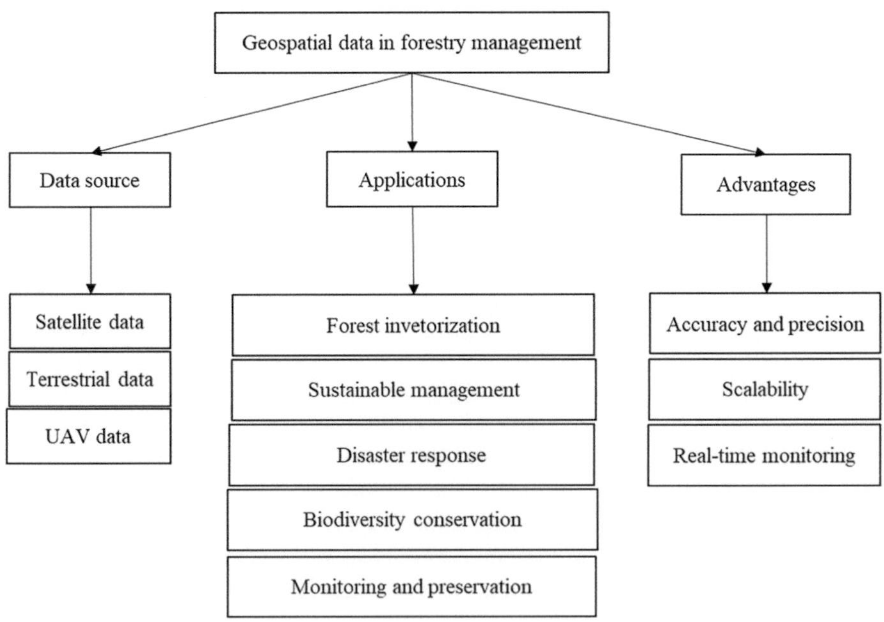

Fig. 3 Geospatial data in forest management

insights into vegetation monitoring, land cover classification, and biomass estimation. These optical sensors detect visible and near-infrared light, facilitating comprehensive evaluations of forest structure, species distribution, and deforestation (Almeida et al. 2025). Synthetic Aperture Radar (SAR) data, exemplified by Sentinel-1, is particularly adept in forest monitoring in all weather circumstances due to its ability to penetrate cloud cover. SAR data is utilized to evaluate forest structure, tree height, and biomass estimation (Mastro et al. 2022). Moreover, hyperspectral imagery offers comprehensive spectral data across an extensive range of wavelengths, facilitating vegetation categorization, species identification, and the detection of specific plant health indicators. This information is crucial for biodiversity evaluations, facilitating the monitoring of forest vitality and species variety (Kacic and Kuenzer 2022).

Airborne data, encompassing LiDAR and UAVs (Unmanned Aerial Vehicles or drones), provide high-resolution information, markedly enhancing the precision of forest assessments. LiDAR technologies, commonly employed in aerial surveys, generate high-resolution digital elevation models (DEMs), quantify canopy heights, and evaluate forest structure (Stone et al. 2016). LiDAR is exceptionally proficient at assessing tree height, crown volume, and canopy density, serving as an essential instrument for comprehensive forest inventory mapping. Unmanned Aerial Vehicles (UAVs), outfitted with diverse sensors including optical, thermal, and multispectral, are progressively employed for forest surveillance, acquiring high-resolution imagery that offers intricate perspectives of forest regions. Drones are highly effective for assessing forest health, tree canopy status, and species diversity, particularly in inaccessible remote regions (Ecke et al. 2022).

Terrestrial systems augment distant sensing technology by delivering precise measurements. Global Navigation Satellite Systems (GNSS) technologies are extensively utilized in field surveys to get accurate geographical coordinates, crucial for georeferencing and mapping forested regions. GNSS is essential for performing forest inventories, delineating forest boundaries, and calibrating aerial and satellite data (Keefe et al. 2019). Terrestrial Laser Scanning (TLS) provides precise three-dimensional assessments of forest structures at ground level. TLS is employed to assess tree heights, canopy volume, and several morphological features, offering an in-depth comprehension of forest structure (Arrizza et al. 2024).

Geospatial data for forestry management is sourced from diverse platforms, including open-access and corporate services, each providing varying degrees of accessibility and quality. Open-access systems like Landsat and Sentinel offer complimentary access to extensive satellite data, facilitating forestry experts' acquisition of essential resources without substantial expenses. Landsat offers optical pictures with a spatial resolution of 30 m, making it especially valuable for the long-term observation of forest alterations. Sentinel satellites provide optical (Sentinel-2) and SAR (Sentinel-1) data with enhanced spatial resolution and increased revisit frequency, rendering them optimal for forestry applications (Mngadi et al. 2021). Google Earth Engine (GEE) improves accessibility by providing a cloud-based platform that grants access to a vast array of geospatial information, including Landsat and Sentinel data, facilitating large-scale forest studies (Vijayakumar et al. 2024). Proprietary data services, provided by private entities such as DigitalGlobe and

Airbus, deliver high-resolution data (sub-meter to 1 m) with swift acquisition times, frequently customized for particular forest management requirements. These services generally incur higher costs than open-access platforms but provide substantial benefits in data resolution, customization, and processing capabilities, which are essential for operational planning, forest resource distribution, and conservation initiatives.

Geospatial data has multiple applications in forestry inventories, facilitating various phases of forest management. A principal application is land cover classification, facilitating the mapping of forested and non-forested regions. Satellite photography, particularly from Landsat and Sentinel, is frequently employed to categorize land cover types and observe deforestation and forest degradation (Hoekman et al. 2020). The analysis of canopy density and forest structure is a vital application in which LiDAR and UAV data are utilized to evaluate forest canopy height, volume, and density, yielding crucial insights on biomass, carbon storage, and overall forest health (Camarretta et al. 2020). Finally, species identification and biodiversity assessment are significantly improved through the utilization of hyperspectral and multispectral data from UAVs, allowing forest managers to monitor species distributions and evaluate forest biodiversity. This aids in identifying possible risks to biodiversity, including disease outbreaks or invasive species (Dainelli et al. 2021).

This table delineates the principal methodologies employed in forestry management, along with their corresponding applications and sources. The amalgamation of these geospatial technologies offers a holistic method for forest monitoring, guaranteeing sustainable management practices and the preservation of biodiversity. Table 1 delineates geographic data types, platforms, and applications pertinent to forest management.

3 Role of Artificial Intelligence in Forestry Applications

Artificial intelligence (AI) is transforming forestry by enhancing the efficiency, precision, and scalability of forest monitoring and management techniques. The amalgamation of AI methodologies with geospatial data has facilitated sophisticated instruments for forest resource management, biodiversity preservation, and climate change alleviation. These AI-driven methodologies possess the capacity to augment decision-making in forestry, rendering it more data-centric and anticipatory. Prominent AI approaches utilized in forestry include machine learning (ML) and deep learning (DL), both of which provide significant benefits for analyzing extensive datasets and deriving valuable insights.

Machine learning (ML) algorithms, including Random Forests and Support Vector Machines (SVMs), have been widely employed in forestry for tasks such as classification, regression, and anomaly detection. Random Forest, a multifaceted ensemble learning technique, is very effective for land cover categorization, vegetation mapping, and tree species identification. This approach utilizes numerous decision trees to manage extensive, high-dimensional datasets and addresses intricate

Table 1 Overview of geospatial methods for forestry management

Method	Data type	Platform	Application	Citation
Optical imagery	Satellite	Landsat, Sentinel-2	Land cover classification, forest health	Soleimannejad et al. (2019,)
SAR	Satellite	Sentinel-1	Forest structure, biomass estimation	Liu and Sorin (2022)
Hyperspectral	Satellite	EO-1 Hyperion, UAV	Species identification, plant health monitoring	Zhang et al. (2020)
LiDAR	Airborne	Airborne LiDAR, UAV	Canopy height, biomass estimation, forest structure	Liu et al. (2018)
UAVs	Airborne	Drones	High-resolution imagery for forest health and structure	Dash et al. (2017)
GNSS	Ground-based	GPS systems	Field survey mapping, tree measurements	Oveland et al. (2018)
TLS	Ground-based	Terrestrial scanners	3D forest structure, tree height estimation	Vaglio Laurin et al. (2019)

interactions among variables. Support Vector Machines (SVMs) are utilized for classifying forest cover types, differentiating between forested and non-forested regions, and recognizing specific tree species using satellite or UAV imagery (Mountrakis and Heydari 2024). Support Vector Machines (SVMs) are especially advantageous in scenarios when the dataset is not linearly separable, as they can discern nonlinear boundaries between classes. Figure 4 presents a summary of artificial intelligence applications, including machine learning and deep learning, in forestry for functions such as tree species identification and biomass calculation.

Conversely, Deep Learning (DL) methodologies, including Convolutional Neural Networks (CNNs) and Transformers, have surfaced as premier strategies for intricate picture analysis and feature extraction. Convolutional Neural Networks (CNNs), due to their capacity to autonomously learn hierarchical features from images, have been

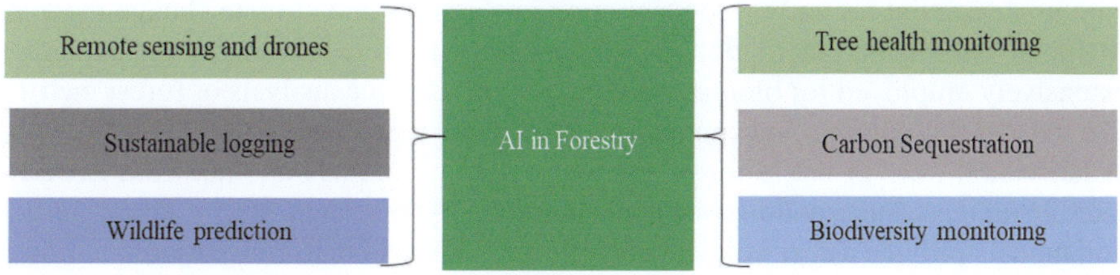

Fig. 4 AI techniques for forestry applications

effectively utilized in remote sensing data, including satellite and drone imaging, for tasks such as land cover categorization, tree species identification, and forest health assessment (Zhang et al. 2020). Convolutional Neural Networks (CNNs) excel in analyzing high-resolution images and identifying nuanced patterns that conventional machine learning algorithms can overlook. Transformers, a contemporary design utilized in natural language processing, have demonstrated potential in geospatial applications owing to their capacity to represent long-range dependencies and manage sequential data proficiently. These designs are especially effective for evaluating time-series data, including multitemporal satellite images, to observe forest changes and dynamics (Nguyen et al. 2018).

The amalgamation of AI with geographical data provides formidable capabilities, especially in object detection and feature segmentation. Object detection for the classification of tree species is a prominent application. Deep learning methods, such as CNNs, enable AI to precisely identify and categorize individual tree species from high-resolution aerial data, aiding in biodiversity evaluations and forest inventory management (Zhong et al. 2024). AI-driven object detection models can efficiently analyze extensive image datasets and precisely identify and categorize tree species, a task that would be arduous and time-consuming with conventional methods. Furthermore, the precision of AI-driven object detection is consistently enhancing, attributable to the accessibility of extensive labeled datasets and the advancement of increasingly complex models.

A notable contribution of AI to forestry is the automated segmentation of forest characteristics from point clouds generated by technologies such as LiDAR. Point clouds encompass extensive three-dimensional data regarding forest structure, such as tree heights, canopy density, and biomass. AI systems, especially Deep Learning models, may partition these point clouds into significant entities such as distinct trees, branches, and trunks. This automated segmentation significantly decreases the manual labor needed for point cloud analysis and delivers precise measurements of forest characteristics. Through the training of models on annotated LiDAR point cloud data, artificial intelligence can identify and delineate tree borders, allowing forest managers to achieve more accurate forest inventories (Liu et al. 2018). Deep learning is being utilized for point cloud segmentation in applications including biomass estimation, carbon stock analysis, and forest health monitoring.

AI-driven predictive analytics is essential for enhancing forestry applications, especially in biomass estimation, carbon stock assessment, and growth pattern forecasting. A significant difficulty in forest management is the estimation of tree biomass, essential for carbon sequestration evaluations and climate change models. Artificial intelligence methodologies, especially machine learning algorithms, are extensively employed for biomass assessment through the analysis of forest metrics like tree height, volume, and canopy structure. AI can precisely forecast the biomass of individual trees or entire forest stands by training models on data from LiDAR, satellite images, and ground-based measurements (Ahmad et al. 2021). These prediction models provide a cost-efficient and effective alternative to conventional biomass estimation techniques, which frequently depend on labor-intensive field surveys.

Likewise, AI-based carbon stock analysis utilizes remote sensing data and machine learning algorithms to quantify the carbon sequestered in forests. Artificial intelligence can utilize forest structural data, like tree diameter, height, and species mix, to forecast carbon stocks and track carbon sequestration over time. These predictive models are essential for assessing the efficacy of forest conservation and afforestation initiatives in alleviating climate change (Rahul et al. 2025).

Besides biomass and carbon stock estimation, AI is utilized for predicting growth patterns and forecasting yields in forestry. Through the examination of historical data regarding forest growth, climatic circumstances, and soil characteristics, AI models can forecast forest development under various environmental scenarios. This forecasting ability is crucial for forest management strategies, including thinning, harvesting, and reforestation planning (Taylor et al. 2009). Additionally, AI models can predict timber yield, assisting forest managers in making data-informed decisions regarding harvesting timelines and resource distribution.

In essence, artificial intelligence possesses the capacity to revolutionize forestry operations through improved data processing, feature extraction, and predictive analytics. The amalgamation of machine learning and deep learning models with geospatial data is enhancing the precision, automation, and scalability of forestry applications. The evolving field of AI is anticipated to enhance its involvement in forestry management, providing creative solutions for biodiversity protection, sustainable resource management, and climate change mitigation.

4 Comprehensive Forestry Inventories: Methodologies and Workflow

4.1 Traditional Versus Modern Approaches

Forestry inventories have seen substantial evolution over the decades, shifting from conventional manual field surveys to advanced approaches that incorporate geospatial technologies and artificial intelligence (AI). Conventional methods, such as manual surveys, entail direct field measurements of arboreal dimensions, species classification, and forest composition. Although these procedures yield excellent precision at localized levels, they are laborious, time-consuming, and frequently unfeasible for extensive inventories. Moreover, manual surveys are susceptible to observer bias and inconsistencies, particularly when performed across varied terrains and isolated areas (Morrison 2016).

Conversely, geospatial and AI-driven approaches provide unmatched benefits in terms of efficiency and scalability. These methodologies include sophisticated instruments such remote sensing, LiDAR, and UAVs, integrated with machine learning algorithms for feature extraction, categorization, and estimate. The amalgamation of geospatial data with artificial intelligence methodologies facilitates high-resolution mapping, real-time analysis, and precise forecasting of forest characteristics, including biomass and carbon sequestration capacity (Wang et al. 2025). Furthermore, contemporary technologies facilitate the evaluation of remote and ecologically delicate regions without disrupting the natural ecosystem.

4.2 Workflow for Comprehensive Inventories

The process for executing thorough forestry inventories includes several essential phases that guarantee precise and effective evaluation of forest resources. The initial phase, data collecting, necessitates careful planning to choose suitable tools and procedures tailored to the unique attributes of the study area. Frequently utilized technologies comprise LiDAR sensors for structural data acquisition, UAVs outfitted with multispectral cameras for spectral analysis, and GNSS systems for accurate georeferencing. Upon data collection, preprocessing is conducted to rectify problems such as noise resulting from sensor constraints and ambient factors. Methods such as statistical filtering and interpolation enhance data quality, while the integration of datasets from diverse sources—such as LiDAR and optical imagery—provides a more comprehensive depiction of forest characteristics (Kellndorfer et al. 2010). During the analysis phase, essential forest metrics such as tree height, canopy density, and species composition are derived through sophisticated image processing and machine learning methodologies. Artificial intelligence models assist in categorizing forest elements, therefore facilitating biodiversity evaluations and resource assessments (Shivaprakash et al. 2022). Moreover, biophysical factors like biomass and carbon storage are assessed using statistical regression models and deep learning algorithms, enhancing the comprehension of forest dynamics and ecosystem services.

4.3 Case Studies

An exemplary instance of regional inventory employing sophisticated technologies is a study in the Amazon basin, where LiDAR technology was combined with artificial intelligence models to estimate tree biomass and delineate canopy structures. This regional forestry inventory illustrated the efficacy of integrating high-resolution LiDAR data with neural networks, attaining an exceptional 95% accuracy in biomass prediction, thus confirming the viability of AI-enhanced LiDAR applications for extensive forest evaluations (Lister et al. 2020). Likewise, another study aimed at mapping biodiversity hotspots in Southeast Asia utilized UAVs outfitted with multispectral sensors to gather comprehensive biological data. Deep learning algorithms were employed to analyze this data, facilitating precise classification of vegetation types and identification of endangered species. This method highlighted the capacity of UAVs and AI to monitor environmentally sensitive areas with high accuracy while reducing environmental disruption.

5 Mapping for Sustainable Forest Management

Mapping is essential for advancing Sustainable Forest Management (SFM) by providing accurate, data-driven insights into forest ecosystems (Khaiter and Erechtchoukova 2022). Advanced mapping techniques enable the identification of regions necessitating conservation, resource management, and restoration. The integration of GIS technologies promotes forest inventory procedures while also facilitating biodiversity conservation and climate change mitigation. Sustainable management is attained by integrating mapping outputs with scientific, social, and policy frameworks to harmonize ecological integrity and human requirements.

5.1 Importance of Forest Mapping in SFM

Forest mapping is essential to sustainable forest management as it offers instruments for tracking deforestation and degradation. High-resolution images and geospatial analytics monitor alterations in forest cover, aiding in the mitigation of illicit logging, habitat degradation, and carbon stock depletion (Pandey et al. 2022). Furthermore, mapping facilitates zoning for conservation and resource management, designating areas of ecological importance for protection, while establishing sustainable extraction zones to mitigate environmental impacts. This zoning facilitates a balance between resource utilization and ecosystem conservation, promoting long-term forest sustainability.

5.2 Innovations in Mapping Techniques

Technological improvements have transformed forest mapping, especially through the application of 3D models for canopy and terrain research (Dassot et al. 2011). These models, obtained from LiDAR and photogrammetry, furnish comprehensive structural data, facilitating precise evaluations of biomass, carbon sequestration, and topographical effects on forest development. Furthermore, time-series analysis for forest change detection utilizes satellite imagery and remote sensing datasets to monitor temporal variations in forest cover (Decuyper et al. 2022). This technique identifies trends of deforestation and afforestation, offering vital insights for adaptive management strategies.

5.3 AI-Driven Decision Support Systems

The incorporation of artificial intelligence (AI) in forest mapping has resulted in the creation of real-time monitoring dashboards that amalgamate spatial data, sensor inputs, and machine learning algorithms for immediate evaluations of forest health (Ali et al. 2025). These technologies provide proactive measures against deforestation, wildfires, or pest infestations. Moreover, AI-driven predictive instruments for policy formulation and catastrophe management improve decision-making by modeling diverse scenarios, anticipating results, and optimizing resource distribution. These advances substantially advance the attainment of SFM objectives by facilitating prompt, informed, and efficient management interventions.

6 Challenges and Limitations

The incorporation of advanced geospatial technology and Artificial Intelligence (AI) has transformed forest mapping and sustainable management; yet, various hurdles and constraints hinder the extensive adoption and efficacy of these approaches (Katharria et al. 2025). These difficulties are classified into technical, socio-economic, policy, ethical, and environmental categories. Resolving these concerns is essential for guaranteeing the effective, fair, and sustainable utilization of geospatial tools and technologies.

6.1 Technical Challenges

A major technological problem in forest mapping is the quality and resolution of data. High-resolution spatial data is essential for accurate study of forest characteristics, including canopy density, species variety, and topographical aspects (Wulder et al. 1996). Acquiring such data necessitates costly sensors, sophisticated equipment such as LiDAR, and continuous satellite coverage. Fluctuations in data quality stemming from meteorological factors, sensor constraints, or inconsistencies in spatial resolution can lead to erroneous interpretations and impede decision-making (Xu et al. 2023). Moreover, the amalgamation of data from several sources might intensify discrepancies, complicating the establishment of a cohesive dataset for analysis.

A significant concern is the interoperability of tools and systems. Geospatial technologies encompass various software, hardware, and analytical methodologies, each characterized by distinct standards and formats (Breunig et al. 2020). Guaranteeing interoperability and effortless data exchange across many platforms continues to be a challenging endeavor. Integrating outputs from remote sensing instruments with AI-driven predictive models frequently necessitates substantial customization and technical proficiency. The absence of standardization hinders collaboration and scalability, restricting the extensive utilization of geospatial technology in forest management.

6.2 *Socio-Economic and Policy Barriers*

The expense and availability of technology provide considerable socio-economic obstacles, especially in low-income and developing areas (Kanagawa and Nakata 2008). Premium geospatial tools, sensors, and processing software frequently necessitate significant financial inputs, rendering them unattainable for small-scale forestry projects, local communities, and non-governmental groups. Despite the availability of such technologies, ongoing expenses associated with software updates, data collecting, and maintenance further limit their utilization (Taylor et al. 2020).

The necessity for a proficient personnel and training constitutes a significant constraint. The successful implementation of sophisticated technologies necessitates experts skilled in geospatial analysis, remote sensing, AI integration, and data interpretation (Dritsas and Trigka 2025). Numerous regions exhibit a deficiency in training programs and institutional capacity to cultivate these skills. This disparity frequently results in the underutilization of current technologies or dependence on external specialists, which is not a viable solution for sustainable long-term forest management.

Policy obstacles also impede the efficient implementation of forest mapping technologies. Insufficient legislative frameworks, restricted governmental backing, and bureaucratic inefficiencies may hinder the execution of technology-driven solutions. Policymakers frequently possess an inadequate comprehension of the capabilities of geospatial tools, leading to insufficient financial support and strategic planning.

6.3 *Ethical and Environmental Concerns*

With the growing dependence of geospatial technologies on AI, data privacy and security have surfaced as critical ethical issues. The acquisition and analysis of geographical data frequently encompass sensitive information pertaining to land ownership, resource utilization, and indigenous territory (Olson et al. 2016). In the absence of stringent privacy policies and secure data storage systems, there exists a danger of misuse, unauthorized access, or data breaches. Moreover, AI models employed in forest management may unintentionally perpetuate biases inherent in the training data, resulting in an inequitable allocation of resources or erroneous decision-making.

The possible environmental effects of geospatial tools also require consideration (Satapathy et al. 2008). The utilization of drones, satellites, and other remote sensing technology requires substantial energy, hence contributing to carbon emissions. Moreover, the disposal of electrical waste from obsolete sensors and equipment might adversely affect nearby ecosystems. Although these technologies are vital for advancing sustainability, their manufacturing, utilization, and disposal require meticulous management to mitigate adverse environmental impacts.

There is an increasing worry regarding the ethical ramifications of implementing modern technologies in populations reliant on forests (Hill 2002). These tribes frequently possess profound cultural and spiritual links to their forests, which may conflict with the data-centric methodologies of GIS tools. Harmonizing technology interventions with reverence for traditional knowledge and traditions is crucial for cultivating trust and collaboration among stakeholders.

6.4 Addressing the Challenges

A multifaceted strategy is required to address these difficulties. Investment in research and development can enhance the quality and accessibility of geospatial tools while fostering interoperability through defined frameworks. Governments and international organizations must prioritize financing for capacity-building efforts, ensuring that local communities and institutions with the capabilities to utilize innovative technologies effectively (Abiodun 1998). Comprehensive ethical rules and stringent privacy policies must be instituted to safeguard sensitive data and avert misuse.

Furthermore, including renewable energy sources into the functionality of geospatial tools can diminish their ecological impact. Cooperative initiatives among technology developers, legislators, and local communities can synchronize technological progress with socio-cultural and environmental imperatives. By tackling these obstacles, geospatial technologies can significantly contribute to sustainable forest management and the alleviation of global environmental issues (Clark 2007).

7 Future Perspectives and Recommendations

The amalgamation of innovative technologies with sustainable forest management (SFM) possesses significant potential to tackle urgent worldwide environmental issues (Valente et al. 2015). Emerging technological developments, policy changes, and inclusive practices will be crucial in determining the future of forest conservation and resource management.

7.1 Emerging Trends in Technology

Edge computing and cloud integration are poised to revolutionize geospatial analytics for SFM. Edge computing facilitates data processing near the source, such as sensors or drones, hence minimizing latency and enabling instantaneous decision-making.

This functionality is especially advantageous for assessing forest vitality and identifying hazards such as illicit logging or wildfires (Pandey et al. 2022). In conjunction with cloud integration, edge computing enables effortless data sharing, storage, and analysis across platforms, hence improving collaboration and scalability.

A notable advancement is the promise of quantum computing in geospatial analytics. Quantum computing, with its unmatched processing capability, can evaluate extensive datasets more effectively than conventional techniques (Pandey et al. 2022). This advancement may facilitate precise simulations of forest ecosystems, enhance resource allocation, and refine forecast models for climate and biodiversity effects. Although in its early development, quantum computing has the potential to transform geospatial analytics and provide solutions to intricate forest management issues.

7.2 *Policy and Collaboration Needs*

To leverage the promise of emerging technologies, a conducive policy environment and cooperative strategies are necessary. Promoting open data and technology sharing can democratize access to sophisticated technologies and stimulate innovation (Gao 2022). Governments, academic institutions, and business entities should collaborate to establish accessible repositories of geospatial data for all stakeholders, thereby promoting openness and inclusion in forest management initiatives.

Moreover, public–private partnerships for extensive projects might facilitate the execution of technology-driven solutions. These collaborations can consolidate resources, experience, and financing to advance and implement geospatial technologies on a broader scale. Collaborations between technology companies and forestry departments can yield cost-effective solutions customized to local requirements, while also guaranteeing widespread adoption and sustainable maintenance of these systems.

7.3 *Vision for Sustainable Forest Management*

Realizing a holistic vision for SFM necessitates a combination of technological advancement and human-centered methodologies (Ystgaard and Moor 2023). Incorporating community involvement is essential for guaranteeing that forest management techniques are just and efficient. Local populations, especially those reliant on forests, possess significant traditional knowledge that can enhance technical insights. Involving these communities in mapping, monitoring, and decision-making processes empowers them and promotes sustainable practices grounded on cultural and ecological understanding.

Utilizing technology for international environmental objectives is fundamental to future sustainable forest management activities. Technologies such as artificial intelligence, remote sensing, and geospatial analytics might facilitate the attainment of the United Nations Sustainable Development Goals (SDGs), especially those pertaining to climate action, biodiversity preservation, and sustainable resource utilization (Ziemba et al. 2024). AI-driven predictive models can enhance reforestation initiatives, whereas remote sensing can monitor advancements in diminishing deforestation rates.

In summary, the future of SFM depends on the harmonious combination of advanced technology, cooperative policy frameworks, and community engagement. By cultivating innovation, advancing inclusivity, and synchronizing initiatives with global environmental goals, we can guarantee the sustainable stewardship of forests for future generations.

8 Conclusion

The amalgamation of geospatial data with Artificial Intelligence (AI) in forestry has transformed the methodologies for monitoring, managing, and conserving forests. These technologies offer unparalleled accuracy and efficiency in forestry inventories, facilitating the evaluation of critical factors such as canopy density, species richness, and biomass. Utilizing tools such as remote sensing, LiDAR, and AI-driven analytics, forest managers can obtain actionable insights to combat deforestation, habitat degradation, and climate change. The capabilities of these technologies surpass inventory management, playing a crucial role in policy formulation, disaster response, and sustainable resource management.

The progress in geospatial data and artificial intelligence highlights the necessity of interdisciplinary collaboration to attain Sustainable Forest Management objectives. Effective forest management necessitates the involvement of various stakeholders, including scientists, policymakers, local communities, and technology developers. By cultivating collaborations among various entities, the potential of breakthrough technologies can be enhanced, facilitating their wider implementation and adoption. Collaboration between AI researchers and ecologists can enhance predictive models, while relationships with local populations can incorporate traditional knowledge into technology frameworks. Moreover, public–private partnerships and open-data efforts help democratize access to sophisticated technologies and stimulate innovation in the forestry sector.

Integrating technology innovations with sustainable practices is essential for the future of forest management. Although technologies like AI and GIS analytics provide significant advantages, their use must conform to ecological, social, and ethical standards. Data privacy and security must be addressed to safeguard sensitive information regarding indigenous territories and natural resources. Likewise, the ecological impact of these technologies, encompassing energy usage and electronic waste, must be mitigated through the implementation of sustainable practices and renewable energy sources.

Sustainability necessitates a people-centric approach, acknowledging the crucial role of local communities in forest protection. Communities reliant on forests frequently possess profound cultural and spiritual attachments to their surroundings, with traditional knowledge that enhances scientific approaches. Involving these communities in mapping, monitoring, and decision-making processes guarantees that forest management plans are fair and culturally attuned. This method not only empowers communities but also improves the sustainability and efficacy of conservation initiatives.

The integration of geospatial data and artificial intelligence with forestry practices signifies a pivotal advancement in forest management. These technologies provide the necessary instruments to tackle urgent global issues such as deforestation, climate change, and biodiversity decline. Nonetheless, their complete potential can solely be actualized through interdisciplinary collaboration, inclusive policymaking, and a dedication to sustainable practices. By integrating technical progress with environmental conservation and community involvement, we can guarantee that forests persist as essential ecosystems and resources for future generations.

The pursuit of SFM presents hurdles; but, through a unified vision, cutting-edge technologies, and collaborative efforts, achieving a harmonious equilibrium between conservation and development is feasible. Forests are not merely resources; they are essential to the vitality of our planet. Maintaining them through sustainable practices and technology advancements is both a necessity and a responsibility we owe to ourselves and future generations.

References

Abiodun AA (1998) Human and institutional capacity building and utilization in science and technology in Africa: An appraisal of Africa's performance to-date and the way forward. Afr Dev Rev 10(1):10–51. https://doi.org/10.1111/j.1467-8268.1998.tb00095.x

Ahmad A, Gilani H, Ahmad SR (2021) Forest aboveground biomass estimation and mapping through high-resolution optical satellite imagery—a literature review. Forests 12(7):914. https://doi.org/10.3390/f12070914

Ali F et al (2024) Spectral intelligence: AI-driven hyperspectral imaging for agricultural and ecosystem applications. Agronomy 14(10):2260. https://doi.org/10.3390/agronomy14102260

Ali G, Mijwil MM, Adamopoulos I, Ayad J (2025) Leveraging the internet of things, remote sensing, and artificial intelligence for sustainable forest management. Babylonian J Internet Things 2025:1–65. https://doi.org/10.58496/BJIoT/2025/001

Arrizza S, Marras S, Ferrara R, Pellizzaro G (2024) Terrestrial laser scanning (TLS) for tree structure studies: a review of methods for wood-leaf classifications from 3D point clouds. Remote Sens Appl 36:101364. https://doi.org/10.1016/j.rsase.2024.101364

Breunig M et al (2020) Geospatial data management research: progress and future directions. ISPRS Int J Geoinf 9(2):95. https://doi.org/10.3390/ijgi9020095

Buchelt A et al (2024) Exploring artificial intelligence for applications of drones in forest ecology and management. For Ecol Manage 551:121530. https://doi.org/10.1016/j.foreco.2023.121530

Camarretta N et al (2020) Monitoring forest structure to guide adaptive management of forest restoration: a review of remote sensing approaches. New For (Dordr) 51(4):573–596. https://doi.org/10.1007/s11056-019-09754-5

Clark WW (2007) Partnerships in creating agile sustainable development communities. J Clean Prod 15(3):294–302. https://doi.org/10.1016/j.jclepro.2006.02.008

Coleman S, Göb R, Manco G, Pievatolo A, Tort-Martorell X, Reis MS (2016) How can SMEs benefit from big data? Challenges and a path forward. Qual Reliab Eng Int 32(6):2151–2164. https://doi.org/10.1002/qre.2008

Coppin P, Lambin E, Jonckheere I, Muys B (2002) Digital change detection methods in natural ecosystem monitoring: a review. In: Analysis of multi-temporal remote sensing images. World Scientific, pp 3–36. https://doi.org/10.1142/9789812777249_0001

Dainelli R, Toscano P, Di Gennaro SF, Matese A (2021) Recent advances in unmanned aerial vehicles forest remote sensing—a systematic review. Part II: research applications. Forests 12(4):397. https://doi.org/10.3390/f12040397

Dash JP, Watt MS, Pearse GD, Heaphy M, Dungey HS (2017) Assessing very high resolution UAV imagery for monitoring forest health during a simulated disease outbreak. ISPRS J Photogramm Remote Sens 131:1–14. https://doi.org/10.1016/j.isprsjprs.2017.07.007

Dassot M, Constant T, Fournier M (2011) The use of terrestrial LiDAR technology in forest science: application fields, benefits and challenges. Ann For Sci 68(5):959–974. https://doi.org/10.1007/s13595-011-0102-2

de Almeida DRA et al (2025) Remote sensing approaches to monitor tropical forest restoration: current methods and future possibilities. J Appl Ecol 62(2):188–206. https://doi.org/10.1111/1365-2664.14830

Decuyper M et al (2022) Continuous monitoring of forest change dynamics with satellite time series. Remote Sens Environ 269:112829. https://doi.org/10.1016/j.rse.2021.112829

Dritsas E, Trigka M (2025) Remote sensing and geospatial analysis in the big data era: a survey. Remote Sens (Basel) 17(3):550. https://doi.org/10.3390/rs17030550

Ecke S et al (2022) UAV-based forest health monitoring: a systematic review. Remote Sens (Basel) 14(13):3205. https://doi.org/10.3390/rs14133205

Gao (2022) Spatial scientometrics. In: Encyclopedia of big data. Springer, Cham, pp 872–874. https://doi.org/10.1007/978-3-319-32010-6_493

Gardner T (2010) Monitoring forest biodiversity. Routledge, London. https://doi.org/10.4324/9781849775106

Hill CM (2002) Primate conservation and local communities—ethical issues and debates. Am Anthropol 104(4):1184–1194. https://doi.org/10.1525/aa.2002.104.4.1184

Hoekman D et al (2020) Wide-area near-real-time monitoring of tropical forest degradation and deforestation using Sentinel-1. Remote Sens (Basel) 12(19):3263. https://doi.org/10.3390/rs12193263

Kacic P, Kuenzer C (2022) Forest biodiversity monitoring based on remotely sensed spectral diversity—a review. Remote Sens (Basel) 14(21):5363. https://doi.org/10.3390/rs14215363

Kanagawa M, Nakata T (2008) Assessment of access to electricity and the socio-economic impacts in rural areas of developing countries. Energy Policy 36(6):2016–2029. https://doi.org/10.1016/j.enpol.2008.01.041

Katharria A, Rajwar K, Pant M, Velasquez JD, Snasel V, Deep K (2025) Information fusion in smart agriculture: machine learning applications and future research directions. https://doi.org/10.2139/ssrn.5187083

Keefe RF et al (2019) Positioning methods and the use of location and activity data in forests. Forests 10(5):458. https://doi.org/10.3390/f10050458

Keenan RJ (2015) Climate change impacts and adaptation in forest management: a review. Ann For Sci 72(2):145–167. https://doi.org/10.1007/s13595-014-0446-5

Kellndorfer JM, Walker WS, LaPoint E, Kirsch K, Bishop J, Fiske G (2010) Statistical fusion of lidar, InSAR, and optical remote sensing data for forest stand height characterization: a regional-scale method based on LVIS, SRTM, Landsat ETM+, and ancillary data sets. J Geophys Res Biogeosci. https://doi.org/10.1029/2009JG000997

Khaiter PA, Erechtchoukova MG (2022) Advanced scientific methods and tools in sustainable forest management: a synergetic perspective. In: Forest dynamics and conservation. Springer Nature Singapore, Singapore, pp 279–309. https://doi.org/10.1007/978-981-19-0071-6_14

Knoke T, Kindu M, Schneider T, Gobakken T (2021) Inventory of forest attributes to support the integration of non-provisioning ecosystem services and biodiversity into forest planning—from collecting data to providing information. Curr For Rep 7(1):38–58. https://doi.org/10.1007/s40725-021-00138-7

Lister AJ et al (2020) Use of remote sensing data to improve the efficiency of national forest inventories: a case study from the United States National Forest Inventory. Forests 11(12):1364. https://doi.org/10.3390/f11121364

Liu M, Sorin P (2022) Estimation of biomass burning emissions by integrating ICESat-2, Landsat 8, and Sentinel-1 data. Remote Sens Environ 280:113172

Liu K, Shen X, Cao L, Wang G, Cao F (2018) Estimating forest structural attributes using UAV-LiDAR data in Ginkgo plantations. ISPRS J Photogramm Remote Sens 146:465–482. https://doi.org/10.1016/j.isprsjprs.2018.11.001

Luo Q, Li S, Wang H, Cheng H (2024) Mapping human pressure for nature conservation: a review. Remote Sens (Basel) 16(20):3866. https://doi.org/10.3390/rs16203866

Mastro P, Masiello G, Serio C, Pepe A (2022) Change detection techniques with synthetic aperture radar images: experiments with Random Forests and Sentinel-1 observations. Remote Sens (Basel) 14(14):3323. https://doi.org/10.3390/rs14143323

Mngadi M, Odindi J, Peerbhay K, Mutanga O (2021) Examining the effectiveness of Sentinel-1 and 2 imagery for commercial forest species mapping. Geocarto Int 36(1):1–12. https://doi.org/10.1080/10106049.2019.1585483

Morrison LW (2016) Observer error in vegetation surveys: a review. J Plant Ecol 9(4):367–379. https://doi.org/10.1093/jpe/rtv077

Mountrakis G, Heydari SS (2024) Effect of intra-year Landsat scene availability in land cover land use classification in the conterminous United States using deep neural networks. ISPRS J Photogramm Remote Sens 212:164–180. https://doi.org/10.1016/j.isprsjprs.2024.04.027

Ngome CS, Yeom C (2024) Comparative analysis of participatory forest management in Kenya: Embaringo and Gathiuru Community forest associations. Asia-Pac J Converg Res Interchange 10(1):227–245. https://doi.org/10.47116/apjcri.2024.01.19

Nguyen TH, Jones SD, Soto-Berelov M, Haywood A, Hislop S (2018) A spatial and temporal analysis of forest dynamics using Landsat time-series. Remote Sens Environ 217:461–475. https://doi.org/10.1016/j.rse.2018.08.028

Olson R, Hackett J, DeRoy S (2016) Mapping the digital terrain: towards indigenous geographic information and spatial data quality indicators for indigenous knowledge and traditional land-use data collection. Cartogr J 53(4):348–355. https://doi.org/10.1080/00087041.2016.1190146

Oveland I, Hauglin M, Giannetti F, Kjørsvik NS, Gobakken T (2018) Comparing three different ground based laser scanning methods for tree stem detection. Remote Sens (Basel) 10(4):538. https://doi.org/10.3390/rs10040538

Pandey S, Kumari N, Dash SK, Al Nawajish S (2022) Challenges and monitoring methods of forest management through geospatial application. In: Advances in remote sensing for forest monitoring. Wiley, New York, NY, pp 289–328. https://doi.org/10.1002/9781119788157.ch13

Paul KI et al (2013) Development and testing of allometric equations for estimating above-ground biomass of mixed-species environmental plantings. For Ecol Manage 310:483–494. https://doi.org/10.1016/j.foreco.2013.08.054

Rahul W, Seema G, Prakash KC, Vandana S (2025) Community participation, technical solutions, and adaptive management: keys to successful compensatory afforestation fund management and planning authority (CAMPA) plantation initiatives in Maharashtra'. J Resour Ecol. https://doi.org/10.5814/j.issn.1674-764x.2025.03.009

Sahle M, Saito O, Fürst C, Yeshitela K (2018) Quantification and mapping of the supply of and demand for carbon storage and sequestration service in woody biomass and soil to mitigate

climate change in the socio-ecological environment. Sci Total Environ 624:342–354. https://doi.org/10.1016/j.scitotenv.2017.12.033
Satapathy DR, Katpatal YB, Wate SR (2008) Application of geospatial technologies for environmental impact assessment: an Indian Scenario. Int J Remote Sens 29(2):355–386. https://doi.org/10.1080/01431160701269002
Sharma S et al (2024) Remote sensing and GIS in natural resource management: comparing tools and emphasizing the importance of in-situ data. Remote Sens (Basel) 16(22):4161. https://doi.org/10.3390/rs16224161
Shivaprakash KN et al (2022) Potential for artificial intelligence (AI) and machine learning (ML) applications in biodiversity conservation, managing forests, and related services in India. Sustainability 14(12):7154. https://doi.org/10.3390/su14127154
Soleimannejad L, Ullah S, Abedi R, Dees M, Koch B (2019) Evaluating the potential of Sentinel-2, Landsat-8, and IRS satellite images in tree species classification of hyrcanian forest of Iran using random forest. J Sustain For 38(7):615–628. https://doi.org/10.1080/10549811.2019.1598443
Stone C, Webster M, Osborn J, Iqbal I (2016) Alternatives to LiDAR-derived canopy height models for softwood plantations: a review and example using photogrammetry. Aust For 79(4):271–282. https://doi.org/10.1080/00049158.2016.1241134
Taylor AR, Chen HYH, VanDamme L (2009) A review of forest succession models and their suitability for forest management planning. For Sci 55(1):23–36. https://doi.org/10.1093/forestscience/55.1.23
Taylor R, Davis C, Brandt J, Parker M, Stäuble T, Said Z (2020) The rise of big data and supporting technologies in keeping watch on the world's forests. Int For Rev 22(1):129–141. https://doi.org/10.1505/146554820829523880
Vaglio Laurin G et al (2019) Tree height in tropical forest as measured by different ground, proximal, and remote sensing instruments, and impacts on above ground biomass estimates. Int J Appl Earth Observ Geoinform 82:101899. https://doi.org/10.1016/j.jag.2019.101899
Valente S, Coelho C, Ribeiro C, Marsh G (2015) Sustainable forest management in Portugal: transition from global policies to local participatory strategies. Int For Rev 17(3):368–383. https://doi.org/10.1505/146554815815982620
Vijayakumar S, Saravanakumar R, Arulanandam M, Ilakkiya S (2024) Google Earth Engine: empowering developing countries with large-scale geospatial data analysis—a comprehensive review. Arab J Geosci 17(4):139. https://doi.org/10.1007/s12517-024-11948-x
Wang T, Zuo Y, Manda T, Hwarari D, Yang L (2025) Harnessing artificial intelligence, machine learning and deep learning for sustainable forestry management and conservation: transformative potential and future perspectives. Plants 14(7):998. https://doi.org/10.3390/plants14070998
Wolfslehner B, Seidl R (2010) Harnessing ecosystem models and multi-criteria decision analysis for the support of forest management. Environ Manage 46(6):850–861. https://doi.org/10.1007/s00267-009-9414-5
Wulder MA, Franklin SE, Lavigne MB (1996) High spatial resolution optical image texture for improved estimation of forest stand leaf area index. Can J Remote Sens 22(4):441–449. https://doi.org/10.1080/07038992.1996.10874668
Xu Y, Bai T, Yu W, Chang S, Atkinson PM, Ghamisi P (2023) AI security for geoscience and remote sensing: challenges and future trends. IEEE Geosci Remote Sens Mag 11(2):60–85. https://doi.org/10.1109/MGRS.2023.3272825
Yin D, Wang L (2016) How to assess the accuracy of the individual tree-based forest inventory derived from remotely sensed data: a review. Int J Remote Sens 37(19):4521–4553. https://doi.org/10.1080/01431161.2016.1214302
Ystgaard KF, De Moor K (2023) Envisioning the future: a multi-disciplinary approach to human-centered intelligent environments. Qual User Exp 8(1):11. https://doi.org/10.1007/s41233-023-00064-5
Zhang N, Yang G, Pan Y, Yang X, Chen L, Zhao C (2020) A review of advanced technologies and development for hyperspectral-based plant disease detection in the past three decades. Remote Sens (Basel) 12(19):1–34. https://doi.org/10.3390/rs12193188

Zhang N, Yang G, Pan Y, Yang X, Chen L, Zhao C (2020) A review of advanced technologies and development for hyperspectral-based plant disease detection in the past three decades. Remote Sens (Basel) 12(19):3188. https://doi.org/10.3390/rs12193188

Zhong L, Dai Z, Fang P, Cao Y, Wang L (2024) A review: tree species classification based on remote sensing data and classic deep learning-based methods. Forests 15(5):852. https://doi.org/10.3390/f15050852

Ziemba EW et al (2024) Leveraging artificial intelligence to meet the sustainable development goals. J Econ Manage 46:508–583. https://doi.org/10.22367/jem.2024.46.19

Prioritization of Erosion Prone Watersheds of the Song River Basin Using Morphometric Indices Coupled with AHP-TOPSIS and Geospatial Technology

Sunil Singh, Kabysruti Bora, Sanju Purohit, and Jitendra Binwal

1 Introduction

Erosion is a significant challenge for watershed management, posing risks to ecosystem stability, community livelihoods and economic development (Kumar et al. 2022a; Singh et al. 2022). Particularly in regions with rugged terrain and high rainfall, erosion can exacerbate soil degradation, diminish biodiversity, reduce water quality and increase vulnerability to natural disasters such as landslides and floods (Sarma et al. 2024). The complexity and variability of erosion processes are frequently not adequately addressed by traditional watershed management methods. In order to combat these constraints, it is imperative to implement sophisticated scientific methodologies that integrate morphometric analysis with decision-making frameworks and geospatial technologies. Morphometric indices are essential for comprehending the erosion susceptibility of a watershed, as they quantify its physical and geometric characteristics (Nasir et al. 2023). These indices, including parameters such as linear, areal and relief aspect provide quantitative insights into the watershed's structure and its potential for erosion (Shekar et al. 2023). However, a singular reliance on these indices may prove inadequate in the absence of a methodical strategy to identify and prioritize the most vulnerable regions.

S. Singh (✉) · J. Binwal
School of Liberal Arts and Sciences, Mody University of Science and Technology (MUST), Lakshmangarh, Rajasthan, India
e-mail: sunilgeo.hnbgu@gmail.com

K. Bora
Department of Geography, D.B.S. PG College, Dehradun, India

S. Purohit
Department of Environmental/Ecological Studies and Sustainability, Akamai University, Waimea, HI, USA

M. O. Sarif and A. Sharifi (eds.), *Advanced Geospatial Intelligence and AI for Environmental Resilience and Sustainable Development*, Advances in Geographic Information Science, https://doi.org/10.1007/978-3-032-03714-5_11

The Analytical Hierarchy Process (AHP) and the Technique for Order of Preference by Similarity to an Ideal Solution (TOPSIS) provide effective frameworks for prioritizing erosion-prone watersheds (Kumar et al. 2022a; Shekar et al. 2025). By decomposing the problem into a hierarchy and allocating relative weights to each criterion, AHP enables the evaluation of multiple criteria (Saaty and Wind 1980). This approach facilitates the systematic evaluation of morphometric indices in accordance with their influence on erosion risk (Kumar et al. 2022b). TOPSIS enhances this by arranging the watersheds in order of their proximity to a sustainable solution identifying the most require intervention. The incorporation of AHP-TOPSIS with geospatial technology, predominantly Geographic Information Systems (GIS), significantly boosts the analysis of erosion-prone watersheds. GIS offers a spatially explicit platform for the visualization and combination of morphometric parameters which enables a thorough estimation of watershed situation.

It enables the identification of critical areas and the support in decision-making by overlaying morphometric data with spatial layers of erosion risk. By incorporating these with advanced decision-making contexts and geospatial technologies, numerous scholars have conducted extensive research on the prioritization of erosion-prone watersheds. Singh and Jain (2002) offered foundational insights into the ways in which morphometric parameters, such as drainage density and slope, influence erosion risk in the Indian Himalayas. Meshram and Sharma (2017) demonstrated the efficacy of the analytical hierarchy process (AHP) in managing complex decision-making scenarios by integrating expert opinions into watershed prioritization. Sharma and Mahajan (2020) illustrated the potential of geographic information systems (GIS) to improve visualization and targeted interventions in erosion management by spatially analyzing morphometric data. Kumar and Sarkar (2022) presented how to use AHP and TOPSIS together to rank watersheds based on a number of factors. Singh and Singh (2018) further examine the incorporation of GIS with morphometric parameters, emphasising its importance in improving the precision of erosion risk mapping. Kumar et al. (2022b) and Sarkar et al. (2022) presented the utilisation of MCDM for the determination of holistic watershed prioritization. They analyse the robustness of this approach to erosion risk management. Patel et al. (2022) examined the synergy between GIS and AHP-TOPSIS, highlighting the potential of these combined methodologies to enhance erosion risk assessment and management strategies in Ami River basin. Based on these previous research and methodologies, suggested a comprehensive framework for ranking erosion-prone watersheds within the Song River Basin. This incorporates geospatial technologies for spatial analysis and visualization; AHP-TOPSIS for multi-criteria decision-making and morphometric indices from DEM and RS data. The study aims to provide insights into the optimization of erosion control efforts and the enhancement of the resilience of vulnerable watersheds by employing advanced approach for sustainable soil conservation.

2 Data and Methodology

2.1 Study Area

The Song River basin is located in the Dehradun district of Uttarakhand and lies between 30°32′ 37″ N latitude and 78°06′ 28″ E longitude (Fig. 1). It encompasses an area of 1047.120 km^2 and drains the central and eastern part of the Doon Valley and make confluence with the Ganga River near Virbhadra. It originates from the southern slope of the Surkanda mountain in the Mussoorie range of the Garhwal Himalayas and flows in the north-west direction. Its major tributaries are Kali Gad, Bindal River and Rispana River. With its tributaries and associated watercourses, the basin supports various ecological functions, farming activities and anthropological settlements.

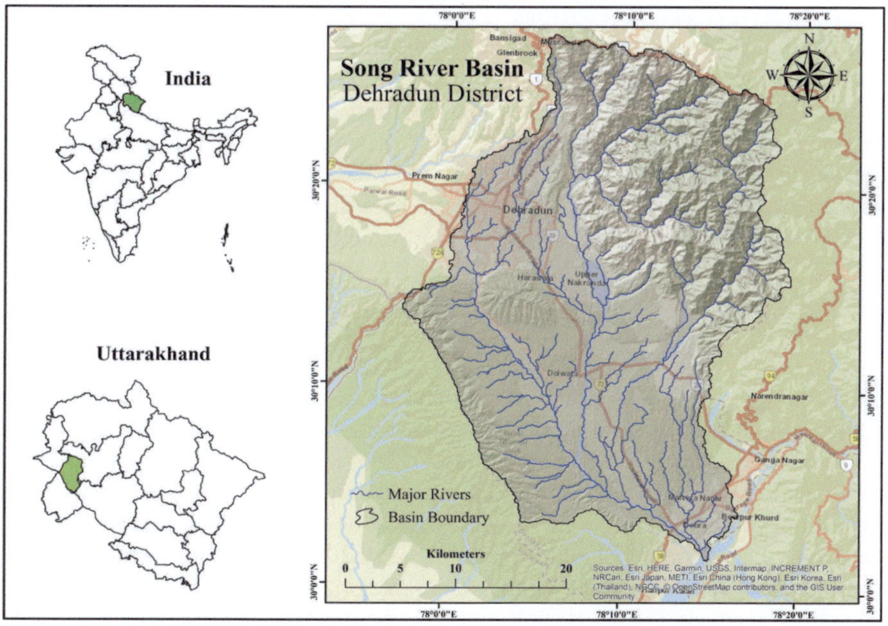

Fig. 1 Location map of the study area showing the Song River Basin in the Dehradun District, Uttarakhand, India. The map highlights the basin boundary and major river network within the basin. Insets on the left show the location of Uttarakhand within India and the location of the Dehradun district within Uttarakhand

Table 1 Data sources and scale used in the study

Data	Scale	Source
Topographical sheets	1:50,000	Survey of India (https://soinakshe.uk.gov.in/)
CartoDEM	(30 m resolution) Streams numbers, stream length, slope, aspect and relief	National Remote Sensing Centre (NRSC) (https://bhuvan.nrsc.gov.in/home/index.php)

2.2 Data Sources

The data used in the examination were derived from various geospatial portals (Table 1). The topographical sheets (1:50,000) were extracted from the Survey of India (https://soinakshe.uk.gov.in/) and utilized to extract the drainage network and various hydrological characteristics of the basin. The digital elevation (DEM) data were derived from the Bhuvan portal of National Remote Sensing Centre (NRSC), ISRO (https://bhuvan.nrsc.gov.in/home/index.php) on 30 m spatial resolution. The CartoDEM were then used for assessing the drainage network, frequency, density, texture, slope, absolute relief, relative relief, circulatory ration, elongation ratio, hypsometry and various other linear, areal and relief aspects of the river basin. The assessment and mapping have been done in QGIS and ArcGIS 10.8 software.

2.3 Method

The methodology employs a multi-step approach, beginning with the pre-processing and analysis of the spatial datasets using Geographic Information Systems (GIS) and Remote Sensing (RS) techniques. The DEM were used in the GIS environment to delineate the song river basin and extract the various indices (Fig. 2). The basin was further subdivided into nine sub watersheds, designated SW1 to SW9, then the morphometric indices (including linear, aerial and relief aspects) were calculated using standard mathematical formulas to characterize the geomorphological properties of watersheds (Table 2). Subsequently determining the morphometry, ranks have been assigned to individual criteria for each sub-watershed using Analytical Hierarchy Process (AHP) in the hierarchical structure of criteria and sub-criteria, integrating expert knowledge and stakeholder preferences. Thereafter, Technique for Order Preference by Similarity to Ideal Solution (TOPSIS) method was then employed to rank the sub-watersheds, based on their proximity to the ideal solution, considering multiple criteria simultaneously.

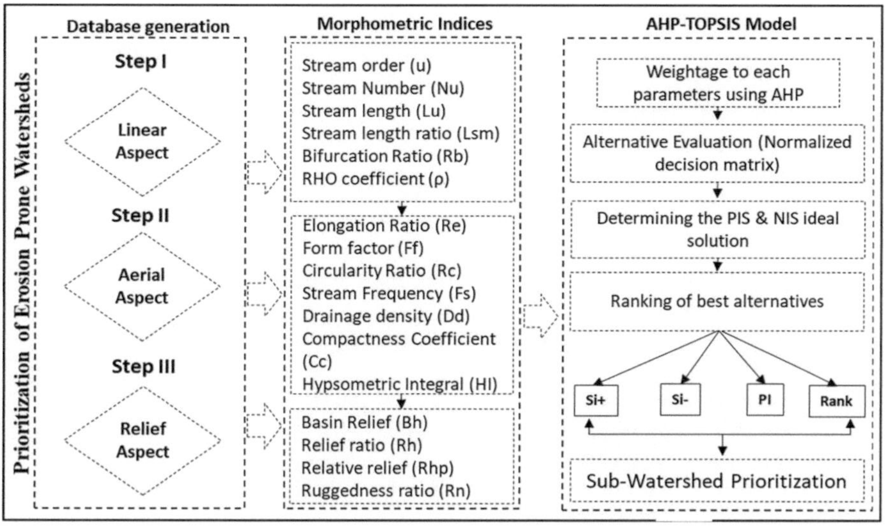

Fig. 2 Methodological framework of the study illustrating the step-by-step process followed for analyzing the Song River Basin. The flowchart outlines data collection, pre-processing, analysis and interpretation stages used to achieve the research objectives

2.3.1 AHP-TOPSIS Model

The study employs the AHP-TOPSIS model, which combines the Technique for Order Performance by Similarity to Ideal Solution (TOPSIS), introduced by Hwang and Yoon in 1981, with the Analytic Hierarchy Process (AHP) developed by Saaty in 1980. This approach assigns weights to different criteria based on insights from highly indexed journals, expert opinions and relevant literature. To determine the factors influencing watershed prioritization in mountainous regions, a structured questionnaire was used to consult 10 experts. The TOPSIS method begins by calculating the Euclidean distance between negative ideal solutions (NIS) and positive ideal solutions (PIS) among decision alternatives. The optimal solution has the shortest distance to the PIS and the longest distance from the NIS, leading to a calculated proximity coefficient. The alternative with the highest proximity coefficient is preferred. The process involves several steps: normalizing the decision matrix (NDM), calculating the weighted normalized decision matrix (WNDM), and identifying both PIS and NIS. Distances to these solutions are computed, followed by the calculation of the closeness coefficient (CC), which ranges from 0 to 1. The alternatives with higher closeness coefficients are considered superior (Kannan et al. 2009). The AHP-TOPSIS model was executed in five steps, detailed as follows:

1. Normalizing the Decision Matrix (NDM):

$$x = \frac{a_{ij}}{\sqrt{\sum_{i=1}^{m} a_{ij}}} \tag{1}$$

Table 2 Mathematical expressions used for the evaluation of various morphometric indices of the Song River Basin, categorized into linear, areal, and relief aspects, along with their respective formulas and references

	Parameters	Formula	References
Linear aspect	Stream order (Su)	Hierarchical order	Strahler (1964)
	Stream length (Lu)	Length of the streams	Horton (1945)
	Mean stream length (Lsm)	Lsm = Lu/Nu where, Lu = Stream length of order 'U'	Horton (1945)
	Stream number (Nu)	Nu = Total number of stream segments of order 'U'	
	Stream length ratio (Rl)	RL = Lu/Lu − 1; where Lu = Total stream length of order 'U', Lu1 = Stream length of next lower order	Horton (1945)
	Bifurcation ratio (Rb)	Rb = Nu/ Nu + 1; where, Nu = Total number of stream segment of order 'u'; Nu + 1 = Number of segment of next higher order	Schumm (1956)
	Mean bifurcation ratio (Rbm)	Rbm = Average of bifurcation ratios of all orders	Strahler (1964)
Areal aspect	Basin area (A)	Area of watershed in (km^2)	
	Basin length (Lb)		
	Basin width		
	Basin perimeter (P)	Perimeter of watershed in (km^2)	
	Form factor (Rf)	Rf = $A/(Lb)^2$; where, A = Area of watershed, Lb = Basin length	Horton (1932)
	Elongation ratio (Re)	Re = $2\sqrt{(A/\pi)}/Lb$; where, A = Area of watershed, π = 3.14, Lb = Basin length	Schumm (1956)
	Circulatory ratio (Rc)	Rc = $4\pi A/P^2$; where, A = Area of watershed, π = 3.14, P = Perimeter of watershed	
	Drainage density (Dd)	Dd = L/A where, L = Total length of streams; A = Area of watershed	Horton (1945)
	Stream frequency (Fs)	Fs = Nu/A, where, N = Total number of streams; A = Area of watershed	Horton (1945)
	Texture ratio (T)	Texture ratio (T) T = N1/P where, N1 = Total number of first order streams; P = Perimeter of watershed	Horton (1945)
Relief aspect	Relative relief (Rh)	Difference between maximum and minimum elevation of watershed	Schumm (1956)
	Relief ratio (Bh)	Rr = H/Lb, where Rh = relief ratio, H = total relief (relative relief) of the basin in km, Lb = basin length	Schumm (1956)

(continued)

Table 2 (continued)

	Parameters	Formula	References
	Basin slope (Sb)	Sb = VI/HE (in degree), where Sb = basin slope, VI = vertical interval, HE = horizontal equivalent	Strahler (1950)
	Hypsometric Integral (HI)	HI = (Elev_mean − Elev_min)/(Elev_mean − Elev_min)	Pike and Wilson (1971)

Here, R_{ij} are elements of the normalized decision matrix, and a_{ij} represents the performance of the ith alternative in the jth criteria.

2. Weighted Normalized Decision Matrix (WNDM):

$$V_{ij} = R_{ij} * W_j \tag{2}$$

V_{ij} are elements of the weighted normalized matrix, R_{ij} are elements of the normalized matrix, and W_j represents the weight of criteria j.

3. Calculating PIS (positive ideal solutions) and NIS (negative ideal solutions):

$$\begin{aligned} A^+ &= \left\{\left(\left(\max_i v_{ij} \lfloor j \in J\right), \min_i v_{ij} \lfloor j \in J'\right) | i = 1, 2, 3, \ldots, n\right\} \\ &= \left\{V_1^+, V_2^+, \ldots V_j^+ \ldots V_m^+\right\} \end{aligned} \tag{3}$$

$$\begin{aligned} A^- &= \left\{\left(\left(\min_i v_{ij} \lfloor j \in J\right), \max_i v_{ij} \lfloor j \in J'\right) | i = 1, 2, 3, \ldots, n\right\} \\ &= \left\{V_1^-, V_2^-, \ldots V_j^- \ldots V_m^-\right\} \end{aligned} \tag{4}$$

4. Computing solution distances and negative ideal solutions (NIS):

$$S_i^+ = \sqrt{\sum_{j=1}^{n} \left(\mathrm{v}_\mathrm{i}^+ - \mathrm{v}_\mathrm{ij}\right)^2}, \quad \mathrm{i} = 1, 2, \ldots, \mathrm{m} \tag{5}$$

$$S_i^- = \sqrt{\sum_{j=1}^{n} \left(\mathrm{v}_\mathrm{i}^- - \mathrm{v}_\mathrm{ij}\right)^2}, \quad \mathrm{i} = 1, 2, \ldots, \mathrm{m} \tag{6}$$

5. The final step involves calculating the closeness coefficient (CC) for each alternative, as shown in Eq. (7). This coefficient ranges from 0 to 1, with higher values indicating better alternatives.

$$Cl_+^i = \frac{d_{i-}}{d_{i+} + d_{i-}}; \quad 0 \le 1; \ i = 1, 2, \ldots, m \tag{7}$$

whereas, Cl_{+}^{i} represents closeness coefficient, d_{i+} indicates PIS and d_{i-} denotes NIS.

3 Results and Discussion

3.1 Morphometric Indices of the Basin

The analysis of morphometric indices is vital to understanding the hydrological and geomorphological characteristics of a river basin. Morphometry involves the quantitative study of the Earth's surface, landforms and the processes shaping them. It provides critical insights into the basin's drainage characteristics, runoff potential, soil erosion susceptibility and groundwater recharge capability. To prioritize watersheds for management and sustainable development of natural resources, assessing morphometric indices is indispensable to understand watershed dynamics. The computation involves linear, areal and relief aspects, as detailed below.

3.1.1 Linear Aspects of the Basin

Stream order (u) refers to the hierarchical classification of river channels within a drainage system (Singh et al. 2022). It was primarily proposed by R.E. Horton in 1945 and further refined by Strahler in 1957. In this system, a 2nd order stream forms when two 1st order streams or unbranched finger-tip tributaries meet. Subsequently, at junctions, two second-order streams merge to create a third-order stream and this pattern continues. The differentiation of stream orders within a basin is influenced by the underlying physiographic and structural characteristics of the area. The WS8 has the highest stream order of 6th order followed by the WS1, WS2, WS3, WS7, WS4, WS6, WS5 and WS9 (Fig. 3; Table 3). In a watershed, the total number of streams includes those from every stream order. According to Horton's (1932) law of stream number (Nu), the number of streams forms a geometric sequence, decreasing as stream order increases. The northern half of the basin, characterized by steep terrain, contains the highest concentration of lower-order stream segments and fingertip tributaries. It has been noted that as stream order rises, the number of streams declines across all watersheds. The Song River Basin has a total of 1017 streams, consisting of 772 first order, 186 s order, followed by 3rd, 4th and 5th order streams (Table 3). The cumulative distance traversed by streams in a watershed is ascertained by summing the lengths of all individual stream segments. Horton's Law of 1945 states that the cumulative stream length (Lu) decreases sequentially. This metric is essential for comprehending runoff characteristics: shorter stream segments denote steeper gradients, whereas longer segments generally indicate gentler slopes. A cumulative length of 1489.72 km of streams has been documented across the basin, with the longest lengths found in first-order rivers (763.22 km), succeeded

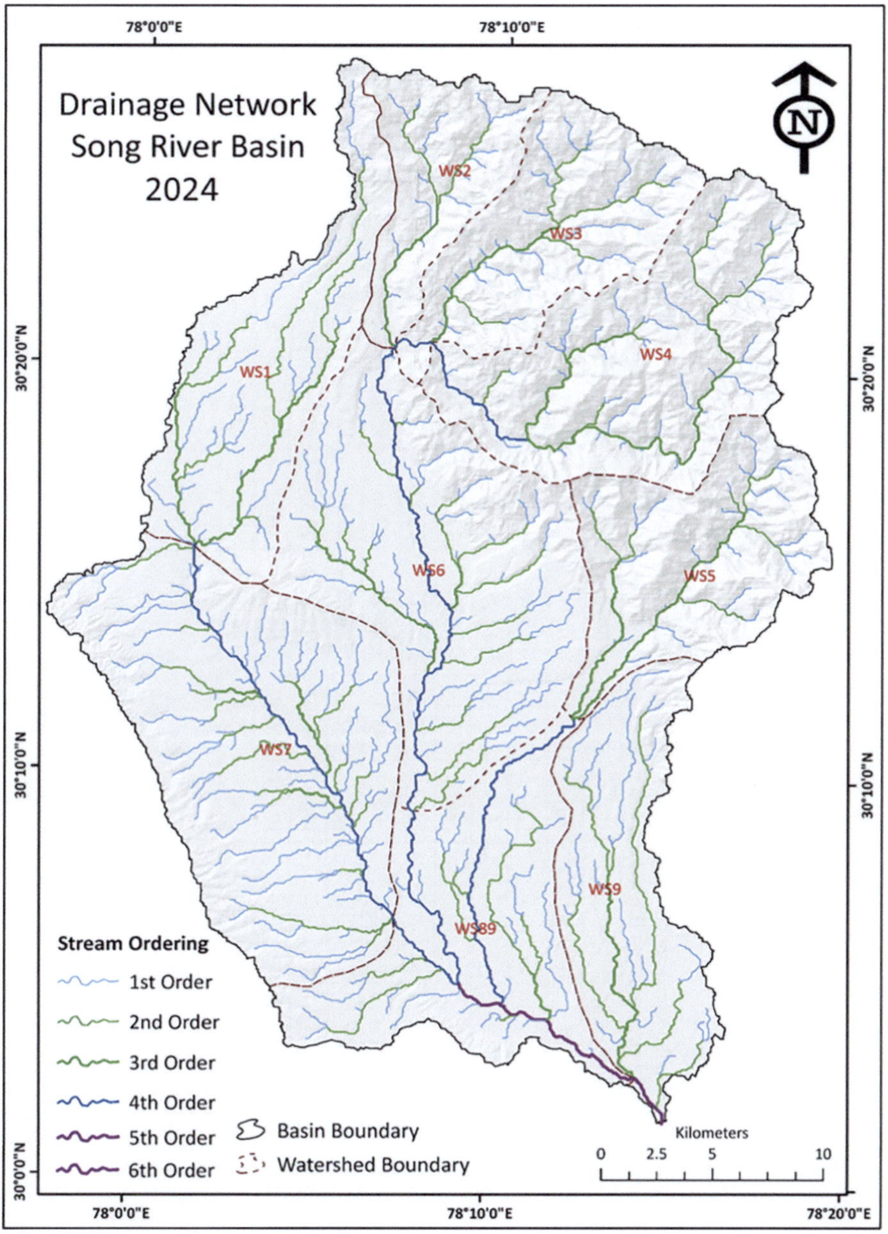

Fig. 3 Drainage network map of the Song River Basin (2024) displaying the stream ordering (based on Strahler's method), delineated watershed boundaries (WS1–WS9), and the overall basin boundary. The map categorizes streams from 1st to 6th order and provides a detailed spatial representation of the basin's hydrological structure

Table 3 Stream ordering and corresponding stream lengths (Lu) of the Song River Basin using Strahler's method. The table presents the number of streams by order, total number of streams (Nu), total stream length (Lu in km), and mean stream length (Lsm) for each sub-watershed (WS1–WS9)

Watersheds	Stream orders (Strahler's method)							Stream length (Lu in km)					Total	
	I	II	III	IV	V	VI	Nu	I	II	III	IV	V	(Lu)	Lsm
WS1	21	4	2	1	–	–	28	72.12	15.16	13.15	0.08	0	100.51	20.10
WS2	39	9	2	1	–	–	51	27.34	14.28	9.4	7.38	0.051	58.451	11.69
WS3	66	16	3	1	–	–	86	53.1	26.04	11.76	11.71		102.61	25.65
WS4	94	23	4	1	–	–	122	77.63	22.57	20.54	26.95		147.69	36.92
WS5	75	16	4	1	–	–	96	45.3	28.05	19.38	13.02		105.75	26.43
WS6	142	34	8	2	1	–	187	132.89	79.16	26.38	12.62	26.67	277.72	55.544
WS7	147	38	8	2	1	–	196	180.93	98.33	20.33	2	23.3	324.89	64.97
WS8	128	34	10	3	1	1	177	96.59	64.02	46.02	10.7	8.21	225.54	45.10
WS9	59	12	2	1	–	–	74	77.32	44.01	22.4	2.83		146.56	36.64

by second-order (391.62 km), third-order (189.36 km), fourth-order (87.29 km) and fifth-order (58.23 km) streams. The mean lengths of streams in successive orders are compared to establish the stream length ratio (Lsm). In this basin, sub-watershed stream length ratios range from 11.69 (WS2) to 64.98 (WS7), indicating that the watershed is in the late juvenile geomorphic stage.

The bifurcation ratio (Rb) is defined as the ratio of the number of streams of order Nu to the number of streams of order Nu + 1 (Singh et al. 2022). Geological and structural characteristics of the basin's subsurface primarily influence this ratio. Typically ranging from 2 to 5 in watersheds where geological structures do not alter the drainage pattern (Strahler 1957), a higher Rb value indicates a more gradual and extended peak flow, whereas a lower Rb value suggests a more sudden and pronounced peak flow. The watershed exhibits a bifurcation ratio ranging from 2 to 6 across all sub-watersheds (Table 4), indicating high infiltration capacity and significant groundwater potentiality. Moreover, the mean bifurcation ratio (Rbm) value varies from 2.31 (SW1) to 3.65 (WS7) within the study area, highlighting the effective structural control on drainage development. The RHO coefficient (ρ), which compares stream length (Lu) to bifurcation ratio (Rb), serves to assess the water storage capacity of the basin. The examination indicates that SW7 shows the highest RHO value, whereas SW9 demonstrates the lowest.

3.1.2 Aerial Aspects of the Basin

The sub-watersheds (SW) of the basin are characterized by varying elongation ratios (Re), which reflect their shape and hydrological behaviour. An elongation ratio (Re) is a measure of how overextended or circular a watershed is, with values closer to 1 indicating a more circular shape, and values closer to 0 indicating an elongated shape. The sub-watersheds of the Song River vary in shape as indicated by their elongation ratios (Re). WS4, with a Re of 0.79, is the most circular, suggesting efficient drainage and lower flood risk. WS3, WS7, and WS8 have Re values around 0.70–0.73, indicating moderate elongation (Table 4). WS2, WS5, and WS6, with Re values of 0.67–0.68, are slightly more elongated. WS1 and WS9, with the lowest Re values of 0.47 and 0.51, are the most elongated, potentially leading to slower drainage and higher flood risk (Table 4). The sub-watersheds of the Song River exhibit varying form factors (Rf), indicating their shapes and potential hydrological behaviour. The form factor, a dimensionless value, is calculated by dividing the watershed area by the square of its length. Higher values suggest more circular watersheds, leading to faster runoff and higher peak discharge rates after rainfall. WS7, with the highest form factor of 9.10, is the most circular, indicating quicker runoff. In contrast, WS1, with a form factor of 4.27, is more elongated, suggesting slower runoff and lower peak flows (Table 4).

The circulatory ratio (Rc) is a key indicator in hydrology, reflecting the shape and compactness of watersheds. In the sub-watersheds of the Song River, the Rc values vary, indicating differences in their geomorphological characteristics. WS1 (0.26) and WS9 (0.29) have lower Rc values, suggesting elongated shapes that may result

Table 4 Morphometric analysis of the Song River Basin depicting linear, areal, and relief aspects for each sub-watershed (WS1–WS9). The table includes values for stream frequency (ρ), drainage density (Dd), stream frequency (Fs), texture ratio (Rt), overland flow length (Lo), mean bifurcation ratio (Rbm), compactness coefficient (C), relative relief (Rh), relief ratio (Bh), hypsometric integral (HI), ruggedness number (Rn), elongation ratio (Re), form factor (Rf), circulatory ratio (Rc) and constant of channel maintenance (Cc)

WS	Linear aspect							Aerial aspect				Relief aspect			
	ρ	Dd	Fs	Rt	Lo	Rbm	C	Rh	Bh	HI	Rn	Re	Rf	Rc	Cc
WS1	0.23	0.97	0.27	0.39	2.06	2.31	1.03	1533	63.16	0.49	1.49	0.47	4.27	0.26	0.24
WS2	0.18	1.13	0.99	1.21	1.77	2.71	0.89	1693	140.03	0.48	1.91	0.67	4.28	0.37	0.26
WS3	0.16	1.21	1.01	1.77	1.65	3.11	0.83	1869	126.63	0.48	2.26	0.70	5.75	0.45	0.22
WS4	0.18	1.10	0.91	1.96	1.82	3.46	0.91	1956	117.83	0.49	2.14	0.79	8.11	0.44	0.19
WS5	0.16	1.14	1.04	1.88	1.75	3.17	0.87	1956	122.48	0.49	2.24	0.68	5.79	0.45	0.21
WS6	0.24	1.58	1.06	3.10	1.27	3.61	0.63	1048	47.29	0.48	1.66	0.68	7.93	0.61	0.17
WS7	0.85	1.64	0.99	2.86	1.22	3.65	0.61	473	21.67	0.49	0.77	0.73	9.10	0.53	0.17
WS8	0.18	1.99	1.56	2.64	1.01	3.38	0.50	385	23.32	0.49	0.77	0.73	6.87	0.32	0.22
WS9	0.09	1.60	0.81	1.18	1.25	3.23	0.63	1006	47.57	0.48	1.61	0.51	4.34	0.29	0.23

in slower drainage and prolonged runoff. WS2 (0.37) and WS8 (0.32) have moderate Rc values, indicating slightly more compact forms. In contrast, WS3 (0.45), WS4 (0.44), and WS5 (0.45) show relatively higher Rc values, suggesting more efficient drainage due to their compact shapes. Notably, WS6 (0.61) and WS7 (0.53) have the highest Rc values, indicating the most circular and compact forms, which facilitate faster runoff and efficient water flow (Table 4). The evaluation of stream frequency (Fs) in the Song River Basin provides insights into the drainage characteristics of its watersheds. Watershed 1, with a low stream frequency of 0.27, suggests a sparse drainage network, potentially due to less rugged terrain or higher soil infiltration. In contrast, watersheds WS2, WS3, WS4, WS7 and WS9 have moderate stream frequencies ranging from 0.81 to 1.01, indicating a balanced drainage network that suggests moderate surface runoff and soil permeability. High stream frequencies in watersheds WS5, WS6, and WS8 ranging from 1.04 to 1.56, reflect a denser drainage network, possibly due to steep terrain, less permeable soils, or higher precipitation levels. These higher values suggest increased erosion potential and a higher risk of flooding, making them crucial areas for conservation efforts (Table 4).

The drainage density (Dd) of the Song River Basin varies across different watersheds, reflecting diverse topographical and hydrological characteristics. WS1 has the lowest Dd of 0.97, indicating a less dissected terrain with higher potential for infiltration and reduced runoff. In contrast, Watersheds WS6, WS7, WS8 and WS9 exhibit higher Dd values (ranging from 1.58 to 1.99), suggesting steeper slopes or a more dissected landscape, which may lead to quicker runoff and increased erosion potential. Watersheds 2, 3, 4, and 5 have moderate Dd values (between 1.10 and 1.21), indicating a balanced hydrological regime with moderate drainage channel development. These variations highlight the importance of considering local geological and land use factors in watershed management and flood risk assessment (Table 4). The Compactness Coefficient (Cc) for the sub-watersheds of the Song River reveals their shape efficiency and drainage characteristics. A lower Cc value indicates a more elongated shape, while a higher value suggests a shape closer to a circle, which generally promotes more efficient drainage. The WS4, WS6, and WS7, with Cc values of 0.19, 0.17 and 0.17 respectively, exhibit more elongated or irregular shapes, which may result in less efficient drainage and longer water flow paths (Table 4). Conversely, WS1, WS2 and WS8, with Cc values of 0.24, 0.26, and 0.22, respectively, have somewhat more compact shapes, potentially enhancing their drainage efficiency compared to the more elongated sub-watersheds.

3.1.3 Relief Aspects of the Basin

The relief of a basin is a crucial determinant in influencing its denudational characteristics and, therefore, its geomorphic processes. The basin relief (Bh) is the vertical distance between its maximum and minimum elevations. This investigation recorded a maximum Bh of 140.03 m and a minimum of 21.67 m (Table 4). The resultant relief stipulates the gradient of the river basin, which in consequently influences soil erodibility. The upper section of the basin displays a steep gradient, resulting in reduced infiltration and increased runoff. In contrast, the southern and central areas feature gentle gradients and flat topography, resulting in decreased runoff and improved infiltration. The analysis indicates that WS2 is the highest-ranked, whereas WS7 is the lowest (Table 4). The relief ratio (Rh) of the indicates a spectrum of topographical gradients, demonstrating the diversity in steepness and height throughout the area. Sub-watersheds WS1 and WS2, possessing relief ratios of 1533 and 1693 respectively, demonstrate moderate slopes, signifying a moderately steep topography. The WS3, exhibiting a relief ratio of 1869, exhibits a little higher gradient. Conversely, WS4 and WS5 exhibit the largest relief ratios at 1956, indicating extremely steep gradients and significant topographical relief. In contrast, WS6 shows a relief ratio of 1048, signifying a less steep gradient, while WS7 and WS8, with ratios of 473 and 385 respectively, denote the gentle.

These variations reflect the diverse topography of the basin, impacting hydrological and geomorphological processes. Understanding these differences is crucial for effective watershed management, flood risk assessment and sustainable land use planning. The regions with high relative humidity might require erosion control measures, whereas places with low relative humidity may be more conducive to agricultural activity. The relief ratio, or ruggedness number (Rn), of the sub-watersheds in the Song River Basin reflects varying degrees of topographical roughness throughout the area. The 1st Watershed (WS1), possessing an Rn of 1.49, features moderately rugged topography, whereas the WS2, with an Rn of 1.91, reveals significantly steeper terrain. The WS3 exhibits the greatest Rn of 2.26, indicating the most rugged and steep terrain among the watersheds. Likewise, WS4 and WS5, showing Rn values of 2.14 and 2.24 respectively, exhibit substantial elevation variations, which signify rugged topography. On the other hand, WS6, exhibiting an Rn of 1.66, shows a relatively moderate rugged terrain, although less pronounced than that of WS2 to WS5. The WS7 and WS8 showed the lowest Rn of 0.77, representing mostly flat or gently sloped terrains. Eventually, WS9, with an Rn of 1.61, represents a moderate degree of ruggedness, demonstrating a balance between steep and flat terrains (Table 4). Langbein (1947) introduced the hypsometric integral (HI) to describe the area-altitude relationship. Later Strahler (1952) modified this method to use relative heights and areas for comparing developmental phases. The hypsometric integral (HI) represents the area below the hypsometric curve, ranging from 0 (highly eroded regions) to 1 (less eroded regions) (Pike and Wilson 1971). The HI and its curve indicate the relative age and degree of basin dissection. The hypsometric curve plots the proportion of total watershed height against the proportion of watershed area (Fig. 4). Concave curves with low HI values represent old, eroded landscapes; smooth, S-shaped curves indicate mature, moderately eroded landscapes; and convex curves with high HI values denote young, slightly eroded landscapes (Singh et al. 2022). In the sub-watersheds of the Song River basin, HI values vary between 0.48 and 0.49, indicating consistent topography throughout sub-watersheds WS1 to WS9 (Fig. 5; Table 4).

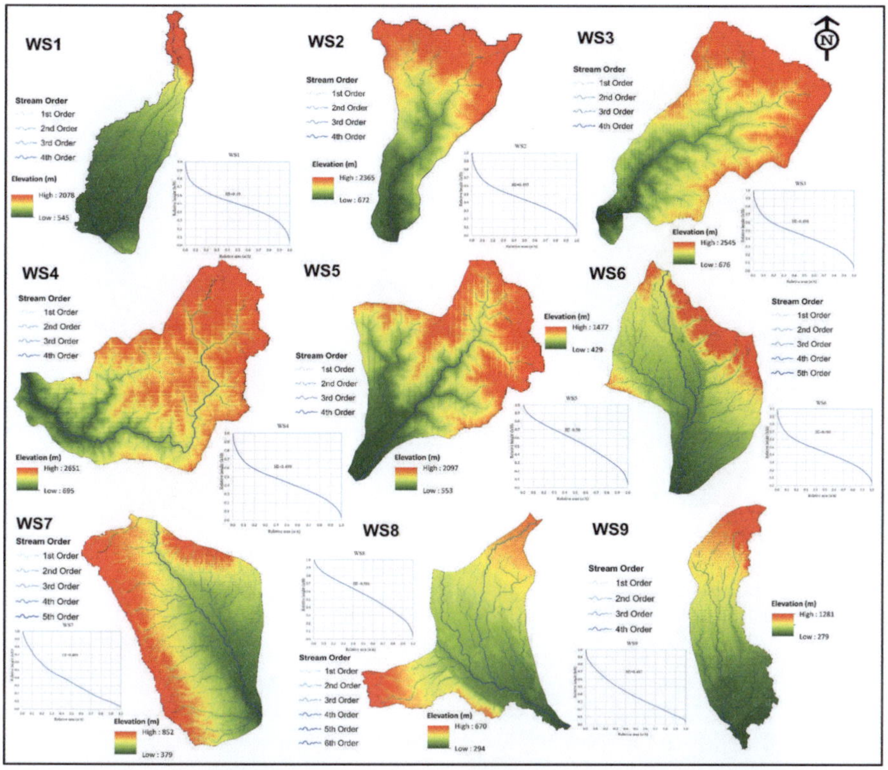

Fig. 4 Sub-watershed delineation and geomorphic analysis of the Song River Basin. The figure presents nine sub-watersheds (WS1–WS9), showing stream ordering (1st to 6th order), elevation range (absolute relief), and corresponding hypsometric curves. Elevation is represented using a color gradient from green (low) to red (high), and the hypsometric curves illustrate the terrain characteristics and erosional stages of each sub-watershed

3.2 Correlation Between Morphometric Indices and AHP-TOPSIS Results

The correlation among the morphometric indices of the Song River Basin has been assessed in R-Studio software using R programming and the Python language. The Rf (indicator of basin compactness) shows positive correlation with the Re (0.739*) and the Rc (0.743*), indicating adequate drainage network that favours oval or elongated shape of the basin. This phenomenon is frequently observed in basins that were formed through specific geomorphological processes (Strahler 1964). Nevertheless, Rf demonstrates a negative correlation with Rh (− 0.375), indicating that basins with high compactness are not significantly associated with steep relief. Likewise, a slight positive correlation between Rf and HI (0.362) suggests that compact basins may marginally favour higher elevation profiles. The Re has a positive correlation

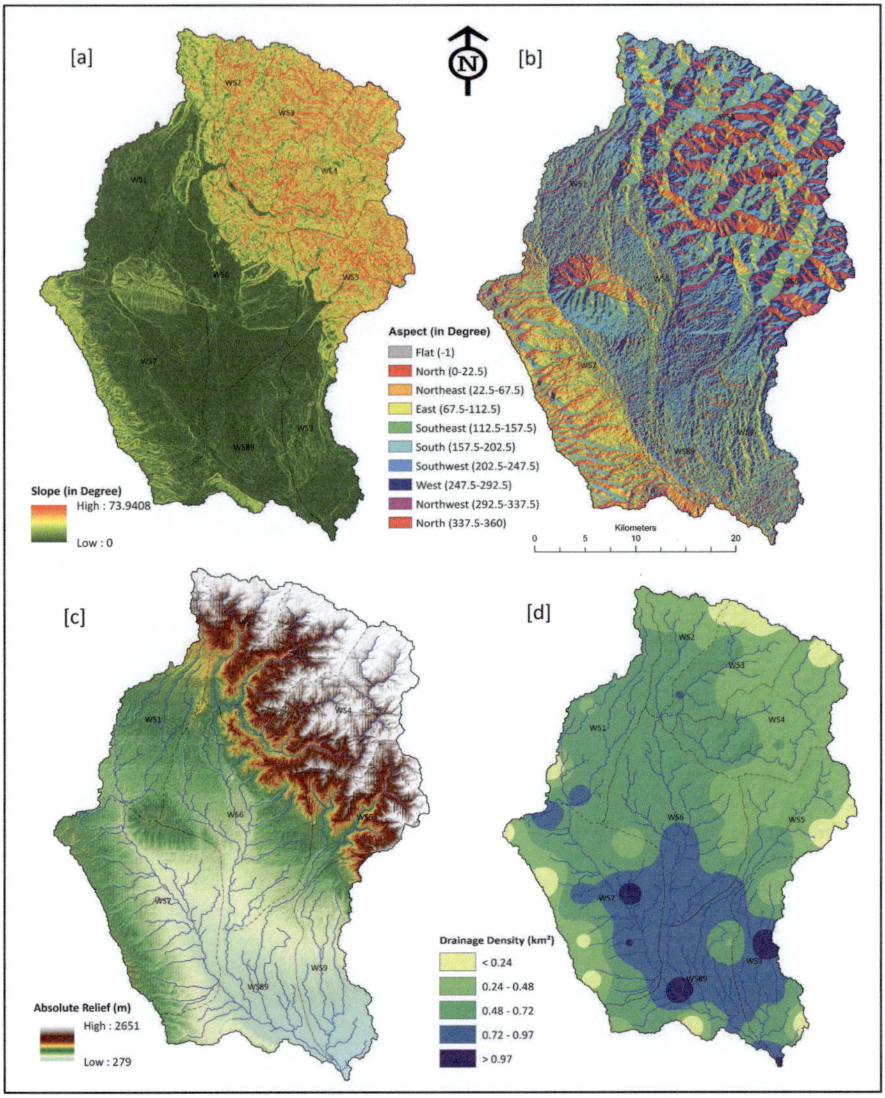

Fig. 5 Spatial representation of key terrain parameters for the Song River Basin: (**a**) Slope (in degrees), (**b**) Aspect (slope orientation), (**c**) Absolute Relief (elevation difference), and (**d**) Drainage Density (km/km^2) across the basin

with Rc (0.609), suggesting that more extended basins are likely to be more circular, potentially as a result of their geomorphological processes. Re exhibits a weak positive correlation with Dd (0.209), demonstrating that elongation does not substantially affect the density of the basin's river network. Moreover, its associations with indices such as Fs and Rh are moderate, suggesting a minimal influence of elongation on these characteristics. The Rc, an indicator of basin morphology, exhibits a weak positive connection with Fs (0.308) and a negligible association with Dd (0.118) (Fig. 6; Table 5). The insufficient correlations reveal that the basin's morphology does not significantly affect stream density or the quantity of streams per unit area. The negative association of Rc with HI (− 0.240) suggests that a circular basin shape may not significantly relate to the basin's elevation profile, which represents the overall topographic features.

Dd demonstrates a robust correlation with Fs (0.690*), indicating that basins characterised by a high density of drainage systems typically possess a higher number of streams. The association suggests regions with a greater abundance of streams sustain a more extensive drainage network (Hack 1957). The moderate positive association of Dd with Rf (0.406) indicates that compact basins generally possess a more intricate drainage network. Conversely, Dd demonstrates a substantial negative association with Bh (− 0.778*), indicating that basins with greater relative

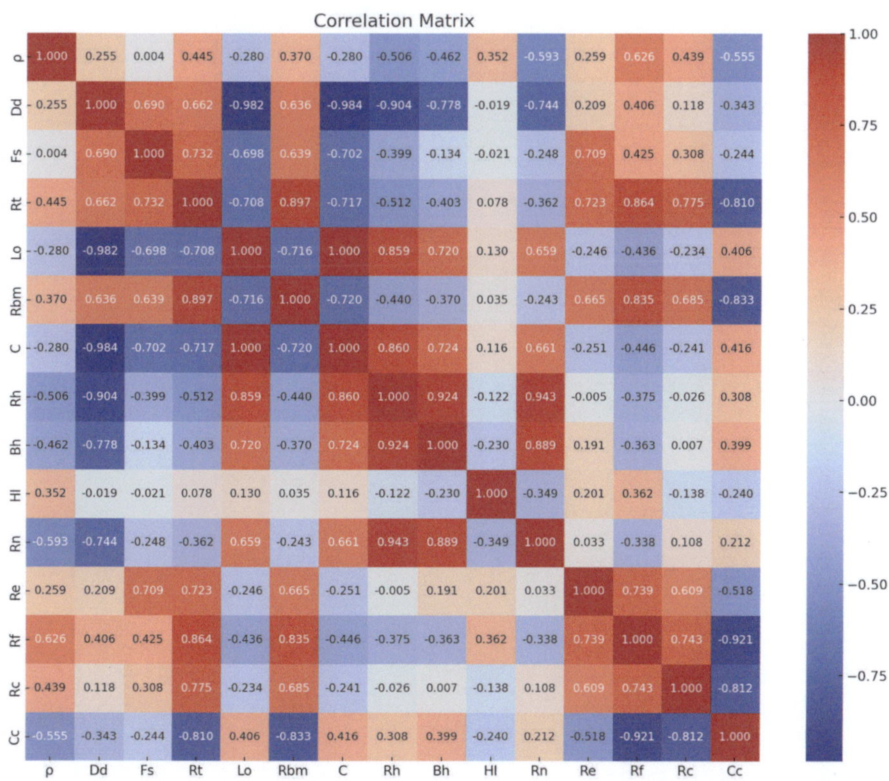

Fig. 6 Correlation matrix illustrating the relationships between various morphometric parameters of the Song River Basin. The color scale represents the strength and direction of correlations, ranging from −1 (strong negative) to +1 (strong positive), providing insights into the interdependence of hydrological and geomorphological variables

Table 5 Correlation matrix among selected morphometric indices of the river basin. Statistically significant correlations are indicated at the 0.05 (*) and 0.01 (**) levels. The table also presents the sum of correlations for each parameter, the grand total (GT), and the computed weights used for further analysis

MP	ρ	Dd	Fs	Rt	Lo	Rbm	C	Rh	Bh	HI	Rn	Re	Rf	Rc	Cc
ρ	1	0.255	0.004	0.445	−0.280	0.370	−0.280	−0.506	−0.462	0.352	−0.593	0.259	0.626	0.439	−0.555
Dd	0.255	1	0.690*	0.662	−0.982**	0.636	−0.984**	−0.904**	−0.778*	−0.019	−0.744*	0.209	0.406	0.118	−0.343
Fs	0.004	0.690*	1	0.732*	−0.698*	0.639	−0.702*	−0.399	−0.134	−0.021	−0.248	0.709*	0.425	0.308	−0.244
Rt	0.445	0.662	0.732*	1	−0.708*	0.897**	−0.717*	−0.512	−0.403	0.078	−0.362	0.723*	0.864**	0.775*	−0.810**
Lo	−0.280	−0.982**	−0.698*	−0.708*	1	−0.716*	1.000**	0.859**	0.720*	0.130	0.659	−0.246	−0.436	−0.234	0.406
Rbm	0.370	0.636	0.639	0.897**	−0.716*	1	−0.720*	−0.440	−0.370	0.035	−0.243	0.665	0.835**	0.685*	−0.833**
C	−0.280	−0.984**	−0.702*	−0.717*	1.000**	−0.720*	1	0.860**	0.724*	0.116	0.661	−0.251	−0.446	−0.241	0.416
Rh	−0.506	−0.904**	−0.399	−0.512	0.859**	−0.440	0.860**	1	0.924**	−0.122	0.943**	−0.005	−0.375	−0.026	0.308
Bh	−0.462	−0.778*	−0.134	−0.403	0.720*	−0.370	0.724*	0.924**	1	−0.230	0.889**	0.191	−0.363	0.007	0.399
HI	0.352	−0.019	−0.021	0.078	0.130	0.035	0.116	−0.122	−0.230	1	−0.349	0.201	0.362	−0.138	−0.240
Rn	−0.593	−0.744*	−0.248	−0.362	0.659	−0.243	0.661	0.943**	0.889**	−0.349	1	0.033	−0.338	0.108	0.212
Re	0.259	0.209	0.709*	0.723*	−0.246	0.665	−0.251	−0.005	0.191	0.201	0.033	1	0.739*	0.609	−0.518
Rf	0.626	0.406	0.425	0.864**	−0.436	0.835**	−0.446	−0.375	−0.363	0.362	−0.338	0.739*	1	0.743*	−0.921**
Rc	0.439	0.118	0.308	0.775*	−0.234	0.685*	−0.241	−0.026	0.007	−0.138	0.108	0.609	0.743*	1	−0.812**
Cc	−0.555	−0.343	−0.244	−0.810**	0.406	−0.833**	0.416	0.308	0.399	−0.240	0.212	−0.518	−0.921**	−0.812**	1
Sum	1.075	2.925	1.330	0.909	0.998	2.293	0.975	−1.077	−0.364	1.155	0.540	2.147	0.861	1.951	1
GT	16.55	16.55	16.55	16.55	16.55	16.55	16.55	16.55	16.55	16.55	16.55	16.55	16.55	16.55	16.55
Weight	0.065	0.177	0.080	0.055	0.060	0.139	0.059	−0.065	−0.022	0.070	0.033	0.130	0.052	0.118	0.051

*Correlation is significant at the 0.05 level (2-tailed)
**Correlation is significant at the 0.01 level (2-tailed)
Sum Sum of correlation; *GT* Grand total of correlation

relief generally display less drainage densities. This association could be ascribed to the heightened challenge of stream flow in steep terrains, resulting in less dense drainage networks (Schumm 1956). Fs exhibits a slight positive association with HI (0.425) and slight negative correlations with Dd (− 0.248) and Rh (− 0.399) (Fig. 6; Table 5). The positive correlation with HI indicates that basins with greater elevation ranges may possess a more extensive network of streams. The negative association with Dd suggests that a higher frequency of streams does not inherently correspond with enhanced drainage density, while the modest negative connection with Rh implies that basins characterised by frequent streams may not consistently exhibit high relative relief. Rh, an indicator of the basin's topographic steepness, demonstrates a robust positive connection with Bh (0.924**), indicating that regions with more relative relief are likely to display elevated relief ratios (Fig. 6; Table 5). However, the association of Rh with other morphometric indices such as HI (− 0.122) and Fs (− 0.399) is minimal, suggesting that relative relief is not significantly associated with the overall elevation profile of the basin or the characteristics of the stream network (Strahler 1964). The Bh has a substantial negative association with Dd (− 0.778*), indicating that basins characterised by elevated relief generally possess less drainage densities. This indicates that steep terrains may restrict the formation of a dense drainage network (Schumm 1956). The negative correlation of Bh with HI (− 0.230) indicates that the relief ratio is not significantly associated with the overall configuration of the basin's elevation profile, illustrating that relief and elevation profiles do not consistently correspond in a direct manner. Eventually, the HI, reflecting the overall topographic relief of a basin, has weak positive associations with Rf (0.362) and weak negative associations with Rc (− 0.240) and Bh (− 0.230) (Fig. 4; Table 5). This suggests that HI operates autonomously from other morphometric indices, exerting limited impact on the basin's shape or relief characteristics, highlighting the intricacy of basin topography.

After computing the morphometric indices for sub-watershed (SW) prioritization, a decision matrix was generated to normalize the data for the MCDM model. To standardize the parameters and eliminate incompatibility due to different units of measurement, normalization was performed using the linear method in the TOPSIS model. The weights for prioritization were derived using the AHP scale, based on pair-wise comparison of parameters concerning their contribution to identification and are shown in Tables 6 and 7. Among all the evaluated morphometric indices, the RHO coefficient (ρ), drainage density (Dd), stream frequency (Fs), drainage texture (Rt), length of overland flow (Lo), mean bifurcation ratio (Rbm), constant of channel maintenance (C) and hypsometric integral (HI) exhibited the most significant influence on watershed erodibility, with scores ranging from 0.09 to 0.85, 0.97 to 1.99, 0.27 to 1.56, 0.39 to 3.10, 1.01 to 2.06, 2.31 to 3.65, 0.50 to 1.03 (Table 4). The determined inconsistency rate for the final matrix was 0.081 which is considered acceptable as it is below 0.1 (Saaty 1977). After determining the weights of each parameter, the weighted normalized decision matrix (NDM) was computed (Eqs. 1–6), succeeded by the recognition of the worst and best values, PIS and NIS, for the TOPSIS model. The results demonstrated that WS6, WS7, WS8 showcased highest

Table 6 Showing the Normalized matrix (NM) of the TOPSIS model used in the study. The table presents the normalized values of selected morphometric parameters for each sub-watershed (SW1 to SW9), forming the basis for multi-criteria decision analysis

SW	ρ	Dd	Fs	Rt	Rbm	Lo	C	Hl	Rn	Rf	Re	Rc	Cc
SW1	0.231	0.229	0.089	0.063	0.240	0.438	0.438	0.336	0.286	0.219	0.236	0.200	0.369
SW2	0.182	0.267	0.326	0.196	0.281	0.376	0.376	0.330	0.367	0.219	0.334	0.288	0.401
SW3	0.156	0.286	0.335	0.285	0.324	0.351	0.351	0.330	0.434	0.295	0.351	0.352	0.337
SW4	0.182	0.259	0.299	0.316	0.359	0.387	0.387	0.336	0.412	0.416	0.393	0.341	0.303
SW5	0.156	0.270	0.343	0.304	0.330	0.371	0.371	0.336	0.430	0.297	0.338	0.349	0.331
SW6	0.244	0.374	0.352	0.501	0.375	0.269	0.269	0.330	0.318	0.406	0.336	0.475	0.261
SW7	0.856	0.387	0.326	0.462	0.380	0.260	0.260	0.336	0.149	0.466	0.363	0.417	0.261
SW8	0.176	0.470	0.516	0.427	0.351	0.214	0.214	0.336	0.147	0.352	0.362	0.249	0.342
SW9	0.093	0.378	0.267	0.191	0.336	0.266	0.266	0.330	0.309	0.222	0.254	0.231	0.368

Table 7 Calculation of the Positive Ideal Solution (A^+), Negative Ideal Solution (A^-), and AHP-derived weights for selected morphometric parameters used in the TOPSIS model. These values are essential for ranking sub-watersheds based on their relative morphometric performance

Parameters	ρ	Dd	Fs	Rt	Rbm	Lo	C	HI	Rn	Rf	Re	Rc	Cc
A^+	0.094	0.080	0.082	0.070	0.046	0.040	0.029	0.017	0.013	0.009	0.007	0.007	0.006
A^-	0.010	0.039	0.014	0.009	0.029	0.019	0.014	0.016	0.004	0.004	0.004	0.003	0.004
AHP-Weight	0.11	0.17	0.16	0.14	0.12	0.091	0.066	0.050	0.030	0.019	0.019	0.014	0.014

Table 8 Weighted normalized matrix for the TOPSIS model, showing the weighted values of morphometric parameters for each sub-watershed (SW1–SW9). The table also includes the calculated distances from the Positive Ideal Solution (Si^{+}), Negative Ideal Solution (Si^{-}), the Performance Index (PI) and the resulting priority ranking of the sub-watersheds

WS	ρ	Dd	Fs	Rt	Rbm	Lo	C	HI	Rn	Rf	Re	Rc	Cc	Si +	Si-	PI	Rank
SW1	0.025	0.039	0.014	0.009	0.029	0.040	0.029	0.017	0.009	0.004	0.004	0.003	0.005	0.12	0.030	0.195	9
SW2	0.020	0.045	0.052	0.027	0.034	0.034	0.025	0.016	0.011	0.004	0.006	0.004	0.006	0.10	0.048	0.329	7
SW3	0.017	0.049	0.054	0.040	0.039	0.032	0.023	0.016	0.013	0.006	0.007	0.005	0.005	0.09	0.055	0.371	6
SW4	0.020	0.044	0.048	0.044	0.043	0.035	0.026	0.017	0.012	0.008	0.007	0.005	0.004	0.09	0.056	0.377	5
SW5	0.017	0.046	0.055	0.043	0.040	0.034	0.025	0.017	0.013	0.006	0.006	0.005	0.005	0.093	0.058	0.384	4
SW6	0.027	0.064	0.056	0.070	0.045	0.024	0.018	0.016	0.010	0.008	0.006	0.007	0.004	0.08	0.082	0.518	3
SW7	0.094	0.066	0.052	0.065	0.046	0.024	0.017	0.017	0.004	0.009	0.007	0.006	0.004	0.04	0.112	0.736	1
SW8	0.019	0.080	0.082	0.060	0.042	0.019	0.014	0.017	0.004	0.007	0.007	0.003	0.005	0.08	0.096	0.545	2
SW9	0.010	0.064	0.043	0.027	0.040	0.024	0.018	0.016	0.009	0.004	0.005	0.003	0.005	0.11	0.044	0.295	8

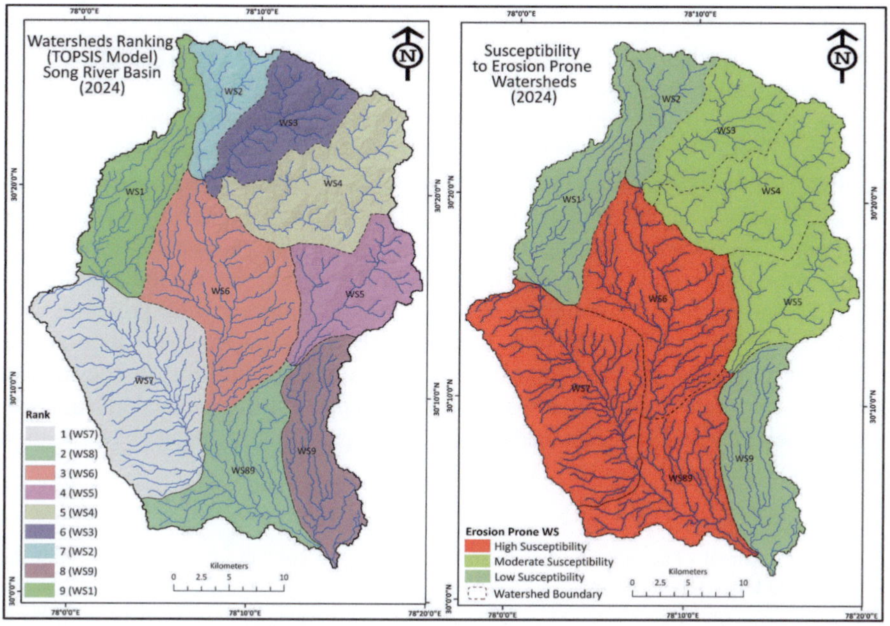

Fig. 7 Watershed prioritization and erosion susceptibility analysis of the Song River Basin (2024). The left map shows watershed rankings based on the AHP-TOPSIS multi-criteria decision model. The right map illustrates the erosion susceptibility of each watershed, classified into high, moderate, and low susceptibility zones, aiding in targeted watershed management and conservation planning

scores (0.518, 0.736, 0.545) and were ranked 1 to 3, which indicates high susceptibility and sensitivity to erosion (Table 8). Whereas WS3, WS4, WS5 (with score 0.371, 0.377 and 0.384) shows the moderate susceptibility and WS1, WS2 and WS9 (with scores 0.195, 0.329 and 0.295), indicate lower sensitivity to erosion (Fig. 7 and Table 8). The basin is mostly composed of silt, clay, sand, fine loam, coarse loam, loamy skeletal materials, together with gravel and pebbles, encounters considerable erosion issues. Shallow soils in the northern region, characterized by sparse vegetation are particularly susceptible to fluvial processes. Soil erosion is intensified by improper land management and environmental degradation, such as deforestation and agricultural growth.

4 Conclusion

The present work emphases on the prioritization of erosion prone watersheds to assess the erodibility based on the morphometric indices (linear, areal and relief aspect) derived from digitizing the survey of India (SOI) toposheets and CartoDEM with geospatial technology. The integration of morphometric indices with AHP-TOPSIS and GIS technology has demonstrated efficacy in sustainable watershed management. This comprehensive methodology facilitates an in-depth investigation of many morphometric characteristics, permitting the identification of high-risk zones prone to erosion. The study employs AHP-TOPSIS, a multi-criteria decision-making technique, to effectively evaluate watersheds according to their susceptibility, thereby enabling targeted intervention measures. Geospatial technology improves data accuracy and visualisation, offering a thorough spatial perspective. These hybrid methodology enhances decision-making by combining qualitative assessments with quantitative evaluations, making it particularly useful for addressing complex hydrological and environmental challenges. The prioritization outcomes assist policymakers, planners and resource managers in identifying critical sub-watersheds that require immediate conservation measures, such as afforestation, soil erosion control and water conservation strategies. By adopting this approach for watershed management, stakeholders can allocate resources more efficiently, mitigate land degradation and promote sustainable water resource planning.

Conflict of Interest The authors declare that they have no conflict of interest regarding the publication of this research article.

References

Hack JT (1957) Studies of longitudinal stream profiles in Virginia and Maryland. U.S. Geological Survey Professional Paper 294-B

Horton RE (1932) Drainage-basin characteristics. Trans Am Geophys Union 13(1):350–361

Horton RE (1945) Erosional development of streams and their drainage basins; hydrophysical approach to quantitative morphology. Geol Soc Am Bull 56(3):275–370

Kannan G, Pokharel S, Kumar PS (2009) A hybrid approach using ISM and fuzzy TOPSIS for the selection of reverse logistics provider. Resour Conserv Recycl 54(1):28–36

Kumar P, Sarkar P (2022) A comparison of the AHP and TOPSIS multi-criteria decision-making tools for prioritizing sub-watersheds using morphometric parameters' analysis. Modeling Earth Systems and Environment 8(3):3973–3983

Kumar A, Singh S, Pramanik M, Chaudhary S, Negi MS (2022a) Soil erodibility mapping using watershed prioritization and morphometric parameters in conjunction with WSA, SPR and AHP-TOPSIS models in Mandakini basin, India. Int J River Basin Manage 22(2):143–165

Kumar A, Singh S, Pramanik M, Chaudhary S, Maurya AK, Kumar M (2022b) Watershed prioritization for soil erosion mapping in the Lesser Himalayan Indian basin using PCA and WSA methods in conjunction with morphometric parameters and GIS-based approach. Environ Dev Sustain. https://doi.org/10.1007/s10668-021-01586-8

Langbein WB (1947) Topographic characteristics of drainage basins (No. 968-C). USGPO

Meshram SG, Sharma SK (2017) Prioritization of watershed through morphometric parameters: a PCA-based approach. Appl Water Sci 7:1505–1519. https://doi.org/10.1007/s13201-015-0332-9

Nasir MJ, Ahmad W, Jun C, Iqbal J, Bateni SM (2023) Soil erosion susceptibility assessment of Swat River sub-watersheds using the morphometry-based compound factor approach and GIS. Environ Earth Sci 82(12):315

Patel A, Singh MM, Singh SK, Kushwaha K, Singh R (2022) AHP and TOPSIS based sub-watershed prioritization and tectonic analysis of Ami River Basin, Uttar Pradesh. J Geol Soc India 98(3):423–430
Pike RJ, Wilson SE (1971) Elevation-relief ratio, hypsometric integral, and geomorphic area-altitude analysis. Geol Soc Am Bull 82(4):1079–1084
Saaty TL (1977) A scaling method for priorities in hierarchical structures. J Math Psychol 15(3):234–281
Saaty TL, Wind Y (1980) Marketing applications of the analytic hierarchy process. Manag Sci 26(7):641–658
Sarkar P, Kumar P, Vishwakarma DK, Ashok A, Elbeltagi A, Gupta S, Kuriqi A (2022) Watershed prioritization using morphometric analysis by MCDM approaches. Eco Inform 70:101763
Sarma B, Sarmah N, Malakar A (2024) Degradation of water and land resources in the Himalayan mountain ecosystems. In: Addressing the climate crisis in the Indian Himalayas: Can traditional ecological knowledge help?, pp 31–52
Schumm SA (1956) Evolution of drainage systems and slopes in badlands at Perth Amboy, New Jersey. Geol Soc Am Bull 67(5):597–646
Sharma S, Mahajan AK (2020) GIS-based sub-watershed prioritization through morphometric analysis in the outer Himalayan region of India. Appl Water Sci 10(7):1–11. https://doi.org/10.1007/s13201-020-01243-x
Shekar PR, Mathew A, Hasher FFB, Mehmood K, Zhran M (2025) Towards sustainable development: ranking of soil erosion-prone areas using morphometric analysis and multi-criteria decision-making techniques. Sustainability 17(5):2124
Shekar PR, Mathew A, Abdo HG, Almohamad H, Abdullah Al Dughairi A, Al-Mutiry M (2023) Prioritizing sub-watersheds for soil erosion using geospatial techniques based on morphometric and hypsometric analysis: a case study of the Indian Wyra River basin. Appl Water Sci 13(7):160
Singh S, Kumar A, Negi MS (2022) Hydro-morphological investigations of Neeru watershed using DEM and geospatial techniques. Int J Geography Geol Environ 4(2):13–23. https://doi.org/10.22271/27067483.2022.v4.i2a.110
Singh R, Jain SK (2002) Assessment of erosion-prone watersheds using morphometric indices: A case study in the Indian Himalayas. J Environ Manage 64(2):109–124
Singh O, Singh J (2018) Soil erosion susceptibility assessment of the lower Himachal Himalayan watershed. J Geol Soc India 92:157–165. https://doi.org/10.1007/s12594-018-0975-x
Strahler AN (1950) Davis' concepts of slope development viewed in the light of recent quantitative investigations. Ann Am Assoc Geogr 40(3):209–213
Strahler AN (1952) Hypsometric (area-altitude) analysis of erosional topography. Geol Soc Am Bull 63(11):1117–1142
Strahler AN (1957) Quantitative analysis of watershed geomorphology. EOS Trans Am Geophys Union 38(6):913–920
Strahler AN (1964) Part II. Quantitative geomorphology of drainage basin and channel networks. In: Chow VT (ed) Handbook of applied hydrology

Coastal and Environmental Risk Management

Application of Geospatial Technologies and AI to Detect and Analyze the Shoreline Change of Visakhapatnam, a Coastal District

Sravanthi Karnata and Namrata Kumari

1 Introduction

Coastal areas are features of the dynamic and complex biophysical environment that influence many issues about the support of biodiversity, economic activities, and habitation. However, these areas are becoming more sensitive to climate change, relative sea level variation, and anthropogenic influences that cause appreciable changes in shorelines in the long term. Shoreline changes, which include erosion and deposition, directly affect the stability of the coastal environment. To understand and manage these changes, sustainable coastal management needs to influence the decision-making process. Shoreline change detection is a wide concept that can result from a combination of different methods and techniques, such as geospatial sciences, remote sensing streams, and complex algorithms. In the past, the observations of shoreline change have been conducted through field surveys and historical maps, which provide good data, although they are time-consuming and sometimes contain restricted spatial and temporal data. As the technologies developed, the assistance provided by the integration of satellite remote sensing and GIS started playing a huge role in analyzing shorelines by acquiring extremely high spatial resolution of shorelines, monitoring shoreline change for long periods, automatic detection, and processing huge amounts of Shoreline data sets conveniently (Yasir et al. 2020). These technologies, when adopted alongside AI and ML solutions, can provide more accurate, automated, and largely scalable methodologies for detecting and analyzing shoreline shifts. The Random Forest, Support Vector Machines, and Convolutional Neural Networks have shown increased capability for extracting key patterns and features from satellite data (Bayram et al. 2017; Bengoufa et al. 2021). These methods improve the ability based on which the satellite imagery

S. Karnata (✉) · N. Kumari
Department of Geography, Lovely Professional University, Phagwara, Punjab, India
e-mail: sravanthikarnata7@gmail.com

M. O. Sarif and A. Sharifi (eds.), *Advanced Geospatial Intelligence and AI for Environmental Resilience and Sustainable Development*, Advances in Geographic Information Science, https://doi.org/10.1007/978-3-032-03714-5_12

can be classified or segmented to enable the detection of land–water boundaries. On the same note, the use of time series analysis and AI-developed models to estimate future positions of shorelines provides planners with foresight on future positions of the coasts, hence aiding in risk prevention (Bengoufa et al. 2021; Abd-Elhamid et al. 2023; Chen et al. 2021). Tsai, in their study, used edge detection techniques to monitor the changes in the shoreline in Zengwun with the help of Landsat data, with which they observed that more than 80% of Taiwan's shoreline is facing severe recession (Tsai 2022).

This chapter provides information about the geospatial technologies and artificial intelligence (AI) used in shoreline change detection and analysis with special reference to Visakhapatnam-India, which is an economic and environmental hub with a long shoreline along the Bay of Bengal. However, the district is facing major shoreline changes due to dynamic factors such as tidal actions and storm surges, besides human-induced activities, including the expansion of the area to accommodate commercial and industrial port activities. The region presents an interesting case to showcase the role of applying AI and GIS in monitoring and mitigating coastal transformations. The methodological approach proposed in this chapter employs geospatial resources such as QGIS and ArcMap, in conjunction with advanced ML algorithms and plugins to process satellite images identified across a ten-year timeline (2014–2024). The process involves data collection and preparation, shoreline mapping by Image segmentation, Random Forest classification, and change identification through statistical data analytic tools, thereby incorporating DSAS techniques. In other words, the results seek to define shoreline behavior, capture areas of erosion and accretion, and offer information that policymakers and coastal planners may utilize. As such, this research aims to bring together geospatial technologies and AI for shoreline attribute extraction, which can also be employed to identify shoreline changes affecting other coastal areas of the world. It also underscores the role of technology in the mitigation of emerging hazards such as climate change and coastal distress. This chapter aids in the progression of the field of geospatial science and stresses the importance of AI in improving knowledge about active coastal processes.

2 Recent Developments

Over the past decades, coastal studies have been revolutionized with the help of geospatial technologies and artificial intelligence (AI). Originally, evaluating the shoreline shift was a slow and highly qualitative process conducted manually using aerial imagery, field surveys, and basic GIS tools—a very time-consuming and hardly scalable approach to the problem. However, with the advancement in remote sensing technology and improved satellite imagery of the coastal features, assessing the coastal zones has remarkably gotten easier and more accurate (Sharma et al. 2024). First of all, the synergy of AI with geographic tools has led to substantial enhancements in the possibility of applying data mining techniques and pattern recognition for extracting and categorizing the shoreline features; this implies effective methods and tools based on AI for the detection and analysis of shoreline variation. Based on current developments, there is a growing indication that machine learning (ML) and deep learning (DL) are increasingly being incorporated into geospatial analysis (Bengoufa et al. 2021a; 2021b; Boussetta et al. 2023). It is now increasingly common to apply AI models for examples of work such as the categorization of land and water, demarcation of shorelines, and even forecasting future shifts in coastal regions. Together with the new technological tools and cloud storage systems, including Google Earth Engine (GEE) and GIS programs like ArcGIS Pro, these technologies can significantly improve the rate and quality of large-scale data analysis and monitoring of coastal ecosystems. A study by Santra et al. (2023) used the SVM approach with GEE to quantify the shoreline dynamics, which had 99.5% accurate results. The focus of this section is on specific recent advancements in the application of geospatial technologies and AIS in the evaluation of shoreline changes and innovation in the satellite data sample collection to enhance machine learning algorithms and GIS-based systems. Object-based image Classification (OBIA), a technique that classifies image objects rather than individual pixels to DSAS (Digital Shoreline Analysis System), a GIS tool developed by the USGS for shoreline metrics computation have allowed for a more precise and comprehensive assessment of shoreline dynamics, such as erosion, accretion, and land–water boundary shifts (Dronova et al. 2011; Kaplan and Avdan 2017; Cenci et al. 2018).

2.1 *Advancements in Geospatial Technologies*

Improved Satellite Data: One of the significant developments in recent years has been the availability of high-resolution satellite imagery. Datasets such as those from **Landsat 8** (launched in 2013) and **Sentinel-2** (launched in 2015) have significantly enhanced the ability to monitor coastal areas. The **Sentinel-2** mission, in particular, provides high-frequency imagery (every 5–10 days) with a spatial resolution of 10–20 m, which is ideal for detecting even small shoreline movements and fluctuations. This data aids in the creation of detailed maps of shorelines, which helps to analyze both the erosion and accretion zones with greater accuracy (Kaplan and Avdan 2017; Tian et al. 2020).

GIS Tools: Technology advancements led to the development of modern GIS software such as ArcGIS Pro and QGIS, which have become central to the analysis of shoreline changes over the last decade. These platforms offer a wide array of tools for image classification, extraction of features, and spatial analysis, which play a huge role in shoreline mapping. These tools include machine learning, such as Random Forest training and Support Vector Machine models, that enable the classification of land and water boundaries automatically. Integrating these tools with cloud-based platforms can handle large volumes of geospatial data, making it easier to analyze temporal changes in coastal zones (Bayram et al. 2017; Bengoufa et al. 2021a; Sharma et al. 2024; Cenci et al. 2018; Chen et al. 2005).

DSAS (Digital Shoreline Analysis System): DSAS is a widely used tool for shoreline change metrics developed by the U.S Geological Survey (USGS) specifically designed to calculate various shoreline change statistics including the Net Shoreline Movement (NSM), End Point Rate (EPR) to quantify and analyze shifts in shoreline positions over time including essential information on the rate and direction of shoreline movement, which plays a crucial role in coastal management and risk assessment. DSAS underwent several updates in recent years to improve its functionality and integration with modern GIS platforms, such as the ArcMap extension of DSAS that supports high-resolution satellite imagery (Nassar et al. 2019; Gopinath et al. 2023; Thinh and Hens 2017).

2.2 *AI and Machine Learning in Coastal Studies*

Supervised and Unsupervised Learning Techniques: Machine learning (ML) algorithms have increasingly been employed to automate the process of classifying land–water boundaries in satellite imagery. Supervised learning techniques, such as Random Forest (RF) and Support Vector Machines (SVM), are commonly used for this task. With the help of manually trained sample datasets, these models can predict the features in new, unseen images, significantly reducing the manual work for large data sets. On the other hand, **unsupervised learning** techniques such as **K-means clustering** have also been used for shoreline classification, particularly when labeled datasets are unavailable. Unsupervised methods can identify patterns in data based solely on the inherent structure of the data itself (Bayram et al. 2017; Innocent Abbas 2013; Talukdar et al. 2020; Pham et al. 2024).

Hybrid Models: A growing trend is the use of hybrid models, which combine machine learning techniques with traditional geospatial methodologies. One such approach is Object-Based Image Analysis (OBIA), which integrates spectral and spatial features to perform segmentation and classification. OBIA uses both machine learning algorithms (such as RF or SVM) and spatial information (such as shape, size, and adjacency) to detect shoreline features, unlike other methods, such as pixel-based methods. This method assigns a class label to each pixel in an image that shares similar properties (Prasetyo et al. 2016). This technique improves classification accuracy and has proven to be particularly useful for coastal applications where both spectral and spatial characteristics are crucial for proper classification. QGIS, an open-source GIS software, plays a crucial role in the implementation of OBIA for coastal studies. The results can be combined with machine learning algorithms, such as Random Forest (RF) and Support Vector Machines (SVM), to enhance classification results. This significantly improves the accuracy of land–water boundary delineation in coastal zones, especially when working with satellite data (Bayram et al. 2017; Dronova et al. 2011; Kaplan and Avdan 2017; Mohammed et al. 2024).

2.3 Integration of AI with Geospatial Technologies

The integration of AI and machine learning with GIS tools has been a major advancement in the field of shoreline change detection. Machine learning algorithms, such as RF, SVM, and CNNs, can now be directly implemented within GIS software, providing users with an end-to-end solution for geospatial analysis (Bengoufa et al. 2021a). For example, the recent developments in OBIA, machine learning, and DSAS offer a powerful suite of tools for detecting and analyzing shoreline features, while ML algorithms such as Random Forest and Support Vector Machines enhance the classification by automating the detection of land–water boundaries. Once the shorelines are classified and extracted, **DSAS** is used to calculate shoreline change metrics, which can be visualized and analyzed within GIS platforms like **ArcGIS** or **QGIS**. The integration of Object-Based Image Classification (OBIA) with machine learning algorithms, combined with the application of DSAS for shoreline change analysis, represents a powerful advancement in the field of coastal monitoring. These tools allow for the efficient detection, classification, and analysis of shoreline changes, offering a more precise and automated approach to the study of coastal dynamics (Boussetta et al. 2023; Tian et al. 2020; Niculescu and Lam 2019; Sreekesh et al. 2020).

3 Study Rationale

Shoreline change detection is important in ascertaining the status of coastal areas, which are areas of dynamic natural features, including erosion, deposition of sediments, and anthropogenic activities. This calls for an efficient and reliable monitoring system of these changes towards efficient coastal management, hazards, and risk assessment. The current research is interested in extending the use of advanced geospatial technologies, which include AI & ML, together with the conventional GIS tools, in the study of shoreline fluctuation in Visakhapatnam, which is frequently affected by storm surges and tidal range. With the availability of high-resolution satellite data from sources like Landsat 8 (2014–2015) and Sentinel-2 (2016–2024), this research aims to employ advanced techniques such as Object-Based Image Analysis (OBIA) and Random Forest (RF) for automated classification of shoreline features. Additionally, DSAS will be used to calculate key shoreline change metrics such as Net Shoreline Movement (NSM), End Point Rate (EPR), and weighted linear regression rates (WLR). The AI integration with GIS platforms like QGIS and ArcGIS makes it possible to identify movements of the shoreline over the past decade with high precision, which turns out to be a helpful resource for the identification of trends and tendencies in coastal dynamics necessary for sustainable coastal management in Visakhapatnam.

4 Materials and Methods

This section outlines the methodology adopted for detecting and analyzing shoreline changes in Visakhapatnam using Sentinel-2 satellite imagery (2015–2024) Level-2A data from the Copernicus Data Space ecosystem. The process encompassed data acquisition, preprocessing, image classification, shoreline extraction, and change analysis, conducted using QGIS for machine learning-based classification and ArcMap for shoreline change computation through the Digital Shoreline Analysis System (DSAS).

4.1 Study Area

Visakhapatnam is a coastal district located between 17° 43′ N and 83° 17′ E on the Northeastern coast of Andhra Pradesh (Fig. 1). It is nestled between the Eastern Ghats and the Bay of Bengal, making it a key coastal area. Visakhapatnam serves as a major industrial hub, hosting various industries like steel plants, refineries, and pharmaceuticals. As of the 2011 census of India, Visakhapatnam had a population of 4.3 lakhs occupying approximately 682 km^2. It is home to the Visakhapatnam Port, one of the busiest and largest cargo-handling ports in India, playing a pivotal role in international trade. The city's economy thrives on diverse sectors such as fishing, information technology, tourism, and small-scale industries. Vizag's strategic location as a coastal city contributes significantly to its economic growth and trade connections. Visakhapatnam boasts diverse ecosystems, including coastal plains, mangrove forests, and marine biodiversity along its coastline. The region is also known for its picturesque beaches and scenic beauty, attracting tourists and environmental enthusiasts. The mangrove ecosystems in Vizag act as natural buffers against coastal erosion and provide habitats for various marine species, contributing to the overall ecological balance of the region. This particular district was selected due to its dynamic shoreline changes driven by natural processes and anthropogenic activities. The region experiences significant coastal erosion, accretion, and periodic cyclonic events, making it an ideal case study for shoreline monitoring and analysis (Gireesh 2023; Kannan et al. 2015).

4.2 Data Collection and Preprocessing

The Sentinel-2 Level-2A imagery and Landsat-8 Level 2 (Table 1), which are atmospherically corrected and provide surface reflectance data, were downloaded. Data selection focused on images with less than 5% cloud cover to ensure clarity and precision. Uniform periods were chosen each year to maintain seasonal consistency and minimize variability due to tidal or climatic factors. The selected images were clipped

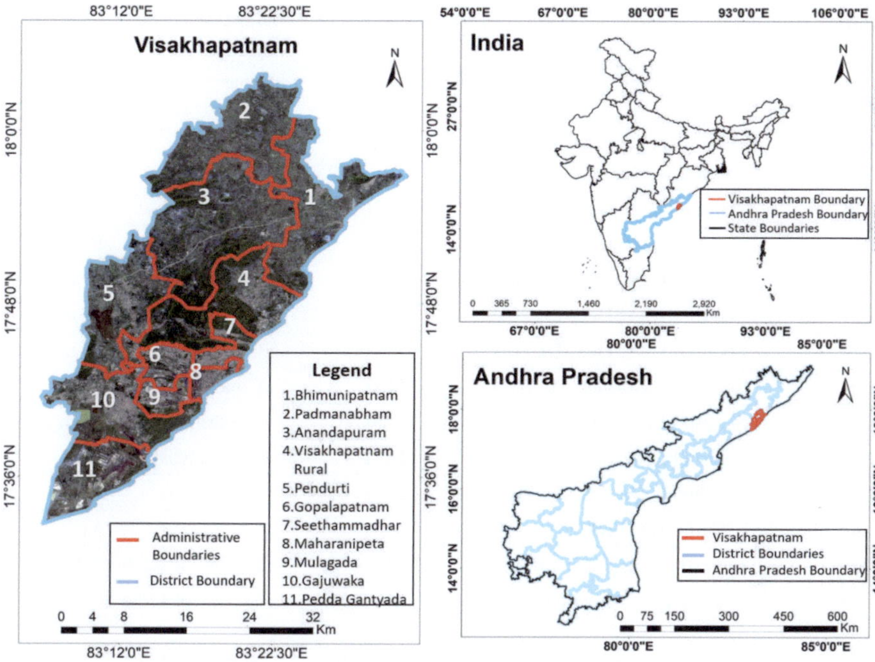

Fig. 1 Map of the study area, Visakhapatnam District

to the study area using a predefined polygon shapefile delineating the Visakhapatnam coastline. Administrative boundary data at the State level, District level, and Taluka level are collected to understand the study area and to extract the satellite imagery accordingly. Tide gauge data were collected from the PSMSL to calculate the uncertainty values for DSAS analysis.

4.3 *Object-Based Image Analysis*

Object-based image analysis (OBIA) was applied to classify the satellite imagery into land and water categories. The images were segmented using the Orfeo Toolbox plugin in QGIS, which groups pixels into homogeneous objects based on spectral and spatial properties with the help of the Mean Shift Algorithm (Comaniciu and Meer 2002). The Mean Shift algorithm was chosen for image segmentation due to its non-parametric nature, as it does not require prior knowledge of the number of clusters. Unlike k-means or other supervised techniques, Mean Shift automatically identifies clusters based on the underlying data distribution, making it well-suited for complex and heterogeneous satellite imagery. It also considers both spectral and spatial information simultaneously, which enhances the accuracy of segmenting land and water features. Its ability to handle arbitrary cluster shapes and adapt to the local

Table 1 Data collection

Year	Latest data (cloud < 5%)	Type/Sensor	Source	Resolution
Remote sensing data				
2014	Landsat 8; Dec 7, 2014	OLI/TIRS	USGS	30 m
2015	Landsat 8; Dec, 10, 2015	OLI/TIRS	USGS	30 m
2016	Sentinel L2A; Dec, 26, 2016	MSI	Copernicus	10 m
2017	Sentinel L2A; Dec, 29, 2017	MSI	Copernicus	10 m
2018	Sentinel L2A; Dec, 29, 2018	MSI	Copernicus	10 m
2019	Sentinel L2A; Dec, 29, 2019	MSI	Copernicus	10 m
2020	Sentinel L2A; Dec, 28, 2020	MSI	Copernicus	10 m
2021	Sentinel L2A; Oct, 18, 2021	MSI	Copernicus	10 m
2022	Sentinel L2A; Oct, 29, 2022	MSI	Copernicus	10 m
2023	Sentinel L2A; Nov, 14, 2023	MSI	Copernicus	10 m
2024	Sentinel L2A; Oct, 27, 2016	MSI	Copernicus	10 m
Administrative boundary data				
2024	India Boundary-State level	Shapefile	Survey of India	NA
2024	Andhra Pradesh Boundary- District Level	Shapefile	Survey of India	NA
2024	Visakhapatnam Boundary – Taluka Level	Shapefile	Survey of India	NA
Tidal data				
2014 to 2022	Tide Gauge Data of Visakhapatnam	Metric values	PSMSL	NA

structure of the data makes it particularly effective for object-based image analysis (OBIA) in remote sensing.

Mean Shift Algorithm: The Mean Shift algorithm iteratively shifts data points towards regions of higher density using a weighted average of neighboring points within a kernel window

$$m(x) = \sum_{i=1}^{n} \frac{K(x_i - x)x_i}{K(x_i - x)}$$

where m(x) = new shifted mean position, x = current position of the data point, x_i = neighboring data points within the kernel window, $K(x_i - x)$ = kernel function (e.g., Gaussian kernel) determining weight

Kernel Function in Mean Shift: A common choice for the kernel function is the **Gaussian kernel**:

$$K(x) = e^{-\frac{||x||^2}{2h^2}}$$

where h = bandwidth (radius of the search window), $\|x\|^2$ = squared Euclidean distance between data points.

Convergence Condition: The algorithm stops when the shift in mean location is below a small threshold ε.

$$\|m(x) - x\| < \varepsilon$$

where ϵ is a small tolerance value for convergence.

Mean Shift in Image Segmentation: In image segmentation, Mean Shift operates in the feature space (e.g., spatial + color space). Each pixel is treated as a data point, and clustering is performed to group similar pixels together.

The feature vector for each pixel x includes: Spatial coordinates (x, y) and Color values (R, G, B), the combined distance metric is

$$D = \sqrt{D = \lambda_s(x_1 - x_2)^2 + \lambda_c(c_1 - c_2)^2}$$

where λ_s = spatial weight, λ_c = color weight, $x_1,\ x_2$ = spatial coordinates of pixels, $c_1,\ c_2$ = color values of pixels. This segmentation (Fig. 2) output was assigned its zonal statistics with mean and standard deviation calculations, and within the QGIS environment. This function helps to produce an attribute table with the values of each object in the segmented image (Winsemius and Braaten 2023). This helps to analyze land cover classes based on spectral and spatial properties. Each segment is treated as an individual zone, and the raster pixel values of each zone are extracted. For each segmented object (zone), the Mean (μ) and Standard Deviation (SD, σ) are calculated using the pixel values within the segment.

Mean Calculation: The mean pixel value for a given segmented region (zone) is:

$$\mu = \frac{1}{N} = \sum_{i=1}^{n} x_1$$

where μ = mean pixel value for the object, x_i = individual pixel values within the segment, and N = total number of pixels in the segment.

Standard Deviation (SD) Calculation: The SD measures the variability of pixel values within each segment:

$$\sigma = \sqrt{\frac{1}{N}\sum_{i=1}^{n}(x_i - \mu)\text{^}2}$$

where σ = standard deviation of pixel values in the segment, μ = mean pixel value of the segment, and x_i = individual pixel values.

Training data were manually digitized within QGIS by creating point features into different classes of LULC level 1 description (Built-up, Agriculture, Forest, Barren, and Waterbody) (National Remote Sensing Centre 2014). These samples

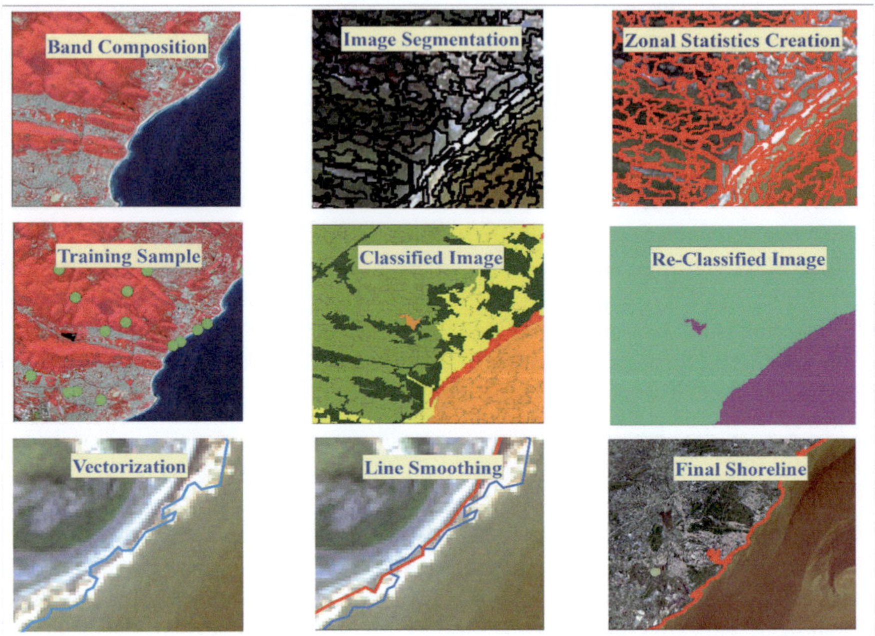

Fig. 2 OBIA, RF, and shoreline extraction process

were labeled based on spectral characteristics, including reflectance values from key bands such as coastal, blue, green, and near-infrared (NIR). The Random Forest (RF) algorithm, a robust ensemble learning technique, was employed for classification.

Random Forest (RF) Classifier: The Random Forest (RF) algorithm, a robust ensemble learning technique, was employed for classification. Random Forest classification was selected due to its high accuracy, robustness against overfitting, and ability to handle large, high-dimensional datasets, making it well-suited for complex land–water classification tasks in satellite imagery. RF operates by constructing multiple decision trees and aggregating their results. The classification is based on the majority voting principle, such as:

$$P(c) = \frac{1}{N}\sum_{i=1}^{n} h_i(x)$$

where P(c) = probability of class c, N = number of trees, $h_i(x)$ = decision of the ith tree for input x.

The RF algorithm was employed to train the sample that was created earlier with the segmentation to create a training model. The created model is used on the segmented imagery with zonal statistics in generating classified raster outputs that delineate different land use and land cover classes. All the above processes are operated in the QGIS environment, and the classified image is operated in ArcMap to reclassify and extract the shoreline. The classified raster image has five classes (Built-up, Agriculture, Forest, Barren, and Waterbody) and is reclassified into two classes such as land and water body, in ArcGIS. Figure 2 illustrates the step-by-step workflow of the OBIA and shoreline extraction process, from image segmentation and classification to vectorization and final shoreline delineation for further analysis.

4.4 Shoreline Extraction

The classified raster was subsequently converted into vector format to extract the shoreline. The Raster to Vector Conversion tool in ArcGIS facilitated this transformation, isolating the land–water boundary as a polyline representation of the shoreline. To improve the accuracy and visual consistency of the shoreline, smoothing was applied using the Smooth line tool in ArcGIS, which is suitable for reducing noise in extracted shorelines while maintaining overall shape accuracy. This step removed minor irregularities introduced during the classification process and ensured a clear delineation of the coastline.

4.5 Shoreline Change Analysis

For the analysis of shoreline changes, the vector shorelines for each year were imported into ArcMap and analyzed using the DSAS extension. A baseline was manually created landward of the shorelines to serve as a reference for change measurement. Transects were generated perpendicular to the baseline at consistent intervals of 100 m to quantify the movement of the shoreline over time. DSAS computed key metrics, including Net Shoreline Movement (NSM), which measures the total distance of shoreline change, End Point Rate (EPR), which calculates the rate of change between the earliest and latest shorelines, and Weighted Linear Regression (WLR), which assesses long-term trends in shoreline dynamics (Barik et al. 2021). The detailed analysis can be seen in Fig. 4, which represents the methodological flowchart outlining the sequential steps followed in this chapter.

Net Shoreline Movement (NSM): NSM calculates the absolute movement of the shoreline between the oldest and most recent shorelines, and it can be expressed as below:

$$NSM = S_f - S_i$$

where NSM: Net Shoreline Movement (m), S_f = shoreline position at the final year, S_i = shoreline position at the initial year, while a positive NSM value indicates accretion, whereas a negative value represents erosion.

Endpoint Rate (EPR): EPR estimated the annual rate of shoreline change by dividing NSM by the elapsed time between two shoreline positions, and it can be expressed as below:

$$EPR = \frac{NSM}{t_f - t_i}$$

where EPR = End point Rate (m/year), $t_{f = \text{final year}}$, t_i = initial year. The assumption in EPR is a linear trend in shoreline change, but it does not consider intermediate shoreline positions, making it sensitive to anomalies.

Weighted Linear Regression (WLR): To overcome the limitations of NSM and EPR, WLR provides a statistically robust estimation of shoreline trends by assigning weights to shoreline positions. This is expressed as below:

$$Y = a + bt + e$$

where Y = shoreline position (m), a = intercept, b = regression slope (change rate), t = time (years), e = residual error.

The rate of change using WLR is calculated as:

$$WLR = \frac{\Sigma w_i \left(x_i - x^{-}\right)\left(y_i - y^{-}\right)}{\Sigma w_i (x_i - x^{-})^2}$$

where w_i = weight assigned to each shoreline, $x_{i,}$ y_i = time and shoreline position respectively, x^{-},y^{-} = mean values of time and shoreline. WLR is less affected by outliers and short-term fluctuations, making it ideal for long-term trend analysis.

4.6 Validation

Several validation steps were incorporated to ensure the reliability of shoreline change analysis with the integration of AI and GIS. The Random Forest (RF) classification applied to the segmented image effectively distinguished between different classes. Further, the zonal statistics-based training approach improved classification accuracy based on calculations of mean and standard deviations of the segmented objects. The extracted shorelines were cross-validated against the Google Earth imagery. The uncertainty values shown in Table 2 play an important role in the DSAS environment and were derived using PSML tide gauge data, allowing for a more precise representation of shoreline dynamics. Other parameters considered in uncertainty calculations include seasonal errors, which are zero in this case as all the data is taken from the same seasons each year. Shoreline proxy offset, measurement errors, toposheet offset, and rectification errors are zero and not applicable in this study due to the unique methodology used for the research. The digital error is taken as half of the spatial resolution of the satellite data.

Table 2 Uncertainty calculations

Positional error (m)	2014	2015	2016	2017	2018	2019	2020	2021	2022	2023	2024
Seasonal error (E_s)	NA	NA	NA	NA	NA	NA	NA	NA	NA	NA	NA
Tidal fluctuation (E_{td})	7	7.1	7	7.2	7	7.2	7.3	7.2	7.2	7.3	7.2
Shoreline proxy offset (E_o)	NA	NA	NA	NA	NA	NA	NA	NA	NA	NA	NA
Measurement errors (m)	0	0	0	0	0	0	0	0	0	0	0
Rectification error (E_r)	0	0	0	0	0	0	0	0	0	0	0
Digitizing error (E_d)	15	15	15	5	5	5	5	5	5	5	5
Toposheet survey offset (E_t)	NA	NA	NA	NA	NA	NA	NA	NA	NA	NA	NA
Pixel error (E_p)	0	0	0	0	0	0	0	0	0	0	0
Total shoreline position error (E_{sp}) (m)	22.00	22.10	22.00	12.20	12.00	12.20	12.30	12.20	12.20	12.30	12.20
Year	2014–2015	2015–2016	2016–2017	2017–2018	2018–2019	2019–2020	2020–2021	2021–2022	2022–2023	2023–2024	2014–2024
Uncertainty	**6.64**	**5.85**	**4.92**	**4.92**	**4.95**	**4.95**	**4.94**	**4.95**	**4.95**	**3.49**	**0.32**

Uncertainty Estimation: Shoreline analysis is subject to various sources of uncertainty, including tidal variations, sensor resolution, and classification errors. The total uncertainty (u_t) was calculated as:

$$Esp = \sqrt[2]{E_s + E_{td} + E_o + E_m + E_r + E_d + E_t + E_p}$$

$$U_t = \frac{\sqrt[2]{Esp_i + Esp_f}}{no.\ of\ years}$$

U_t = Uncertainty.

Esp_i = Total shoreline Positional error of the initial year.

Esp_f = Total shoreline Positional error of the final year.

The DSAS model was configured with a 300-m onshore baseline, ensuring that the measured shoreline changes were consistent and not influenced by small-scale fluctuations. The positioning of the Baseline, Transects, and extracted shoreline can be seen in Fig. 3. The baseline positioning helped improve the assessment of long-term accretion and erosion trends. For the final analysis, **Weighted Linear Regression (WLR)** was applied to shorelines from 2014 to 2024, while **End Point Rate (EPR)** and **Net Shoreline Movement (NSM)** were calculated using only 2014 and 2024 shorelines to focus on decadal change. The derived results for different parameters are cross-checked with different historical data available through literature, showing high accuracy overall. For example, a study conducted by Gireesh et al. (Gireesh et al. 2023) employed the maximum classification method and used DSAS analysis, which showed erosion and accretion rates of − 6.78 m/year, and 3.8 respectively, for the years 2014 to 2019 (Gireesh et al. 2023). Another study conducted by Kannan and Murthy (2016) showed that 39 km of shoreline is under erosion and 75 km of shoreline is under accretion, while stable land is found to be 41 km, where they have taken 155 km of coastline of the old Visakhapatnam district (Kannan et al. 2015). The study by Shaik et al. (2024) used DSAS analysis to understand the response of shorelines to rising sea levels, and their results show − 50.43 m/year EPR for the years 2011–2014, and WLR shows the mean accretion and erosion as − 1.02 m/year and 1.01 m/year respectively (Shaik et al. 2024). Altogether, the results derived from this approach of integrating GIS and ML algorithms in this chapter show an accuracy of more than 95% overall, highlighting the scope of GIS and AI integrations for shoreline change detection and monitoring in the future.

4.7 *Visualization and Interpretation*

The results of the shoreline change analysis were visualized using a combination of maps, statistical plots, interpolated spatial models, and tables. Shoreline evolution maps were generated for all the years from 2014 and 2024, highlighting the spatial

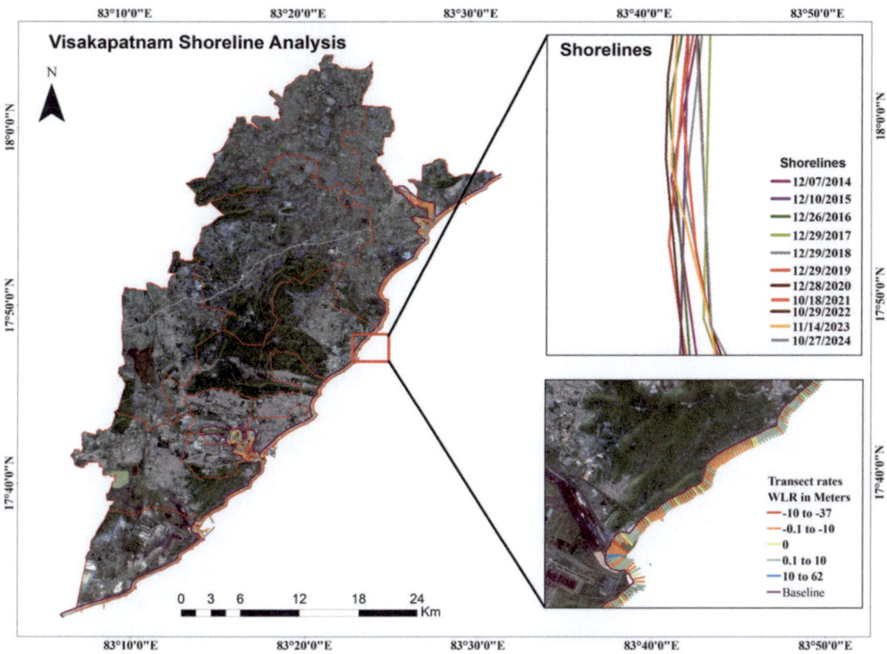

Fig. 3 Map of extracted shoreline and transects

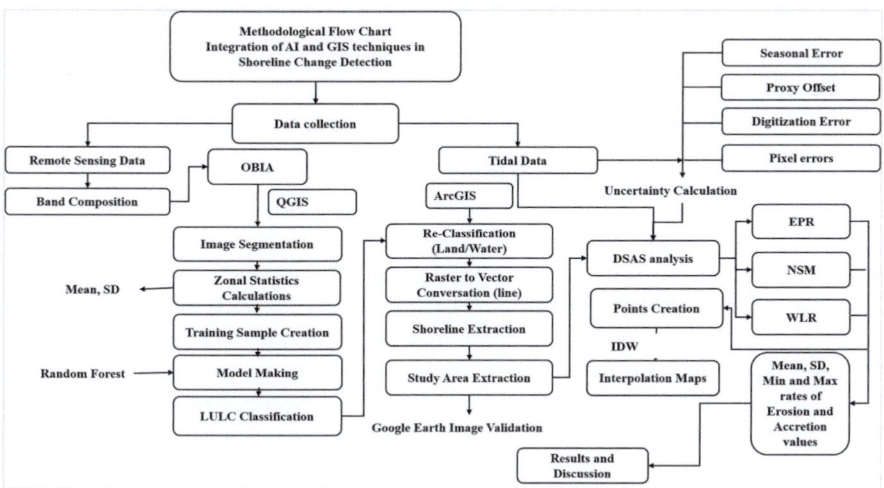

Fig. 4 Methodological flow chart

extent of erosion and accretion along the Visakhapatnam coastline. The extracted shorelines were overlaid to provide a visual representation of change, along with the transects with erosion-prone areas marked in red and accretion zones in blue. To enhance interpretability, interpolated heat maps were created to show the spatial distribution of EPR, NSM, and WLR values along the study area. Using gradient color schemes (e.g., dark red for high erosion and dark blue for high accretion), these maps effectively conveyed areas of concern, such as high erosion hotspots or stable regions. For interpolation mapping, the shoreline is divided into three zones to give better visuals. Zone one consists of the Bheemunipatnam shoreline. Zone two consists of Visakhapatnam Rural, Seethammadhara, and Maharanipeta. Zone three consists of the shorelines of Gajuwaka and Peddagantyada in spatial. Additionally, statistical tables were employed to present key metrics, while charts depicted the transects experiencing erosion, accretion, and stability. The statistical analysis revealed that 39% of transects exhibited accretion under WLR, with a mean accretion rate of 3.83 m/year, whereas 37% of transects showed erosion, averaging − 2.17 m/year. The maximum shoreline retreat (− 37.5 m) and progradation (62.9 m) were identified using WLR, while NSM recorded extreme values of − 557 m for erosion and + 372 m for accretion. These results indicate that while accretion dominates certain segments of the coastline, significant erosion is also occurring, likely due to sediment deposition, anthropogenic influences, or nearshore hydrodynamics. The visualization approach effectively communicates the shoreline changes, offering critical insights for coastal management and planning.

5 Results and Discussions

5.1 Shoreline Change Analysis

The analysis of shoreline changes in Visakhapatnam from 2014 to 2024 using Object-Based Image Analysis (OBIA), Random Forest (RF) classification, and the Digital Shoreline Analysis System (DSAS) reveals a dynamic coastal environment influenced by both natural processes and human interventions. The study quantified shoreline changes using three key metrics: Weighted Linear Regression (WLR), End Point Rate (EPR), and Net Shoreline Movement (NSM), each of which offers a different perspective on the rate and magnitude of coastal change.

The shoreline of Visakhapatnam stretches over 135 km, with a 125 km shoreline covered by DSAS, as the 50 to 100 transects derived no data values due to no intersection between one or more shorelines of different years. The WLR analysis covered 1250 transects, while EPR and NSM covered 1219 transects, and the area distribution can be observed in Table 3. The overall results indicate a mixed pattern of erosion and accretion across the study area, with accretion dominating in certain sections while significant erosion was observed in others. A total of 1250 transects (each transect is of a 100 m uniform interval) were analyzed, covering a total of

125 km of the Visakhapatnam shoreline, among which 55.5 km of shoreline exhibited accretion, 41.3 km of shoreline exhibited erosion, with 26.1 km of shoreline remained stable. The Spatial distribution analysis indicates that Bheemunipatnam, Maharanipeta, and Gajuwaka are facing high Erosion, while Visakhapatnam rural, Seethammadhara, and Peddagantyada are facing high Accretion of − 10 and + 10 m/year. On the other hand, high erosion rates of more than − 10 m found in Bheemunipatnam and Maharanipeta are mainly due to the presence of the river mouth and port, respectively, while the spotted high accretion zones with more than 10 m are due to wave actions.

End Point Rate (EPR) Analysis: The End Point Rate (EPR) was computed using shorelines from 2014 and 2024, allowing an assessment of long-term shoreline dynamics. A total of 1,219 transects were generated along the Visakhapatnam coast. Of these, 413 transects (33.89%) showed erosion, 554 transects (45.45%) showed accretion, and 262 transects (21.48%) remained stable (Table 4). The mean shoreline change rate using EPR was + 0.2 m/year, with a maximum accretion of + 37.25 m/year and maximum erosion of − 57.14 m/year. The mean erosion rate was − 1.02 m/year with a standard deviation of 4.2 m, whereas the mean accretion rate was + 1.2 m/year with a standard deviation of 3.3 m. This indicates both erosion and accretion rates are not uniform across the coastline where some areas experience high erosion while some experience high accretion (Fig. 5). Out of the analyzed 121.9 km shoreline, 55.5 km long is undergoing accretion with an average of + 1.2 m/year, and 41.3 km shoreline is facing erosion with an average of − 1.02 m/year, while the 26 km shoreline is stable with no major changes.

The spatial distribution maps (Fig. 6) indicated that high erosional activities are concentrated in certain locations, particularly Bheemunipatnam and Gajuwaka, where the erosional activities are higher than − 3 m/year, and the other places show accretion of more than 3 m/year, suggesting sediment deposition and shoreline advancement. with a dominant stable area found in the Maharanipeta, Seethammadhara, and south of Bheemunipatnam, Gajuwaka. This trend aligns with previous studies on coastal sediment transport in Visakhapatnam, which suggest that coastal protection structures, harbor developments, and sediment deposition patterns influence these changes.

Net Shoreline Movement (NSM) Analysis: The Net Shoreline Movement (NSM) provides an absolute measure of shoreline displacement by comparing the 2014 and 2024 shoreline positions. A total of 1219 transects were analyzed, revealing that 413 transects (33.89%) show erosion, 554 transects (45.45%) indicate accretion, and 262

Table 3 Shoreline analysis: transects area distribution (135 km)

Analysis type	Total covered (km)	Erosion (km)	Accretion (km)	Stable (km)	No-value area (km)
EPR	121.9	41.3	55.5	26.1	13.1
NSM	121.9	41.3	55.5	26.1	13.1
WLR	125	47.4	48.8	28.8	10

Table 4 Statistics of the EPR, NSM, and WLR

Parameters	NSM	EPR	WLR
Total transects	1219	1219	1250
Total erosional transects	413	413	474
Total accretional transects	554	554	488
Total stable transects	262	262	288
Erosional transects %	33.88	33.88	37.92
Accretional transects %	45.45	45.45	39.04
Total stable transects% %	21.49	21.49	23.04
Mean shoreline change	1.98	0.2	0.49
Mean erosion	−9.20	−1.02	−2.18
SD erosion	41.54	4.20	3.83
Maximum erosion	−556.8	−57.14	−37.41
Mean accretion	11.18	1.2	3.26
SD accretion	32.83	3.3	5.54
Maximum accretion	371.73	37.25	62.84

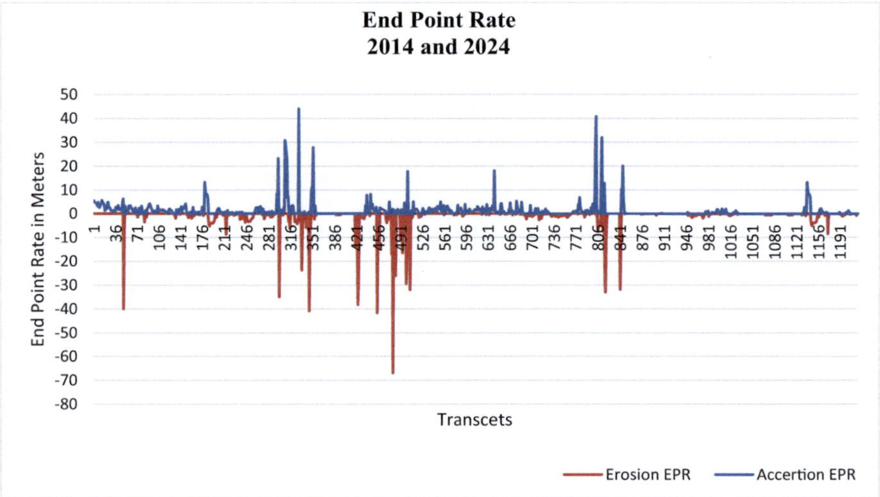

Fig. 5 End point rates of shoreline change (2014–2024)

transects (21.48%) remain stable. The mean shoreline change rate using NSM was + 1.9 m, with a maximum accretion of + 371 m and maximum erosion of -557 m (Table 4). The mean erosion rate was − 9.2 m with a standard deviation of 41.5 m, whereas the mean accretion rate was + 11.17 m with a standard deviation of 32.8 m indicating severe changes along the shoreline with some areas experiencing high erosion and other regions see substantial accretion, and the spikes can be observed in the Fig. 7. The dominant observed phenomenon was accretion as out of the analyzed

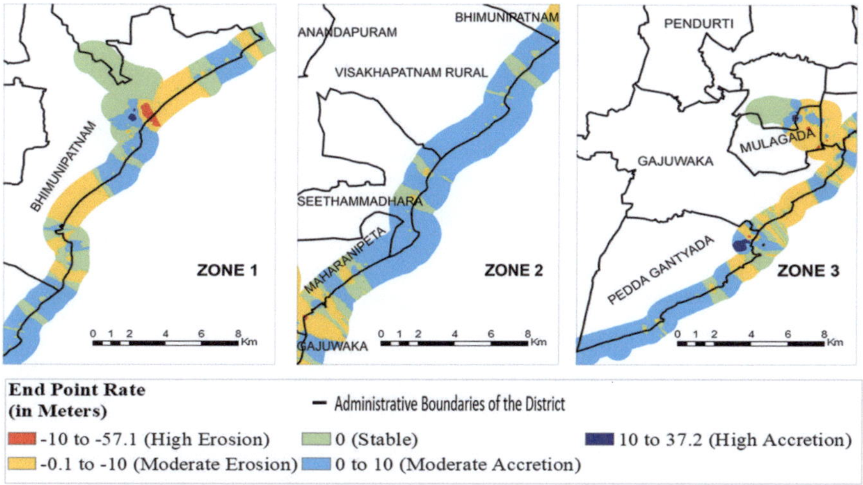

Fig. 6 Spatial analysis of endpoint rates

121.9 km shoreline, 55.5 km long shoreline advanced from the last 11 years with an average of + 11.17 m net movement, while 41.3 km shoreline was eroded with an average of − 9.2 m net movement while 26 km long shoreline is stable is no major changes.

The spatial distribution maps (Fig. 8) indicated the same findings that are seen in EPR, highlighting substantial spatial variations in shoreline movements, suggesting a mix of natural coastal dynamics and human interventions. The similarity between EPR and NSM values suggests that erosion and accretion hotspots in both methods are in the same regions. The NSM method is particularly useful for identifying the areas

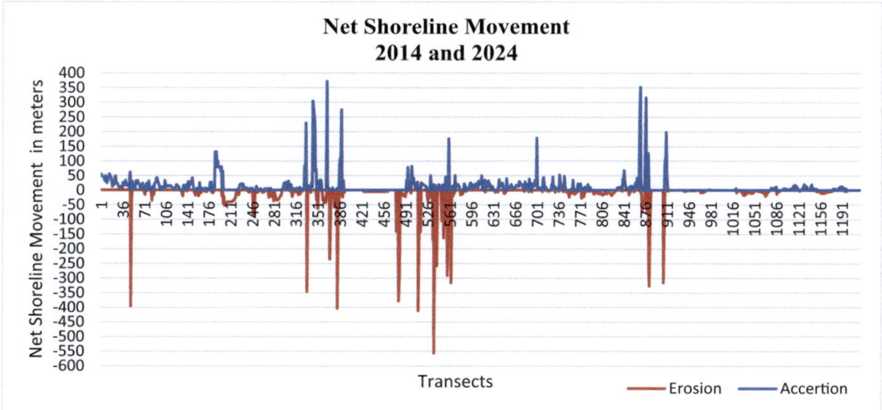

Fig. 7 Net shoreline movement (2014–2024)

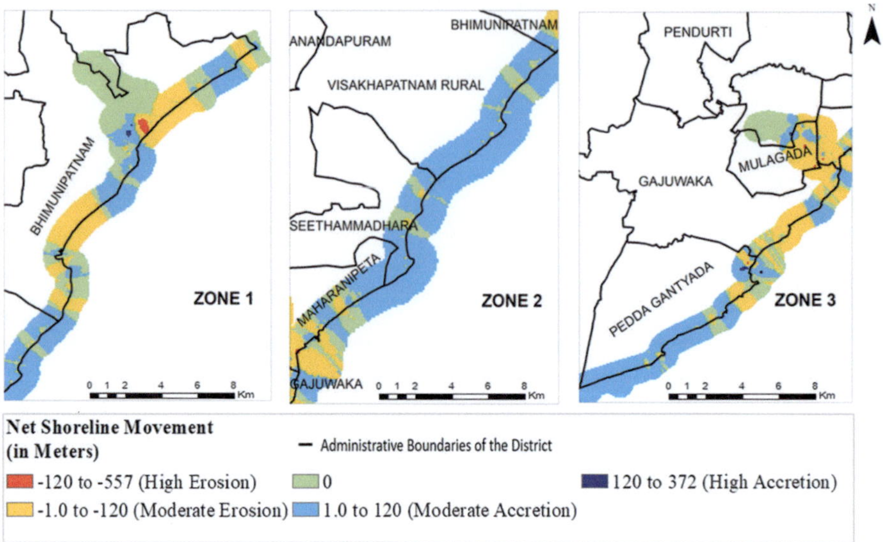

Fig. 8 Spatial analysis of net shoreline movement

of severe coastal vulnerability and planning mitigation strategies, as it represents the total movement of the shoreline from one period to another.

Weighted Linear Regression (WLR) Analysis: The WLR analysis was conducted to incorporate shoreline position uncertainty and provide a more robust estimate of shoreline change trends. This method considers all available shoreline positions and assigns a weighted significance to each, allowing for a more statistically robust estimation of shoreline change trends over time. A total of 1,250 transects were generated, covering various coastal zones including **Bheemunipatnam, Visakhapatnam Rural, Seethammadhara, Maharani Peta, Gajuwaka, and Pedda Gantyada.** The analysis reveals that out of 474 transects (37.92%) showed erosion, 488 transects (39.04%) showed accretion, and 288 transects (23.04%) were stable (Table 4). The mean shoreline change calculated through WLR is 0.49 m/year, indicating an overall trend of shoreline advancement. The maximum accretion observed is + 62.84 m/year indicating advancement of land in some areas, while the maximum erosion recorded is − 37.41 m/year indicating land loss in a few regions due to strong wave action, human interference, and reduced sediment supply. The mean erosion rate is − 2.18 m/year, with a standard deviation of 3.83 m, indicating that erosion rates are relatively dispersed but not excessively extreme, whereas the mean accretion rate is + 3.26 m/year, with a standard deviation of 5.54 m, showing greater fluctuations in the extent of land gain (Fig. 9).

The spatial distribution analysis (Fig. 10) shows that the northern parts of Bheemunipatnam Maharani Peta and Gajuwaka exhibit high erosion rates with values more than − 3 m/year, and the high accretion rates are found in the southern parts of Visakhapatnam Rural, Seethammadhara, and parts of Pedda Gantyada. While most

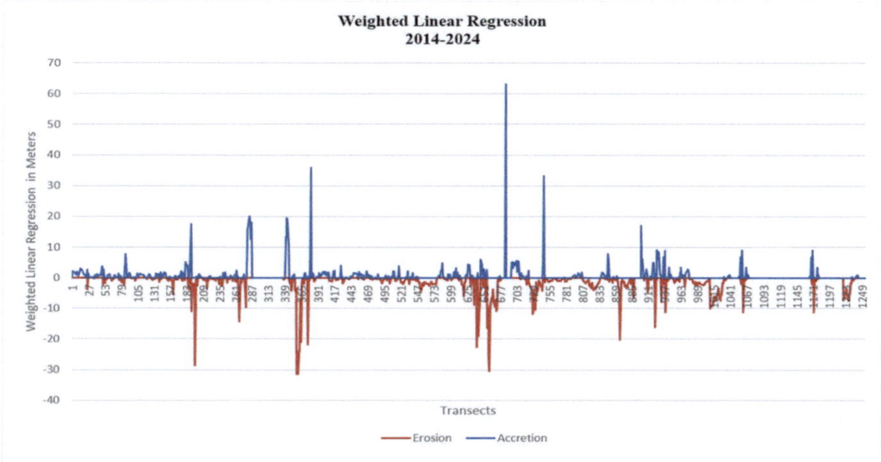

Fig. 9 Weighted linear regression of shoreline change (2014–2024)

of the parts of Pedda Gantyada and Maharani Peta exhibit stable land. Compared to the End Point Rate (EPR) and Net Shoreline Movement (NSM) methods, the results of WLR show a more reliable and accurate estimate as it considers multiple shorelines of different uniform periods, reducing the short-term fluctuation, giving long-term analysis, and offering a clear picture of long-term coastal dynamics.

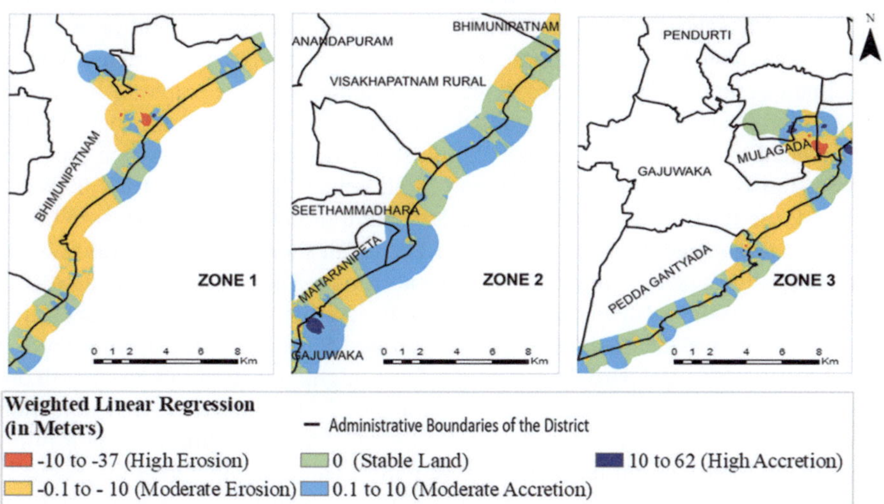

Fig. 10 Spatial analysis of weighted linear regression

5.2 *Comparative Analysis of Shoreline Change Methods: EPR, NSM, and WLR*

The analysis of shoreline dynamics along the Visakhapatnam coast using three different methodologies—End Point Rate (EPR), Net Shoreline Movement (NSM), and Weighted Linear Regression (WLR)—reveals distinct spatial and statistical trends. Each of the above-adopted methods provides valuable information while highlighting the complexities of coastal erosion with their differences.

The EPR method, calculated using 2014 and 2024 shoreline positions, provides a rate-based perspective on shoreline changes. With 1,219 transects, the results indicate that 33.88% of transects experienced erosion, 45.45% showed accretion, and 21.49% remained stable. The average and standard deviation rates of erosion and accretion show dominant change on the shoreline with high fluctuation. The maximum erosion recorded was − 57.14 m/year, while the maximum accretion reached + 37.25 m/year. The spatial distribution shows severe erosion around the river mouth at Bheemunipatnam, where strong hydrodynamic forces and sediment transport cause instability. In other regions, such as Maharani Peta, high erosion rates can be observed, which can be linked to port activities that disrupt natural sediment deposition patterns. The NSM method, which measures absolute shoreline displacement between 2014 and 2024, presents a different perspective. However, extreme variations were observed, with maximum erosion reaching − 556.8 m and maximum accretion at + 371.73 m. The mean erosion value of − 9.20 m, with a high standard deviation of 41.54 m, suggests significant spatial inconsistencies, reflecting localized influences such as wave action, sediment availability, and coastal structures. The mean accretion was + 11.18 m, with a standard deviation of 32.83 m, reinforcing the variability of shoreline advancement across the region. NSM focuses on analyzing the magnitude of shoreline change rather than rates, which helps understand the extent of long-term coastal modifications rather than gradual trends over time. On the other hand, WLR is calculated by integrating multiple shoreline positions over time, which provides a more statistically robust representation of shoreline trends. With 1250 transects analyzed, the results indicate that 37.92% of the transects were erosional, 39.04% were accretional, and 23.04% were stable. The mean shoreline change was 0.49 m/year, suggesting a general trend of moderate accretion. The mean erosion value of − 2.18 m, with a standard deviation of 3.83 m, indicates that WLR captures more stable trends compared to the extreme values seen in NSM. Similarly, the mean accretion of + 3.26 m, with a standard deviation of 5.54 m, highlights its ability to smooth out short-term fluctuations. The maximum erosion (− 37.41 m) and maximum accretion (+ 62.84 m) values are lower than those observed in NSM, confirming that WLR minimizes the influence of outlier events and provides a more reliable long-term assessment of shoreline evolution Comparing these three methods, NSM shows the largest variations in erosion and accretion due to its absolute measurement approach, making it effective for understanding drastic shoreline shifts but less suited for capturing consistent trends over time. EPR provides a clearer rate of change but is limited by its reliance on only two shoreline positions, which may not account for seasonal or episodic changes. WLR offers more robust results, a balanced view with minimizes short-term fluctuations, reduces anomalies, and highlights persistent trends. Thus, WLR is the most reliable approach for coastal management and long-term planning as it focuses on both erosion and accretion without being overly influenced by short—term deviations. Together, the EPR, NSM, and WLR methods provide a comprehensive understanding of shoreline dynamics. While EPR offers a simple and effective rate-based assessment using two shoreline positions, NSM delivers insight into the absolute spatial displacement of the coastline. WLR, by integrating multiple shoreline positions, captures long-term trends and minimizes short-term fluctuations. Used collectively, these methods allow for a robust and multi-faceted analysis, enabling coastal managers to identify both sudden and gradual changes in shoreline position and prioritize mitigation efforts accordingly.

6 Perspectives

The major limitations of the study include the time range and size of the areas, as this is only based on the last 11 years of data, and is limited to the Visakhapatnam district. Technical limitations include OBIA analysis, during which a few of the shoreline intersections are missed due to which in DSAS analysis up to 100 transects being derived with no data values. While this study adopted ML algorithms, it can also use multiple other ML algorithms in classifications, such as support vector machine learning, which can provide more accurate results. Future scope of the study should focus on long-term period data for a larger area with the help of deep learning and predictive models of AI, ML, and cloud-based systems like GEE (Google Earth Engine), which will provide effective and accurate results for future prediction for coastal planning and management.

7 Conclusion

The Visakhapatnam coast shoreline changes were evaluated through a study examining Sentinel-2 Level-2A satellite imagery from 2014 to 2024 by applying Object-Based Image Analysis (OBIA), Random Forest (RF) classification, and Digital Shoreline Analysis System (DSAS). The shoreline evolution was assessed using three key metrics: Net Shoreline Movement (NSM), End Point Rate (EPR), and Weighted Linear Regression (WLR). The WLR established itself as the most reliable method for studying long-term trends along with shoreline alterations by minimizing the effects of short-term complications and inconsistencies. The methodological approach adopted in this research integrates AI techniques such as ML algorithms and GIS techniques, which enable automated classification and precise feature extraction. AI-driven **Object-Based Image Analysis (OBIA)** methods, such as **Random Forest**

(RF) and Mean-Shift Clustering, have enhanced the accuracy of shoreline delineation by distinguishing between land and water more effectively than traditional pixel-based methods. Machine learning algorithms allow for handling large datasets easily in pattern recognition, which means human errors can be reduced and improve classification efficiency. The role of AI models in future prediction helps in coastal management and planning with the help of historical trends and patterns. Along with that, the integration and AI and GIS can unlock robust approaches for lasting coastal monitoring because of its efficient scalability. The analysis revealed that 55.5 km of shoreline experienced accretion, 41.3 km underwent erosion, and 26.1 km remained stable. The most significant erosion was observed in Bheemunipatnam and Maharanipeta, with WLR indicating an average retreat rate of − 1.8 to − 2.5 m/year. In contrast, Seethammadhara and Peddagantyada exhibited significant accretion, with rates ranging from 1.5 to 3.2 m/year. These findings can assist local planners in identifying vulnerable coastal zones, guiding the implementation of shoreline protection measures, and sustainable development strategies. While the study successfully quantified shoreline changes, certain limitations could be acknowledged, as the coastline is 135 km, the transects should be 1350, while DSAS only calculated 1250 transects due to a few zones where one or more shorelines are not intersecting properly. Future studies could include in-depth calculation of the parameters, such as NSM, EPR, as well as the integration of UVA-based shoreline mapping and real-time tide gauge data to enhance accuracy with deep learning models. The scope of the study could be widened by adopting the predictive models with the help of time-series forecasting and AI-driven predictive analytics for better planning and recommendations.

References

Abd-Elhamid HF, Zeleňáková M, Barańczuk J, Gergelova MB, Mahdy M (2023) Historical trend analysis and forecasting of shoreline change at the Nile delta using RS data and GIS with the DSAS tool. Remote Sens 15(7):1737. https://doi.org/10.3390/rs15071737

Barik G, Guru B, Sangma F (2021) Shoreline changes analysis and forecast using digital shoreline assessment system 5.0: evidences from parts of East Coast of India. J Indian Soc Remote Sens 49(11):2815–2830. https://doi.org/10.1007/s12524-021-01424-4

Bayram B et al (2017) The efficiency of random forest method for shoreline extraction from Landsat-8 And Gokturk-2 imageries. ISPRS Ann Photogramm Remote Sensing Spatial Inform Sci IV-4/W4:141–145. https://doi.org/10.5194/isprs-annals-iv-4-w4-141-2017

Bengoufa S, Niculescu S, Mihoubi MK, Belkessa R, Abbad K (2021b) Rocky shoreline extraction using a deep learning model and object-based image analysis. Int Arch Photogramm Remote Sens Spatial Inform Sci XLIII-B3-2021:23–29. https://doi.org/10.5194/isprs-archives-xliii-b3-2021-23-2021

Bengoufa S et al (2021a) Machine learning and shoreline monitoring using optical satellite images: case study of the Mostaganem shoreline, Algeria. J Appl Remote Sens 15(2). https://doi.org/10.1117/1.jrs.15.026509

Boussetta A, Niculescu S, Bengoufa S, Zagrarni MF (2023) Deep and machine learning methods for the (semi-)automatic extraction of sandy shoreline and erosion risk assessment basing on remote sensing data (case of Jerba island-Tunisia). Remote Sens Appl Soc Environ 32:101084. https://doi.org/10.1016/j.rsase.2023.101084

Cenci L, Disperati L, Persichillo MG, Oliveira ER, Alves FL, Phillips M (2018) Integrating remote sensing and GIS techniques for monitoring and modeling shoreline evolution to support coastal risk management. GIScience Remote Sens 55(3):1376370. https://doi.org/10.1080/15481603.2017.1376370

Chen SS, Chen LF, Liu QH, Li X, Tan Q (2005) Remote sensing and GIS-based integrated analysis of coastal changes and their environmental impacts in Lingding Bay, Pearl River Estuary, South China. Ocean Coast Manage 48(1):65–83. https://doi.org/10.1016/j.ocecoaman.2004.11.004

Chen D, Wang Y, Shen Z, Liao J, Chen J, Sun S (2021) Long time-series mapping and change detection of coastal zone land use based on google earth engine and multi-source data fusion. Remote Sens 14(1):1. https://doi.org/10.3390/rs14010001

Comaniciu D, Meer P (2002) Mean shift: a robust approach toward feature space analysis. IEEE Trans Pattern Anal Mach Intell 24(5):603–619. https://doi.org/10.1109/34.1000236

Dronova I, Gong P, Wang L (2011) Object-based analysis and change detection of major wetland cover types and their classification uncertainty during the low water period at Poyang Lake, China. Remote Sens Environ 115(12):3220–3236. https://doi.org/10.1016/j.rse.2011.07.006

Gireesh B et al (2023) Extraction and mapping of shoreline changes along the Visakhapatnam-Kakinada coast using satellite imageries. J Earth Syst Sci 132(2):52. https://doi.org/10.1007/s12040-023-02052-x

Gopinath G et al (2023) Long-term shoreline and islands change detection with digital shoreline analysis using RS data and GIS. Water 15(2):244. https://doi.org/10.3390/w15020244

Innocent Abbas I (2013) An assessment of land use/land cover changes in a section of Niger Delta, Nigeria. Front Sci 2(6):137–143. https://doi.org/10.5923/j.fs.20120206.02

Kannan R et al (2015) Detection of shoreline changes Visakhapatnam Coast, Andhra Pradesh from multi-temporal satellite images. J Remote Sens GIS 05(01):1000157. https://doi.org/10.4172/2469-4134.1000157

Kaplan G, Avdan U (2017) Object-based water body extraction model using Sentinel-2 satellite imagery. Eur J Remote Sens 50(1):137–143. https://doi.org/10.1080/22797254.2017.1297540

Mohammed MW, Nasir SM, Khdir SH (2024) Using object-based image analysis approach for analyzing the impact of annual average change of climate elements on LULC change in northern of Sulaimani province. In: 1st international conference on climate changes and their environmental risks from a geographical perspective, pp 218–236. https://doi.org/10.14500/ccergp2022.gen466

Nassar K, Mahmod WE, Fath H, Masria A, Nadaoka K, Negm A (2019) Shoreline change detection using DSAS technique: case of North Sinai coast, Egypt. Mar Georesour Geotechnol 37(1):1448912. https://doi.org/10.1080/1064119X.2018.1448912

National Remote Sensing Centre (2014) Land use/land cover database on a 1:50,000 scale. Natural Resources Census Project, LUCMD, LRUMG, RSAA, NRSC, ISRO. https://bhuvan-app1.nrsc.gov.in/2dresources/thematic/2LULC/lulc1112.pdf

Niculescu S, Lam CN (2019) Geographic object-based image analysis of changes in land cover in the coastal zones of the Red River Delta (Vietnam). J Environ Prot 10(03):413–430. https://doi.org/10.4236/jep.2019.103024

Pham QB et al (2024) Multi-spectral remote sensing and GIS-based analysis for decadal land use land cover changes and future prediction using random forest tree and artificial neural network. Adv Space Res 74(1):17–47. https://doi.org/10.1016/j.asr.2024.03.027

Prasetyo Y, Gunawan SA, Maksum ZU (2016) Determination of the water catchment area in Semarang City Using a combination of object-based Image analysis (OBIA) classification, INSAR and Geographic Information System (GIS) methods Based on a High-Resolution SPOT 6 image and radar imagery. IOP Conf Series Earth Environ Sci 47:012027. https://doi.org/10.1088/1755-1315/47/1/012027

Santra M, Dwivedi CS, Pandey AC (2023) Quantifying shoreline dynamics in the Indian Sundarban delta with Google Earth Engine (GEE)-based automatic extraction approach. Trop Ecol. https://doi.org/10.1007/s42965-023-00321-w

Shaik I, Suhail M, Venkata N (2024) Shoreline delineation and change analysis in response to sea level rise and coastal bathymetry along the coast of Visakhapatnam, India using high-resolution optical imagery. Acta Geophys 72:4453–4472. https://doi.org/10.1007/s11600-024-01341-3

Sharma S et al (2024) Remote sensing and GIS in natural resource management: comparing tools and emphasizing the importance of in-situ data. Remote Sens 16(22):4161. https://doi.org/10.3390/rs16224161

Sreekesh S, Kaur N, Naik SRS (2020) An OBIA and rule algorithm for coastline extraction from high- and medium-resolution multispectral remote sensing images. Remote Sens Earth Syst Sci 3(1–2):24–34. https://doi.org/10.1007/s41976-020-00032-z

Talukdar S et al (2020) Land-use land-cover classification by machine learning classifiers for satellite observations—a review. Remote Sens 12(7):1135. https://doi.org/10.3390/rs12071135

Thinh NA, Hens L (2017) A Digital Shoreline Analysis System (DSAS) applied on mangrove shoreline changes along the Giao Thuy coastal area (Nam Dinh, Vietnam) during 2005–2014. Vietnam J Earth Sci 39(1):87–96. https://doi.org/10.15625/0866-7187/39/1/9231

Tian Y, Jia M, Wang Z, Mao D, Du B, Wang C (2020) Monitoring invasion process of *Spartina alterniflora* by seasonal Sentinel-2 imagery and an object-based random forest classification. Remote Sens 12(9):1383. https://doi.org/10.3390/rs12091383

Tsai YS (2022) Monitoring 23-year of shoreline changes of the Zengwun Estuary in Southern Taiwan using time-series Landsat data and edge detection techniques. Sci Total Environ 839:156310. https://doi.org/10.1016/j.scitotenv.2022.156310

Winsemius S, Braaten J (2023) Zonal statistics. In: Springer eBooks, pp 463–485. https://doi.org/10.1007/978-3-031-26588-4_24

Yasir M et al (2020) Automatic coastline extraction and changes analysis using remote sensing and GIS technology. IEEE Access 8:3027881. https://doi.org/10.1109/ACCESS.2020.3027881

A Systematic and Structured Review on Groundwater Contamination, Sources, Prevalence, Health Risk and Mitigation Strategies in Relation with Land Use/Land Cover in Bihar State of India

Sushmita Kumari and Md. Omar Sarif

1 Introduction

Given the indispensability of water for life and the current era of globalization. it is imperative that we coexist harmoniously with nature and establish enduring goals (Naujokas et al. 2013; Doncaster and Bullock 2024). Arising arsenic toxicity affects more than 200 million individuals across approximately 105 (Fig. 1) nations globally (Murcott 2012). Examining the influence of escalating human activity and population growth on groundwater is crucial. Arsenic has had detrimental effects on human health worldwide through diverse pathways (IARC 2004). The largest recorded concentration of arsenic was in La Pampa, Argentina, reaching 5280 g/L. Other notable concentrations were found in Bangladesh at 4730 g/L, Taiwan at 3590 g/L, South 24 Parganas, India at 3700 g/L, Antofagasta, Chile, and Brazil at 2100 g/L, in that order (Chakraborti et al. 2018a; Smedley and Kinniburgh 2002) (Fig. 2). In this Paper Study Samuel and Adeloju (2021) said that identifies areas where there has been documented evidence of arsenic contamination in surface or groundwater (Adeloju et al. 2021). There have been reports of abnormally high

Supplementary Information The online version contains supplementary material available at https://doi.org/10.1007/978-3-032-03714-5_13.

S. Kumari · Md. O. Sarif (✉)
Department of Geography, Lovely Professional University, Phagwara 144411, Punjab, India
e-mail: sarif@hiroshima-u.ac.jp

Md. O. Sarif
Centre for Peaceful and Sustainable Futures (CEPEAS), The IDEC Institute, Hiroshima University, Higashihiroshima, Hiroshima 7398529, Japan

M. O. Sarif and A. Sharifi (eds.), *Advanced Geospatial Intelligence and AI for Environmental Resilience and Sustainable Development*, Advances in Geographic Information Science, https://doi.org/10.1007/978-3-032-03714-5_13

amounts of arsenic in groundwater in several European nations, including Croatia, Hungary, and Serbia. The WHO standard of 10 μg/L and the EC guideline of 7.5 μg/L (EC Directive 2006/118/EC) were found to be repeatedly exceeded by groundwater arsenic levels in Hungary (Adeloju et al. 2021). Groundwater samples taken from Romania and Hungary have arsenic concentrations ranging from less than 0.5 to 240 μg/L (Rowland et al. 2011). A summary of the variations in groundwater arsenic levels across the globe is shown in Table 1. The two affected locations with the greatest number of individuals at danger of arsenic exposure are West Bengal, India, and Bangladesh (Shaji et al. 2021). Bangladesh was chosen as a case study in order to highlight the seriousness of groundwater poisoning by arsenic. People without access to integrated urban water delivery systems, mostly in rural areas, are the most impacted in West Bengal and Bangladesh. Millions of rural humans in these areas thus frequently drink water tainted with arsenic. Many people in this region have already been identified as having indications of diseases linked to arsenic (Shankar and Shanker 2014a). About 1.4 million, or 30% of Bangladesh's 4.75 million tube wells, have arsenic levels higher than 50 μg/L, according to UNICEF. In Bangladesh, 12.6% of the water samples tested exceeded the national drinking water arsenic threshold, according to a UNICEF survey on household drinking water quality (Cecilia and Sargsyan Hrachya 2018). This means that around 22 million people in Bangladesh could be exposed to dangerous levels of arsenic. Furthermore, around 6 million people in West Bengal, India, have been exposed to high amounts of arsenic, which range from 50 to 3200 μg/L (Hamzaoui-Azaza et al. 2020). Significantly, methanogenesis-influenced fluids had the highest concentrations of arsenic (23–208 μg/L, with an average of 123 μg/L), whereas sulphate reduction-dominated waters had lower values (< 0.5 to 58 μg/L, with an average of 11.5 μg/L). The nations in the Americas most affected are the United States, Argentina, Chile, and Mexico, as seen in Fig. 2. It is predicted that at least four million individuals in Latin America are impacted by arsenic levels higher than 50 μg/L (Tapia et al. 2022). It has been discovered that some wells in Argentina, Bolivia, and Peru have abnormally high quantities of arsenic—up to several milligrams per liter (Mukherjee et al. 2019).

Arsenic pollution in groundwater currently affects twenty states and four Union Territories throughout India (Shaji et al. 2021; Raza et al. 2023). Arsenic, known for its carcinogenic properties, presents a major global concern, especially in the unconsolidated aquifers of the world's most recent organic formations (Li et al. 2021; Santra et al. 2013). More than 90% of arsenic pollution stems from geogenic sources (Shaji et al. 2021). Arsenic naturally enters soil and groundwater as rocks and minerals erode, and human activities can also introduce it, leading to contamination of soil and groundwater (Bhunia et al. 2014; Siddiqui and Chaudhry 2017). The main reason for arsenic contamination in groundwater is frequently linked to old alluvial sediments, affected by phenomena like plate tectonics, mountain building, erosion, and sediment accumulation (Mukherjee et al. 2019). Arsenic pollution in groundwater has become a notable issue in Bihar, India, affecting millions of inhabitants in rural and semi-urban areas (Saha and Sahu 2015). The extensive Himalayan mountain range provides water to the Ganga River basin system, which then feeds into the Bay of Bengal, supporting Bihar's agricultural requirements (Singh and

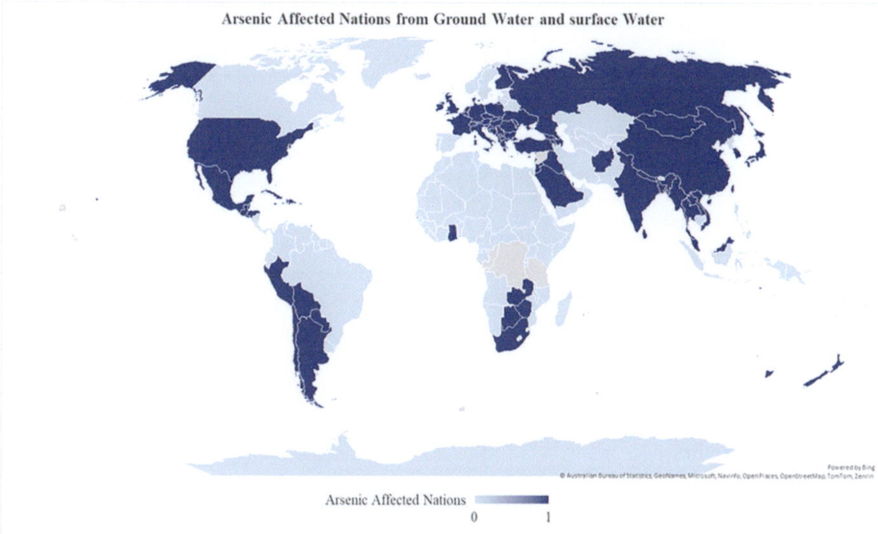

Fig. 1 A worldwide map of countries/territories experiencing surface and groundwater contamination by arsenic exposure (Adeloju et al. 2021)

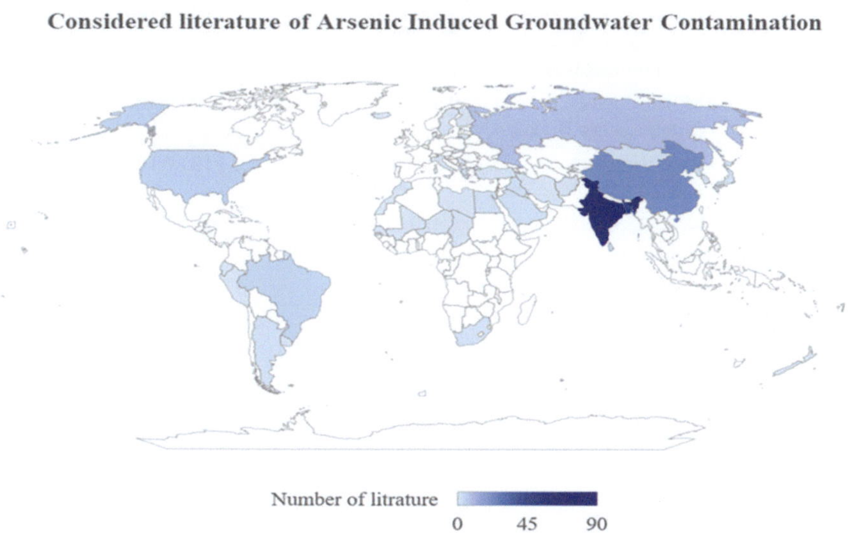

Fig. 2 The literature based global affected areas by arsenic contamination in ground water

Table 1 Variations in groundwater arsenic levels across the globe

Country	Region	(As), μg/L	Permissible limit, μg/L
India	West Bangal (Uttar Pradesh)	10–3200	50 (WHO)
Bangladesh	Noakhali	< 1 to 4730	50 (WHO)
Pakistan	Muzaffargarh (southern Punjab)	Up to 906	50
Nepal	Rupandehi	Up to 2620	50
Japan	Fukuoka Prefecture (southern region)	1–293	10 (WHO)
Peru	–	500	50 (WHO)
Romania	–	10–176	10 (WHO)
Slovakia	Banska Bistrica and Nitra	37–39	10 (WHO)
Spain	–	< 1 to 100	10 (WHO)
Taiwan	–	10–1820	10 (WHO)
Thailand	Ron Phibun	1 to > 5000	10 (WHO)
UK	Southwest England	< 1 to 80	10 (WHO)
USA	Tulare Lake	Up to 2600	10 (USEPA)
Australia	Perth Northern New South Wales Victoria (gold mining regions)	7000 52–337 < 1 to 300,000	10 (WHO)
Brazil	Minas Gerais (Southeastern Brazil)	0.4–350 (surface water)	10 (WHO)
Cambodia	Prey Veng and Kandal-Mekong delta	Up to 9001-1610	10 (WHO)
Canada	Nova Scotia (Halifax Country)	1.5–738.8	10 (WHO)
Afghanistan	Ghanzi	10–500	10 (WHO)
Argentina		10–1000	50 (WHO)
Chile		900–1040	50 (WHO)
China		50–440	50 (WHO)
Chaina Inner Magnolia		1–2400	50 (WHO)
Croatia (Eastern)	Osijek area	27 (shallow groundwater 205–240 (deep groundwater)	10 (WHO)
Finland	Southwest Finland	17–980	10 (WHO)
Germany	–	10–150	10 (WHO)
Ghana	–	< 1 to 175	50 (WHO)
Greece	Fairbanks (mine tailings)	Up to 10,000	10 (WHO)

(continued)

Table 1 (continued)

Country	Region	(As), μg/L	Permissible limit, μg/L
Hungary	Pannonian Basin	10–176 0.5–240	10 (WHO)
Italy	Volcano island and Phlegrean fields	0.1–6940	10 (WHO)
Japan	Fukuoka prefecture (southern region)	1–293	10 (WHO)
Lao PDR	Laos	277.8	50 (WHO)
Mexico	Lagunera	8–620	25
New Zealand	TaupobVolcanic zone	21	10 (WHO)
Vietnam	Red River Delta (North Vietnam) Mekong Delta (Southern Vietnam)	< 1 to 3050	10 (WHO)

Sources Adeloju et al. (2021) and Shankar and Shanker (2014a)

Choudhary 2010). The sediment carried by tributaries of the Ganga River originates from its starting point and is transported along its entire course (Ahmad et al. 2018; Chandra et al. 2018). Arsenopyrite, a compound present in the sediment, is the primary reason behind arsenic contamination in the groundwater of the Gangetic basin (Chakraborti et al. 2018a; Shaji et al. 2021; Singh and Choudhary 2010). The existence of arsenopyrite in the sediments contributes to arsenic pollution in the groundwater of the Gangetic basin. Arsenic concentrations have increased in the Gangetic plains, impacting 18 out of Bihar's 38 districts with notable groundwater pollution (Ghosh et al. 2019). It's probable that more than nine million individuals in Bihar are ingesting water with arsenic levels surpassing safe threshold (Chakraborti et al. 2018a). The scenario in rural Bihar is troubling, as more than 80% of the populace depends on groundwater for drinking, frequently surpassing the WHO/BIS threshold of 10 g/L for arsenic content. This puts a substantial portion of the population at risk of arsenic contamination. Furthermore, a notable number of cancer patients with elevated arsenic levels in their bloodstream are located in regions close to the Ganges River (Kumar et al. 2021a). A study from 2007 found that 10.8% of water samples, equivalent to 7164 out of 66,623, gathered from 11 districts in Bihar, surpassed the permissible limit of 50 μg/L for arsenic concentration in India (Kumar Jha and Chandra Gupta 2017). According to a 2018 study, 17.8% of the 19,961 drinking water samples analyzed showed arsenic levels above 50 μg/L (Chakraborti et al. 2018a). The problem of arsenic-contaminated groundwater was first noticed in 2002 in Semaria Ojhapatti village, situated in the Bhojpur district of Bihar (Singh 2015; Ghosh and Singh 2010). The inhabitants of the area experienced significant impacts from arsenic poisoning, showing various visible signs of arsenicosis (Thomas and Bradham 2016a; Saha 2003). A recent study in the Boxer district of the state examined health problems resulting from arsenic contamination, revealing the highest recorded groundwater arsenic level at 1929 μg/L (Thakur et al. 2015). In this tiny village, a troubling finding emerged: blood samples from the residents exhibited the highest arsenic levels ever documented in the state, reaching 664.7 μg/L (Thakur et al. 2015). Long-term and consistent intake of toxic inorganic arsenic through ingestion and drinking for about five to ten years can lead to arsenicosis (Aullón Alcaine et al. 2020). This condition may manifest as peripheral vascular disease, typically affecting the blood vessels in the lower limbs and feet, along with various skin disorders and cancers affecting organs such as the bladder, kidneys, and lungs (Hopenhayn 2006) Additionally, individuals may experience diabetes, hypertension, and reproductive issues (Bhowmick et al. 2007a; Singh and Ghosh 2012; Jhariya et al. 2021; Ravindra and Mor 2019; Kumar et al. 2016a). Therefore, mitigation strategy is an essential element to overcome this critical problem because of arsenic contamination. Various techniques such as oxidation (Shankar and Shanker 2014b), co-precipitation adsorption processes (Khalid 2017), Bucket Treatment Units (Rahaman et al. 2022), Solar Oxidation (Smedley and Kinniburgh 2002), Nano filtration (Hashim et al. 2019) can effectively remove arsenic from water, ensuring its safety for consumption. This research proposed the presence of arsenic contamination in the nearby groundwater, a potential hazard that has been discussed in detailed discourse. This study examines the health hazards linked with arsenic intake, the concentrations of arsenic in groundwater, soil, and food supplies in the Bihar area, and evaluates the vulnerability of the local populace to arsenic pollution. This review provides an understanding of the suitable framework for identifying research deficiencies in the domain and examining a structured method for further investigation into arsenic contamination in groundwater. Information on the effects of groundwater pollution on human health and agricultural systems, methods for treating groundwater pollution, and possible mitigation measures is insufficient. This gap arises mainly because many studies on arsenic contamination concentrate on broader research topics (Santra et al. 2013; Brammer and Ravenscroft 2009; Bhowmick et al. 2018). Therefore, the aim of this review paper is to accomplish the following objectives to systematically investigate the correlation between groundwater pollution, its repercussions, and potential strategies for mitigation by considering 142 articles at global level (Fig. 2): (i) To delve into the importance of mapping and monitoring land use/ land cover (LULC) changes dynamically with groundwater contamination; (ii) To present the quantification approach(s) of arsenic induced groundwater pollution; (iii) To demonstrate how arsenic poisoning affects human health; (iv) To investigate the various methods for reducing and mitigating arsenic pollution.

2 Data and Methods

In Fig. 3, the strategy for writing a systematic and structured review on groundwater contamination, sources, prevalence, health risk, and mitigation strategies in relation to land use/land cover in Bihar State of India appears to be a comprehensive approach encompassing multiple facets. Here's a breakdown of the strategy:

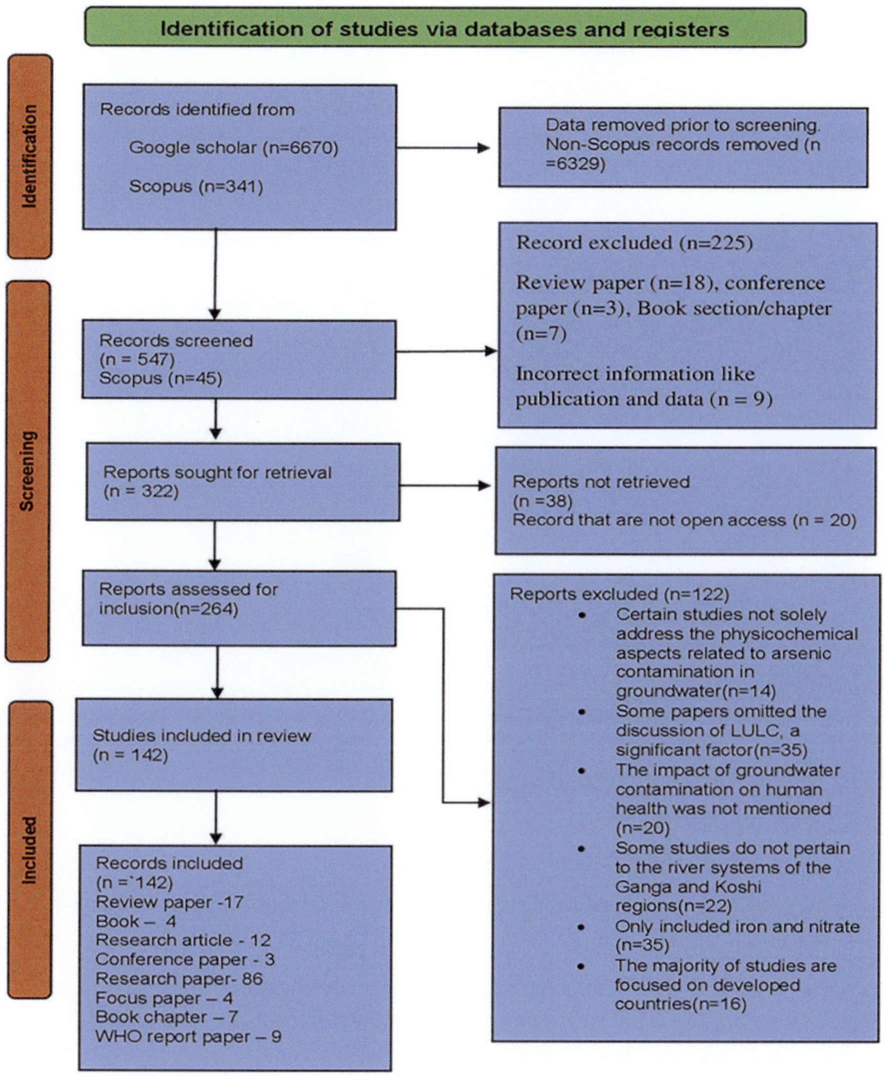

Fig. 3 A schematic representation outlining the systematic approach for composing this review work

1. Systematic Approach: This involves employing a methodical and rigorous process for literature review and data synthesis. It likely includes defining research questions, establishing inclusion/exclusion criteria for studies (Fig. 3), and systematically searching [derived directly from the keywords inherent based on common keywords (Fig. 4)] and selecting relevant literature.

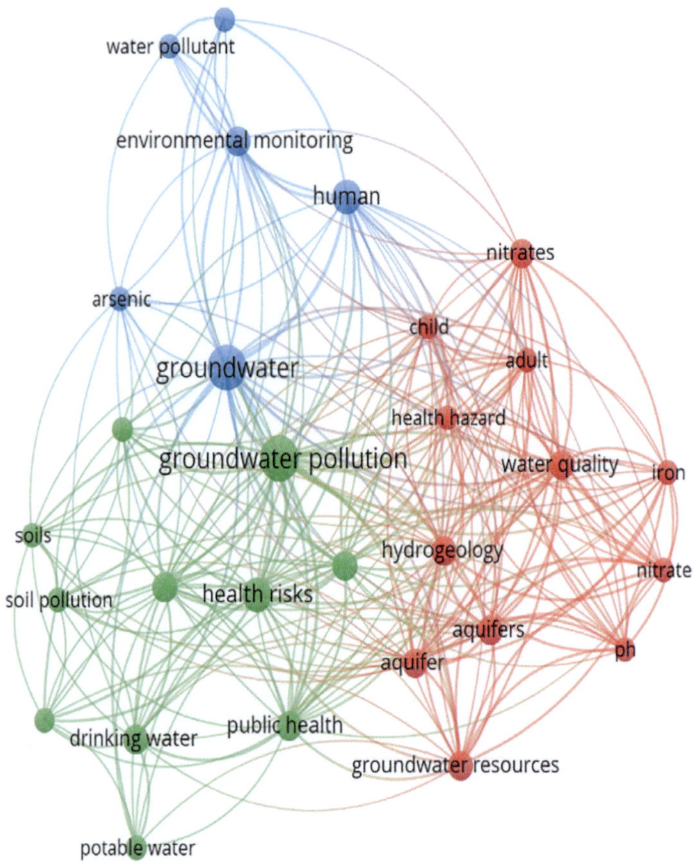

Fig. 4 Keywords connections in the literature in the context of groundwater contamination

2. Structured Review: This suggests organizing the paper in a clear and logical manner, possibly following a predefined structure such as Introduction, Methods, Results, Discussion, and Conclusion (IMRaD) format. Each section would serve a specific purpose, contributing to the overall coherence and readability of the paper.
3. Focus on Groundwater Contamination: The primary focus of this study is on groundwater contamination, encompassing aspects such as its sources, prevalence, associated health risks, and potential mitigation strategies. This indicates a deep dive into understanding the various factors contributing to groundwater pollution.
4. Consideration of LULC: Our strategy involves exploring the interplay between LULC patterns and groundwater contamination. This implies examining how human activities, urbanization, agriculture, industrialization, etc., influence the quality of groundwater, thereby addressing the broader environmental context.

5. Regional Context (Bihar State, India): By narrowing down the geographical scope to Bihar State in India, we are focusing on a specific region, which allows for a more targeted analysis of groundwater issues tailored to the local context. This approach acknowledges the unique socio-economic, geographical, and environmental factors influencing groundwater dynamics in Bihar.

Overall, our chosen strategy integrates systematic methods with a structured approach to comprehensively explore groundwater contamination in Bihar State, considering its sources, prevalence, health implications, and mitigation strategies within the context of LULC dynamics (Fig. 5).

3 Explore Significance of Land Use/Land Cover Mapping and Dynamics in the Context of Groundwater Contamination

Mapping land use and land cover (LULC) is an essential method for analysing the distribution of natural resources and human activities across different areas. It is essential to comprehend how different land use patterns affect the quality of the environment, especially groundwater supplies. Groundwater quality is greatly impacted by changes in LULC, particularly in areas experiencing fast industrialization, agricultural growth, and urbanization (Supplementary Table 1). It is crucial to comprehend these processes in order to manage water resources sustainably.

3.1 Effect on the Contamination of Groundwater

Deforestation, urbanization, intensification of agriculture, and industrialization are examples of changes in LULC that can drastically change the natural hydrological cycle (Malakootian and Nozari 2020). For example, the conversion of forest land to urban or agricultural regions diminishes groundwater recharge, lowers infiltration rates, and increases surface runoff (Chaudhary and Kumar 2018). When pesticides, herbicides, and fertilizers are used in agriculture, dangerous chemicals can seep into the soil and eventually contaminate groundwater. In a similar vein, cities produce a lot of wastewaters, which can contaminate aquifers if it is not adequately handled.

3.2 LULC Dynamics and the Quality of Groundwater

The term "LULC dynamics" describes how land cover varies over time as a result of either natural or man-made forces Areas susceptible to groundwater contamination can be identified by keeping an eye on these tendencies.

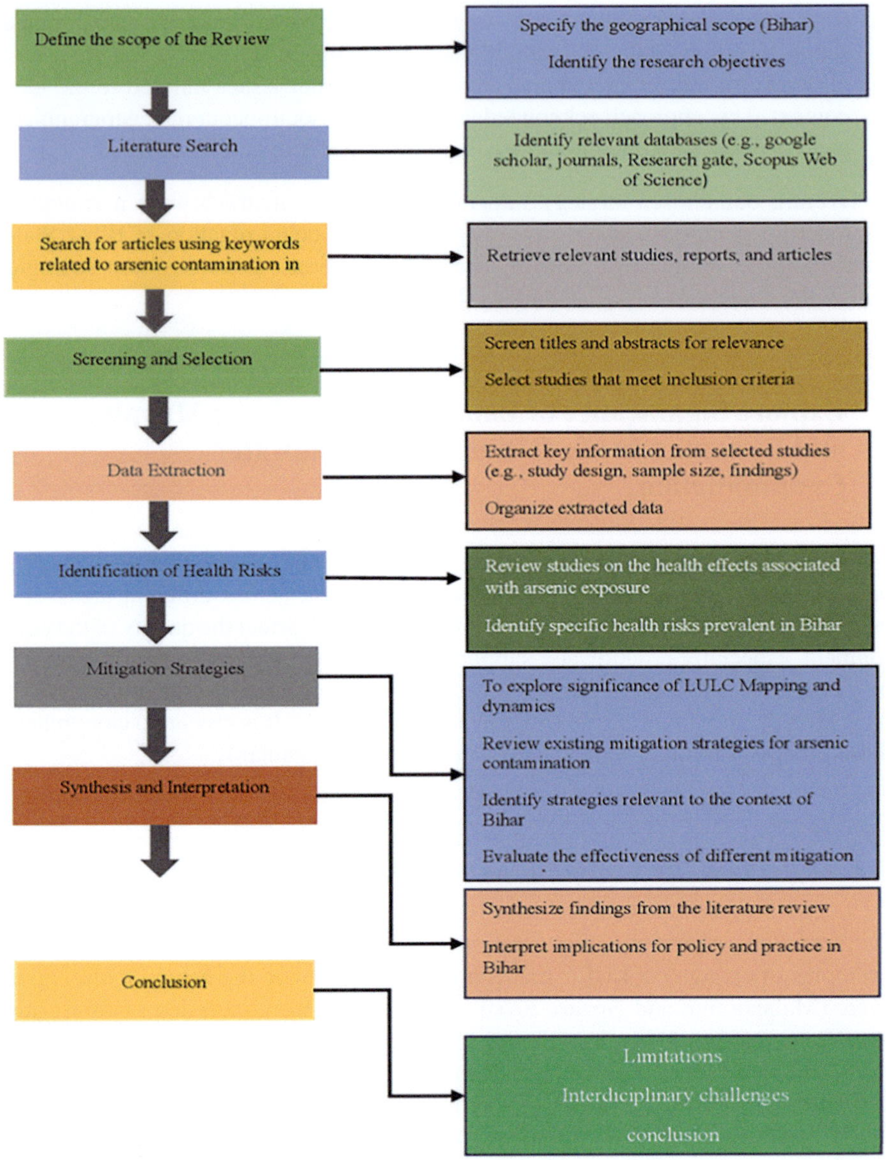

Fig. 5 Flowchart of the review work

Adhikari et al. (2020) and Mahanta et al. (2016)—For instance, as agricultural fields grow, the widespread application of nitrogen-based fertilizers frequently results in higher nitrate concentrations in groundwater.

Dandge and Patil (2022)—Similarly, heavy metals including lead, cadmium, and arsenic are found in groundwater sources as a result of industrial activity. Understanding the dynamics of land cover change is especially important in semi-arid and arid regions, where groundwater is the main supply of drinking water.

Karale et al. (2024) and Gómez et al. (2006)—In these regions, excessive groundwater extraction for irrigation frequently causes salinization, which lowers the quality of the water. Furthermore, the substitution of impermeable surfaces for natural vegetation in urban settings diminishes the soil's inherent filtration capacity, raising the possibility of groundwater contamination (Rizeei et al. 2018; Ju et al. 2018).

3.3 LULC Mapping's Function in Mitigation Techniques

The spatially detailed understanding of land use patterns that LULC mapping offers is essential for creating groundwater resource protection plans. Several studies have investigated the relationship between land use, Review of Studies on Groundwater and Land Use/Land Cover (LULC) Relationships. Several studies have explored the intricate connections between land use and land cover (LULC) changes and their effects on groundwater resources, employing a range of methodologies, data sources, and analytical tools. Siddik et al. (2022) have used fieldwork data and multispectral imagery from the USGS Earth Explorer (Landsat 5 TM and Landsat 8 OLI), the impact of seasonal fluctuations in LULC on groundwater recharge was evaluated. Urbanization, vegetation cover, exposed soil, water bodies, and flooded regions were among the variables that were examined (Siddik et al. 2022). Wetspot analysis and image resolution changes were among the methods used; the main tools were ArcGIS (10.5) and Q-GIS (3.6). Their results highlighted the importance of land management techniques and showed that seasonal LULC fluctuations had a substantial impact on groundwater recharge. Adhikari et al. (2020) investigated the effects of changes in the built-up area on Ho Chi Minh City's (HCMC) groundwater recharge. Both secondary data from UNESCO reports and primary data from fieldwork were used. Crop varieties, metropolitan areas, woods, and bodies of water were among the variables. Their research, which used models like DYNA-CLUE and methods like SWAT, found that urbanization has a major impact on groundwater sustainability and replenishment. Chaudhary and Kumar (2018) used 30-m-resolution Landsat 5 TM (8-bit) and Landsat 8 OLI/TIRS (12-bit) photos to examine how urbanization affects high-potential groundwater recharge areas. Urban growth causes these zones to shrink, which lowers groundwater levels, according to field data, satellite images, and Central Ground Water Board (CGWB) inputs. LULC analysis and GIS-based classification with ArcGIS and Erdas Imagine were among the methods used. Roshan et al. (2024) suggested a novel method for assessing how LULC changes affect drinking water quality by fusing remote sensing (RS) and geographic information systems (GIS). ERDAS IMAGINE and ArcGIS tools were used to link fieldwork data with Sentinel-2 satellite images (10-m spatial resolution).

The study demonstrated how flexible RS and GIS are in evaluating the relationship between LULC and water quality factors, providing an approach that may be applied to similar research projects. Mondal et al. (2020) investigated how human activity affects the volume and quality of groundwater. They examined factors like habitation, farming, industrial activity, and vegetation using field work data, USGS Landsat pictures, and laboratory analysis. Their findings showed that human activity dramatically changes the volume and quality of groundwater. Elmahdy et al. (2020) The study offered a thorough grasp of how variations in groundwater levels and quality over time impact environmental and geotechnical situations. The study used sophisticated techniques like change detection analysis and support vector machines (SVMs). These techniques made use of satellite imagery, particularly Landsat TM and ETM+, laboratory analysis, and field data. Furthermore, the data processing and analysis required the use of programs like ArcGIS and ENVI. Reyes Gomez (2017) examined the relationship between LULC and water quality in semi-arid regions, focusing on arsenic, fluoride, and nitrate contamination. The study identified actionable strategies to address the linkages between LULC changes and deteriorating water quality. Karale et al. (2024) emphasized sustainable land management and conservation efforts to mitigate LULC-induced challenges. Utilizing USGS Landsat data, their analysis employed supervised classification and tools like ArcGIS and Q-GIS to evaluate variables including forests, agriculture, and settlements. Tola and Deyassa (2024) analysed hydrological changes associated with deforestation and agricultural expansion. Using CA Markov Chain Analysis and supervised classification, they linked LULC changes to declining groundwater recharge rates. Dandge and Patil (2022) explored groundwater quality using WQI techniques. They identified elevated levels of Total Hardness, Electrical Conductivity, alkalinity, and other indicators as critical factors degrading groundwater quality. Ju et al. (2018) integrated principal component analysis (PCA), dual isotope analysis, and LULC data to trace nitrate contamination sources in complex geochemical processes. Rizeei et al. (2018) investigated groundwater hazards and nitrate levels in areas undergoing rapid urban development. Their findings highlighted the urgent need for pollution mitigation in the context of rising water demand. These studies collectively underscore the profound influence of LULC changes on groundwater systems, advocating for integrated management strategies and advanced analytical tools to ensure sustainable resource utilization. Each study's methodological rigor and use of cutting-edge tools such as GIS, RS, and advanced statistical models contribute valuable insights to the field of hydrogeology. These studies collectively highlight the varied approaches used to assess groundwater quality, land use changes, and the critical role of arsenic contamination across different regions. Researchers and policymakers can pinpoint crucial areas where land use practices endanger groundwater quality by combining remote sensing and Geographic Information System (GIS) technology. In order to reduce the dangers of contamination, this knowledge can direct the deployment of strategies like buffer zones, sustainable farming methods, and regulated urban development (Camacho et al. 2011). Additionally, LULC maps are useful for forecasting patterns in land use changes in the future, allowing for preventative actions to shield groundwater from possible contamination. Local governments and environmental organizations can create land management strategies that strike a balance between environmental preservation and development requirements by keeping an eye on these trends (Naz et al. 2024). In conclusion, knowledge of the connection between groundwater quality and land use patterns requires a grasp of LULC mapping and dynamics. Stakeholders can more accurately determine the causes and dangers of groundwater contamination by examining changes in land cover, opening the door to sustainable management and conservation of water resources (Supplementary Table 1).

4 Assessment of Groundwater Quality

4.1 Rivers Water Quality in Bihar

According to the 2019 and 2021 Central Pollution Control Board (CPCB) reports, water quality monitoring was carried out at 95 locations across 21 rivers in Bihar. Among these, 45 sites situated across eighteen rivers were discovered to be failing to meet the Water Quality Criteria for BOD (Table 2). The list includes eighteen rivers affected by contamination: Daha, Dhous, Gandak, Ganga, Gangi, Ghaghara, Harbora, Kamala, Kohra, Lakhandei, Manusmar, Parmar, Punpun, Ramrekha, Sirsiya, Sone, Burhi Gandak/Sikrahna, and Bagmati. Further information regarding the polluted sections of these rivers is provided in the subsequent table.

4.2 Criteria for the Identification Groundwater Contaminants and Classification of Persistent Reservation Sites

The term Persistent Reservation Sites describes portions of rivers or other bodies of water that have been severely contaminated by different types of pollutants, such as sewage discharge, agricultural runoff, or industrial waste. These areas frequently have poor water quality, endangering aquatic life and putting public health at risk. These area identification aids in the prioritization of restoration projects and the implementation of pollution mitigation strategies with the goal of preserving water resources and restoring ecological balance for the benefit of society and the environment.

Thakur et al. (2015) have said that contaminants like arsenic, fluoride, iron, and nitrate in drinking water pose significant health risks to populations, particularly in regions like Bihar, India, where these pollutants are prevalent (Table 3). To create efficient mitigation strategies and safeguard the public's health, it is imperative to comprehend their effects on health.

Table 2 The count of rivers with polluted stretches in Bihar

Sl. No.	Rivers	Polluted river length and/or location	Maximum measured BOD (mg/L)	Class of priority
1	Sone	Koelwar	4.0	V
2	Sirsiya	Rasual	30.0	II
3	Ramrekha	Harinagar	12.0	III
4	Punpun	Over Punpun	10.00	IV
5	Parmar	Along Jog Bani	3.4	V
6	Manusamar	Along Asitamadhi	6.0	IV
7	Lakhandei	Along Sitamarhi	11.0	III
8	Kohara	Along the Manjhaulia	8.0	IV
9	Kamla	Along the Darbhanga route	5.0	V
10	Harbora	In nearby of Narkatiaganj	8.0	IV
11	Gagra	In along with Revelganj	3.6	V
12	Gangi	At Ara	8.0	IV
13	Ganga	Between Patna, Bhagalpur, Buxar, and Fatwah	7.0	IV
14	Gandak	Along with Rewaghat	3.8	V
15	Dahos	From Gopalganj to Siwan	5.6	V
16	Daha	Along the Madhuvapur route	10.0	IV
17	Burdhi Gandakj/ Sikhrana	To Pakridaya from Narkatiaganj	10.0	IV
18	Bagmati	Along Sirnia	3.6	V

It is alarming that groundwater in some parts of Bihar has levels of arsenic pollution that are higher than WHO guidelines. Prolonged exposure to arsenic can exacerbate socioeconomic problems in the impacted populations and cause a variety of cancers.

Particularly in the Maner and Shahpur blocks, with excessively high levels of arsenic, the risk of cancer incidence is heightened. Additionally, arsenic contamination disrupts the economic stability of these regions by impacting agricultural productivity and increasing healthcare costs associated with treating arsenic-related health conditions.

Fluoride contamination, especially pronounced in Navinger, Jamui block, poses health challenges such as skeletal fluorosis. This condition, characterized by stiffness in the back and joints, significantly impairs mobility and reduces the quality of life for those affected. High fluoride levels in water sources also contribute to dental and skeletal fluorosis, further burdening healthcare systems and diminishing productivity within communities.

Iron contamination in groundwater, notably in block 31 of district 20, is another health concern in Bihar. While iron is an essential nutrient, excessive intake through contaminated water sources can lead to adverse health effects such as gastrointestinal

Table 3 Groundwater contaminants

Contaminants	Permissible limit	Impact on health	Locations and circumstances
Arsenic	10 ppb by WHO and > 50 ppb by BIS indicates a health issue	Human health is at risk from water that has more than 50 parts per billion	There are serious dangers associated with arsenic pollution in groundwater in 57 blocks of District 15 of Bihar. Because arsenic is poisonous, prolonged exposure to it through drinking water can cause a number of malignancies. Additionally, it exacerbates the socioeconomic situation of those affected. In the Maner and Shahpur blocks, the districts of Patna and Bhojpur had the greatest levels of pollution, with 1810 and 1630 parts per billion, respectively
Floride	Compared to 1 mg/L, arsenic levels in water up to 1.5 mg/L are seen more tolerable	> 1.5 mg/L produced a stanning effect on tooth enamel	The maximum concentration was found in 14 blocks of 9 districts in the Navinger (2.85 mg/L) of the Jamui block. This area has higher fluoride levels (5–10 mg/L per liter), which might cause stiffness in the back and make it difficult to move naturally
Iron	As much as 1 mg/L in potable water	1.5 mg/L is dangerous	In Bihar, iron is a prevalent water pollutant. There is iron in the ground water in block 31 from district 20
Nitrate	45 mg/L or higher in drinking water	For children drinking water with 45 mg/L is dangerous	Distinguishing between natural and artificial sources of nitrogen in groundwater poses difficulties. Nitrogen contamination is widespread in groundwater, commonly originating from nonpoint sources. In Bihar, iron is found in 15 blocks across 9 districts, with Mokama exhibiting the highest concentration. The Fatuah block in Patna records 228 mg/L

Source Thakur et al. (2015)

issues and liver damage. Addressing iron pollution requires not only improving water quality but also educating communities on appropriate water treatment methods to minimize health risks.

Nitrate contamination, widespread across Bihar, originates from various nonpoint sources, making it challenging to distinguish between natural and artificial sources. Elevated nitrate levels in drinking water, as seen in blocks like Fatuah in Patna, In babies, it can cause methemoglobinemia, sometimes referred to as "blue baby syndrome," which reduces the blood's ability to carry oxygen and may cause serious health problems or even death. Furthermore, aquatic ecosystems and biodiversity are at risk when nitrate poisoning in groundwater spreads to surface water bodies.

Table 4 Classification of persistent reservation sites

Priority class	Group	Threshold
First priority	I	Monitoring point with BOD concentration higher than 30.0 mg/L
Second priority	II	BOD levels at the monitoring site range from 20.0 to 30.0 mg/L
Third priority	III	Observing areas where the BOD is between 10.0 and 20.0 mg/L
Fourth priority	IV	Observing areas where the BOD is between 6.0 and 10 mg/L
Fifth priority	V	Monitoring Locations having BOD between 3 and 6 mg/L

Source Central Pollution Control Board River Stretches for Restoration of Aquatic Resources (2015)

River segments are categorized into groups (I to V) mentioned in Table 4 based on their pollution rating score, as detailed in Table 4. The prioritization of river stretches is determined by their observed BOD levels, following the criteria outlined below.

5 Quantification Approach for Arsenic in Ground Water

Using analytical techniques like inductively coupled plasma mass spectrometry and atomic absorption spectrometry, the table illustrates the arsenic concentrations found in different water sources. Utilizing mathematical models and sampling techniques, this approach captures spatial and temporal variations. This quantification method ensures precise risk evaluation and effective mitigation strategies. Quantifying arsenic in groundwater is crucial for assessing its presence and potential health risks to human populations. Various quantitative methods have been developed and utilized for this purpose (Supplementary Table 2), each with its own advantages and limitations. One common approach is colorimetric analysis, which involves the reaction of arsenic with a reagent to produce a colored complex. This method works well for field applications and is simple and reasonably priced.

Okonuzzaman et al. (2022)—However, it may lack precision and accuracy compared to more advanced techniques. Atomic absorption spectrometry (AAS) and atomic fluorescence spectrometry (AFS) are widely used analytical techniques for arsenic quantification. AAS measures the absorption of light by arsenic atoms in a sample, while AFS detects the fluorescence emitted by arsenic atoms following their excitation. These methods offer high sensitivity and accuracy but require specialized equipment and skilled operators (Li et al. 2021). Inductively coupled plasma mass spectrometry (ICP-MS) is another powerful technique for arsenic quantification. It offers excellent sensitivity, allowing detection at trace levels, and can simultaneously analyze multiple elements. However, it requires expensive instrumentation and expertise in sample preparation and analysis (Richards et al. 2022). Ion chromatography (IC) coupled with conductivity detection is a method commonly used for analyzing arsenic species in groundwater. It separates arsenic species based on their ionic properties and measures their concentrations using conductivity. This approach provides good selectivity and sensitivity for arsenic speciation but requires meticulous sample preparation and calibration (Rowland et al. 2011). In recent years, biosensors and nanomaterial-based sensors have emerged as promising tools for arsenic quantification. These devices utilize biological receptors or functionalized nanomaterials to selectively capture arsenic ions, producing a signal that can be quantified. While these methods offer potential for rapid, on-site analysis, their performance in complex environmental matrices requires further validation. Hopenhayn (2006) investigated arsenic contamination in groundwater samples from the Kosi region using a combination of survey methods, questionnaires, and interviews as primary data sources. Secondary data were obtained from the Central Ground Water Board (CGWB) and Mahavir Cancer Sansthan, Patna. The study focused on iron and nitrate variables, employing advanced analytical tools like the UV double-beam spectrophotometer PHARO 300 and the Merkoquant arsenic kit, which corroborated the spectrophotometric findings. Results revealed that while groundwater samples were contaminated with arsenic, water samples from the Kosi River remained uncontaminated, highlighting a critical distinction in contamination sources and impacts in the region (Kumar Jha and Chandra Gupta 2017).

Overall, the choice of quantification approach for arsenic in groundwater depends on factors such as analytical requirements, resource availability, and sample characteristics. A combination of techniques, including both traditional and emerging methods, may be necessary to comprehensively assess arsenic contamination and ensure water safety. Aullón Alcaine et al. (2020) determined the maximum levels of pollution by studying the content of arsenic in the sediments of dry lake surfaces. Using primary fieldwork and experimental techniques, the study concentrated on physicochemical parameters that were examined using conductivity detection in conjunction with ion chromatography (IC). The results, which included shallow groundwater, loess sediment, and volcanic ash as highlights, connected arsenic mobilization to geochemical processes, hydrogeological characteristics of neighbouring aquifers, and climatic conditions. Prabhakar and Tiwari (2015) investigated the dynamics of groundwater storage and recharge utilizing secondary data sources, including Survey of India maps, USGS Landsat TM, DEM data, and topographic maps. The study, which used ArcGIS, AIO and signature editor, and ERDAS IMAGINE (9.2), found that building canals and increasing vegetation greatly improved groundwater reserves in some villages, helping to control local water demand. Shaji et al. (2021) gave a worldwide overview of areas impacted by arsenic pollution using secondary data from UNEP, WHO, CGWB, and earlier studies. Geological mapping and scatter plot analysis were two analytical techniques that indicated a relationship between pH levels and arsenic presence but did not show any linkage between electrical conductivity and arsenic levels. The study discovered that, after sedimentary formations, mining is the second-largest source of arsenic worldwide. Arsenic pollution is common in Asia's Indo-Gangetic alluvium and groundwater belts linked to orogenic tectonics. Paul et al. (2021) used secondary data from the USGS Earth Explorer to examine how human activity and geology affect water quality. Both natural and man-made factors were found to have a considerable impact on water quality, even though there were only slight temperature differences between the research areas. Conductivity, alkalinity, magnesium, and salt were among the

Table 5 Quantification(s) approaches of identify groundwater pollution

Sl. No.	Data type	Methodology	Findings	Source
1	Primary data source—field methods Secondary data source—resource Environment, ESRI, US	Variables—fluoride (F) and arsenic (As) Analytical tools an analysis of the difference between F and As was conducted using the t-test Next, a one-way ANOVA	Findings—the study findings align with previous research indicating that exposure to fluoride (F)). At specific locations, arsenic levels surpassed the recommended thresholds established by the World Health Organization (WHO)	Li et al. (2021)

factors that were assessed using colorimetric analysis. The study established a link between changes in land use and land cover (LULC) and arsenic contamination using multivariate statistics, principal component analysis, and hydro geochemistry (Table 5).

5.1 *Discovery of Arsenicosis*

Arsenic (As) is the 20th most prevalent element in the Earth's crust and the 53rd most abundant element on the planet's surface. It possesses an atomic mass of 74.9216 and a melting point of 816.80 °C (Ranjan 2019a). In 1982, the author discovered instances of arsenicosis in West Bengal following a patient's visit to a dermatological clinic at the Calcutta School of Tropical Medicine. The author initially observed the condition in a patient residing in district 107 in Khulna (Saha 2003).

5.2 *Sources of Arsenic*

There are multiple avenues for encountering arsenic, a hazardous metalloid. Arsenic can naturally seep into groundwater from rocks and minerals via geological mechanisms. Industrial activities and mining represent human actions that release arsenic into the environment (Smedley and Kinniburgh 2002; Li et al. 2021; Siddik et al. 2022; Ram et al. 2021; Kumar et al. 2019; Omwene et al. 2019a). Soil pollution mainly originates from agricultural activities, notably the use of arsenic-containing pesticides and fertilizers. Arsenic can also be released into the atmosphere through the combustion of wood and fossil fuels, resulting in atmospheric deposition. Furthermore, some industries discharge waste containing arsenic into water sources, elevating the potential for contamination. A comprehensive strategy to reduce arsenic exposure should encompass both natural and human-induced sources across various environmental routes.

5.2.1 Natural Sources

Arsenic exists in significant quantities in the atmosphere, water sources, and terrestrial environments, occurring naturally in the Earth's crust. Both natural processes and human activities can introduce arsenic into drinking water and groundwater, leading to contamination. The main natural sources of inorganic arsenic are geologic formations, including sedimentary and volcanic rocks, soils, geothermal zones, coal reserves, volcanic eruptions, and geothermal water. The primary source of the GBM river system is vast sediment accumulations. Through the ages, the mighty GBM river has carried arsenic from the ancient Himalayas downstream into these deposits during certain periods of the Pleistocene and Holocene epochs. Arsenic then leaches out from these sediments into Holocene alluvium and groundwater aquifers within the fan deposit zones (Shaji et al. 2021; Ghosh et al. 2019; Gómez et al. 2006; Ranjan 2019a; Gounot 1994).

5.2.2 Anthropogenic Sources

Instances of human-induced arsenic pollution in groundwater have been recorded in fifty-four nations. Human activities, including metal refining through burning, coal mining, petroleum extraction, and operations in various industries such as chemical, glass, cement, and leather, are the main contributors to these pollutants. These sectors encompass the manufacturing of fertilizers, oils, pigments, herbicides, fungicides, chemicals, and the management of acid mine drainage (Shaji et al. 2021; Rodriǵuez-Lado et al. 2013). Because these minerals are mined and smelted, arsenic seeps into surface and groundwater through garbage holes waste dumps, extraction basins, and mines (Rowland et al. 2011; Ranjan 2019b).

5.3 *Sources of Arsenic Exposure*

There are multiple pathways for arsenic exposure, but the primary concern arises from consuming contaminated drinking water, especially in regions where naturally occurring arsenic levels are elevated. The risk of exposure increases when consuming foods containing high levels of arsenic, such as rice and certain types of seafood (Li et al. 2021; Ghosh and Kumar 2014; Shankar and Shanker 2014b; Kumar et al. 2021b; Mohammed Abdul et al. 2015). Increased exposure to arsenic can result from various factors such as the utilization of arsenic-based pesticides, mining activities, and industrial processes, all of which can release this element into the environment, including the air, soil, and water. It's crucial for effective mitigation approaches to address these multiple sources comprehensively in order to safeguard public health and minimize the hazards linked with arsenic exposure. Human exposure to arsenic primarily occurs through dietary intake and water consumption, particularly groundwater from both natural and human-made origins. Arsenic-containing foods such as rice, vegetables, meat, eggs, and milk contribute to this exposure. Subsequent sections will explore the factors that contribute to arsenic exposure in greater detail (Thomas and Bradham 2016b; Morales-Simfors et al. 2020; Alarcón-Herrera et al. 2020).

5.4 Arsenic (As) Intake by Human

Contaminated food and water sources are primary contributors to arsenic (As) poisoning, posing significant health risks to individuals. Consuming food or water contaminated with arsenic from polluted soil can result in serious health issues, including skin lesions and cancer (Fig. 6). To address these risks, the World Health Organization (WHO) has established standards for safe levels of arsenic in drinking water as a measure to mitigate these health hazards. Monitoring and controlling arsenic consumption are essential to safeguard human health, water and food sources, and other valuable resources.

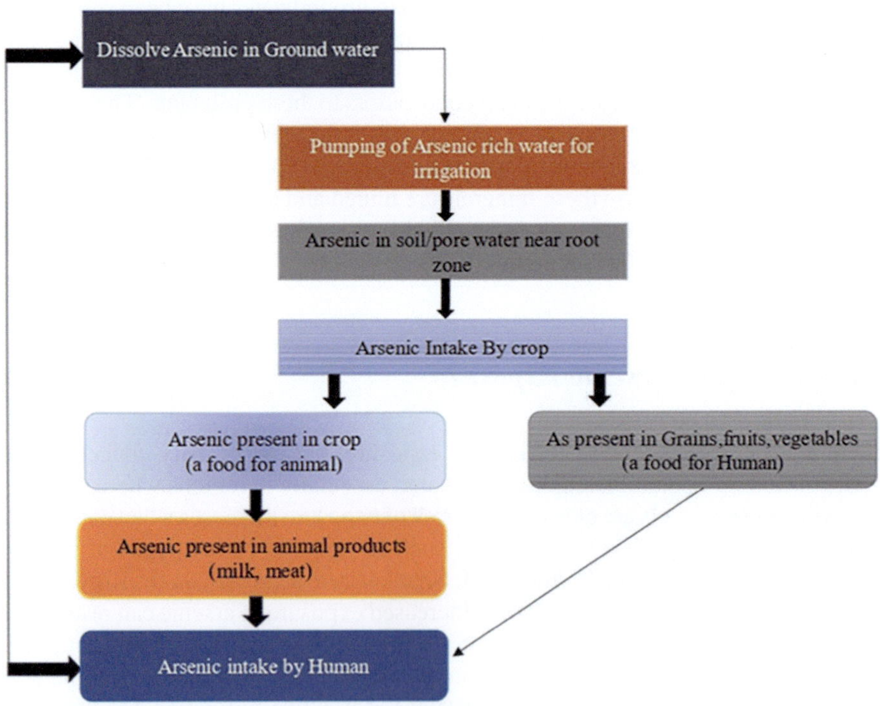

Fig. 6 Mechanism of arsenic intake by human from soil and water

5.5 The Process of Arsenic Leaching from Soil

There are two dominant opinions regarding arsenic leaching in the earth, Theory 60 of Pyrite Oxidation and Theory Ox Hydroxide-Reduction (Borgoño et al. 1977; Warhurst and Nickson 2007). The first explanation contends that excessive groundwater extraction for residential and agricultural purposes is continuously oxidizing pyrites (FeS_2), Around 60,000 years ago, arsenic deposits were deposited in the soil of affected regions and contributed to a decrease in the aquifer level. The second theory states that iron oxychloride dissolves reductively, releasing arsenic. But both of these elements contribute to the leaching of arsenic from soil rock (Das et al. 2001). Before arsenic to be leached from soil, its compounds need to breakdown and be transferred from the solid phase into the soil solution. This is due to a confluence of physical and chemical processes. Initially, when water from irrigation or rainfall passes through the soil, soluble arsenic molecules within the soil structure dissolve. Secondly, the acidity of the environment plays a crucial role in promoting the leaching of arsenic, making soil pH an important factor. Another factor affecting leaching is the presence of organic matter in the soil, which reacts with arsenic to create compounds and enhances its movement. Microbial activity can also aid in the release of arsenic by promoting the breakdown of minerals containing arsenic Finally, particular ions like phosphate may induce arsenic to compete with it for binding sites on soil particles, thereby contaminating groundwater.

5.6 Release of Arsenic in Water

Arsenic leakage into water is mostly caused by natural geological processes, industrial operations, and agricultural practices. The decomposition of arsenic-containing minerals is a geological phenomenon that can lead to elevated levels of arsenic in groundwater (Smedley and Kinniburgh 2002; Rowland et al. 2011). Other human activities that introduce arsenic into water sources include burning and mining (Fig. 6) (Ranjan 2019a; Pal et al. 2009). Arsenic-based pesticides and fertilizers found in agricultural runoff can also contaminate water (Dandge and Patil 2022; Foley et al. 2005). Reduced oxygen levels, such as those found in certain environments, can speed up the release of arsenic and prompt less harmful forms of the metal to convert into more soluble and harmful variants. This process exacerbates the degradation of water quality (Foley et al. 2005; Fendorf et al. 2010). Groundwater can get contaminated by arsenic in three different ways (Ranjan 2019b): (i) Arsenic released by oxidizing pyrite that is rich in arsenic, (ii) Groundwater contamination resulting from the acidic solution of iron hydroxide and the dissolved release of arsenic, (iii) When arsenic and phosphate from fertilizers exchange ions, arsenic is liberated.

5.7 *Groundwater Quality: Indian Scenario*

Recent research conducted by the CGWB indicates that over 50 million people in India are at risk of arsenic contamination from groundwater sources (Table 6). A significant research endeavour examining arsenic contamination of groundwater in India, particularly concentrated within the Ganga basin (Chakraborti et al. 2018a). The Ganga River basin, home to over 500 million people, comprises more than a quarter of India's population, accounting for more than 26%.

The Ganga River basin stands as one of the most densely populated and productive regions globally (Khan et al. 2016). Groundwater is vital for meeting the diverse water requirements of the country, including irrigation, residential use, commercial activities, and agriculture (Saha and Ray 2018). Presently, the Ganga is among the most contaminated rivers globally, containing various pollutants, including chemicals and harmful bacteria, exceeding the WHO's recommended levels by over three thousand times (Shaji et al. 2021; Kumar et al. 2016a; Tandon and Sinha 2018). Although the Ganga and Brahmaputra alluvial plains boast the country's most abundant groundwater reservoirs, they are also severely tainted by arsenic contamination (Table 7). Most of the groundwater extraction occurs in the Indo-Gangetic region of Northern and Northwestern India, leading to considerable depletion and reduction of the water table in many regions (Chakraborti et al. 2018a; James et al. 2009; MacDonald et al. 2016). The first reports of groundwater contamination with arsenic emerged from the Chandigarh district in northern India (Datta and Kaul 1976). Furthermore, the second occurrence was documented in the lower Gangetic plains of West Bengal (Saha 2003; Saha and Ray 2017). Following this, reports surfaced from various states including West Bengal, Bihar, Uttar Pradesh, Jharkhand, Assam, Manipur, Meghalaya, Mizoram, Nagaland, Sikkim, Tripura, and Himachal Pradesh (Fig. 7). Additionally, reports were also received from Punjab, Himachal Pradesh, Andhra Pradesh, and West Bengal in India (Shaji et al. 2021). Elevated levels of arsenic in groundwater are observed in two types of terrains: alluvial and hard rock. Alluvial aquifers contribute to 90% of arsenic occurrence in India, with the remaining 10% attributed to hard rock aquifers. This encompasses regions such as Karnataka and Chhattisgarh (Central Ground Water Board Arsenic Hot Spot in Ground Water in India 2019; Omwene et al. 2018, 2019b; Pokhrel et al. 2009; Sorg et al. 2014; Vieira et al. 2017; Yang et al. 2020).

6 The Impact of Arsenic Contamination on Human Health

Inorganic arsenite (As^{3+}) and its derivatives are considered to present a higher risk compared to inorganic arsenate (As^{5+}), the body excretes absorbed inorganic arsenic through urine. In the past, excretion and methylation were considered detoxification methods (Adeloju et al. 2021). Currently, it is understood that these methylated forms of arsenic may present greater risks in terms of human metabolism (Gebel

Table 6 Certain regions within the district have groundwater containing arsenic levels ranging from 0.01 to 0.05 mg/L

Sl. No.	State	District
1	Andhra Pradesh	Kurnool, Nellore, and Guntur
2	Assam	Sonitpur, Jorhat, Golaghat, Lakhimpur, Sibsagar, Nalbari, and Nagaon
3	Bihar	Gopalganj, Katihar, Khagaria, Lakhisarai, Purnea, Madhepura, Muzaffarpur, W. Champaran, Bagursarai, Bhagalpur, Bhojpur, Buxar, Darbhanga, and so on
4	Chhattisgarh	Rajanandgaon
5	Delhi	North east Delhi
6	Daman and Diu	Diu
7	Gujrat	Surendra Nagar, Vadodara, Kachchh, Patan, Rajkot, Gandhinagar, Amreli, Anand, Bharuch, Bhavnagar, Dahod, and Kachchh
8	Haryana	Palwal, Rohtak, Sirsa, Bhiwani, Soni Path, and Mahendragarh
9	Himachal Pradesh	Kangra
10	Jammu and Kashmir	Jammu, Kathua, and Rajouri
11	Jharkhand	Saheb Ganj
12	Karnataka	Raichur
13	Madhya Pradesh	The following places: Umaria, Neemach, Khandwa, Mandsaur, Batul, Dhar, and Chhindwara
14	Odisha	Ganapati
15	Punjab	Sangrur, Tarn Taran, Hoshiarpur, Gurdaspur, and Faridkot
16	Rajasthan	Ganga Nager
17	Tmailnadu	Cuddalore, Ramnathpuram, Peramblur, Thirunelveli, Thiruvallur, Tuticorin, Dindigul, Nagapattinam
18	Telangana	Nalgonda
19	Uttar Pradesh	Jhansi, Kaushambi, Pilibhit, Shahjahanpur, Badaun, Bahraich, Basti, Deoria, and Gorakhpur
20	West bangal	Hooghly, Howrah, Cooch Behar, Murshidabad, Nadia, South 24 Parganas, and North 24 Parganas

Source CGWB, "Groundwater Quality in Shallow Aquifers in India," Faridabad, 2018

2002). Continued intake of water with elevated arsenic concentrations can result in the onset of different forms of cancer (Lalwani et al. 2004). The adverse effects and symptoms of arsenic poisoning can differ depending on the location and, to some extent, the individual affected (Ghosh and Kumar 2014). Reports suggest that individuals with a balanced diet may experience fewer adverse effects compared to those with poor dietary habits. The initial identification of arsenic-induced skin

Table 7 Districts exhibiting heightened levels of arsenic contamination (greater than 0.05 mg/L)

Sl. No.	State	District
1	Assam	Cahar, Dhamija, Jorahat, and Nagaon
2	Bihar	The subsequent locations: Begusarai, Bhagalpur, Bhojpur, Buxar, Darbhanga, Katihar, Khagaria, Kishanganj, Lakhisarai, Munger, Samastipur, Vaishali, Saran, Purnia
3	Chhattisgarh	Rajanandgaon
4	Hariyana	Jhajjar, Ambala
5	Karnataka	Raichur
6	Manipur	Bishnupur, Thouba
7	Punjab	Amritsar, Roper, Tarn Taran
8	Uttar Pradesh	Agra, Aligarh, Azamgarh, Balia, Balarampur, Deoria, Gonda, Gorakhpur, Lakhimpur kheri, Maunath Bhanjnam, Matura, and Moradabad
9	West Bangal	Hooghly, Malda, Murshidabad, Nadia, South 24 Parganas, and North 24 Parganas

Source Report of the central Ground Water Board, 2010 and 2011

lesions was noted in 1983 at the Department of Dermatology, School of Tropical Medicine in Calcutta, India (Saha 2003). Prolonged exposure to arsenic results in a condition termed "arsenicosis," which exhibits skin symptoms including melanosis, keratosis, and leucomelanosis (Bhowmick et al. 2007b). The World Health Organization defines arsenicosis as a chronic condition resulting from the consumption of arsenic at levels surpassing safety standards for a minimum of six months. Typically, it is identified by observable skin abnormalities such as melanosis and keratosis, with or without affecting internal organs (Naujokas et al. 2013). The initial detection of arsenic contamination in tubewell water occurred in northern Bangladesh in 1993 (Ahmad et al. 2018). Ingesting groundwater tainted with arsenic can result in significant health issues, including skin conditions, kidney and bladder ailments, as well as bronchiectasis and coronary artery disease. In Supplementary Table 3 and Fig. 8, we have portrayed the major health problems induced by arsenic contamination through groundwater. Some of the problems include hyperkeratosis, arsenicosis, darkening of the palms and soles, high blood pressure, liver and heart damage, diabetes, and Bowen's disease (Chakraborti et al. 2018b). In all these research papers, the impacts of arsenic are discussed using different methods, as follows. Aullón Alcaine et al. (2020) emphasized the interdependence of natural processes including shallow groundwater contact and sedimentary features, as well as the impact that geochemical reactions, hydrogeological features, and climate play in mobilizing arsenic in dry lake deposits. To go along with this, Prabhakar and Tiwari (2015) illustrated the potential for anthropogenic activities such as canal construction and vegetation enhancement to alleviate groundwater scarcity by mapping the effects of drainage density and changes in land use using GIS-based technologies. Adding

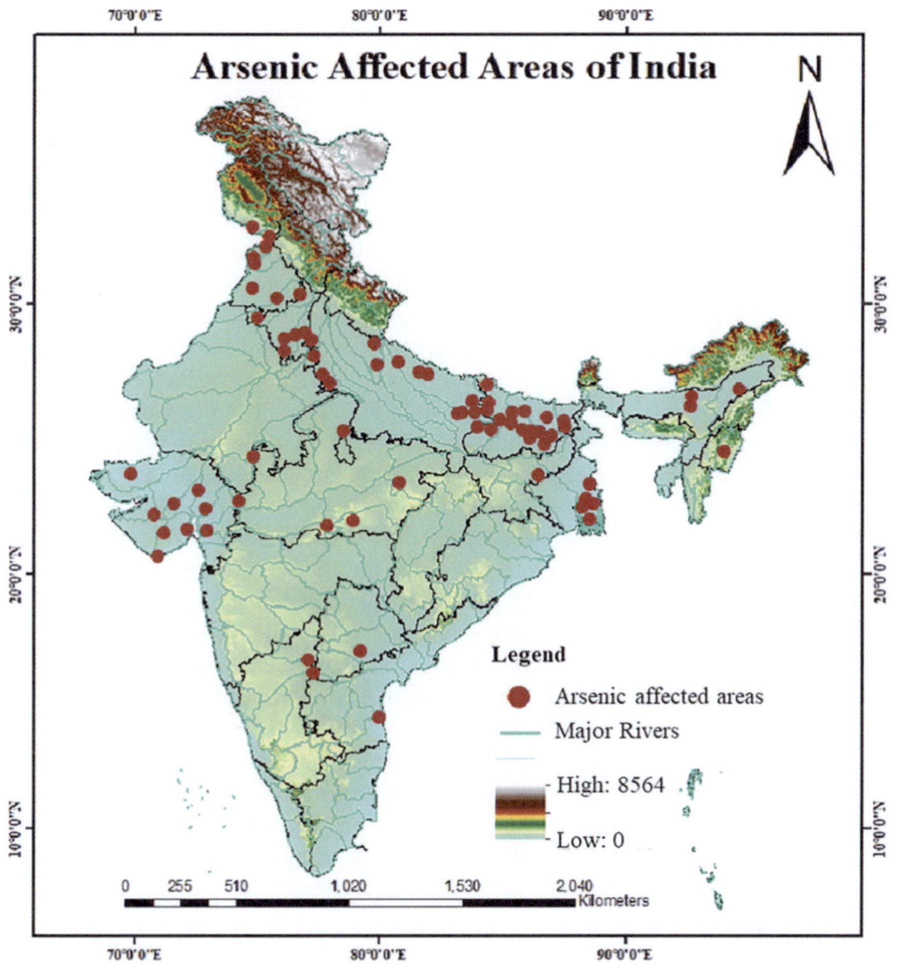

Fig. 7 Arsenic pollution has impacted multiple states and union territories throughout India, affecting a total of twenty states and four union territories

a global perspective, Shaji et al. (2021) correlated pH levels with arsenic prevalence and used international databases to assess patterns of arsenic pollution. Mining was shown to be the second most important global source of contamination, after sedimentary formations. Their findings reinforce the geological and tectonic influences observed in arsenic distribution, particularly in Asia's Indo-Gangetic alluvium. Finally, Paul et al. (2021) investigated how LULC changes affect arsenic pollution in groundwater by utilizing hydrogeochemical and multivariate statistical techniques to assess the interaction between geological factors and human activity. Li et al. (2021) conducted a comprehensive study on fluoride and arsenic contamination in groundwater, utilizing field sampling from 417 sites as the primary data source.

Secondary data were analyzed using the inverse distance weighting (IDW) technique and processed in ArcGIS 10.2 (ESRI Inc., US). The study employed statistical tools such as T-tests, ANOVA, and Microsoft Excel for data analysis, adhering to the World Health Organization (WHO) parameters for water quality. The findings highlighted how changes in land use, influenced by human activities, can elevate the health risks associated with groundwater contamination by fluoride and arsenic. These insights underscore the critical need for sustainable land management practices to mitigate such risks. Thakur and Gupta (2019) conducted a study focusing on the prevalence of arsenicosis, particularly in Bihar, analyzing its impact through variables such as household drinking water samples and related health outcomes. Their findings revealed a higher prevalence of arsenicosis among females and children compared to males. Key factors, including per capita income, sanitation, awareness levels, and the depth of water sources, showed a significant positive correlation with protective measures like water purification. This study highlights the critical role of socioeconomic and environmental factors in addressing health damage caused by arsenic contamination. Saha and Ray (2018) in a book chapter, explored the Indo-Ganga–Brahmaputra plain using secondary data sources from CGWB, UNESCO, and WHO. Their study delves into the broader implications of groundwater issues in this region, emphasizing the role of global and national standards in understanding and managing water quality challenges. Andualem and Demeke (2019) conducted a study integrating fieldwork and secondary data sources, such as USGS Earth Explorer, to assess groundwater potential. They analyzed variables including land cover, drainage density, lineament density, soil, geology, slope, and geomorphology. Utilizing ArcGIS and a multi-criteria decision-making (MCDM) approach, they developed thematic layers to delineate groundwater potential zones. Validation of the resulting groundwater potential map, based on existing pumping wells, revealed a prediction accuracy of 70.5%. This high accuracy underscores the reliability of the delineated zones, highlighting the effectiveness of the MCDA technique in groundwater potential mapping.

Together, These studies demonstrate the intricate relationships that shape groundwater reserves and water quality between natural and man-made variables. To solve water management issues, they stress the value of combining fieldwork, sophisticated analytical methods, and geospatial tools, especially in areas vulnerable to arsenic poisoning.

6.1 Impact of Arsenic Contamination on Human Health: Indian Scenario

As per a report issued on September 19, 2016, by the Ministry of Drinking Water and Sanitation of the Government of India, around 1.32 million individuals in India experience negative consequences due to arsenic-contaminated groundwater. In West Bengal, the affected population exceeds 8.9 million, while Bihar has over 1.6 million

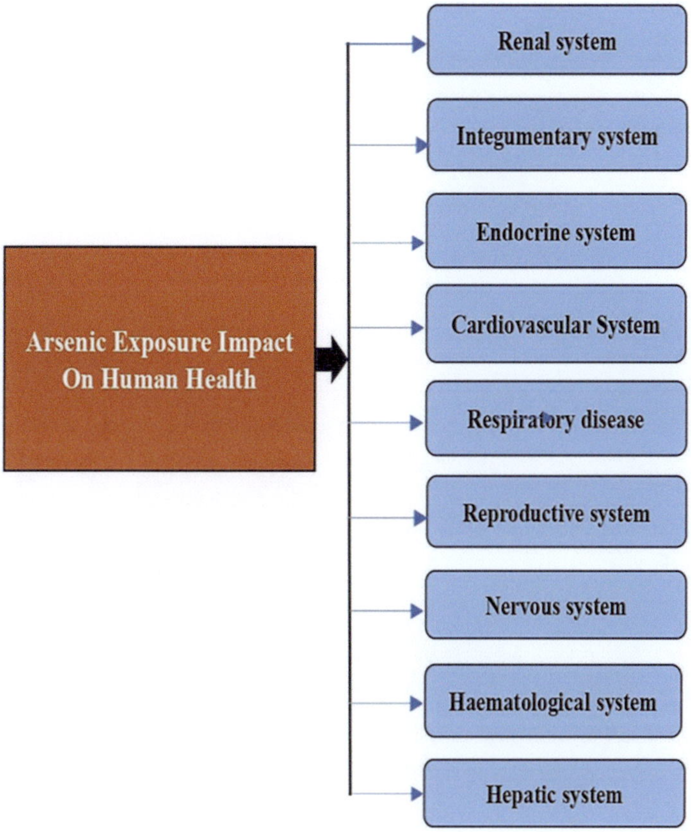

Fig. 8 Effects of arsenic exposures on human health

affected individuals, Uttar Pradesh has 1.5 million, and Assam has over 1.2 million people affected by arsenic contamination in the state's groundwater (Fig. 9).

The extensive Ganga river basin, originating in the Himalayas and draining into the Bay of Bengal, plays a vital role in providing sustenance for the state of Bihar (Singh and Choudhary 2010). Throughout its entire course, the tributaries of the Ganga river transport sediment from its origin (Ahmad et al. 2018; Kumar Jha and Chandra Gupta 2017). Arsenopyrite, discovered in the sedimentary deposits of the Gangetic basin, acts as the primary origin of arsenic pollution in the region's groundwater (Fig. 4) (Chakraborti et al. 2018a; Goyal et al. 2010). As per the Indian Ministry of Water Resources, arsenic contamination has significantly affected 1600 settlements across 67 blocks and 18 districts within the states (Kumar et al. 2021a). Nowadays, the approximately 50 million residents of the state, of whom 13.85 million are exposed to arsenic levels in water exceeding 10 μg/L, are encountering increased hazards. As a result, the current population faces a higher likelihood of arsenic exposure

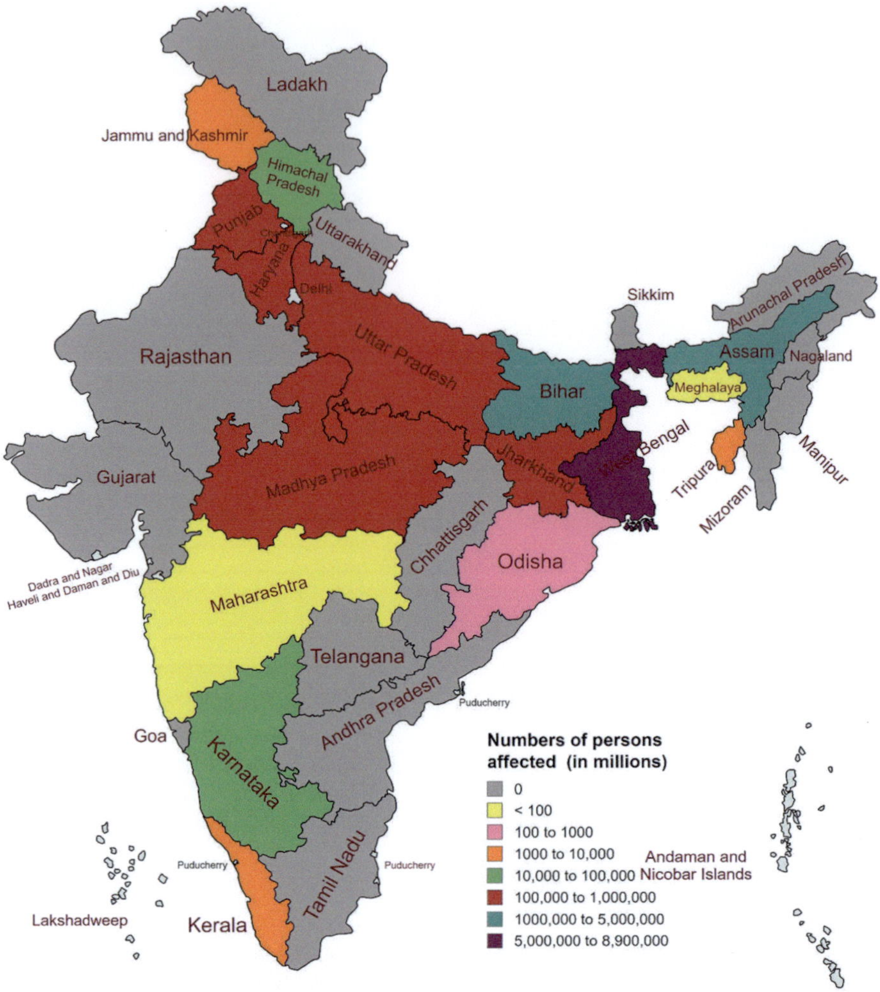

Fig. 9 Persons affected by groundwater-arsenic problems (*Source* Inter-Ministerial Group (IMG) on Arsenic Mitigation, 2015)

compared to previous survey estimations. The first instance of arsenic contamination in groundwater in Bihar was identified in Semaria Ojhapatti village, located in the Bhojpur district. Afflicted individuals exhibited characteristic symptoms of arsenicosis along with additional internal health issues.

Recent studies indicate that arsenic contamination poses a significant problem in several districts of Bihar, this includes Buxar, Bhojpur, Patna, Saran, Vaishali, Samastipur, Begusarai, Khagaria, Munger, and Bhagalpur. Arsenic primarily enters the body through drinking water Subsequently, arsenic travels through the digestive system, enters the bloodstream, and reaches vital organs, Eroding and toxifying cells, resulting in illness and disturbances in metabolic functions. According to Globo Can's assessment, there were 18.0 million fresh cancer diagnoses, 9.5 million fatalities, and 43.8 million recurrences (within five years of surviving) in 2018. In 2016, India documented 3.9 million cancer instances according to statistics from the Indian Council of Medical Research's (ICMR) National Cancer Registry Program. Arsenic penetrates the body via ingestion, then progresses through the digestive tract, bloodstream, and essential organs, inducing organ toxicity that hampers metabolism and results in illness. The states registering the highest numbers of cancer cases were Uttar Pradesh (674,386), Maharashtra (364,997), and Bihar (359,228) Conversely, Tamil Nadu (222,748), Karnataka (202,156), Andhra Pradesh (159,696), Telangana (115,333), and Kerala (115,511) reported lower rates of cancer. This study aims to examine the factors influencing cancer incidence in the Bihar Ganga basin and evaluate the influence of arsenic on it.

6.2 Reported Articles on Ground Water Contamination in Bihar

It was reported that drinking water sources in 3,791 rural areas across 11 districts are affected by fluoride contamination. Within the Kosi basin, there are nine sites with increased iron levels. Consumption of contaminated water can lead to various health issues, including skin problems, liver ailments, kidney disorders, and other bodily complications. Included in the afflicted district are the following: Buxar, Darbhanga, Katihar, Samastipur, Saran, Sitamarhi, Patna, Vaishali, Araria, Nawada, Sheikhupura, Kaimur, Munger, Nalanda, Rohtas, Katihar, Khagaria, Lakhisarai, Begusarai, Bhagalpur, and Bhojpur. In certain areas of the districts of Saharsa and Madhepura, health problems caused by arsenic have also been reported (Kumar et al. 2016b, 2021b; Central Ground Water Board Arsenic Hot Spot in Ground Water in India 2019; Shankar and Rakshit 2015).

6.2.1 Contaminated Water Report by NHRC

According to the National Human Rights Commission (NHRC) report and Bihar Economic Survey (2021–2022), heightened levels of iron, fluoride, and arsenic were detected in groundwater across 31 out of the 38 districts in the state (Fig. 10). As per the NHRC, investigations revealed chemical contamination of groundwater in 30,272 rural areas. Among them, 4742 rural regions adjacent to the Ganges demonstrate the highest incidence of arsenic among all 30,742 areas. Conversely, the Koshi Basin encompasses 3791 rural wards spanning 11 districts affected by fluoride, with certain areas also experiencing elevated iron levels. The human rights committee states that these chemicals may present serious health hazards, including problems with the liver and kidneys. According to the Commission, the report should include information

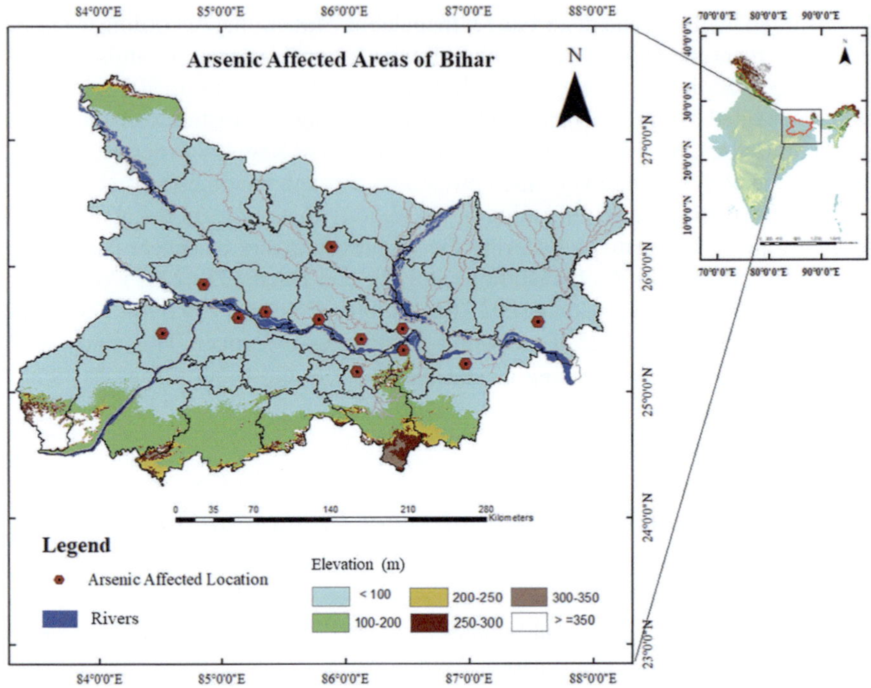

Fig. 10 Map of arsenic affected areas of Bihar *Sources* Bihar Economic Survey (2022)

on how the "Har-Ghar-Jal-Nal-Yojana" initiative is being implemented as well as the safety measures put in place to guarantee that those districts have access to potable water. The Commission has also requested a report on the groundwater random sample from the State Pollution Control Board (Bihar Economic Survey 2022).

Groundwater contamination induced by Lead, iron, nitrate, fluoride, and arsenic have many health problems, like as cancer and gallbladder cancer, are being brought on by it for those living in the Gangetic belt. In Bihar's rural areas, there is a shortage of clean drinking water along with social and economic problems. Along with serious health issues, it is dangerous for the ecosystem as well. One's capacity to participate in social and economic activities is restricted by contaminated drinking water. Water contamination has been shown to have negative effects on individual health, as well as wider health effects. Human sickness and mortality have been linked to increased levels of lead, nitrate, fluoride, iron, and arsenic contamination in drinking water (Thakur et al. 2015; Kumar et al. 2021b). Skin, bladder, and lung cancer risks may be elevated in those with the skin illness linked to arsenic exposure (Richards et al. 2022; Mendoza-Chávez et al. 2021; Gute 2004). For the impacted parties, there were monetary, social, and environmental repercussions to these health problems. The societal and economic burdens on affected households and communities are reduced when arsenic-contaminated water is ingested. The bulk of arsenic studies are of an epidemiological nature.

Studies in epidemiology have shown that long-term exposure to arsenic in drinking water leads to serious illness. The cross-sectional studies emphasize the illness's present exposure more than anything else. Pollution-related health issues have a financial cost associated with them because of lost production and medical bills (Santra et al. 2013; Adhikari et al. 2020). Advanced forms of keratosis pose a significant public health risk in Bangladesh and West Bengal, India, due to their severe nature and the potential for social isolation in affected areas if not addressed (Ahmad et al. 2018). Skin abnormalities develop after extended (5–10 years) exposure to drinking water contaminated with arsenic (Keratosis, Melanosis), hyperpigmentation and a higher chance of birth defects, cancers of the skin, bladder, and lungs, among other things. Arsenic exposure lowers the social and financial costs for the afflicted individual. There is limited research that simultaneously examines social and economic aspects to evaluate the socio-economic impacts on households exposed to arsenic in their drinking water.

Arsenic present in drinking water can lead to various types of cancers (Smedley and Kinniburgh 2002; Kumar et al. 2016b; Mazumder 2003). The socio-economic structure is significantly impacted by the arsenic issue as well. The three main kinds of socioeconomic issues are those related to agriculture, health, and other issues (Thakur and Gupta 2019). A high concentration of contaminated water reduces soil fertility, agricultural productivity, and health issues by entering the food chain (Brammer and Ravenscroft 2009). In Bangladesh and Nepal, arsenic-contaminated groundwater utilized for agricultural purposes infiltrates into the food chain. Each of these three issues contributes to social and economic issues. The skin sores, bladder issues, cancer, and death are only a few of the health issues that arise; the few social issues include social ignorance, sadness, and suicidal thoughts (Kumar et al. 2021a).

7 The Available Techniques to Minimize and Mitigate Arsenic Contamination

The only established solutions for addressing long-term arsenic poisoning are sufficient physical activity, a healthy diet, and access to safe drinking water (Maeda 2009). Efficient watershed management, promoting water conservation, and raising public awareness are essential tactics in tackling the arsenic (Frankenberger and Losi 2015). Biological transformation of inorganic arsenic entails substituting the hydroxyl group of arsenic acid with a CH_3 group, thereby altering it into a less toxic form (Central Ground Water Board Arsenic Hot Spot in Ground Water in India 2019; Berg et al. 2007). The first step in the methylation process of As^{5+} involves reducing it to As^{3+}. Following this, coenzyme S-adenosyl methionine methylates As^{3+} to produce dimethyl arsine (Pierce and Moore 1982). Bacteria that decrease sulfate levels often boost metabolism Moreover, multiple types of fungi have demonstrated the capacity to reduce arsenic levels (United States Department of Health and Human Services 2000). Across the globe, different technologies are under investigation for eliminating arsenic contamination from groundwater. Electrocoagulation is acknowledged as an effective method for achieving drinking water quality standards and treating arsenic-contaminated water (Mendoza-Chávez et al. 2021). A cement-based filter medium (CBFM) can also be utilized to remove heavy metals from groundwater (Holmes et al. 2019). Other commonly used techniques for treating arsenic-contaminated water include reverse osmosis, adsorption, ion-exchange, coagulation, chemical oxidation, and advanced oxidation processes (AOPs) (Omwene et al. 2019b). The summary of each of these processes' essential elements mentioned in Supplementary Table 4.

7.1 Oxidation

Arsenite can undergo oxidation upon exposure to various substances like oxygen, ozone, hydrogen peroxide, permanganate, and free chlorine (Khalid 2017). In developing countries, atmospheric oxygen, hypochlorite, and permanganate are frequently employed as oxidizing agents (Khalid 2017). While arsenic oxidation occurs slowly in the atmosphere, certain chemicals such as permanganate and chlorine can rapidly convert arsenite to arsenate in various conditions (Wegelin et al. 2000).

7.2 Co-precipitation and Adsorption Processes

Activated alumina, ferric chloride, ferric sulphate, aluminium alum, and other coagulants can be used to extract arsenic from water (Chakraborti et al. 2019). It's been found that ferric salts remove arsenic more effectively than alum on a weight basis and that they work across a larger pH range (Pal et al. 2009). Pentavalent According to, arsenic has the potential to dissolve more readily than trivalent arsenic in each (Pierce and Moore 1982). The Bucket Treatment Unit (BTU) utilizes coagulation, co-precipitation, and adsorption principles, specifically designed for household use. Consisting of two 20-L buckets stacked on top of each other, the procedure involves manually mixing chemicals with arsenic-contaminated water in one bucket by vigorous shaking, followed by gentle mixing to promote flocculation (Pallier et al. 2010). Subsequently, the blended water is left undisturbed for one to two hours to settle. After this duration, the water from the first bucket is carefully directed through a sand filter positioned in the second bucket, ensuring that any sediment collected is not transferred. At this stage, the water in the second bucket is effectively purified (Mazumder 2003).

7.3 Solar Oxidation

To diminish arsenic concentrations in drinking water, a straightforward technique is utilized, which involves exposing arsenic to sunlight in transparent containers (MacDonald et al. 2016). In the presence of extra oxidizing agents, ultraviolet radiation can accelerate the oxidation of arsenite. Various substances have been utilized to eliminate arsenic compounds, including both arsenite (As^{3+}) and arsenate (As^{5+}) However, converting arsenite (As^{3+}) to arsenate (As^{5+}) is essential for eliminating arsenite (As^{3+}) (Vieira et al. 2017; Pierce and Moore 1982; Jiang et al. 2013). As a consequence, the process becomes costly and may sometimes yield undesirable byproducts (Rodriǵuez-Lado et al. 2013). Thus, a number of solid materials having oxidative characteristics have been created to skip the pre-oxidation stage utilizing expensive oxidizing chemicals (Siddiqui and Chaudhry 2017).

7.4 Biological Oxidation

Studies show that specific microorganisms, like Leptothrix ochracea and Gallionella ferruginea, enhance the biological oxidation of iron, creating conditions conducive to the adsorption of arsenic (Pallier et al. 2010).

Another technique for removing arsenic from groundwater involves using a nanofiltration (NF) membrane system (Song et al. 2015). In Vietnam, utilizing laterite as an economical adsorbent in a sustainable filtration system aimed at eliminating arsenic from groundwater also offers the advantage of increasing groundwater levels through managed aquifer recharge (Postma et al. 2016). Finding clay-rich laterite in India is relatively easy, and it can be employed as a filtration technique to eliminate arsenic from groundwater (Bagyaraj et al. 2013). Arsenic found in groundwater can be efficiently removed by utilizing clay-rich laterite, which is abundantly available in India (Shaji et al. 2021; Mendoza-Chávez et al. 2021).

8 Limitations

The absence of comprehensive and reliable data on arsenic pollution in groundwater throughout Bihar's regions could potentially limit the review's depth. Making direct comparisons and reaching broad conclusions may be complicated by variations in data gathering techniques and reporting requirements.

8.1 Temporal Dynamics

The empirical dynamics of groundwater arsenic contamination show variations in concentration over time (Brammer and Ravenscroft 2009; Mondal et al. 2021; Sinha et al. 2018). Many variables, such as hydrogeological conditions, anthropogenic activities, and seasonal oscillations, contribute to the dynamic nature of arsenic levels (Sorg et al. 2014). Effective mitigation initiatives depend on an understanding of these temporal patterns, which help identify high-risk periods and implement targeted interventions to manage and minimize arsenic exposure in impacted locations. Developing effective and enduring strategies is imperative to tackle the persistent problem of arsenic contamination in groundwater (Ghosh 2021). This necessitates vigilant monitoring of changes over various time frames. Keeping up with the ever-changing landscape of arsenic pollution proves to be a challenge, especially when it comes to staying updated on the latest methods for mitigating it, understanding its health effects, and gauging its prevalence. Due to gaps in timely data, providing an up-to-the-minute assessment of the arsenic contamination in Bihar's groundwater presents a considerable hurdle.

Bihar presents a range of hydrogeological conditions, encompassing differences in the types of aquifers and geological formations. The diverse variations observed could complicate the understanding of contamination origins and the movement of arsenic. Due to this heterogeneity, mitigation strategies may not have universal applicability, as interventions effective in one location may not yield the same results elsewhere (Yawar et al. 2016).

8.2 Limited Focus on Social and Economic Factors

The review may have overlooked important social and economic elements of arsenic because it focuses mostly on technical and scientific aspects. Components of arsenic pollution that are potentially overlooked by the review are the socio-economic differences in exposure and access to mitigation methods, as well as the technical and scientific components of the problem.

8.3 Limited Evaluation of Health Risks

The research seeks to evaluate the health risks linked to arsenic exposure, but a comprehensive understanding of the full range of health consequences was hindered by gaps in epidemiological data or differences in study methodologies.

8.4 *Effectiveness of Mitigation Strategies: It's Possible that the Review Overlooked Long-Term Effects or Newly Emerging Health Issues*

The efficacy of mitigation techniques is subject to variation contingent upon regional conditions, community involvement, and technology progress. It's possible that the study missed recent information on the accomplishments or shortcomings of ongoing mitigation projects. The depth of analysis about the effectiveness of particular mitigation strategies may be limited by difficulties obtaining comprehensive project assessments or monitoring data. Because pertinent studies in local languages or unpublished reports might not be included, relying solely on published literature could introduce language bias. Publication bias, which holds that successful outcomes are more likely to be published, may have an impact on how successful mitigation techniques are judged overall.

8.5 *Interdisciplinary Challenge*

Collaboration across scientific, social, and policy sectors is necessary due to the interdisciplinary character of the problem. The review's comprehensiveness could be impacted, nonetheless, by shortcomings in the synthesis of data from several fields. The study might not adequately convey the subtleties of multidisciplinary exchanges and teamwork in relation to Bihar's arsenic poisoning. Although the review offers a useful summary of the body of knowledge, these limitations emphasize the need for care in interpreting results and point to areas that want more investigation and study.

9 Need of This Study

In Bihar, India, groundwater is an essential resource for industrial, agricultural, and drinking purposes. However, there are now major environmental and public health problems due to the growing contamination from both natural and man-made causes. There have been numerous reports of arsenic, fluoride, nitrate, and heavy metals in groundwater, which can cause long-term health problems like cancer, bone problems, and neurological impairments. Groundwater quality is greatly impacted by Bihar's land use and land cover (LULC) patterns, which are marked by intense agriculture, fast urbanization, and industrial growth. Groundwater contamination is made worse by changes in land cover, such as deforestation and uncontrolled trash disposal. A thorough, organized assessment that links contamination sources, prevalence, health hazards, and mitigation measures in relation to LULC changes in Bihar is lacking despite the abundance of research on groundwater pollution.

By methodically examining the body of research and offering insights into the major variables influencing groundwater quality, this review seeks to close this knowledge gap.

10 Discussion

This review offers a thorough examination of Bihar's groundwater contamination, highlighting the relationship between it and changes in land cover and use. The study identifies the main determinants of groundwater quality, such as geogenic sources, agricultural runoff, and industrial outflow. The health hazards posed by arsenic, fluoride, and nitrate contamination—which impact millions of inhabitants—are given particular consideration. In order to protect public health, the conversation emphasizes the critical necessity for coordinated land and water management policies, long-term mitigation techniques, and community education initiative.

11 Conclusion

Groundwater pollution is a significant worldwide issue, with arsenic occurring naturally in old alluvial soil, water, and biota. People mainly rely on groundwater for drinking and household use, leading to increasing concerns about various health problems like cancer, skin disorders, dental issues, and hair loss that have been documented in scientific research over time. To ensure effective and sustainable management of this finite natural resource as outlined in SDG6, a comprehensive analysis is essential in the study region. Hence, the objective of the research is to develop a strategy for the optimal utilization of groundwater reservoirs to meet human needs, agricultural demands, and other industrial sectors. There hasn't been any ground water pollution research in the field of my study where I'm investigating, despite the fact that people are already eating extremely dangerous water owing to a lack of understanding and awareness. Since there are an increasing number of harmful diseases available every day. Hence, my research will serve as a more robust effort to ascertain the necessity for further investigation in my field of study. The groundwater within the Ganga, Brahmaputra, and Meghna basins is highly susceptible to arsenic contamination. A significant number of individuals in this region have experienced skin ailments, black foot disease, neurological complications, reproductive challenges, hair thinning, cancer-related issues, and multiple cases of youth mortality due to illnesses. These incidents have occurred in various areas where elevated levels of arsenic have been detected in drinking and irrigation water, as well as in food items. The persistent issue of arsenic pollution in Bihar's groundwater has detrimental impacts on human health. Social marginalization, community rejection, and job loss are critical social and economic challenges, respectively. Poisoning has been reported in the GBM's arsenic-contaminated sections. Millions of people's lives are currently at risk because community involvement and arsenic mitigation programs have failed. This review paper, aimed at these research goals, seeks to add to the body of knowledge on arsenic pollution in Bihar and serve as a useful tool for those searching for long-term fixes and ways to improve the region's water quality. Despite government efforts to construct water tanks as a political objective, there is a chronic shortage of potable water in Bihar, where contaminated water sources continue to cause arsenic toxicity. I suggest that the World Health Organization (WHO) and the United Nations Environment Programme (UNEP) undertake a thorough assessment of global arsenic contamination levels and launch an international awareness campaign to educate the public, healthcare providers, medical experts, and policymakers about this matter. Also, this will assist us in achieving our SDG.

References

Adeloju SB, Khan S, Patti AF (2021) Applied sciences arsenic contamination of groundwater and its implications for drinking water quality and human health in under-developed countries and remote communities—a review

Adhikari RK, Mohanasundaram S, Shrestha S (2020) Impacts of land-use changes on the groundwater recharge in the Ho Chi Minh City, Vietnam. Environ Res 185:109440. https://doi.org/10.1016/j.envres.2020.109440

Ahmad SA, Khan MH, Haque M (2018) Arsenic contamination in groundwater in bangladesh: implications and challenges for healthcare policy. Risk Manag Healthc Policy 11:251–261. https://doi.org/10.2147/RMHP.S153188

Alarcón-Herrera MT, Martin-Alarcon DA, Gutiérrez M, Reynoso-Cuevas L, Martín-Domínguez A, Olmos-Márquez MA, Bundschuh J (2020) Co-occurrence, possible origin, and health-risk assessment of arsenic and fluoride in drinking water sources in Mexico: geographical data visualization. Sci Total Environ 698:134168. https://doi.org/10.1016/J.SCITOTENV.2019.134168

Andualem TG, Demeke GG (2019) Groundwater potential assessment using gis and remote sensing: a case study of Guna Tana landscape, Upper Blue Nile Basin, Ethiopia. J Hydrol Reg Stud 24:100610. https://doi.org/10.1016/j.ejrh.2019.100610

Aullón Alcaine A, Schulz C, Bundschuh J, Jacks G, Thunvik R, Gustafsson JP, Mörth CM, Sracek O, Ahmad A, Bhattacharya P (2020) Hydrogeochemical controls on the mobility of arsenic, fluoride and other geogenic co-contaminants in the shallow aquifers of Northeastern La Pampa Province in Argentina. Sci Total Environ 715:136671. https://doi.org/10.1016/j.scitotenv.2020.136671

Bagyaraj M, Ramkumar T, Venkatramanan S, Gurugnanam B (2013) Application of remote sensing and GIS analysis for identifying groundwater potential zone in parts of Kodaikanal Taluk, South India. Front Earth Sci 7:65–75. https://doi.org/10.1007/s11707-012-0347-6

Berg M, Stengel C, Trang PTK, Hung Viet P, Sampson ML, Leng M, Samreth S, Fredericks D (2007) Magnitude of arsenic pollution in the Mekong and Red River Deltas—Cambodia and Vietnam. Sci Total Environ 372:413–425. https://doi.org/10.1016/j.scitotenv.2006.09.010

Bhowmick SK, Dasari G, Levens KL, Rettig KR (2007a) The prevalence of elevated serum thyroid-stimulating hormone in childhood/adolescent obesity and of autoimmune thyroid diseases in a subgroup. J Natl Med Assoc 99:773–776

Bhowmick S, Pramanik S, Singh P, Mondal P, Chatterjee D, Nriagu J (2018) Arsenic in groundwater of West Bengal, India: a review of human health risks and assessment of possible intervention options. Sci Total Environ 612:148–169. https://doi.org/10.1016/j.scitotenv.2017.08.216

Bhowmick SK, Dasari G, Levens KL, Rettig KR (2007) The prevalence of elevated serum thyroid-stimulating hormone in childhood / adolescent obesity and of autoimmune thyroid diseases Ina subgroup 99:773–776

Bhunia GS, Chatterjee N, Pal DK (2014) Identification of groundwater potential zone of Nawada District, Bihar (India)—a study based on remote sensing and GIS platform. Hydrol Res 45:631–644. https://doi.org/10.2166/nh.2013.258

Bihar Economic Survey (2022) GB. Bihar Economic Survey 2021–22, Patna, India. https://www.adriindia.org/images/report/1646034053BiharEconomicSurvey21-22English_opt.pdf

Borgoño JM, Vicent P, Venturino H, Infante A (1977) Arsenic in the drinking water of the city of Antofagasta: epidemiological and clinical study before and after the installation of a treatment plant. Environ Health Perspect 19:103–105. https://doi.org/10.1289/ehp.7719103

Brammer H, Ravenscroft P (2009) Arsenic in groundwater: a threat to sustainable agriculture in south and South-East Asia. Environ Int 35:647–654. https://doi.org/10.1016/j.envint.2008.10.004

Camacho LM, Gutiérrez M, Alarcón-Herrera MT, Villalba ML, Deng S (2011) Occurrence and treatment of arsenic in groundwater and soil in Northern Mexico and Southwestern USA. Chemosphere 83:211–225. https://doi.org/10.1016/j.chemosphere.2010.12.067

Cecilia S, Sargsyan Hrachya PF (2018) Mitigating arsenic in drinking water. Unicef Policy Br Mitig Arsen Drink Water 2018:1–6

CPCB (2021) Central pollution control board annual report 2019–20. Central Pollution Control Board. https://cpcb.nic.in/annual-report.php

CPCB (2019) Central pollution control board annual report 2019. www.cpcb.nic.in

Chakraborti D, Singh SK, Rahman MM, Dutta RN, Mukherjee SC, Pati S, Kar PB (2018a) Groundwater arsenic contamination in the Ganga river basin: a future health danger. Int J Environ Res Public Health 15:1–19. https://doi.org/10.3390/ijerph15020180

Chakraborti D, Singh SK, Rahman MM, Dutta RN, Mukherjee SC, Pati S, Kar PB (2018b) Groundwater arsenic contamination in the ganga river basin: a future health danger. Int J Environ Res Public Health 15:1–19

Chakraborti S, Banerjee A, Sannigrahi S, Pramanik S, Maiti A, Jha S (2019) Assessing the dynamic relationship among land use pattern and land surface temperature: a spatial regression approach. Asian Geogr 36:93–116. https://doi.org/10.1080/10225706.2019.1623054

Chandra R, Kumari S, Jain SK, Singh AK (2018) Assessment of ground water resources for irrigation in Aurngabad District of South Bihar. Int J Curr Microbiol Appl Sci 7:1606–1617. https://doi.org/10.20546/ijcmas.2018.704.181

Chaudhary BS, Kumar S (2018) Identification of groundwater potential zones using remote sensing and GIS of K-J watershed, India. J Geol Soc India 91:717–721. https://doi.org/10.1007/s12594-018-0929-3

Dandge KP, Patil SS (2022) Spatial distribution of ground water quality index using remote sensing and GIS techniques. Appl Water Sci 12:1–18. https://doi.org/10.1007/s13201-021-01546-7

Das D, Chatterjee A, Samanta G, Chowdhury TR (2001) A simple household device to remove arsenic from groundwater and two years performance report of arsenic removal plant for treating groundwater with community participation. *International Workshop on Technologies for Arsenic Removal from Drinking Water* organized by Bangladesh University of Engineering and Technology (BUET), Dhaka, Bangladesh and The United Nations University (UNU), Tokyo, Japan. 5th–7th May 2001, Dhaka, Bangladesh; Chapter 19; page: 231–250. https://archive.unu.edu/env/Arsenic/Das.pdf

Datta DV, Kaul MK (1976) Arsenic content of drinking water in villages in Northern India: a concept of arsenocosis. J Assoc Phys India 24:599-604

Doncaster CP, Bullock JM (2024) Living in harmony with nature is achievable only as a non-ideal vision. Environ Sci Policy 152:103658. https://doi.org/10.1016/j.envsci.2023.103658

Elmahdy S, Mohamed M, Ali T (2020) Land use/land cover changes impact on groundwater level and quality in the northern part of the United Arab Emirates. Remote Sens 12. https://doi.org/10.3390/rs12111715

Fendorf S, Michael HA, Van GA (2010) Support Online Mater for 1123:1–4. https://doi.org/10.1126/science.1172974

Foley JA, DeFries R, Asner GP, Barford C, Bonan G, Carpenter SR, Chapin FS, Coe MT, Daily GC, Gibbs HK et al (2005) Global consequences of land use. Science 309:570–574. https://doi.org/10.1126/science.1111772

Frankenberger WT, Losi ME (2015) Applications of bioremediation in the cleanup of heavy metals and metalloids. Bioremed Sci Appl 173–210 https://doi.org/10.2136/sssaspecpub43.c11

Gebel TW (2002) Arsenic methylation is a process of detoxification through accelerated excretion. Int J Hyg Environ Health 205:505–508. https://doi.org/10.1078/1438-4639-00177

Ghosh AK (2021) Strategic management, decision theory, and decision science: contributions to policy issues. In: Strategic management, decision theory, and decision science, pp 195-209. https://doi.org/10.1007/978-981-16-1368-5

Ghosh A, Sansthan MC, Pradesh U, Bengal W (2019) Environmental exposures and human health. Challenges. https://doi.org/10.4018/978-1-5225-7635-8.ch005

Ghosh A, Kumar A (2014) Human and ecological risk assessment: an international journal health risk assessment due to groundwater arsenic contamination: children are at high risk

Ghosh NC, Singh RD (2010) Groundwater arsenic contamination in India: vulnerability and scope for remedy. Environ Sci 1–23. api.semanticscholar.org/CorpusID:5773264

Ghosh AK, Singh SK (2012) Health Risk Assessment Due to Groundwater Arsenic Contamination: Children Are at High Risk. Hum. Ecol. Risk Assess. Int. J.; 18, 751–766. https://doi.org/10.1080/10807039.2012.688700

Gómez JJ, Lillo J, Sahún B (2006) Naturally occurring arsenic in groundwater and identification of the geochemical sources in the Duero Cenozoic Basin, Spain. Environ Geol 50:1151–1170. https://doi.org/10.1007/s00254-006-0288-z

Gounot AM (1994) Microbial ecology of groundwaters. In: Gibert J, Danielopol DL, Stanford JA (eds) Groundwater ecology. Elsevier, pp 189–215. https://doi.org/10.1016/b978-0-08-050762-0.50014-0

Goyal SK, Chaudhary BS, Singh O, Sethi GK, Thakur PK (2010) Variability analysis of groundwater levels—AGIS-based case study. J Indian Soc Remote Sens 38:355–364. https://doi.org/10.1007/s12524-010-0024-8

Gute DM (2004) Drinking water and cancer. Epidemiology 15:378. https://doi.org/10.1097/01.ede.0000122632.26957.60

Hamzaoui-Azaza F, Ameur M, Chaouch R, Cheikha LB, Gueddari M, Carrillo-Rivera JJ (2020) Assessment of groundwater quality based on GIS and geochemical methods: coastal aquifer of Bouficha (North-Eastern Tunisia). J Coast Conserv 2020:24. https://doi.org/10.1007/s11852-020-00762-8

Hashim KS, Hussein AH, Zubaidi SL, Kot P, Kraidi L, Alkhaddar R, Shaw A, Alwash R (2019) Effect of initial pH value on the removal of reactive black dye from water by electrocoagulation (EC) Method. J Phys Conf Ser 1294:2–8. https://doi.org/10.1088/1742-6596/1294/7/072017

Holmes RR, Hart ML, Kevern JT (2019) Removal of arsenic from synthetic groundwater using sulfur-enhanced cement-based filter media. J Hazard Toxic Radioact Waste 23:1–9. https://doi.org/10.1061/(asce)hz.2153-5515.0000440

Hopenhayn C (2006) Arsenic in drinking water: impact on human health. Elements 2:103–107. https://doi.org/10.2113/gselements.2.2.103

IARC (2004) Some drinking-water disinfectants and contaminants, including arsenic. IARC monographs on the evaluation of carcinogenic risks to humans, vol 84. World Health Organization, International Agency for Research on Cancer. https://publications.iarc.who.int/_publications/media/download/2661/be62bf756a4a6eaafe4ca5599d2bf8c73427d788.pdf

James S, Rodell M, Velicogna I, Famiglietti JS (2009) Satellite-based estimates of groundwater depletion in India. https://doi.org/10.1038/nature08238

Jhariya DC, Mondal KC, Kumar T, Indhulekha K, Khan R, Singh VK (2021) Assessment of groundwater potential zone using GIS-based multi-influencing factor (MIF), multi-criteria decision analysis (MCDA) and electrical resistivity survey techniques in Raipur City, Chhattisgarh, India. Aqua Water Infrastruct Ecosyst Soc 70:375–400. https://doi.org/10.2166/aqua.2021.129

Jiang J, Ashekuzzaman SM, Jiang A, Sharifuzzaman SM, Chowdhury SR (2013) Arsenic contaminated groundwater and its treatment options in Bangladesh. https://doi.org/10.3390/ijerph10010018

Ju Y, Kaown D, Lee KK (2018) A three-pronged approach for identifying source and extent of nitrate contamination in groundwater. J Soil Water Conserv 73:493–503. https://doi.org/10.2489/jswc.73.5.493

Karale OS, Gavit BK, Bhat AG, Paradkar V, Mukherjee S, Gupta A (2024) Analysing decadal land use land cover dynamics in the sub-upper Krishna Basin of Maharashtra, India using remote sensing and GIS. J Exp Agric Int 46:87–95. https://doi.org/10.9734/jeai/2024/v46i12297

Khalid F (2017) An overview of arsenic removal technologies in India. Invertis J Renew Energy 7:5. https://doi.org/10.5958/2454-7611.2017.00002.9

Khan MYA, Gani KM, Chakrapani GJ (2016) Assessment of surface water quality and its spatial variation. A case study of Ramganga River, Ganga Basin, India. Arab J Geosci 9:1–9. https://doi.org/10.1007/s12517-015-2134-7

Kpiebaya P, Amuah EEY, Shaibu AG, Baatuuwie BN, Avornyo VK, Dekongmen BW (2022) Spatial assessment of groundwater potential using Quantum GIS and multi-criteria decision analysis (QGIS-AHP) in the Sawla-Tuna-Kalba district of Ghana. J Hydrol: Reg Stud 43(April):101197. https://doi.org/10.1016/j.ejrh.2022.101197

Kumar M, Rahman MM, Ramanathan AL, Naidu R (2016a) Arsenic and other elements in drinking water and dietary components from the Middle Gangetic Plain of Bihar, India: health risk index. Sci Total Environ 539:125–134. https://doi.org/10.1016/j.scitotenv.2015.08.039

Kumar M, Chandola VK, Kumar S (2019) Groundwater arsenic contamination and its distribution in Bhojpur District, Bihar, India. Int J Curr Microbiol Appl Sci 8:1421–1431. https://doi.org/10.20546/ijcmas.2019.810.167

Kumar A, Rahman S, Iqubal A, Ali M, Niraj PK, Anand G, Kumar P, Ghosh AK (2016) Ground water arsenic contamination: a local survey in India. https://doi.org/10.4103/2008-7802.188085

Kumar A, Ali M, Kumar R, Kumar M, Sagar P, Pandey RK, Akhouri V, Kumar V, Anand G, Niraj PK (2021) Arsenic exposure in indo gangetic plains of bihar causing increased cancer risk. Sci Rep 11:1–16. https://doi.org/10.1038/s41598-021-81579-9

Kumar A, Kumar R, Rahman S, Ali M, Kumar R, Nupur N, Gaurav A, Raj V, Anand G, Niraj PK et al (2021b) Assessment of arsenic exposure in the population of Sabalpur Village of Saran District of Bihar with mitigation approach https://doi.org/10.1007/s11356-021-13521-5

Kumar Jha A, Chandra Gupta Y (2017) Se study of arsenic in the Koshi Region of Khagaria District. Chem Sci Rev Lett 6:2120–2126

Lalwani S, Dogra TD, Bhardwaj DN, Sharma RK, Murty OP, Vij A (2004) Study on arsenic level in ground water of delhi using hydride generator accessory coupled with atomic absorption spectrophotometer. Indian J Clin Biochem 19:135–140. https://doi.org/10.1007/BF02894273

Li Y, Bi Y, Mi W, Xie S, Ji L (2021) Land-use change caused by anthropogenic activities increase fluoride and arsenic pollution in groundwater and human health risk. J Hazard Mater 406:1–9. https://doi.org/10.1016/j.jhazmat.2020.124337

MacDonald AM, Bonsor HC, Ahmed KM, Burgess WG, Basharat M, Calow RC, Dixit A, Foster SSD, Gopal K, Lapworth DJ et al (2016) Groundwater quality and depletion in the indo-Gangetic basin mapped from in situ observations. Nat Geosci 9:762–766. https://doi.org/10.1038/ngeo2791

Maeda S (1994) Safety and environmental effects. In: The chemistry of organic arsenic, antimony and bismuth compounds, pp 725–759. https://doi.org/10.1002/0470023473.ch19

Mahanta R, Chowdhury J, Nath HK (2016) Health costs of arsenic contamination of drinking water in Assam, India. Econ Anal Policy 49:30–42. https://doi.org/10.1016/j.eap.2015.11.013

Malakootian M, Nozari M (2020) GIS-based DRASTIC and composite DRASTIC indices for assessing groundwater vulnerability in the Baghin Aquifer, Kerman, Iran. Nat Hazards Earth Syst Sci 20:2351–2363. https://doi.org/10.5194/nhess-20-2351-2020

Mazumder DNG (2003) Chronic arsenic toxicity: clinical features, epidemiology, and treatment: experience in West Bengal. J Environ Sci Heal Part A Toxic/Hazardous Subst Environ Eng 38:141–163. https://doi.org/10.1081/ESE-120016886

Mendoza-Chávez CE, Carabin A, Dirany A, Drogui P, Buelna G, Meza-Montenegro MM, Ulloa-Mercado RG, Diaz-Tenorio LM, Leyva-Soto LA, Gortáres-Moroyoqui P (2021) Statistical optimization of arsenic removal from synthetic water by electrocoagulation system and its application with real arsenic-polluted groundwater. Environ Technol 42:3463–3474. https://doi.org/10.1080/09593330.2020.1732472

Mohammed Abdul KS, Jayasinghe SS, Chandana EPS, Jayasumana C, De Silva PMCS (2015) Arsenic and human health effects: a review. Environ Toxicol Pharmacol 40:828–846. https://doi.org/10.1016/j.etap.2015.09.016

Mondal D, Rahman MM, Suman S, Sharma P, Siddique AB, Rahman MA, Bari ASMF, Kumar R, Bose N, Singh SK et al (2021) Arsenic exposure from food exceeds that from drinking water in endemic area of Bihar, India. Sci Total Environ 754:142082. https://doi.org/10.1016/j.scitotenv.2020.142082

Mondal KC, Rathod KG, Joshi HM, Mandal HS, Khan R, Rajendra K, Mawale YK, Priya K, Jhariya DC (2020) Impact of land-use and land-cover change on groundwater quality and quantity in the Raipur, Chhattisgarh, India: a remote sensing and GIS approach. IOP Conf Ser Earth Environ Sci 597. https://doi.org/10.1088/1755-1315/597/1/012011

Morales-Simfors N, Bundschuh J, Herath I, Inguaggiato C, Caselli AT, Tapia J, Choquehuayta FEA, Armienta MA, Ormachea M, Joseph E et al (2020) Arsenic in Latin America: a critical overview on the geochemistry of arsenic originating from geothermal features and volcanic emissions for solving its environmental consequences. Sci Total Environ 716:135564. https://doi.org/10.1016/j.scitotenv.2019.135564

Mukherjee A, Gupta S, Coomar P, Fryar AE, Guillot S, Verma S, Bhattacharya P, Bundschuh J, Charlet L (2019) Plate tectonics influence on geogenic arsenic cycling: from primary sources to global groundwater enrichment. Sci Total Environ 683:793–807. https://doi.org/10.1016/j.scitotenv.2019.04.255

Murcott S (2012) Arsenic contamination in the world: an international sourcebook, vol 11. London. ISBN 9781780400389

Naujokas MF, Anderson B, Ahsan H, Vasken Aposhian H, Graziano JH, Thompson C, Suk WA (2013) The broad scope of health effects from chronic arsenic exposure: update on a worldwide public health problem. Environ Health Perspect 121:295–302. https://doi.org/10.1289/ehp.1205875

Naz I, Ahmad I, Aslam RW, Quddoos A, Yaseen A (2024) Integrated assessment and geostatistical evaluation of groundwater quality through water quality indices. Water 16. https://doi.org/10.3390/w16010063

Okonuzzaman R, Chin LIW, Bon MANY, Fai TSY, Zhihong YE (2022) ScienceDirect arsenic accumulation in rice: sources , human health impact and probable mitigation approaches 29:309–327. https://doi.org/10.1016/j.rsci.2022.02.002

Omwene PI, Çelen M, Öncel MS, Kobya M (2018) PT. Process Saf Environ Prot. https://doi.org/10.1016/j.psep.2018.10.003

Omwene PI, Çelen M, Öncel MS, Kobya M (2019a) Arsenic removal from naturally arsenic contaminated ground water by packed-bed electrocoagulator using Al and Fe scrap anodes. Process Saf Environ Prot 121:20–31. https://doi.org/10.1016/j.psep.2018.10.003

Omwene PI, Meltem C, Öncel MS, Kobya M (2019) Arsenic removal from naturally arsenic contaminated ground water by packed-bed electrocoagulator using Al and Fe scrap anodes 121:20–31. https://doi.org/10.1016/j.psep.2018.10.003

Pal P, Sen M, Manna A, Pal J, Pal P, Roy P (2009) Contamination of groundwater by arsenic: a review of occurrence , causes , impacts , remedies and membrane-based purification 8168. https://doi.org/10.1080/19438150903185077

Pallier V, Feuillade-Cathalifaud G, Serpaud B, Bollinger JC (2010) Effect of organic matter on arsenic removal during coagulation/flocculation treatment. J Colloid Interface Sci 342:26–32. https://doi.org/10.1016/j.jcis.2009.09.068

Paul V, Vattikuti S, Dash P, Arslan Z (2021) Evaluating hydrogeochemical characteristics of groundwater and surface water in the upper pearl river watershed, USA. Environ Monit Assess 193:1–17. https://doi.org/10.1007/s10661-021-09045-7

Pierce ML, Moore CB (1982) Adsorption of arsenite and arsenate on amorphous iron hydroxide. Water Res 16:1247–1253. https://doi.org/10.1016/0043-1354(82)90143-9

Pokhrel D, Bhandari BS, Viraraghavan T (2009) Arsenic Contamination Of Groundwater in the Terai Region of Nepal: an overview of health concerns and treatment options. Environ Int 35:157–161. https://doi.org/10.1016/J.ENVINT.2008.06.003

Postma D, Pham TKT, Sø HU, Hoang VH, Vi ML, Nguyen TT, Larsen F, Pham HV, Jakobsen R (2016) A model for the evolution in water chemistry of an arsenic contaminated aquifer over the last 6000 years, Red River Floodplain, Vietnam. Geochim Cosmochim Acta 195:277–292. https://doi.org/10.1016/j.gca.2016.09.014

Prabhakar A, Tiwari H (2015) Land use and land cover effect on groundwater storage. Model Earth Syst Environ 1:1–10. https://doi.org/10.1007/s40808-015-0053-y

Rahaman S, Mise N, Ichihara S (2022) Arsenic contamination in food chain in Bangladesh: a review on health hazards, socioeconomic impacts and implications. Hyg Environ Heal Adv 2:100004. https://doi.org/10.1016/j.heha.2022.100004

Ram A, Tiwari SK, Pandey HK, Chaurasia AK, Singh S, Singh YV (2021) Groundwater quality assessment using water quality index (WQI) under GIS framework. Appl Water Sci 11:1–20. https://doi.org/10.1007/s13201-021-01376-7

Ranjan A (2019) Spatial analysis of arsenic contamination of groundwater around the world and India 4931:6–15

Ranjan A (2019) International journal of innovative studies in sociology and humanities (IJISSH) spatial analysis of arsenic contamination of groundwater around the World and India 4931:6–15

Ravindra K, Mor S (2019) Distribution and health risk assessment of arsenic and selected heavy metals in groundwater of Chandigarh, India. Environ Pollut 250:820–830. https://doi.org/10.1016/j.envpol.2019.03.080

Raza MB, Datta SP, Golui D, Barman M, Das TK, Sahoo RN, Upadhyay D, Rahman MM, Behera B, Naveenkumar A (2023) Synthesis and performance evaluation of novel bentonite-supported nanoscale zero valent iron for remediation of arsenic contaminated water and soil. Molecules 28:1–21. https://doi.org/10.3390/molecules28052168

Reyes Gómez VM, Gutiérrez M, Nájera Haro B, Núñez López D, Alarcón Herrera MT (2017) Groundwater quality impacted by land use/land cover change in a semiarid region of Mexico. Groundwater Sustain Dev 5:160-167. https://doi.org/10.1016/j.gsd.2017.06.003

Richards LA, Parashar N, Kumari R, Kumar A, Mondal D, Ghosh A, Polya DA (2022) Science of the total environment household and community systems for groundwater remediation in Bihar, India: arsenic and inorganic contaminant removal, controls and implications for remediation selection. Sci Total Environ 830:154580. https://doi.org/10.1016/j.scitotenv.2022.154580

Rizeei HM, Azeez OS, Pradhan B, Khamees HH (2018) Assessment of groundwater nitrate contamination hazard in a semi-arid region by using integrated parametric IPNOA and data-driven logistic regression models. Environ Monit Assess 190. https://doi.org/10.1007/s10661-018-7013-8

Rodriǵuez-Lado L, Sun G, Berg M, Zhang Q, Xue H, Zheng Q, Johnson CA (2013) Groundwater arsenic contamination throughout China. Science 341:866–868. https://doi.org/10.1126/science.1237484

Roshan A, Polya DA, Kumar A, Ghosh A, Glenny AM, Sedighi M et al (2024) Comparison of the distribution of groundwater remediation units and contaminant (arsenic, iron, fluoride) distribution in Bihar, India for improved water security and management. J Environ Manag 372(123157):1-14. https://doi.org/10.1016/j.jenvman.2024.123157

Rowland HAL, Omoregie EO, Millot R, Jimenez C, Mertens J, Baciu C, Hug SJ, Berg M (2011) Geochemistry and arsenic behaviour in groundwater resources of the Pannonian Basin (Hungary and Romania). Appl Geochem 26:1–17. https://doi.org/10.1016/j.apgeochem.2010.10.006
Saha KC (2003) Review of Arsenicosis in West Bengal, India—a clinical perspective. Crit Rev Environ Sci Technol 33:127–163. https://doi.org/10.1080/10643380390814514
Saha D, Ray RK (2017) Groundwater resources of India: potential, challenges and management, pp 19–42
Saha D, Ray RK (2018) Groundwater resources of India: potential, challenges and management. Groundw Dev Manag. Issues Chall South Asia 19–42.https://doi.org/10.1007/978-3-319-75115-3_2
Saha D, Sahu S (2015) A decade of investigations on groundwater arsenic contamination in Middle Ganga Plain, India. Environ Geochem Health 2015:1–25. https://doi.org/10.1007/s10653-015-9730-z
Santra SC, Samal AC, Bhattacharya P, Banerjee S, Biswas A, Majumdar J (2013) Arsenic in foodchain and community health risk: a study in gangetic West Bengal. Proc Environ Sci 18:2–13. https://doi.org/10.1016/j.proenv.2013.04.002
Shaji E, Santosh M, Sarath KV, Prakash P, Deepchand V, Divya BV (2021) Geoscience frontiers arsenic contamination of groundwater : a global synopsis with focus on the Indian Peninsula. Geosci Front 12:101079. https://doi.org/10.1016/j.gsf.2020.08.015
Shankar S, Shanker U, Shikha (2014a) Arsenic contamination of groundwater: a review of sources, prevalence, health risks, and strategies for mitigation. Sci World J 2014. https://doi.org/10.1155/2014/304524
Shankar S, Shanker U, Shikha (2014b) Arsenic contamination of groundwater: a review of sources, prevalence health risks, and strategies for mitigation. Sci World J 2014. https://doi.org/10.1155/2014/304524
Shankar P, Rakshit A (2015) Status of ground water quality in selected shallow aquifers of Saharsa District, the Kosi Megafan, the North Bihar Plains. Int J Bioresour Sci 2:173. https://doi.org/10.5958/2454-9541.2015.00013.4
Siddik MS, Tulip SS, Rahman A, Islam MN, Haghighi AT, Mustafa SMT (2022) The impact of land use and land cover change on groundwater recharge in Northwestern Bangladesh. J Environ Manage 315:115130. https://doi.org/10.1016/j.jenvman.2022.115130
Siddiqui SI, Chaudhry SA (2017) Iron Oxide And Its Modified Forms As An Adsorbent For Arsenic Removal: A Comprehensive Recent Advancement. Process Saf Environ Prot 111:592–626. https://doi.org/10.1016/j.psep.2017.08.009
Singh SK (2015) Groundwater arsenic contamination in the middle-Gangetic Plain, Bihar (India): the danger arrived. Int Res J Environ Sci 4:70–76. ISSN 2319-1414
Singh A, Choudhary SK (2010) Arsenic in ground water in five villages under Nathnagar block of Bhagalpur District. BIHAR 4:213–216
Sinha R, Gupta S, Nepal S (2018) Groundwater dynamics in North Bihar plains. Curr Sci 114:2482–2493. https://doi.org/10.18520/cs/v114/i12/2482-2493
Smedley PL, Kinniburgh DG (2002) A review of the source, behaviour and distribution of arsenic in natural waters. Appl Geochem 17:517–568. https://doi.org/10.1016/S0883-2927(02)00018-5
Song J, Jia SY, Ren HT, Wu SH, Han X (2015) Application of a high-surface-area schwertmannite in the removal of arsenate and arsenite. Int J Environ Sci Technol 12:1559–1568. https://doi.org/10.1007/s13762-014-0528-9
Sorg TJ, Chen ASC, Wang L (2014) Arsenic species in drinking water wells in the USA with high arsenic concentrations. Water Res 48:156–169. https://doi.org/10.1016/j.watres.2013.09.016
Tandon SK, Sinha R (2018) The Ganga River: a summary view of a large river system of the Indian sub-continent 61–73. https://doi.org/10.1007/978-981-10-2984-4_5
Tapia J, Mukherjee A, Rodríguez MP, Murray J, Bhattacharya P (2022) Role of tectonics and climate on elevated arsenic in fluvial systems: insights from surface water and sediments along regional transects of Chile. Environ Pollut 314:120151. https://doi.org/10.1016/J.ENVPOL.2022.120151

Thakur BK, Education M, Gupta V (2015) Groundwater arsenic contamination in Bihar: causes , issues and challenges. Manthan J Commerce Manag 2(1):45–60. https://doi.org/10.17492/manthan.v2i1.6434

Thakur BK, Gupta V (2019) Valuing health damages due to groundwater arsenic contamination in Bihar, India. Econ Hum Biol 35:123–132. https://doi.org/10.1016/j.ehb.2019.06.005

Thomas DJ, Bradham K (2016a) Role of complex organic arsenicals in food in aggregate exposure to arsenic. J Environ Sci 49:86–96. https://doi.org/10.1016/J.JES.2016.06.005

Thomas DJ, Bradham K (2016b) Role of complex organic arsenicals in food in aggregate exposure to arsenic. J Environ Sci 49:86–96. https://doi.org/10.1016/j.jes.2016.06.005

Tola B, Deyassa G (2024) A modeling approach for evaluating and predicting the impacts of land use land cover changes on groundwater recharge in Walga Watershed, Upper Omo Basin, Central Ethiopia. J Hydrol: Reg Stud 51(101659):1-34. https://doi.org/10.1016/j.ejrh.2024.101659

Vieira BRC, Pintor AMA, Boaventura RAR, Botelho CMS, Santos SCR (2017) Arsenic removal from water using iron-coated seaweeds. J Environ Manage 192:224–233. https://doi.org/10.1016/j.jenvman.2017.01.054

Warhurst C, Nickson D (2007) Employee experience of aesthetic labour in retail and hospitality. Work Employ Soc 21:103–120. https://doi.org/10.1177/0950017007073622

Wegelin M, Gechter D, Hug S, Mahmud A (2000) Water, sanitation and hygiene: challenges of the millennium soras—a simple arsenic removal process 2:255–258

Yang Q, Flanagan SV, Chillrud S, Ross J, Zeng W, Culbertson C, Spayd S, Backer L, Smith AE, Zheng Y (2020) Reduction in drinking water arsenic exposure and health risk through arsenic treatment among private well households in Maine and New Jersey, USA. Sci Total Environ 738:139683. https://doi.org/10.1016/j.scitotenv.2020.139683

Yawar M, Khan A, Gani KM, Chakrapani GJ (2016) Assessment of surface water quality and its spatial variation. A case study of Ramganga River, Ganga Basin , India. https://doi.org/10.1007/s12517-015-2134-7

Examining the Role of Geospatial Technologies and AI in Disaster Risk Management and Response, Including Hazard Mapping, Vulnerability Assessment, and Real-Time Monitoring

Sumedha Surbhi Singh and C. Jeganathan

1 Introduction

Geospatial technology encompasses a range of tools and techniques used to collect, analyze, and interpret data related to the Earth's surface. This field integrates geographic information systems (GIS), remote sensing, global positioning systems (GPS), and advanced data analytics to map and understand spatial phenomena. It plays a critical role in addressing complex challenges such as urban planning, environmental conservation, disaster management, and climate change adaptation (Cao et al. 2023; Maurya and Kumar 2024; Latorre et al. 2024; Queen et al. 2024). By leveraging satellite imagery, sensor networks, and geospatial databases, this technology provides insights into spatial patterns, trends, and relationships (Nandakumar and Pradeep 2025; Huang 2025; Jiang and Shekhar 2017). Its applications extend across diverse domains, enabling informed decision-making and sustainable development practices.

Artificial Intelligence (AI) has revolutionized geospatial technology by enabling faster and more accurate analysis of spatial data (Vysotskyi and Vysotskyi 2024). AI-powered tools, such as machine learning and deep learning algorithms, can process vast amounts of geospatial data from sources like satellites, drones, and ground-based sensors (Dritsas and Trigka 2025). These algorithms can detect patterns, classify land cover, predict changes, and identify anomalies with minimal human intervention (Stojković et al. 2024). For instance, AI can automate the analysis of satellite imagery

S. S. Singh (✉) · C. Jeganathan
Department of Remote Sensing & Geoinformatics, Birla Institute of Technology, Mesra, Ranchi, India
e-mail: phdrs10009.20@bitmesra.ac.in

C. Jeganathan
Sarala Birla University, Ranchi, Jharkhand, India

M. O. Sarif and A. Sharifi (eds.), *Advanced Geospatial Intelligence and AI for Environmental Resilience and Sustainable Development*, Advances in Geographic Information Science, https://doi.org/10.1007/978-3-032-03714-5_14

to monitor deforestation, map urban sprawl, or track disaster impacts in near real-time (Dritsas and Trigka 2025). It also enhances predictive modelling, allowing geospatial experts to forecast events like floods, wildfires, and droughts by integrating historical data with current environmental conditions.

In addition, AI facilitates the integration and interpretation of diverse datasets, such as climate variables, topography, and socio-economic factors, to generate comprehensive insights. Natural Language Processing (NLP) in AI enables the extraction of relevant information from unstructured geospatial data sources like reports and social media, enriching spatial analyses (Sufi and Kalil 2024). The combination of AI and geospatial technology not only accelerates data-driven decision-making but also enhances the precision and efficiency of solutions for critical global challenges, including climate resilience and sustainable development (Hosen et al. 2023; Lazorenko 2024).

Geospatial technologies and artificial intelligence (AI) have become indispensable tools in disaster risk management and response, providing innovative approaches to understanding, mitigating, and responding to natural and man-made hazards (Pang and Pang 2022; Afolabi et al. 2023; Ivić 2019). These technologies leverage spatial data, satellite imagery, and advanced computational techniques to assess the dynamics of hazards, enabling accurate mapping of risk-prone areas. Hazard mapping, a critical component of disaster preparedness, is now enriched by geospatial technologies like remote sensing and GIS, which provide high-resolution spatial and temporal data. Coupled with AI algorithms, these tools can identify patterns, predict hazard occurrences, and simulate potential impacts, significantly improving the ability to plan and allocate resources effectively (Abid et al. 2021; Beeravelly 2024).

In addition to hazard mapping, geospatial technologies and AI play a pivotal role in vulnerability assessment (Afolabi et al. 2023). By integrating data from various sources—such as population density, socio-economic factors, and infrastructure distribution—these technologies provide insights into the communities and systems most at risk. AI enhances this process by analysing vast datasets, identifying hidden relationships, and generating predictive models. This facilitates the design of targeted interventions and risk reduction strategies, ensuring that the most vulnerable populations are prioritized. Moreover, geospatial analytics enable real-time decision-making by visualizing the spatial extent of vulnerabilities, which is crucial in resource-constrained settings.

Real-time monitoring and response represent another transformative application of geospatial technologies and AI in disaster risk management. Technologies like satellite-based remote sensing, UAVs, and IoT devices continuously collect data on environmental changes, while AI processes this information to detect anomalies, forecast events, and optimize response strategies (Perdicca and Paolanti 2022). For instance, during a flood, geospatial data can be used to monitor water levels, while AI can predict the areas likely to be affected, facilitating timely evacuations and resource deployment (Deepak et al. 2020). Together, these advancements are revolutionizing disaster management, shifting the focus from reactive to proactive approaches that save lives, protect assets, and enhance resilience.

AI plays a transformative role in disaster risk management by enhancing preparedness, response, and recovery efforts. AI-powered predictive models analyse historical disaster data, weather patterns, and real-time sensor readings to forecast potential disasters such as floods, hurricanes, earthquakes, or wildfires. Machine learning algorithms identify patterns and trigger early warnings, enabling authorities to take preventive measures, evacuate people, or allocate resources efficiently (Adikari et al. 2021; Biswas et al. 2024; Pan et al. 2024). AI tools process geospatial and environmental data to identify high-risk zones and create detailed hazard maps. For instance, algorithms can assess flood risk by integrating rainfall data, topography, and river flow information. These insights support urban planning, infrastructure development, and policymaking to mitigate disaster impacts (Alahi et al. 2023). During disasters, AI facilitates real-time monitoring through satellite imagery, drones, and IoT devices. For example, AI algorithms analyze drone footage to detect the extent of damage or monitor wildfire progression. These insights allow emergency responders to prioritise areas requiring immediate attention. Post-disaster, AI-driven image recognition and remote sensing analyze satellite or drone imagery to assess the severity of damage to infrastructure, vegetation, or populated areas. This rapid assessment accelerates insurance claims processing, resource allocation, and recovery planning. AI-enabled Natural Language Processing (NLP) analyzes social media posts during disasters to identify affected regions, understand immediate needs, and track the movement of displaced populations (Islam et al. 2024). This crowd-sourced data complements traditional emergency response mechanisms.

AI optimizes resource allocation, such as determining the best locations for relief camps, efficiently distributing supplies, or deploying emergency personnel. Algorithms account for constraints like transportation logistics, population density, and resource availability to enhance efficiency. AI supports long-term disaster risk reduction by analysing trends in disaster frequency, intensity, and recovery efforts. Predictive models help design resilient infrastructure, recommend sustainable land-use practices, and guide climate adaptation strategies.

By integrating AI into disaster risk management, authorities and organizations can better anticipate, respond to, and recover from disasters, ultimately saving lives and reducing economic losses. This chapter aims to showcase the integration of Artificial Intelligence (AI) and geospatial technology in enhancing disaster risk management. It highlights how the synergy between advanced spatial data processing and intelligent algorithms can significantly improve the accuracy, speed, and efficiency of disaster preparedness, response, and recovery. By exploring real-world applications and case studies, the chapter demonstrates the transformative potential of AI in analyzing vast geospatial datasets, forecasting hazards, assessing risks, and optimizing decision-making processes. The chapter provides valuable insights into the practical use of these cutting-edge technologies for building resilience and reducing the impacts of natural and human-induced disasters.

2 Recent Developments

Artificial intelligence (AI) is transforming disaster risk management by enabling more precise predictions, efficient resource allocation, and real-time responses. Recent advancements in machine learning (ML) models, particularly deep learning, have significantly improved the accuracy of hazard prediction and early warning systems. For example, convolutional neural networks (CNNs) are increasingly used to analyse satellite imagery to detect patterns associated with natural hazards such as floods, landslides, and wildfires. These models can identify at-risk areas with unprecedented precision, facilitating timely alerts and evacuation plans. AI-driven predictive analytics is also being applied to forecast the likelihood of disasters based on historical data, climatic trends, and geophysical parameters, providing critical lead time for preparedness (Biswas et al. 2024; Albahri et al. 2024).

Another major development is the integration of AI with real-time data streams from Internet of Things (IoT) devices and remote sensing technologies. AI algorithms process massive volumes of data from sensors monitoring environmental factors, such as seismic activity, water levels, and wind speed, to detect early warning signs. For instance, real-time flood monitoring systems now use AI to analyze rainfall patterns, river discharge, and terrain models, predicting flood extents with remarkable accuracy. This real-time capability is particularly impactful in urban areas, where even small delays in response can lead to significant loss of life and property. Moreover, AI-powered drones and robotic systems are increasingly being used for rapid damage assessment and search-and-rescue operations, reducing response times and enhancing operational efficiency.

AI is also driving advancements in vulnerability assessment and risk mitigation strategies. By combining data from satellite imagery, socioeconomic factors, and historical disaster records, AI models can identify the most vulnerable communities and infrastructure. Generative AI techniques, such as those used in simulation modelling, are being deployed to assess the potential impacts of disasters under various scenarios, guiding policymakers in developing effective contingency plans. These developments are further supported by the proliferation of open-source AI frameworks and cloud computing platforms, which make sophisticated tools accessible to a broader range of stakeholders, including governments, NGOs, and local communities. Together, these innovations are reshaping disaster risk management, making it more adaptive, data-driven, and inclusive.

3 Use of AI in Different Disaster Risk Management

3.1 Flood

Floods are natural disasters caused by the overflow of water onto land that is typically dry, often resulting from heavy rainfall, river overflow, storm surges, or dam failures. They are classified into various types, such as riverine floods, flash floods, and coastal floods, depending on their origin and behaviour. Floods can devastate communities by damaging infrastructure, homes, and agricultural land, leading to economic losses, displacement, and health hazards due to waterborne diseases.

Artificial Intelligence (AI) plays a crucial role in enhancing flood prediction and early warning systems. Machine learning models integrated with satellite and sensor data now generate highly accurate flood maps even under challenging conditions such as cloud cover, exemplified by platforms like FloodSENS that utilize U-Net models trained on optical satellite imagery (https://incubed.esa.int/portfolio/floodsens/). These systems deliver high-accuracy flood detection across diverse biomes and geographical contexts. By analyzing historical weather patterns, hydrological data, and real-time inputs from remote sensors, machine learning (ML) models can predict the onset, intensity, and duration of floods with high accuracy. Advanced models like recurrent neural networks (RNNs) and long short-term memory (LSTM) networks excel at processing time-series data, enabling precise forecasting of rainfall and river discharge. These predictions help authorities issue timely warnings, giving communities sufficient time to evacuate and take preventive measures. AI-based systems also integrate data from diverse sources, such as satellite imagery, radar, and IoT devices, providing a holistic understanding of flood risks. Rezvani et al. (2024) mapped the flood risk on road networks within Portuguese municipalities using geospatial AI.

Real-time flood monitoring has been revolutionized by the use of AI in conjunction with geospatial technologies. Deep learning models, particularly convolutional neural networks (CNNs), are used to analyze satellite and aerial imagery to identify flood-affected areas. These models can process vast amounts of imagery quickly, providing near-instantaneous updates on flood extent and progression. AI-powered systems also generate high-resolution flood maps that guide emergency response teams in navigating affected areas and allocating resources effectively. In urban settings, these maps are crucial for identifying vulnerable zones, such as low-lying areas or regions with poor drainage, enabling targeted interventions to mitigate flood impacts.

3.1.1 Case Study: Mapping the Flood of May 2022 Using Machine Learning and Google Earth Engine (GEE)

The Lakhimpur district of Assam, India, experienced significant flooding in May 2022, impacting large swathes of land, including forests, agricultural areas, and human settlements. The district is situated on the north bank of river Brahmaputra. Along with Brahmaputra, Subansiri, Ranganadi, and Dikrong rivers flows through it. The population of this district was 10.42 lakh in 2011, and it covers an area of 2277 km^2 (https://lakhimpur.nic.in/Lakhimpur/about.htm). Flood mapping is a critical step in understanding the spatial extent of the flood and planning mitigation measures. Due to cloudy weather, optical datasets are covered with clouds. Figure 1 shows the location of the Lakhimpur district along with the Sentinel-2 image of 23 May 2022. So, this study utilized Sentinel-1 radar imagery to map the flood, leveraging the radar's ability to penetrate cloud cover and provide data during extreme weather conditions (Fig. 2). Figure 3 shows the Sentinel-1 images of 06 May 2022 and 25 May 2022 of Lakhimpur district, Assam.

Sentinel-1, a radar imaging satellite mission by the European Space Agency (ESA), is designed for all-weather, day-and-night Earth observation. The satellite operates in the C-band, providing data in two polarizations: VV (vertical transmit and receive) and VH (vertical transmit, horizontal receive). Its Synthetic Aperture Radar (SAR) enables high-resolution imaging (up to 10 m) regardless of cloud cover,

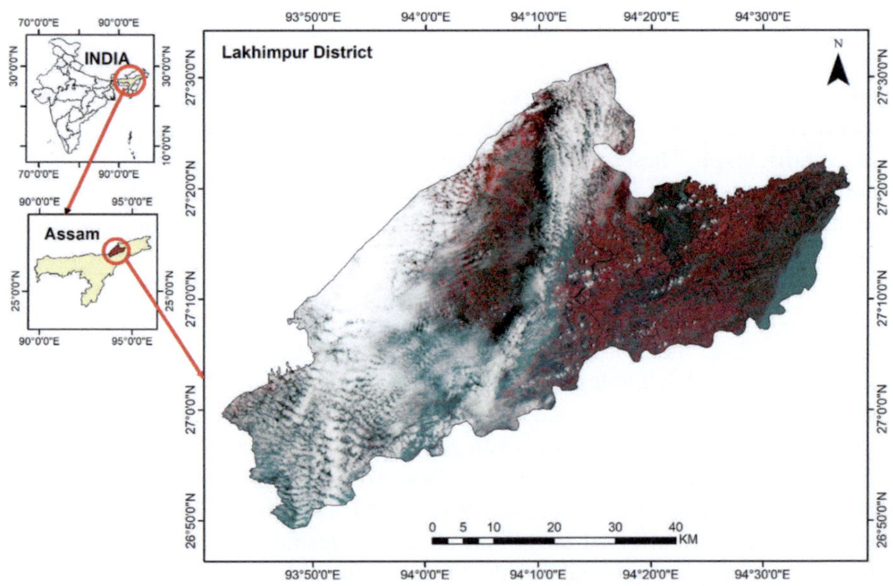

Fig. 1 Location map of Lakhimpur district with Sentinel-2 image of 23-05-2022

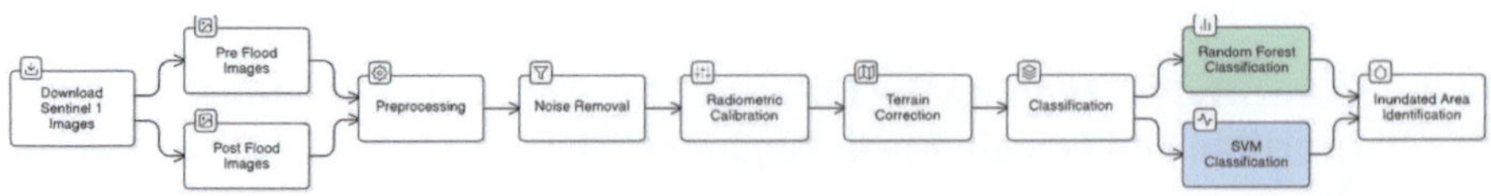

Fig. 2 Methodology flow chart for flood inundation mapping

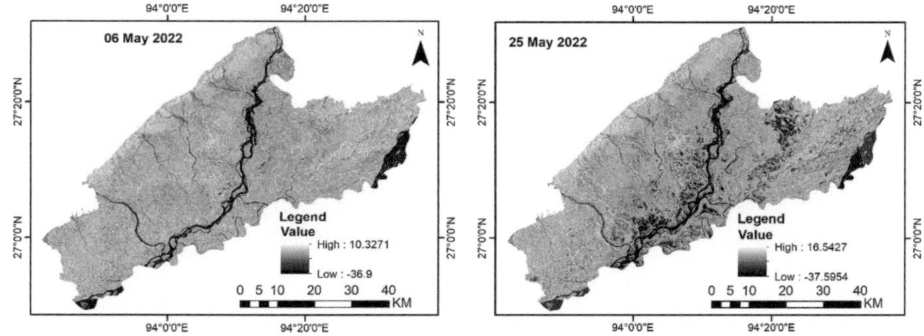

Fig. 3 Sentinel images of the Lakhimpur district

making it particularly effective for flood monitoring. For this study, Sentinel-1 images from 6th May 2022 (pre-flood) and 25th May 2022 (post-flood) were utilized to map the flood extent in Lakhimpur district, Assam. The pre-flood image provided baseline surface conditions, while the post-flood image captured the inundated areas, allowing for the generation of a difference image to highlight changes caused by the flood event. Both images were acquired in the Interferometric Wide (IW) swath mode, ensuring comprehensive coverage and consistent observation geometry. Figure 2 shows the methodology flow chart for this study. Each scene was pre-processed with Sentinel-1 Toolbox, which involves thermal noise removal, radiometric calibration and terrain correction using SRTM 30 or ASTER DEM. After filtering the data collection in GEE for required images, difference image was generated by subtracting the pre-flood backscatter values from the post-flood values. This highlights the areas newly inundated due to the flood. VH/VV ratio was generated for surface roughness analysis. A training set was created to train the data for machine learning classification algorithm with three classes: Flooded area, non-flood area and Water Body.

Two machine learning algorithms, Random Forest (RF) and Support Vector Machine (SVM), were employed to classify the study area into these three classes. The difference image and other extracted features were used as inputs for the classification models. Figure 4 shows the classified images from RF and SVM. RF shows that 402.4 sq. km is flooded, and SVM shows that 410.53 sq. km area is flooded. The classifications identified significant overlap between flooded areas and low-lying terrain, as expected. The slight variation in the results of RF and SVM could be attributed to differences in algorithm sensitivities. SVM showed 86.78% accuracy, and RF showed 84.30% accuracy. The flooded areas were concentrated along the riverbanks and in low-elevation regions, as confirmed by the DEM overlay. This study underscores the importance of machine learning in disaster response and resource management, particularly in flood-prone regions.

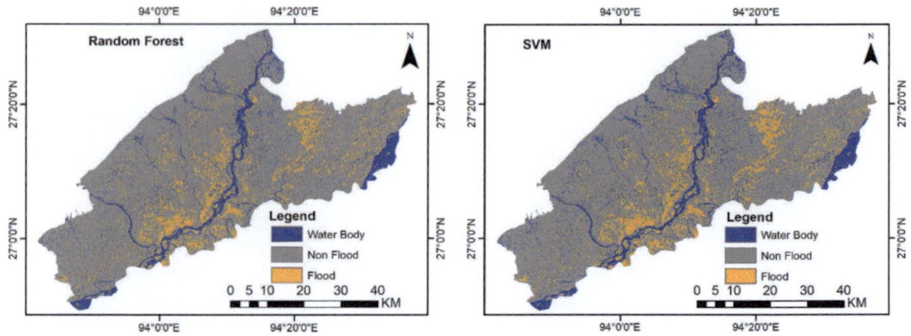

Fig. 4 Classified images using RF and SVM

3.1.2 Post-flood Damage Assessment

AI significantly enhances post-flood damage assessment, enabling faster and more accurate evaluations of affected areas. Machine learning algorithms analyze satellite images, drone footage, and ground reports to estimate the extent of property damage, infrastructure disruption, and agricultural losses. Natural language processing (NLP) tools extract information from social media posts, news reports, and relief requests to identify areas in need of urgent assistance (Wang et al. 2020). These insights allow governments and humanitarian organizations to prioritize their responses, ensuring aid reaches the most severely impacted communities first. Moreover, AI-powered systems can simulate various recovery scenarios, aiding in the planning of long-term reconstruction efforts. Various researchers have mapped inundated areas using remote sensing and GIS techniques (Sivanpillai et al. 2023; Sajjad et al. 2023).

3.1.3 Adaptive Flood Risk Management

Beyond immediate response, AI facilitates adaptive flood risk management by supporting predictive modelling and planning. AI algorithms analyze climate data, land-use changes, and urban growth patterns to assess long-term flood risks under different scenarios. Generative adversarial networks (GANs) and other simulation-based AI models are increasingly used to create virtual scenarios of flooding events, helping policymakers understand potential impacts and devise effective mitigation strategies. These models also support the development of resilient infrastructure, such as optimized drainage systems and flood barriers, tailored to specific risk profiles. By integrating AI into flood risk management frameworks, stakeholders can transition from reactive to proactive approaches, ultimately reducing flood-related losses and enhancing community resilience.

3.2 Cyclone

A cyclone is a large-scale atmospheric disturbance characterized by a low-pressure centre and rotating winds, often accompanied by heavy rainfall and thunderstorms. Cyclones are classified based on their intensity and region, with tropical cyclones forming over warm ocean waters and extratropical cyclones occurring in higher latitudes. In tropical regions, cyclones are known as hurricanes, typhoons, or simply tropical cyclones, depending on their location. These storms can cause widespread destruction through strong winds, torrential rains, and storm surges that lead to coastal flooding. Cyclones pose significant risks to life, property, and infrastructure, particularly in coastal and low-lying areas.

Artificial Intelligence (AI) and Machine learning (ML) models, including deep neural networks (DNNs) and ensemble learning algorithms, process historical weather data, atmospheric conditions, and ocean temperature patterns to forecast cyclone formation, trajectory, and intensity with high accuracy. These models are capable of identifying rapid intensification or weakening phases, which are crucial for real-time decision-making. These AI-driven systems can identify subtle precursors of cyclones, such as changes in sea surface temperature and wind patterns, providing advanced warnings. This early detection allows governments and disaster management authorities to issue timely alerts, initiate evacuation plans, and prepare resources, thereby minimizing loss of life and property. AI systems also integrate data from remote sensing technologies and IoT-based weather sensors, enabling dynamic updates on cyclone characteristics. By combining this information, AI generates detailed spatial models of cyclone paths, helping authorities anticipate affected areas and prepare targeted responses. Table 1 lists some departments which monitor and provide information on cyclones in real-time.

In the aftermath of a cyclone, AI accelerates impact assessment and damage estimation, facilitating efficient disaster response and recovery efforts. Machine learning algorithms analyze satellite and drone imagery to map areas affected by strong winds, heavy rainfall, and storm surges. These systems identify damaged infrastructure, flooded regions, and displaced populations, providing actionable insights for relief operations. Additionally, natural language processing (NLP) tools gather information from social media, news reports, and local communications to validate and supplement geospatial data. AI-powered simulations also estimate economic losses and recovery needs, enabling authorities to prioritize resource allocation effectively. Various studies have used geospatial technology to assess cyclone risk and consequences of cyclone (Hoque et al. 2018, 2019a, b; Mansour 2019; Taznzeen et al. 2021).

AI is integral to building long-term resilience against cyclones through adaptive risk management strategies. Predictive models analyze climate trends, urbanization patterns, and coastal vulnerabilities to assess future cyclone risks under various climate change scenarios. Generative adversarial networks (GANs) and other simulation tools create virtual models of cyclone impacts, helping planners design resilient infrastructure, such as storm-resistant buildings and optimized coastal defences.

Table 1 Departments which monitor and provide information on cyclones in real-time

Platform	Type of data provided	Link
IMD Cyclone Warning Division (India)	• Tropical cyclone warnings for Bay of Bengal and Arabian Sea • Cyclone track and intensity forecasts • Cyclone genesis outlooks • Landfall forecasts and storm surge warnings • Bulletins for fishermen, ports, and disaster managers • High-resolution satellite imagery and radar observations • Seasonal cyclone outlooks and advisory updates	https://mausam.imd.gov.in/imd_latest/contents/cyclone.php https://rsmcnewdelhi.imd.gov.in/
NOAA National Hurricane Center (USA)	• Tropical cyclone forecasts for the Atlantic and Eastern Pacific Oceans • Hurricane track and intensity predictions • Tropical storm advisories and public alerts • Storm surge warnings and outlooks • Wind speed probability maps • Real-time satellite imagery and hurricane hunter aircraft data • Seasonal hurricane activity outlooks	https://www.nhc.noaa.gov

These tools also support land-use planning by identifying areas that should remain undeveloped to mitigate risks from storm surges and flooding. By incorporating AI into cyclone risk management frameworks, stakeholders can transition from reactive measures to proactive strategies, ensuring sustainable development and improved disaster preparedness.

3.3 Drought

Drought is a prolonged period of abnormally low precipitation, leading to water scarcity that affects ecosystems, agriculture, and communities. It is a complex natural hazard categorized into meteorological, agricultural, hydrological, and socio-economic droughts, depending on its impacts and the sectors it affects. Droughts can lead to reduced agricultural yields, food and water insecurity, loss of biodiversity, and economic challenges, particularly in regions dependent on rain-fed farming. Factors such as climate change, deforestation, and unsustainable water management practices have exacerbated the frequency and intensity of droughts globally, underscoring the need for improved monitoring, early warning systems, and sustainable water resource management strategies.

Machine learning (ML) models analyse climate data, soil moisture levels, precipitation trends, and vegetation health. Researchers have used remote sensing data to model drought vulnerability and risk assessment (Alizadeh and Nikoo 2018; Hoque et al. 2019a, b, 2020; Vicente-Serrano et al. 2012; Barzkar et al. 2021; Lai et al. 2019; Arabameri et al. 2021). These models, including support vector machines (SVMs) and recurrent neural networks (RNNs), can detect early warning signs of drought, such as declining rainfall or prolonged dry periods. AI-based systems integrate data from satellite observations, weather stations, and IoT devices to provide region-specific forecasts of drought onset, duration, and intensity. By delivering accurate and timely predictions, AI enables governments and farmers to implement mitigation strategies such as water conservation measures, crop selection adjustments, and early planning for water resource distribution.

During drought and in the post-drought period, Convolutional neural networks (CNNs) and other deep learning models process satellite imagery and remote sensing data to assess soil moisture, vegetation health, surface water availability and other parameters to see the impact and severity of drought. AI can also track drought indicators such as the Standardized Precipitation Index (SPI) or the Normalized Difference Vegetation Index (NDVI) over time, identifying trends that signal worsening conditions. These insights are visualized on interactive dashboards, enabling policymakers, farmers, and other stakeholders to monitor drought conditions at local and regional scales. This dynamic monitoring system ensures that responses are data-driven and aligned with evolving conditions. Table 2 shows some of the agencies which monitor and assess drought conditions and provide data on their websites.

Beyond immediate responses, AI supports long-term drought risk management by guiding the development of adaptive strategies. Generative AI tools simulate future scenarios, enabling stakeholders to assess the effectiveness of interventions such as afforestation, soil conservation, and crop diversification. Optimization algorithms and reinforcement learning models analyze water demand, supply, and usage patterns to recommend efficient water allocation strategies. It can help assist in designing and implementing rainwater harvesting, groundwater recharge, and irrigation scheduling plans. It also facilitates the development of resilient farming systems by recommending drought-tolerant crop varieties and optimizing planting schedules based on climate conditions. By incorporating AI into drought risk management frameworks, societies can reduce vulnerability to droughts, safeguard food security, and ensure sustainable development in water-scarce regions.

Table 2 Agencies which monitor and assess drought conditions and provide data on their websites

Platform	Type of data provided	Link
US Drought Monitor (USA)	• Weekly drought maps for the United States • Classification of drought severity (D0–D4 levels) • State and regional drought summaries • Historical drought data archives • Impacts on agriculture, water resources, and environment • Interactive tools for monitoring drought trends	https://droughtmonitor.unl.edu/
Global Drought Observatory (GDO)	• Global drought monitoring and early warning information • Drought hazard and impact maps • Standardized Precipitation Index (SPI) maps • Vegetation health and soil moisture assessments • Regional drought risk assessments (Africa, Europe, Asia, Americas) • Drought bulletin reports and forecasts	https://drought.emergency.copernicus.eu/

3.3.1 Case Study: Mapping Drought Its Severity in Central Indian States Madhya Pradesh and Chhattisgarh Using Remote Sensing Data

Drought is one of the most significant climatic challenges affecting agriculture, water resources, and socio-economic conditions in India. Accurate and timely monitoring of drought is essential to mitigate its impacts and implement effective management strategies. This case study employs CHIRPS (Climate Hazards Group InfraRed Precipitation with Station Data) rainfall datasets and drought indices such as the Standardized Precipitation Index (SPI) to assess drought severity and trends across a study region in India from 2001 to 2021. The analysis provides insights into spatio-temporal drought patterns and their implications for drought management. The CHIRPS dataset, which integrates satellite imagery and in-situ precipitation data, was used to compute the SPI values. The SPI was calculated for different time scales to assess meteorological drought conditions, categorizing them into seven classes ranging from "Extremely Dry" (< -2) to "Extremely Wet" (> 2). Figure 5 shows the SPI-12 values of Madhya Pradesh and Chhattisgarh. This classification

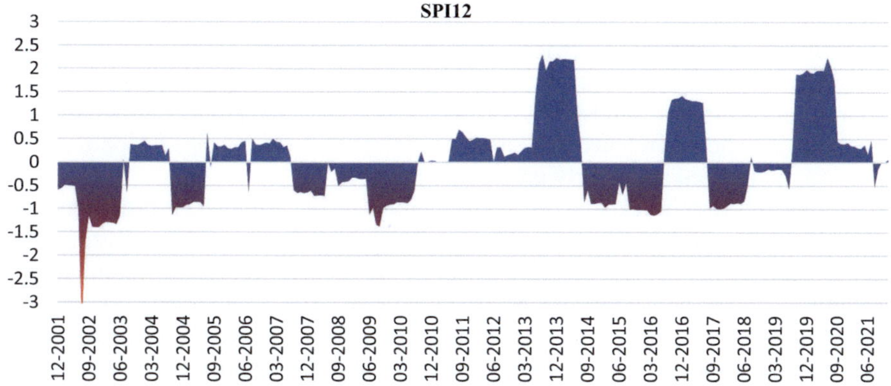

Fig. 5 SPI-12 values of Chhattisgarh and Madhya Pradesh

was applied to identify drought severity for each year in the study period. Data analysis included determining the proportion of years experiencing varying levels of drought and comparing these results against observed rainfall patterns.

Extremely dry conditions (< -2 SPI) were observed in 2002. Very dry conditions (-2 to -1 SPI) were most prevalent in 2004, 2009 and 2015, reflecting widespread drought events. Moderately dry conditions (-1 to -0.5 SPI) were the most frequent drought class, particularly in 2004 and 2017. Normal conditions dominated in years 2003, 2005, 2006, 2007, 2011, 2012 and 2018, indicating recovery during these periods. Wet conditions (SPI > 0.5) were limited, with very wet and extremely wet conditions being sporadic, most notably in 2013, 2016 and 2020.

This study underscores the utility of CHIRPS rainfall data and SPI for effective drought monitoring. The analysis highlights the prevalence of moderate to severe drought conditions in the study area, necessitating proactive measures for water resource management and agricultural planning. The observed trends also emphasize the role of climate variability in influencing drought frequency and intensity. Future work could incorporate additional indices and socio-economic data to develop a more holistic understanding of drought impacts and resilience strategies.

3.4 Earthquake

An earthquake is a sudden and violent shaking of the Earth's surface caused by the release of energy from the Earth's lithosphere. This energy is typically released when tectonic plates move or slip along a fault line, although it can also be triggered by volcanic activity or human activities such as mining and reservoir-induced seismicity (Mahmud 2019; Wallace 1997; Ellsworth 2013). Earthquakes are measured by their magnitude and intensity, with the former quantifying the energy released and the latter describing the observed effects on people, structures, and the environment (Sbarra et al. 2020; Guerrieri et al. 2009). The impact of an earthquake can range from minor tremors to catastrophic destruction, including collapsed buildings, damaged infrastructure, landslides, and tsunamis in coastal regions. Earthquakes are unpredictable and often leave long-term economic and social impacts, particularly in densely populated and poorly prepared regions.

AI-powered systems excel in real-time monitoring of earthquakes, providing critical information during and immediately after seismic events. By analyzing data from seismic networks and satellite imagery, AI models can quickly estimate the magnitude, depth, and epicentre of an earthquake. The ShakeAlert® Earthquake Early Warning System exemplifies this integration, having issued 95 alerts for earthquakes of magnitude 4.5 or higher between 2019 and 2023 with a 94% confirmation rate (https://www.shakealert.org/). Deep learning algorithms process these inputs to generate shake maps, which visualize the intensity and spatial extent of ground shaking. AI is also used to monitor infrastructure damage, analyzing drone imagery and sensor data to identify compromised buildings, roads, and utilities. These insights enable rapid assessment of affected areas, guiding emergency responders in prioritizing rescue efforts and resource allocation. Researchers have used geospatial applications and remote sensing data to study post-earthquake ground deformations and destruction (Markogiannaki et al. 2020; Loos et al. 2020; Sivakumar and Ghosh 2017). Table 3 lists some agencies which are monitoring earthquake in real-time.

AI is crucial in developing long-term strategies for earthquake risk mitigation and resilience. By analyzing historical earthquake data, urban growth patterns, and structural vulnerabilities, AI models identify areas with the highest risk. Predictive analytics support the planning of earthquake-resistant infrastructure by simulating how buildings and lifelines would perform during seismic events. It also aids in land-use planning by recommending zoning policies that minimize construction in high-risk areas. Furthermore, AI-powered tools help educate communities on earthquake preparedness by generating interactive scenarios and training simulations. Kim et al. (2020) used geospatial and remote sensing database to map soil and seismic zones in inland urban areas to reduce earthquake-induced hazards. By integrating AI into earthquake risk management frameworks, societies can improve their resilience, reduce economic losses, and enhance the safety of populations in seismically active regions.

Although predicting earthquakes remains one of the most challenging tasks in disaster management, Artificial Intelligence (AI) is making strides in enhancing early warning capabilities. Machine learning (ML) algorithms, such as neural networks and support vector machines (SVMs), analyze seismic data, tectonic plate movements, and historical earthquake patterns to identify subtle precursors of seismic activity. AI models process real-time data from seismic sensors and geophysical instruments, detecting anomalies that could signal an impending earthquake. While exact prediction of the timing and magnitude of earthquakes is still beyond current technology, AI has improved the ability to issue short-term warnings for aftershocks and secondary hazards, such as landslides and tsunamis. Models process seismic and oceanographic data to identify earthquake-induced changes in sea levels that could

Table 3 Agencies which monitor Earthquake in real-time

Platform	Type of data provided	Link
USGS Earthquake Hazards Program (USA)	• Real-time earthquake detection and reporting worldwide • Earthquake magnitude, location, and depth data • Shake maps (ground shaking intensity maps) • Tsunami alerts related to seismic events • Seismic hazard assessments and fault line mapping • Earthquake catalogs and historical data archives • Public safety resources and educational materials	https://www.usgs.gov/programs/earthquake-hazards
EMSC-CSEM (Europe and Worldwide)	• Real-time earthquake alerts and rapid earthquake location determination • Community Internet Intensity Maps (felt reports from users) • Earthquake magnitude, depth, and epicenter data • Historical earthquake databases • Tsunami information for Mediterranean and nearby regions • Seismological research updates and tools	https://www.emsc-csem.org/

generate tsunamis. Neural networks and ensemble learning techniques analyze the relationship between underwater seismic activity and wave propagation to predict tsunami height, speed, and potential impact zones. These predictions enable coastal communities to evacuate vulnerable areas before tsunamis strike. AI-driven simulations also assist in modelling the effects of tsunamis under various scenarios, helping authorities design better coastal defences and evacuation plans.

3.5 *Fire*

Fire, whether natural or human-induced, is a rapid oxidation process that releases heat, light, and various chemical byproducts, causing significant environmental, economic, and social impacts. Wildfires, which occur in forests, grasslands, and other natural landscapes, are often triggered by lightning, volcanic eruptions, or human activities such as unattended campfires or agricultural practices (Oppong et al. 2024; Keeley and Pausas 2021). Urban fires, on the other hand, occur in populated areas and are typically linked to electrical failures, arson, or industrial accidents. Fires can destroy ecosystems, homes, and infrastructure, disrupt livelihoods, and pose serious health risks due to smoke and air pollution.

Artificial Intelligence (AI) has become a critical tool in predicting wildfires and issuing early warnings to mitigate their devastating impacts. Drone and mobile-based algorithms are improving wildfire risk assessment and early detection by analysing high-resolution data (Lelis et al. 2024; James et al. 2023). Machine learning (ML) models analyse historical fire data, weather patterns, vegetation conditions, and human activities to predict wildfire risk with remarkable accuracy. For instance, algorithms such as random forests and neural networks are used to evaluate factors like temperature, humidity, wind speed, distance from the road and fuel load to estimate the likelihood of ignition and spread. AI-driven systems integrate real-time data from satellites, drones, and IoT-based sensors to detect hotspots and abnormal temperature rises, enabling authorities to issue timely alerts and mobilize resources to prevent fires from escalating.

3.5.1 Case Study: Forest Fire Prediction in Madhya Pradesh, India Using Machine Learning (Singh and Jeganathan 2024)

Madhya Pradesh, with its extensive forest cover, experiences recurrent forest fires that pose significant threats to biodiversity and forest-dependent communities. The forest area of MP is 94,689 km^2, which makes MP the state with the largest forest area in India (https://fsi.nic.in/isfr19/vol2/isfr-2019-vol-ii-madhya-pradesh.pdf). The state's tropical deciduous forests, combined with climatic variability and human activities, create a high risk for fire incidents, particularly during the dry season of March and April. Using advanced geospatial datasets and machine learning techniques, researchers have developed a Random Forest (RF) model to predict forest fire risks, incorporating variables such as NDVI, rainfall, temperature, solar radiation, soil moisture, and litter input. This case study explores the application of these techniques to predict and manage forest fire occurrences in Madhya Pradesh. Figure 6 shows the methodology flow chart for predicting forest fire risk probability.

The study utilized MODIS active fire data (2001–2020) to analyse historical fire occurrences and predict future risks. Climate and ecological datasets, including NDVI, precipitation, and solar radiation, were processed in Google Earth Engine. The RF model was trained on a balanced dataset of 15,000 fire and non-fire points, leveraging data from 2001 to 2017 and validated with 2018–2020 data. The predicted fire risk maps were classified into five categories, ranging from very low to very high risk, and assessed against actual fire incidents. Figure 7 shows the predicted maps of forest fire risk along with actual forest fire locations in those years.

The RF model achieved an accuracy of 86.46% in predicting forest fire risks in Madhya Pradesh. In March 2018, the model accurately identified high-risk zones, with 63.75% of fire events occurring in very high-risk areas. Conversely, in 2020, increased rainfall reduced fire events, with only 8.97% of forested areas under very high risk. The central region of Madhya Pradesh was consistently identified as the most vulnerable due to its dry deciduous forests and lower soil moisture levels.

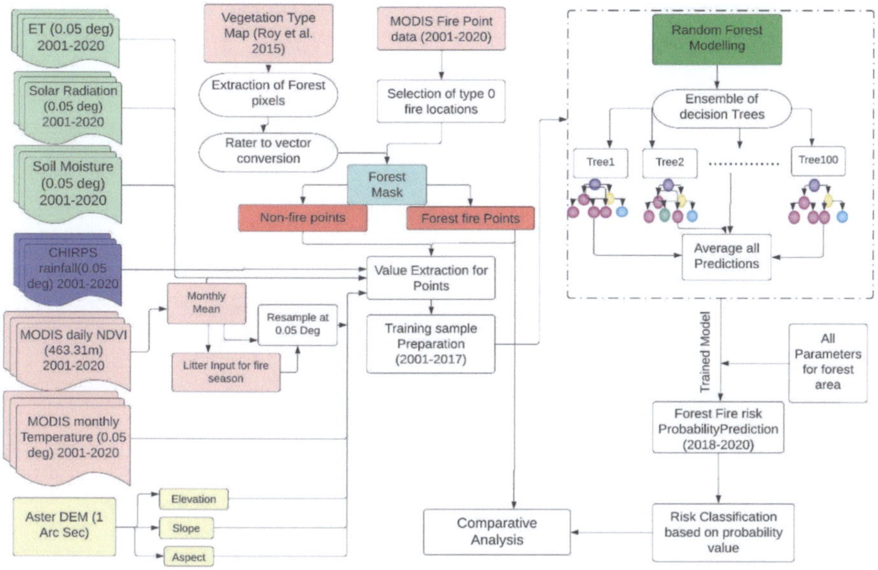

Fig. 6 Methodology flow chat for fire risk probability prediction (*Source* Singh and Jeganathan 2024)

NDVI and rainfall emerged as the most influential variables, emphasizing the critical role of vegetation health and precipitation in mitigating fire risks. This study underscores the effectiveness of integrating AI with geospatial data for forest fire prediction in Madhya Pradesh. The RF model's high accuracy highlights its potential as a tool for proactive fire management and resource allocation. By focusing on high-risk zones during fire-prone months, policymakers can reduce fire incidents and protect the region's biodiversity. Future research should include anthropogenic factors and higher-resolution datasets to enhance predictive capabilities and address the complexities of forest fire dynamics in Central India.

3.5.2 Real-Time Fire Monitoring and Suppression

AI excels in real-time monitoring of wildfires, providing actionable insights during fire events to guide suppression efforts. Deep learning algorithms, particularly convolutional neural networks (CNNs), analyse satellite imagery and drone footage to track the spread and intensity of fires (Thangavel et al. 2023; McCarthy et al. 2021; Rashkovetsky et al. 2021). These models generate dynamic fire maps that indicate high-risk areas, helping firefighters allocate resources efficiently. AI-powered tools also process data from thermal imaging cameras and ground sensors to identify hidden embers or rekindling risks, ensuring thorough containment. Additionally, AI optimizes firefighting strategies by simulating the effectiveness of various tactics,

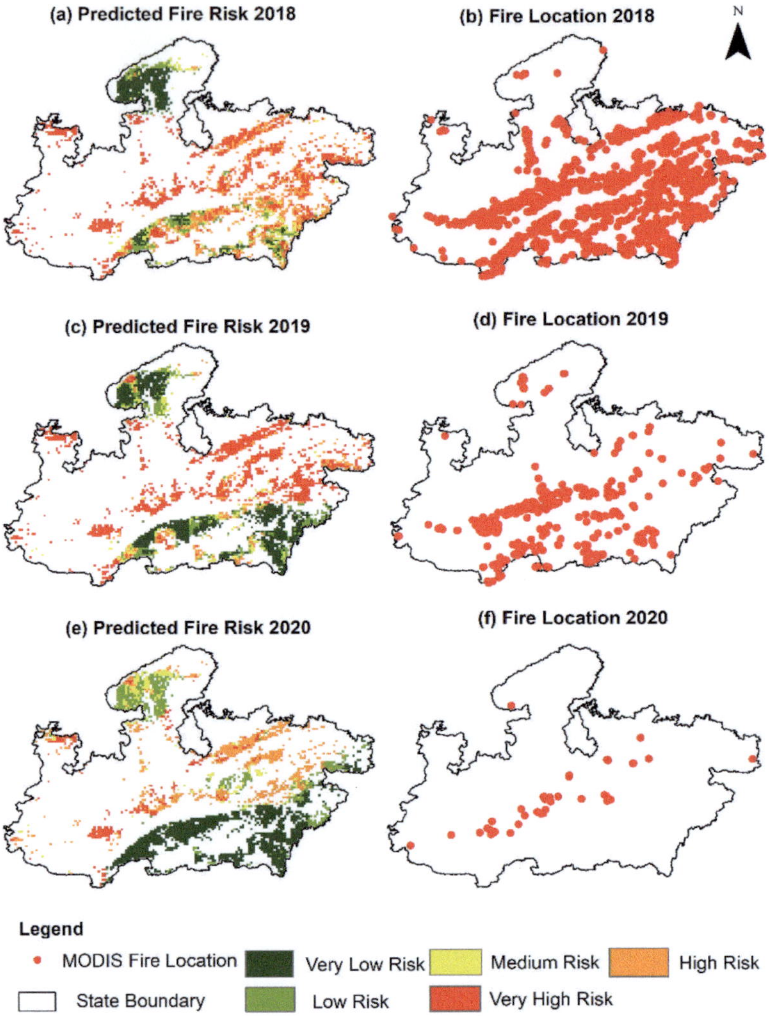

Fig. 7 Predicted fire locations and actual fire points in MP for 2018, 2019 and 2020 (taken from Singh & Jeganathan 2024)

such as aerial water drops or firebreak construction, based on terrain and weather conditions.

3.5.3 Post-fire Impact Assessment and Recovery Planning

In the aftermath of fires, AI aids in damage assessment and recovery planning by analyzing satellite images and aerial surveys to map affected areas. These systems identify burnt vegetation, damaged structures, and disrupted ecosystems, providing precise data for rehabilitation efforts. Natural language processing (NLP) tools further complement this process by extracting reports of fire damage from social media and news articles, offering real-time insights into community needs (Sufi and Khalil 2024). AI also supports the restoration of affected landscapes by recommending strategies for reforestation, soil stabilization, and wildlife habitat recovery. By automating these processes, AI accelerates recovery timelines and promotes sustainable land management practices.

3.5.4 Industrial and Urban Fire Risk Management

AI is increasingly being used to mitigate risks associated with urban and industrial fires. Predictive maintenance algorithms analyze sensor data from manufacturing facilities, power plants, and oil refineries to identify potential fire hazards, such as overheating equipment or gas leaks before incidents occur. Smart fire detection systems equipped with AI algorithms process video feeds and sensor data to detect smoke or flames in real-time, triggering automated responses like sprinkler activation or system shutdowns (Ren et al. 2025; Xiao et al. 2023). In urban areas, AI helps model fire risk based on building materials, density, and emergency access routes, supporting urban planners in designing safer communities. By integrating AI into fire risk management frameworks, both rural and urban areas can significantly reduce fire-related losses and enhance public safety.

3.6 Tornado

A tornado is a rapidly rotating column of air extending from a thunderstorm to the ground, characterized by its funnel-shaped appearance and extreme wind speeds (Snyder et al. 2020). Tornadoes are among the most destructive natural phenomena, capable of uprooting trees, demolishing buildings, and hurling debris at high velocities. They typically form under specific atmospheric conditions, including strong wind shear, instability, and moisture, often associated with severe thunderstorms or supercells (Weinhoff et al. 2023). Tornadoes vary in intensity, with their destructive power measured on the Enhanced Fujita (EF) Scale, ranging from EF0 (light damage) to EF5 (devastating damage) (Wurman et al. 2021). Although they are most common in regions like the central United States, known as "Tornado Alley," they can occur in many parts of the world. Due to their sudden and localized nature, predicting tornadoes remains challenging, making early warning systems, preparedness, and community education vital for minimizing loss of life and property.

Real-time tracking of tornadoes is greatly enhanced by AI-powered systems that analyze live data from weather radars and remote sensing technologies. Deep learning models, such as convolutional neural networks (CNNs), process radar imagery to identify rotational signatures and debris patterns characteristic of tornadoes.

These systems generate detailed maps of a tornado's trajectory, speed, and intensity, enabling emergency response teams to predict its path and prioritize evacuation efforts (Steinkruger et al. 2020). AI also processes crowdsourced data, such as videos and reports from affected individuals, to confirm tornado activity and refine forecasts. This real-time monitoring helps mitigate the impact of tornadoes by guiding resource deployment and improving situational awareness. Table 4 lists some agencies which track tornadoes in real time and provide information about them on their websites.

After a tornado, it plays a crucial role in assessing damage and planning recovery efforts. High-resolution satellite imagery and drone footage are analyzed using

Table 4 Agencies which track tornados in real-time and provide information

Platform	Type of data provided	Link
NOAA Storm Prediction Center (SPC)	• Severe weather forecasts and outlooks (tornadoes, thunderstorms, hail, high winds) • Tornado and severe weather warnings • Mesoscale discussions • Convective outlook maps (day 1 to day 8 forecasts) • Storm reports (hail, wind, tornado events) • Fire weather forecasts • Historical severe weather archives	https://www.spc.noaa.gov/
Tornado HQ	• Real-time tornado warnings and alerts • Live severe thunderstorm warnings • Tornado watch maps • Tornado tracking maps • Archived tornado paths and warning history • Interactive maps showing live and recent warning zones	https://www.tornadohq.com/
India Meteorological Department (IMD)—Nowcast and Thunderstorm Warnings	• Nowcast alerts for thunderstorms, lightning, squalls, hailstorms, and heavy rain • Severe weather bulletins updated every 3 h • Cyclone alerts and warnings • State and district specific storm alerts • Lightning forecasts	https://mausam.imd.gov.in/responsive/districtWiseNowcastGIS.php

machine learning algorithms to map areas of destruction, identifying collapsed buildings, uprooted trees, and damaged infrastructure (Speth et al. 2022). Additionally, AI simulations estimate the economic and social impacts of tornadoes, guiding policymakers in designing recovery strategies and securing necessary funding for rebuilding efforts.

AI is instrumental in developing long-term strategies to mitigate tornado risks and enhance community resilience. Predictive models analyze historical tornado data, land-use patterns, and urban growth to identify high-risk areas and guide zoning regulations. AI-powered simulations, such as generative adversarial networks (GANs), create virtual scenarios of tornado impacts, helping engineers design tornado-resistant buildings and infrastructure (Markowski et al. 2020; Sufi 2022). In agricultural regions, It helps optimize land management practices to minimize the vulnerability of crops and livestock to tornado damage. AI-based educational tools create realistic simulations for public awareness and training programs, ensuring communities are better prepared for future tornado events. By integrating AI into tornado risk management frameworks, society can reduce vulnerabilities and improve recovery capabilities.

Artificial Intelligence (AI) is revolutionizing tornado prediction by analyzing complex meteorological data to enhance early warning capabilities. Machine learning (ML) models process vast datasets, including atmospheric pressure, wind shear, temperature gradients, and humidity levels, to identify conditions conducive to tornado formation. Algorithms like support vector machines (SVMs) and recurrent neural networks (RNNs) excel in analyzing temporal patterns, enabling them to predict tornadoes with increasing accuracy (Lagerquiest et al. 2020). AI systems integrate data from Doppler radar, weather satellites, and ground-based sensors to detect early signs of supercell development, often a precursor to tornadoes. These predictive tools allow meteorologists to issue timely warnings, providing communities with critical minutes to seek shelter and prepare.

3.7 Landslides, Mudslides and Avalanches

Landslides, mudslides, and avalanches are rapid downward movements of soil, rock, or snow, often triggered by natural or human-induced factors. Landslides involve the sliding of rock or debris down a slope caused by factors like heavy rainfall, earthquakes, volcanic activity, or deforestation that destabilizes the terrain (Liu et al. 2015). Mudslides, a specific type of landslide, occur when water saturates the soil, creating a fast-moving flow of mud and debris. They are particularly common in steep, deforested areas following intense rainfall. Avalanches, on the other hand, are the sudden descent of snow, often mixed with ice and debris, down a mountain slope, usually triggered by factors like rapid temperature changes, heavy snowfall, or human activity such as skiing. These events pose significant risks to life, infrastructure, and ecosystems, often cutting off communities and causing economic losses (Xu et al. 2020).

AI-powered systems excel in real-time monitoring of landslides, mudslides, and avalanches, providing dynamic updates to mitigate risks. Deep learning models, particularly convolutional neural networks (CNNs), process satellite imagery and drone footage to identify active slide zones or snowpack instability (Mohan et al. 2020; Zhang and Wang 2024). These models generate detailed hazard maps, highlighting areas at immediate or elevated risk. AI also analyzes data from ground-based sensors, such as inclinometers and snow pressure monitors, to detect early signs of movement or stress. By integrating these data sources, AI creates comprehensive spatial and temporal risk assessments, guiding emergency response teams in prioritizing vulnerable areas.

Following landslides, mudslides, or avalanches, AI accelerates damage assessment and recovery planning. Machine learning algorithms analyze high-resolution imagery and terrain models to map affected areas, identifying blocked roads, damaged infrastructure, and disrupted ecosystems. AI systems can also process data from seismic sensors and social media reports to estimate the extent of destruction and locate stranded individuals (Davison et al. 2020; Pang and Pang 2022). Natural language processing (NLP) tools extract insights from emergency communications, further refining situational awareness. These capabilities help authorities allocate resources effectively for rescue operations and long-term recovery, ensuring aid reaches the most affected regions promptly.

AI supports long-term strategies for mitigating risks associated with landslides, mudslides, and avalanches by analyzing historical events, land-use changes, and environmental conditions (Gomez-Miranda et al. 2025). Predictive models assess the impacts of deforestation, urbanization, and climate change on slope stability and snowpack behaviour. Simulation-based AI tools, such as generative adversarial networks (GANs), create virtual scenarios to evaluate the effectiveness of mitigation measures, such as reforestation, slope reinforcement, or controlled snowpack release. AI also aids in designing resilient infrastructure, including landslide barriers and avalanche control systems, tailored to specific risk profiles (Shahjalal et al. 2025; Xu and Guo 2025). By integrating AI into disaster risk management frameworks, communities can reduce vulnerability and enhance preparedness for these destructive natural hazards.

Artificial Intelligence (AI) is revolutionizing the prediction of landslides, mudslides, and avalanches by analyzing diverse environmental factors to identify potential triggers (Albzaie et al. 2024). Machine learning (ML) models process geological, hydrological, and meteorological data, such as soil saturation, slope stability, precipitation levels, and snowpack conditions, to assess the likelihood of these hazards. Algorithms like decision trees, random forests, and neural networks are particularly effective at detecting patterns that precede slope failure. AI integrates real-time data from sensors, satellites, and remote monitoring stations to issue early warnings, enabling timely evacuation and mitigation efforts (Joshi et al. 2024). These systems are especially valuable in mountainous and hilly regions where sudden events can have catastrophic consequences.

4 Discussion

The integration of artificial intelligence (AI) technologies has revolutionized disaster management frameworks across various hazard types. This transformative approach is reshaping how communities predict, prepare for, respond to, and recover from disasters ranging from floods and earthquakes to wildfires and droughts (Pang and Pang 2022; Gupta et al. 2022). AI-driven predictive capabilities for disasters extend beyond simple impact assessment to customizing relief packages based on demographic and situational variables, ensuring more effective resource allocation during critical response periods (Biswas et al. 2024; Seitzinger and Kalra 2023; Zhang et al. 2022). Advanced hydraulic models augmented by AI can now integrate complex land features, including topography, infrastructure, and urban layouts, enabling ultra-high-resolution simulations at city and regional scales. The computational demands of these sophisticated models are met through high-performance computing systems equipped with specialized hardware, such as NVIDIA A100 Tensor Core GPUs, facilitating rapid processing of complex environmental datasets. This technological convergence has transformed flood risk assessment from a retrospective analysis to a dynamic, predictive framework. The visualization capabilities developed through AI-enhanced geospatial applications have particularly strengthened decision-making processes. Tools like NVIDIA Omniverse provide immersive 3D visualizations that help stakeholders better understand flood patterns and visualize the potential impacts of extreme weather events. These immersive models serve multiple functions beyond technical assessment, including public education, responder training, and facilitating more effective community preparation for disasters (Abid et al. 2021).

Despite significant advancements, AI integration in disaster management faces several limitations that require acknowledgement and mitigation. Data quality and availability represent fundamental challenges, particularly in developing regions where comprehensive historical records and monitoring infrastructure may be limited (Han and Liu 2025). Many AI models depend on extensive historical datasets for training, which may not exist for all locations or disaster types. Computational requirements present another significant limitation, as many advanced AI models require substantial processing power, particularly for real-time applications. Model interpretability remains a persistent challenge, as many advanced AI systems function as "black boxes" where the reasoning behind specific predictions cannot be easily explained. This limitation creates potential barriers to trust and adoption, particularly in high-stakes disaster management contexts where decisions may affect thousands of lives. The ShakeAlert system's experience with false alerts and missed events illustrates the consequences of imperfect model performance (Böse et al. 2022). Integration challenges between different systems and data sources can impede effective implementation. Many disaster management frameworks involve multiple agencies with different technological infrastructures, creating potential interoperability issues. The seamless integration of camera networks, predictive modelling systems, and response planning tools seen in California's wildfire management represents an ideal that many regions struggle to achieve (Govil et al. 2020; Shi et al. 2020).

Ground-truthing and validation requirements present ongoing challenges, particularly for predictive models. Real-time data processing capabilities may be compromised during actual disaster events, precisely when they are most needed. Communication infrastructure damage can disrupt data flows to AI systems, potentially creating blind spots in disaster response. The physical vulnerability of monitoring networks represents an inherent limitation of technology-dependent disaster management approaches. Ethical considerations surrounding AI deployment in disaster contexts raise important questions about privacy, resource allocation, and potential biases in model outputs (Gevaert et al. 2021). Machine learning systems may inadvertently perpetuate existing disparities if training data reflects historical patterns of unequal disaster impacts. These ethical dimensions require careful consideration alongside technical performance metrics (Albahri et al. 2024). Errors such as cloud cover interference in satellite images or sensor malfunctions can reduce the reliability of AI-driven predictions and analyses.

AI models trained on geospatial data for disaster risk management often struggle with generalization and transferability. A model trained in one geographical region may not perform well in another due to differences in terrain, climate, and socio-economic conditions. For instance, a landslide prediction model trained using data from mountainous regions may not account for unique factors influencing landslides in coastal areas. Developing region-specific models can be time-consuming and costly, limiting the scalability of AI solutions. Geospatial data and AI models may fail to account for highly dynamic and extreme scenarios, such as unprecedented climate events or compound disasters (e.g., a hurricane causing both floods and landslides). These rare but catastrophic events often fall outside the scope of training datasets, reducing the accuracy and reliability of AI predictions. As a result, disaster management plans may still need to rely heavily on human expertise and traditional methods.

By addressing these limitations, AI can become a more reliable and equitable tool for enhancing disaster risk management worldwide (Ghaffarian et al. 2023). Future advancements in AI and geospatial technologies promise to significantly improve the accuracy and timeliness of disaster predictions (Vysotskyi and Vysotskyi 2024). As more high-resolution and real-time geospatial data become available from advanced satellites, drones, and IoT networks, AI models will benefit from richer training datasets. This will enable more precise identification of disaster precursors, such as subtle shifts in seismic activity, atmospheric changes, or soil moisture levels, leading to more reliable early warning systems (Meshram et al. 2024). The integration of AI with predictive models could provide longer lead times for evacuation and preparedness, minimizing casualties and property damage.

AI's ability to analyse geospatial data at fine scales opens the possibility for personalized risk assessments and mitigation plans. In the future, individuals and communities may receive tailored disaster alerts and recommendations based on their specific geographic location, infrastructure vulnerability, and socio-economic context. For instance, AI could help farmers adopt crop patterns that are resilient to drought or floods based on localized climate projections. Similarly, urban planners might use AI-driven geospatial analyses to design neighbourhoods with optimized layouts for flood resilience, fire prevention, or evacuation routes.

The convergence of AI, geospatial technologies, and IoT in smart cities will redefine disaster risk management (Elshoukry 2024). Geospatial AI models integrated into urban IoT networks can provide real-time monitoring and automated responses to disasters (Munawar et al. 2021). For example, AI-powered sensors could detect rising water levels in storm drains and automatically activate flood barriers or reroute traffic. Future smart cities may also use AI to simulate disaster scenarios, optimize resource allocation, and coordinate emergency responses, creating resilient urban ecosystems capable of adapting dynamically to evolving threats (Elvas et al. 2021).

AI will drive the deployment of autonomous technologies, such as drones and robots, to assist in disaster risk management. Geospatial data combined with AI algorithms can guide drones for real-time mapping of disaster-affected areas, delivering supplies, or monitoring hazards like wildfires or landslides. Robots equipped with AI and geospatial navigation capabilities could be used for search-and-rescue missions in hazardous environments, reducing risks to human responders. These autonomous systems will become essential tools in disaster management, improving response efficiency and safety.

As the impacts of climate change intensify, AI will play a critical role in guiding adaptation and resilience-building efforts. By analyzing geospatial data on changing weather patterns, sea-level rise, and land-use trends, AI can help design strategies to mitigate climate-related risks. Predictive models will enable long-term planning, such as identifying areas suitable for infrastructure development or natural resource conservation (Young et al. 2024). Additionally, AI can monitor the effectiveness of implemented adaptation measures, iteratively improving strategies for climate resilience (Franchi et al. 2024). AI has the potential to democratize disaster risk management by making geospatial insights more accessible. With cloud-based platforms and open-access datasets, AI-powered tools can be scaled globally, offering affordable solutions to resource-limited regions (Amoah et al. 2023). In the future, even communities without extensive infrastructure can benefit from mobile-based applications that leverage AI for real-time disaster alerts, hazard mapping, and recovery planning. Ensuring equitable access to these technologies will be crucial for reducing global disparities in disaster resilience. Collaborative frameworks supported by AI will enhance global preparedness, ensuring collective responses to shared challenges.

5 Conclusion

The integration of Artificial Intelligence (AI) across various disaster management domains marks a transformative leap in how societies anticipate, prepare for, and respond to natural hazards. From flood mapping and cyclone impact forecasting to drought monitoring, earthquake early warnings, wildfire behaviour modelling, and landslide detection, AI is significantly augmenting traditional disaster response mechanisms. These technologies are shifting the paradigm from reactive to proactive disaster management—offering faster, data-driven insights that can potentially save lives, minimize infrastructure damage, and reduce long-term economic losses.

Despite these advancements, several limitations temper the full realization of AI's potential in disaster risk management. Challenges such as inconsistent or low-quality geospatial data, high computational demands, and limited access to cloud infrastructure in developing regions hinder the widespread application of these tools. Moreover, interpretability remains a key concern—especially for deep learning models. Integration into existing systems, validation of model outputs, real-time processing constraints, and ethical considerations around data use and privacy further complicate the landscape. Therefore, future frameworks must emphasize not only technological development but also capacity building, inclusive policies, and interdisciplinary collaboration to ensure that AI tools serve all communities effectively—especially the most vulnerable.

Looking ahead, the continued evolution of AI and geospatial analytics holds immense promise for more adaptive, intelligent, and responsive disaster risk management systems. With climate change intensifying the frequency and severity of natural hazards, the deployment of AI is not just an option—it is a necessity for building climate-resilient communities. AI will increasingly be integrated with Internet of Things (IoT) networks, real-time sensor data, and citizen science inputs to create dynamic early warning systems. In smart cities, AI will optimize emergency planning and infrastructure resilience. In agriculture, AI-driven drought and flood models will support climate-smart farming. In global frameworks, shared platforms and open data policies will foster cooperation across borders to manage transboundary risks.

The effectiveness of AI in disaster management will depend on our ability to strike a balance—harnessing its speed and precision while remaining attentive to its limitations and ethical implications. By combining AI's analytical power with human judgment, indigenous knowledge, and traditional risk management practices, we can create disaster preparedness systems that are not only technically robust but also socially inclusive and contextually relevant. In doing so, AI and geospatial technologies can pave the way toward a safer, more equitable, and resilient future for all.

Conflict of Interest The authors declare that they have no competing interests whatsoever.

AI Disclosure 'Chatgpt 4.0', 'Consensus'and 'Perplexity AI' have been used to write some parts of the chapter.

References

Abid S, Sulaiman N, Chan S, Nazir U, Abid M, Han H, Ariza-Montes A, Vega-Muñoz A (2021) Toward an integrated disaster management approach: how artificial intelligence can boost disaster management. Sustainability. https://doi.org/10.3390/su132212560

Adikari K, Shrestha S, Ratnayake D, Budhathoki A, Mohanasundaram S, Dailey M (2021) Evaluation of artificial intelligence models for flood and drought forecasting in arid and tropical regions. Environ Model Softw 144:105136. https://doi.org/10.1016/J.ENVSOFT.2021.105136

Afolabi A, Hussain N, Austin-Gabriel B, Ige A, Adepoju P (2023) Geospatial AI and data analytics for satellite-based disaster prediction and risk assessment. Open Access Res J Eng Technol. https://doi.org/10.53022/oarjet.2023.4.2.0058

Alahi M, Sukkuea A, Tina F, Nag A, Kurdthongmee W, Suwannarat K, Mukhopadhyay S (2023) Integration of IoT-enabled technologies and artificial intelligence (AI) for smart city scenario: recent advancements and future trends. Sensors (Basel, Switzerland). https://doi.org/10.3390/s23115206

Albahri A, Khaleel Y, Habeeb M, Ismael R, Hameed Q, Deveci M, Homod R, Albahri O, Alamoodi A, Alzubaidi L (2024) A systematic review of trustworthy artificial intelligence applications in natural disasters. Comput Electr Eng 118:109409. https://doi.org/10.1016/j.compeleceng.2024.109409

Albzaie M, Farghal A, Habib A, Habib M (2024) Sustainability benefits of AI-based engineering solutions for infrastructure resilience in arid regions against extreme rainfall events. Discov Sustain. https://doi.org/10.1007/s43621-024-00500-2

Alizadeh M, Nikoo M (2018) A fusion-based methodology for meteorological drought estimation using remote sensing data. Remote Sens Environ. https://doi.org/10.1016/J.RSE.2018.04.001

Amoah E, McCloskey P, Ouedrago R, Chelal J, Akuleut C, Mwambumba B, Meli B, Oyudi C, Kibwamga E, Cleophas E, Ochola F, Wangiru C, Morang'a K, Mushira L, Maboke M, Titus N, Oltimbao S, Odawa S, Hughes D (2023) A low-cost, AI-powered Measurement Verification and Reporting System for growing trees with smallholder farmers. bioRxiv. https://doi.org/10.1101/2023.11.29.569237.

Arabameri A, Pal S, Santosh M, Chakrabortty R, Roy P, Moayedi H (2021) Drought risk assessment: integrating meteorological, hydrological, agricultural and socio-economic factors using ensemble models and geospatial techniques. Geocarto Int 37:6087–6115. https://doi.org/10.1080/10106049.2021.1926558

Barzkar A, Najafzadeh M, Homaei F (2021) Evaluation of drought events in various climatic conditions using data-driven models and a reliability-based probabilistic model. Nat Hazards. https://doi.org/10.1007/s11069-021-05019-7

Beeravelly S (2024) Smart response: leveraging AI analytics for enhanced disaster resilience. Int J Multidiscip Res. https://doi.org/10.36948/ijfmr.2024.v06i06.32137

Biswas S, Kumar D, Hajiaghaei-Keshteli M, Bera U (2024) An AI-based framework for earthquake relief demand forecasting: a case study in Türkiye. Int J Disaster Risk Reduc. https://doi.org/10.1016/j.ijdrr.2024.104287

Böse M, Andrews J, O'Rourke C, Kilb D, Lux A, Bunn J, McGuire J (2023) Testing the ShakeAlert earthquake early warning system using synthesized earthquake sequences. Seismol Res Lett 94(1):243–259. https://doi.org/10.1785/0220220088

Cao Y, Xu C, Aziz N, Kamaruzzaman S (2023) BIM-GIS integrated utilization in urban disaster management: the contributions, challenges, and future directions. Remote Sens 15:1331. https://doi.org/10.3390/rs15051331

Davison B, Sun W, Bocchini P (2020) Applications of artificial intelligence for disaster management. Nat Hazards. https://doi.org/10.1007/s11069-020-04124-3

Deepak S, Rajan G, Jairaj P (2020) Geospatial approach for assessment of vulnerability to flood in local self governments. Geoenviron Disast. https://doi.org/10.1186/s40677-020-00172-w

Dritsas E, Trigka M (2025) Remote sensing and geospatial analysis in the big data era: a survey. Remote Sens. https://doi.org/10.3390/rs17030550

Ellsworth W (2013) Injection-induced earthquakes. Science. https://doi.org/10.1126/science.1225942

Elshoukry H (2024) Enhancing disaster management by leveraging smart cities technologies. In: 2024 mediterranean smart cities conference (MSCC), pp 1–8. https://doi.org/10.1109/MSCC62288.2024.10697072

Elvas L, Mataloto B, Martins A, Ferreira J (2021) Disaster management in smart cities. Smart Cities 4:819–839. https://doi.org/10.3390/SMARTCITIES4020042

Franchi F, Galassi A, Graziosi F (2024) AI-assisted GIS toward Geo-AI: trends and innovations overview. In: 2024 IEEE 8th forum on research and technologies for society and industry innovation (RTSI), pp 149–153. https://doi.org/10.1109/RTSI61910.2024.10761798.

Gevaert C, Carman M, Rosman B, Georgiadou Y, Soden R (2021) Fairness and accountability of AI in disaster risk management: opportunities and challenges. Patterns. https://doi.org/10.1016/j.patter.2021.100363

Ghaffarian S, Taghikhah F, Maier H (2023) Explainable artificial intelligence in disaster risk management: achievements and prospective futures. Int J Disaster Risk Reduc. https://doi.org/10.1016/j.ijdrr.2023.104123

Gomez-Miranda I, Restrepo-Estrada C, De Albuquerque P, Builes-Jaramillo A (2025) Advanced AI techniques for landslide susceptibility mapping and spatial prediction: a case study in Medellín, Colombia. Appl Comput Geosci https://doi.org/10.1016/j.acags.2025.100226

Govil K, Welch M, Ball J, Pennypacker C (2020) Preliminary results from a wildfire detection system using deep learning on remote camera images. Remote Sens 12:166. https://doi.org/10.3390/rs12010166

Guerrieri L, Vittori E, Comerci V, Esposito E, Porfido S, Michetti AM, Serva L, Silva PG (2009) Mapping and cataloguing earthquake environmental effects for seismic hazard assessment: the contribute of remote sensing techniques, pp 551–554. http://eprints.bice.rm.cnr.it/909/1/376_L.Guerrieri.pdf

Gupta S, Modgil S, Kumar A, Sivarajah U, Irani Z (2022) Artificial intelligence and cloud-based collaborative platforms for managing disaster, extreme weather and emergency operations. Int J Prod Econ. https://doi.org/10.1016/j.ijpe.2022.108642

Han Y, Liu G (2025) The role of GIS in disaster response: improving efficiency and overcoming challenges. Theor Nat Sci. https://doi.org/10.54254/2753-8818/2025.21259.

Hoque M, Phinn S, Roelfsema C, Childs I (2018) Assessing tropical cyclone risks using geospatial techniques. Appl Geogr. https://doi.org/10.1016/J.APGEOG.2018.07.004

Hoque M, Pradhan B, Ahmed N (2019a) Assessing drought vulnerability using geospatial techniques in northwestern part of Bangladesh. Sci Total Environ 705:135957. https://doi.org/10.1016/j.scitotenv.2019.135957

Hoque M, Pradhan B, Ahmed N, Roy S (2019b) Tropical cyclone risk assessment using geospatial techniques for the eastern coastal region of Bangladesh. Sci Total Environ 692:10–22. https://doi.org/10.1016/j.scitotenv.2019.07.132

Hoque M, Pradhan B, Ahmed N, Sohel M (2020) Agricultural drought risk assessment of Northern New South Wales, Australia using geospatial techniques. Sci Total Environ. https://doi.org/10.1016/j.scitotenv.2020.143600

Hosen B, Rahaman M, Kumar S, Sagar L, Akhtar M (2023) Leveraging artificial intelligence and big data for advanced spatial analytics and decision support systems in geography. Malays Appl Geogr. https://doi.org/10.26480/magg.02.2023.62.67

Huang Y (2025) Machine learning for enhanced classification and geospatial distribution analysis. J Comput Technol Appl Math. https://doi.org/10.70393/6a6374616d.323535

Islam M, Rabbi F, Hossain N (2024) Performance evaluation of NLP and CNN models for disaster detection using social media data. Soc Netw Anal Min 14:213. https://doi.org/10.1007/s13278-024-01374-y

Ivić M (2019) Artificial intelligence and geospatial analysis in disaster management. Int Arch Photogramm Remote Sens Spat Inf Sci. https://doi.org/10.5194/isprs-archives-xlii-3-w8-161-2019

James G, Ansaf R, Samahi S, Parker R, Cutler J, Gachette R, Ansaf B (2023) An efficient wildfire detection system for ai-embedded applications using satellite imagery. Fire. https://doi.org/10.3390/fire6040169

Jiang Z, Shekhar S (2017) Spatial big data science, pp 1–131. https://doi.org/10.1007/978-3-319-60195-3

Joshi A, Agarwal S, Kanungo D, Panigrahi R (2024) Integration of Edge–AI Into IoT–cloud architecture for landslide monitoring and prediction. IEEE Trans Industr Inf 20:4246–4258. https://doi.org/10.1109/TII.2023.3319671

Keeley J, Pausas J (2021) Wildfires and global change. Front Ecol Environ. https://doi.org/10.1002/FEE.2359

Kim H, Sun C, Kim M, Cho H, Lee M (2020) GIS-based optimum geospatial characterization for seismic site effect assessment in an Inland Urban Area, South Korea. Appl Sci https://doi.org/10.3390/app10217443

Lagerquist R, McGovern A, Homeyer C, Gagne D, Smith T (2020) Deep learning on three-dimensional multiscale data for next-hour tornado prediction. Mon Weather Rev. https://doi.org/10.1175/mwr-d-19-0372.1

Lai C, Zhong R, Wang Z, Wu X, Chen X, Wang P, Lian Y (2019) Monitoring hydrological drought using long-term satellite-based precipitation data. Sci Total Environ 649:1198–1208. https://doi.org/10.1016/j.scitotenv.2018.08.245

Latorre L, Rego E, De Leo L, Gutierrez M, Zarate J, Garzon A (2024) Tech Report: GIS. https://doi.org/10.18235/0013017

Lazorenko N (2024) Research on the use of artificial intelligence (GeoAI) for tasks of topographic mapping. Spatial Dev. https://doi.org/10.32347/2786-7269.2024.8.430-443

Lelis C, Roncal J, Silveira L, De Aquino R, Marcondes C, Marques J, Loubach D, Verri F, Curtis V, De Souza D (2024) Drone-based AI system for wildfire monitoring and risk prediction. IEEE Access 12:139865–139882. https://doi.org/10.1109/ACCESS.2024.3462436

Liu Z, Zhang H, Guo W, Xu X, Zhao C, Wang W, Yan Q (2015) Which is more hazardous: avalanche, landslide, or mudslide? Nat Hazards 76:1939–1945. https://doi.org/10.1007/s11069-014-1570-0

Loos S, Lallemant D, Baker J, McCaughey J, Yun S, Budhathoki N, Khan F, Singh R (2020) G-DIF: a geospatial data integration framework to rapidly estimate post-earthquake damage. Earthq Spectra 36:1695–1718. https://doi.org/10.1177/8755293020926190

Mahmud S (2019) The combined effects of the gravitational forces on the tectonic plates on earth's surface exerted by the moon, the sun and the other planets are one of the main reasons of the earthquakes vol 6, pp 44–62

Mansour S (2019) Geospatial modelling of tropical cyclone risks to the southern Oman coasts. Int J Disast Risk Reduc. https://doi.org/10.1016/J.IJDRR.2019.101151

Markogiannaki O, Karavias A, Bafi D, Angelou D, Parcharidis I (2020) A geospatial intelligence application to support post-disaster inspections based on local exposure information and on co-seismic DInSAR results: the case of the Durres (Albania) earthquake on November 26, 2019. Nat Hazards. https://doi.org/10.1007/s11069-020-04120-7

Markowski P, Steinkruger D, Young G (2020) An artificially intelligent system for the automated issuance of tornado warnings in simulated convective storms. Weather Forecast. https://doi.org/10.1175/waf-d-19-0249.1

Maurya A, Kumar A (2024) The role of GIS in the study of sustainable development and environmental management. Int J Multidiscip Res. https://doi.org/10.36948/ijfmr.2024.v06i06.32879

McCarthy N, Tohidi A, Aziz Y, Dennie M, Valero M, Hu N (2021) A deep learning approach to downscale geostationary satellite imagery for decision support in high impact wildfires. Forests. https://doi.org/10.3390/F12030294

Meshram N, Mandve P, Vitankar T, Motghare S, Patil S, Katole P (2024) Disaster management system: a machine learning approach to forecasting floods and earthquakes. Int J Res Appl Sci Eng Technol. https://doi.org/10.22214/ijraset.2024.66008

Mohan A, Singh A, Kumar B, Dwivedi R (2020) Review on remote sensing methods for landslide detection using machine and deep learning. Trans Emerg Telecommun Technol. https://doi.org/10.1002/ett.3998

Munawar H, Mojtahedi M, Hammad A, Kouzani A, Mahmud M (2021) Disruptive technologies as a solution for disaster risk management: a review. Sci Total Environ. https://doi.org/10.1016/j.scitotenv.2021.151351

Nandhakumar C, Pradeep D (2025) GeoAI: a systematic review on change detection of geospatial data with different methodologies. In: 2025 6th International Conference on Mobile Computing and Sustainable Informatics (ICMCSI), pp 686–692. https://doi.org/10.1109/ICMCSI64620.2025.10883073

Oppong F, Song Y, Xu C, Li X (2024) Lightning-induced wildfires: an overview. Fire. https://doi.org/10.3390/fire7030079

Pan X, Hou J, Chen G, Li D, Zhou N, Imran M, Li X, Qiao J, Gao X (2024) Rapid urban inundation prediction method based on numerical simulation and AI algorithm. J Hydrol. https://doi.org/10.1016/j.jhydrol.2024.132334

Pang G, Pang G (2022) Artificial intelligence for natural disaster management. IEEE Intell Syst 37:3–6. https://doi.org/10.1109/MIS.2022.3220061

Pierdicca R, Paolanti M (2022) GeoAI: a review of Artificial Intelligence approaches for the interpretation of complex Geomatics data. Geosci Instrum Methods Data Syst. https://doi.org/10.5194/gi-2021-32

Queen Z, Nwokediegwu S, Adesina A, Okwandu A (2024) Geo-information systems in urban planning: a review of business and environmental implications. Magna Sci Adv Res Rev https://doi.org/10.30574/msarr.2024.11.1.0100

Rashkovetsky D, Mauracher F, Langer M, Schmitt M (2021) Wildfire detection from multisensor satellite imagery using deep semantic segmentation. IEEE J Select Topics Appl Earth Observ Remote Sens 14:7001–7016. https://doi.org/10.1109/JSTARS.2021.3093625

Ren Y, Chen H, Zhao C, Zhang K, Zhao L, Sheng Y (2025) Smoke and fire-you only look once: a lightweight deep learning model for video smoke and flame detection in natural scenes. Fire. https://doi.org/10.3390/fire8030104

Rezvani S, Silva M, Almeida N (2024) Mapping geospatial AI flood risk in national road networks. ISPRS Int J Geo Inf. https://doi.org/10.3390/ijgi13090323

Sajjad A, Lu J, Aslam R, Ahmad M (2023) Flood disaster mapping using geospatial techniques: a case study of the 2022 pakistan floods. In: ECWS-7 2023. https://doi.org/10.3390/ecws-7-14312

Sbarra P, Tosi P, De Rubeis V, Sorrentino D (2020) Quantification of earthquake diagnostic effects to assess low macroseismic intensities. Nat Hazards 104(3):1957–1973. https://doi.org/10.1007/S11069-020-04256-6

Seitzinger P, Kalra J (2023) The role of emerging technologies in health emergency planning and preparedness. Emerg Technol Healthc Med. https://doi.org/10.54941/ahfe1004371

Shahjalal M, Rahaman T, Yahia A (2025) AI-driven sustainable infrastructure: smart materials for climate-resilient urban design. Am J Adv Technol Eng Sol. https://doi.org/10.63125/mc5qpg12

Shi J, Wang W, Gao Y, Yu N (2020) Optimal placement and intelligent smoke detection algorithm for wildfire-monitoring cameras. IEEE Access 8:72326–72339. https://doi.org/10.1109/ACCESS.2020.2987991

Singh SS, Jeganathan C (2024) Using ensemble machine learning algorithm to predict forest fire occurrence probability in Madhya Pradesh and Chhattisgarh, India. Adv Space Res 73(6):2969–2987

Sivakumar R, Ghosh S (2017) Earthquake hazard assessment through geospatial model and development of EaHaAsTo tool for visualization: an integrated geological and geoinformatics approach. Environ Earth Sci 76:1–22. https://doi.org/10.1007/s12665-017-6777-4

Sivanpillai R, Oreshkina M, Bear P, Boettcher I, Bradshaw T, Coleman I, Gifford J (2023) Mapping newly inundated areas in post-flood landsat images using thresholding techniques. Int Arch Photogramm Remote Sens Spat Inf Sci. https://doi.org/10.5194/isprs-archives-xlviii-m-3-2023-235-2023

Snyder J, Yan G, Honerkamp R (2020) A review of the characteristics of tornadic wind fields through observations and simulations. J Wind Eng Ind Aerodyn. https://doi.org/10.1016/j.jweia.2020.104195

Speth S, Gonçalves A, Rigault B, Suzuki S, Bouazizi M, Matsuo Y, Prendinger H (2022) Deep learning with RGB and thermal images onboard a drone for monitoring operations. J Field Robot 39:840–868. https://doi.org/10.1002/rob.22082

Steinkruger D, Markowski P, Young G (2020) An artificially intelligent system for the automated issuance of tornado warnings in simulated convective storms. Weather Forecast. https://doi.org/10.1175/waf-d-19-0249.1

Stojković S, Jovanović D, Peulić A (2024) Artificial intelligence in geography: pros or cons?. In: Zbornik radova – VI Kongres geografa Srbije sa medunarodnim ucešcem - zbornik radova. https://doi.org/10.5937/kongef24057s

Sufi F (2022) AI-Tornado: an AI-based Software for analyzing Tornadoes from disaster event dataset. Softw Impacts 14:100357. https://doi.org/10.1016/j.simpa.2022.100357

Sufi F, Khalil I (2024) Automated disaster monitoring from social media posts using ai-based location intelligence and sentiment analysis. IEEE Trans Comput Social Syst 11:4614–4624. https://doi.org/10.1109/TCSS.2022.3157142

Tazneen F, Rahman H, Rahman S, Sultana N, Faisal B (2021) Preliminary application of space-based remote sensing and geospatial technology for investigation on the geo-environmental consequences of cyclone Aila 2009 in the SW Bangladesh. Int J Environ Geoinform. https://doi.org/10.30897/IJEGEO.837770

Thangavel K, Spiller D, Sabatini R, Amici S, Sasidharan S, Fayek H, Marzocca P (2023) Autonomous satellite wildfire detection using hyperspectral imagery and neural networks: a case study on Australian Wildfire. Remote Sens 15:720. https://doi.org/10.3390/rs15030720

Vicente-Serrano S, Beguería S, Gimeno L, Eklundh L, Giuliani G, Weston D, Kenawy A, López-Moreno J, Nieto R, Ayenew T, Konte D, Ardö J, Pegram G (2012) Challenges for drought mitigation in Africa: the potential use of geospatial data and drought information systems. Appl Geogr 34:471–486. https://doi.org/10.1016/J.APGEOG.2012.02.001.

Vysotskyi A, Vysotskyi O (2024) Decentralizing spatial data: the convergence of Geographic Information Systems and Web 3.0 technologies. J Geol Geogr Geoecol. https://doi.org/10.15421/112377

Wallace R (1997) The geology of earthquakes. Earthq Spectra 13:559–560. https://doi.org/10.1193/1.1585967

Wang R, Hu Y, Zhou Z, Yang K (2020) Tracking flooding phase transitions and establishing a passive hotline with AI-enabled social media data. IEEE Access 8:103395–103404. https://doi.org/10.1109/ACCESS.2020.2994187

Wienhoff Z, Nevill J, Rhee D, Poole C, Lombardo F (2023) An approach for assessing misclassification of tornado characteristics using damage. J Appl Meteorol Climatol. https://doi.org/10.1175/jamc-d-22-0197.1

Wurman J, Kosiba K, White T, Robinson P (2021) Supercell tornadoes are much stronger and wider than damage-based ratings indicate. Proc Natl Acad Sci USA. https://doi.org/10.1073/pnas.2021535118

Xiao W, Zhang Z, Zhang X, Xiao S (2023) Flame and smoke detection algorithm based on ODConvBS-YOLOv5s. IEEE Access 11:34005–34014. https://doi.org/10.1109/ACCESS.2023.3263479

Xu X, Zhang H, Gao L, Zhu T (2020) Gravity erosions on the loess gully bank: avalanche, landslide, or mudslide, pp 141–149. https://doi.org/10.1007/978-981-15-3801-8_9

Xu G, Guo T (2025) Advances in AI-powered civil engineering throughout the entire lifecycle. Adv Struct Eng. https://doi.org/10.1177/13694332241307721

Young M, Ehsan U, Singh R, Tafesse E, Gilman M, Harrington C, Metcalf J (2024) Participation versus scale: tensions in the practical demands on participatory AI. First Monday. https://doi.org/10.5210/fm.v29i4.13642

Zhang Q, Wang T (2024) Deep learning for exploring landslides with remote sensing and geo-environmental data: frameworks, progress, challenges, and opportunities. Remote Sens 16:1344. https://doi.org/10.3390/rs16081344

Zhang F, Qiao Q, Wang J, Liu P (2022) Data-driven AI emergency planning in process industry. J Loss Prev Process Ind. https://doi.org/10.1016/j.jlp.2022.104740

GIS-Based Landslide Susceptibility Assessment for Coonoor Taluk, Nilgiris District South India Tamil Nadu Using Machine Learning Algorithm

D. Padmanathan, R. Jegankumar, K. Prakash, and D. Shankar Raja

1 Introduction

Landslides are a natural hazard that primarily affects mountainous regions, posing risks to both property and human life (Panchal and Shrivastava 2022). Landslides can be attributed to a combination of natural and human-induced factors (Bera et al. 2019). Landslide is one of most destructive natural hazards in many countries (Alimohammadlou et al. 2013); with the use of scientific methods, it can be predicted and necessary measures can be implemented to prevent them. Landslides can cause a simple top soil loss to a mass destruction of natural environment including lives. Sometimes, they can change the landscape permanently such as formation of dams post the event of landslide (Highland and Bobrowsky 2008; Eskandarinejad et al. 2025). As of now, numerous landslide classification systems have been developed; among them, the widely accredited classification is done by DJ Varnes. Identification of landslide susceptible areas is vital to safeguard the natural and biological environment in hilly areas (Othman et al. 2012). Landslide Susceptibility Mapping (LSM) using geospatial tools is an efficient method to recognize the susceptible locations (Pourghasemi et al. 2013). Landslides occur due to a combination of factors, such as steep slopes, irregular terrain, deforestation or lack of vegetation, weak geological formations, fractured rock structures, frequent earthquakes, and intense, prolonged rainfall (Anbalagan 1992).

Landslide frequency and magnitude of landslides in recent years have become a major concern across the globe, particularly with undulated landscapes. Understanding the due loss resulted from landslides require susceptibility assessment so

D. Padmanathan · R. Jegankumar (✉) · K. Prakash
Department of Geography, Bharathidasan University, Tiruchirappalli, Tamil Nadu, India
e-mail: jegankumar@bdu.ac.in

D. Shankar Raja
Tata Consultancy Services, San Francisco, USA

M. O. Sarif and A. Sharifi (eds.), *Advanced Geospatial Intelligence and AI for Environmental Resilience and Sustainable Development*, Advances in Geographic Information Science, https://doi.org/10.1007/978-3-032-03714-5_15

that preventive measures can be adopted before the occurrence of landslide (Jaiswal et al. 2010). Landslide susceptibility mapping draws more attention for two reasons: first, increase in the population density and urbanization; second, changes in the climatic conditions such as increased rainfall and temperature variations, contribute to higher risk of landslides (Aleotti and Chowdhury 1999). Although various methods and techniques have been proposed and implemented, no consensus has been reached on the most effective approach for mapping landslide susceptibility (Tyagi et al. 2022). In India, landslides occur in both the Himalayan Mountains and the Western Ghats, particularly along the steep slopes overlooking the Konkan coast. They are also common in the Nilgiri Range, which has a lateritic cap highly susceptible to mass movements. In the Western Ghats, landslides are often triggered by flash floods and heavy rainfall (Prakash 2011). Rainfall in these regions is frequently accompanied by cloudbursts, which last from a few minutes to several hours and leave behind widespread destruction. Over 12% of India's land area is prone to landslides, resulting in approximately 300 fatalities each year.

The Nilgiris district in the Western Ghats of Tamil Nadu has a long history of landslides, which occur almost seasonally, with October and November being particularly prone to such events. Numerous studies have focused on landslide susceptibility mapping using probabilistic methods (Dahal et al. 2008). Several researchers have employed logistic regression for this purpose (Pradhan et al. 2009); while others have applied geotechnical and safety-factor models in different regions (Pradhan et al. 2010). Additionally, modern statistical approaches, including data mining, fuzzy logic, random forest, and artificial neural networks, have been utilized in various case studies.

Landslide susceptibility analysis using data mining, particularly Random Forest models, has become a crucial area of research in recent years (Zhou et al. 2021). To implement the Random Forest approach, a spatial database incorporating various landslide-related factors was designed and developed. The fundamental first step in any study is to construct a dataset tailored to the specific study area. In this research, based on surveys and prior studies, nine key criteria were selected: precipitation, slope, soil texture, aspect, lineament, land use and land cover, elevation, drainage, and density. To generate and analyze data layers, GEE and ArcGIS were used to process all factors in raster format. Essential data were gathered from organizations responsible for natural resource studies and management, then adapted and integrated for optimal results. The study's findings can assist managers and decision-makers in mitigating landslide risks through strategies such as land-use modifications or restricting construction in high-risk areas.

The main aim of the study is to perform a GIS-based landslide susceptibility assessment using random forest machine learning algorithm in Coonoor taluk, Nilgiris districts. The objectives of the study are to

i. Analyse various factors for landslide susceptibility mapping and prepare necessary thematic layers.
ii. Develop a Random Forest machine learning model using selected variables and historical landslide data for landslide susceptibility assessment.

iii. Validate landslide susceptibility using Area under the Curve (AUC) and Receiver Operating Characteristic (ROC).

Achu et al. (2023) emphasized the importance of landslide susceptibility maps for decision-making in disaster management. Their study introduced a novel methodological framework that quantifies uncertainty in landslide prediction using ensembles of eight MLTs. Conducted in the Western Ghats region of Kerala, India, the study analyzed fourteen landslide conditioning factors correlated with 671 historical landslides. Among four ensemble models tested, the weighted mean of probabilities exhibited the highest predictive capability, validated using true skill statistics and area under the curve (AUC) scores. The coefficient of variation was employed for uncertainty analysis, resulting in a confident susceptibility map, wherein approximately 74% of past landslides fell within high-susceptibility, low-uncertainty zones.

Nawazuzzoha et al. (2025) highlighted the growing frequency of landslides due to urbanization and climate change, stressing the need for advanced ML algorithms in LSM. Techniques such as Random Forest (RF), Artificial Neural Networks (ANN), and Support Vector Machines (SVM) demonstrated high predictive accuracy. The study underscored the challenges of compiling comprehensive landslide inventories and proposed ensemble techniques to enhance model robustness. While predicting the temporal and spatial aspects of landslides remains challenging, ML offers promising solutions for landslide prevention. Ballabio et al. (2012) applied SVM for LSM, ensuring model robustness through success and prediction rate curves and receiver operating characteristic (ROC) analysis. Conducted in the Staffora River Basin, Northern Italy, the study found that SVM outperformed logistic regression, linear discriminant analysis, and the Naïve Bayes classifier, exhibiting superior generalization capacity with an 18% higher predictive accuracy.

He et al. (2023) conducted an LSM study in Xining City, integrating GF-2 remote sensing and field survey data. Using frequency ratio (FR), RF, SVM, and ANN models, they mapped landslide susceptibility across slope units. The FR-RF model yielded the highest prediction accuracy, covering 15.48% of the total study area as high susceptibility zones, followed by FR-ANN and FR-SVM models. Verification results confirmed strong predictive performance, establishing FR-RF as the most suitable approach for LSM in the region. Kadavi et al. (2018) developed ensemble-based ML models for LSM in Sacheon-myeon, South Korea. The study utilized AdaBoost, LogitBoost, Multiclass Classifier, and Bagging models to analyze 762 landslides and 20 conditioning factors. The Multiclass Classifier outperformed other models with an AUC of 85.9%, demonstrating high predictive accuracy.

Kalantar et al. (2017) examined the impact of randomly selecting training landslide data on SVM, logistic regression (LR), and ANN models in the Dodangeh watershed, Iran. Their findings showed that random training data selection significantly affected model accuracy, with SVM and LR outperforming ANN, achieving average accuracies of 81.42% and 79.82%, respectively. Kamran et al. (2021) explored GIS-based LSM using different SVM kernel functions in the Tabriz Basin, Iran.

The polynomial-degree-2 (PKF-2) kernel achieved the highest AUC (0.9688), highlighting its superior predictive capability. This model indicated high susceptibility in the northern parts of the region, emphasizing its utility in disaster mitigation planning.

Tran et al. (2024) employed RF, Gradient Boosting (GB), and SVM for LSM in Van Yen District, Vietnam. Utilizing Google Earth Engine, their study analyzed 13 conditioning factors and validated models using AUC scores. The RF model demonstrated the highest accuracy (AUC = 0.883), reinforcing its effectiveness in LSM applications. Lee et al. (2017) applied SVM integrated with GIS for LSM in PyeongChang and Inje, Korea. The study validated susceptibility maps using AUC analysis, achieving 81.36% and 77.49% accuracy for the two study sites, respectively. Sensitivity analysis confirmed the positive effects of topographic, geological, and soil factors in LSM.

Ramzan et al. (2025) reviewed LSM techniques based on ML, categorizing them as regression and classification approaches. Their work summarized key advancements in LSM and recommended future directions for improving model scalability, performance, and interpretability. Pham et al. (2019) introduced a novel hybrid intelligent model (MBSVM) integrating Multi-Boost Ensemble and SVM for LSM in Uttarakhand, India. With an AUC of 0.972 for training and 0.966 for testing, MBSVM outperformed logistic regression, single SVM, and AdaBoost-SVM hybrid models, demonstrating its potential as a reliable landslide hazard assessment tool. Saha et al. (2021) developed a deep learning-based framework for landslide vulnerability mapping (LVM) in Chukha Dzongkhag, Bhutan. The study incorporated physical and social factors, producing nine LVMs using deep learning neural networks (DLNN), ANN, and convolution neural networks (CNN). CNN-RLV exhibited the highest performance (AUC = 0.928), emphasizing the importance of integrating social and physical factors for accurate LVM. Shahabi et al. (2023) reviewed the application of SVM in LSM, comparing it with analytic hierarchy process, LR, ANN, and RF models. Their findings indicated that combining SVM with other techniques could enhance prediction accuracy, with high-quality landslide inventory data being crucial for improved model performance.

Hong (2023) compared SVM, decision tree (DT), and Naïve Bayes (NB) models for LSM in Hoa Binh, Vietnam. Their results confirmed that SVM achieved the highest prediction capability, surpassing DT and NB models. Ullah et al. (2022) applied linear regression (LiR), logistic regression (LoR), and SVM to LSM in Abbottabad, Pakistan. LiR achieved the highest AUC (88%), highlighting its reliability for susceptibility mapping. Xie et al. (2021) proposed a Bayesian optimization-SVM (BO-SVM) model for LSM in Nanping City, achieving superior accuracy (AUC = 0.97) compared to standard SVM (AUC = 0.93). Their study demonstrated the effectiveness of hyperparameter tuning in improving model performance.

These studies collectively underscore the significant role of ML techniques in LSM, highlighting ensemble approaches, deep learning, and advanced optimization strategies as key areas for future research.

2 Significance of the Study

Landslides are a major concern in hilly regions like the Nilgiris, posing risks to human life, infrastructure, and the environment. This study focuses on developing an accurate landslide susceptibility map (LSM) for Coonoor Taluk, a landslide-prone area due to steep slopes, heavy rainfall, and human activities. Using Geographic Information Systems (GIS) and Random Forest (RF) machine learning algorithms, the study integrates hydrological and land-use factors to assess susceptibility. RF improves traditional methods by addressing uncertainties in spatial data and expert judgments, leading to more reliable predictions. GIS enables the spatial visualization of risk zones, aiding planners and policymakers in disaster management.

Recent and historical landslides, such as those in November 2023 that caused severe damage to Coonoor Ghat roads and railways, highlight the urgency of effective risk assessment. Past events, including the 1926 railway disruptions near Katteri station, further emphasize the region's vulnerability. The LSM serves as a crucial tool for identifying high-risk areas and implementing targeted mitigation strategies, including slope stabilization, improved drainage, and land-use planning. It also supports the development of early warning systems and evacuation plans to enhance community resilience. Beyond immediate applications, this research contributes to the broader academic study of landslide susceptibility by demonstrating the effectiveness of RF techniques combined with GIS. The methodological framework can be adapted for other regions with similar environmental conditions, making it a valuable reference for future research. As urban expansion continues into hilly terrains, understanding landslide risks is essential for sustainable development, ensuring safer infrastructure, and reducing long-term disaster costs.

3 Study Area

Coonoor Taluk is situated in the Nilgiris district of Tamil Nadu State. It lies between 12°38′52.4″ N and 12°52′7″ N latitude, and 77°44′11″ E and 77°55′11″ E longitude; flanked by Doddabetta ranges on the south, Hulikal ravine on the southwest, and Kothagiri ridges on the northeast (Fig. 1). The study area covers a total area of about 230.44 square kilometres and is situated at an altitude of 1850 m above mean sea level (MSL). The population of Coonoor taluk was 157,754 as on 2011 census; composed of 78,411 males and 79,343 females. Coonoor is well-connected by road and rail networks through Mettuppalayam, a town at the foothill.

Due to its elevation, the region falls within a tropical zone and experiences a subtropical highland climate. The annual average temperature is around 17 °C, with the highest temperatures recorded in May and the lowest in January. The area is home to indigenous tribes such as the Badugas, Irulas, Kotas, Kurumbas, and Todas, who make up approximately 1.5% of the Taluk's population. Coonoor Taluk has a literacy rate of nearly 88%, and its economy primarily relies on tourism and the

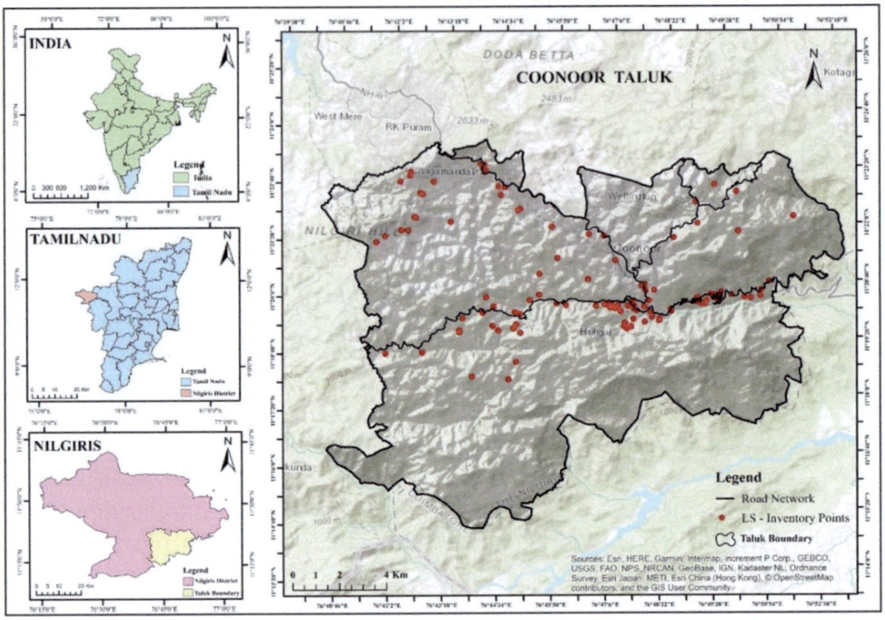

Fig. 1 Location and study area map

tea industry. Major plantation crops include tea, coffee, potatoes, carrots, cabbage, beetroot, garlic, and other vegetables. The region receives most of its rainfall during the southwest monsoon (June to September), with an annual average ranging between 1400 and 2000 mm. According to the landslide inventory reports, occurrences of landslide in the district have been increased during the past three decades. Although there are no severe damages to properties and lives in the district, it has caused a significant damage to infrastructures like road, rail, electricity, telecommunication services, inflicting a cyclic impact on the available services in the area.

Coonoor Taluk features diverse soil types, including Alfisols, Entisols, Inceptisols, and unsurveyed forest areas (Fig. 2). Alfisols, known for their fertility, support various crops and are widely distributed. Entisols, being less developed, are mainly used for grazing or low-intensity farming. Inceptisols, which are moderately developed, accommodate a range of agricultural activities. Unsurveyed forest areas remain unexplored due to dense vegetation, with green zones representing reserved forests crucial for conservation and erosion control. This soil diversity plays a key role in sustaining Coonoor's agriculture and ecology. The taluk's well-connected transport network includes railways linking key locations and major roads like Ooty-Coonoor and CoonoorMettupalayam roads. Village roads support local travel and agriculture. This infrastructure boosts mobility, tourism, and economic growth (Fig. 3).

Fig. 2 Soil types

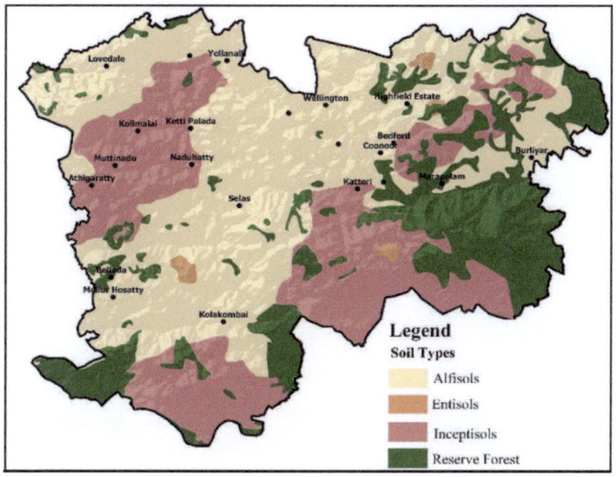

Fig. 3 Transport networks

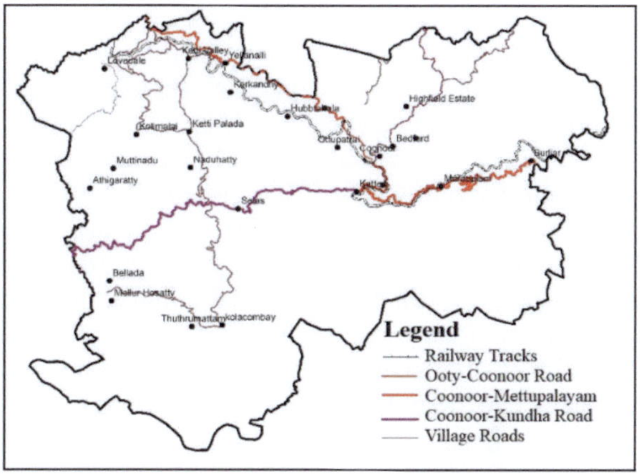

4 Database and Methodology

Landslide inventory reports containing the past landslide occurrences is a crucial data for obtaining influencing factors that induce landslide. This map is then subjected to a random partitioning process, where 70% of the data is reserved for training the model, and the remaining 30% is set aside for testing (Fig. 4). Alongside this, a comprehensive set of geo-environmental factors known to influence landslides is compiled. These factors include rainfall, slope, drainage density, lineament density, soil texture, land use/land cover (LULC), aspect, and elevation. The training dataset, along with these geo-environmental factors, is integrated into Google Earth Engine, a cloud-based platform that facilitates large-scale geospatial analysis. Random Forest

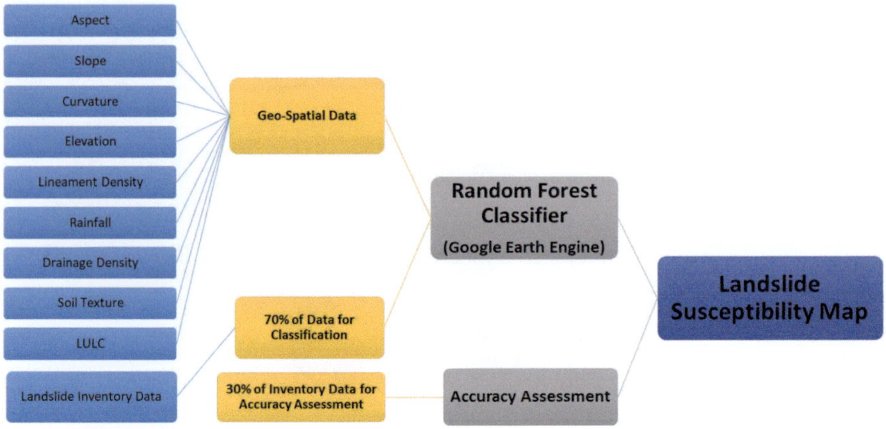

Fig. 4 Methodology framework

(RF) Algorithm is employed to analyse the data. The algorithm, known for its robustness and accuracy, uses multiple decision trees to predict the likelihood of landslides occurring in different areas.

The output of this analysis is a Landslide Susceptibility Map, which highlights areas with varying levels of landslide risk. To ensure the reliability and accuracy of this result, Area under the Curve (AUC) and Receiver Operating Characteristic (ROC) Curve are implemented. They provide a quantitative assessment of the model's performance, indicating how well the model distinguishes between areas prone to landslides and those that are not. This methodology combines historical landslide data with environmental factors, computational power of Google Earth Engine and the predictive capability of Random Forest Algorithm.

5 Results and Discussion

The occurrence of landslides is influenced by a combination of topographic, hydrological, and geological factors (Costanzo et al. 2012). Thus, selecting the appropriate causative factors is a crucial step in developing a landslide susceptibility map.

5.1 Rainfall

Rainfall is a key trigger for landslides. As a natural process, it can lead to devastating loss of life and severe property damage, particularly in mountainous and densely populated regions. Rainfall infiltrates the pore spaces and fractures within rock strata, influencing landslide frequency (Guo et al. 2024). In the study area, mean average

Table 1 Mean annual rainfall

	Rainfall range	Area (sq.km)	Area (%)
1	Very low (<977 mm)	22.83	9.92
2	Low (1000–1016 mm)	36.75	15.96
3	Moderate (1016–1035 mm)	44.48	19.34
4	High (1035–1056 mm)	60.85	26.45
5	Very high (> 1056 mm)	65.53	28.33

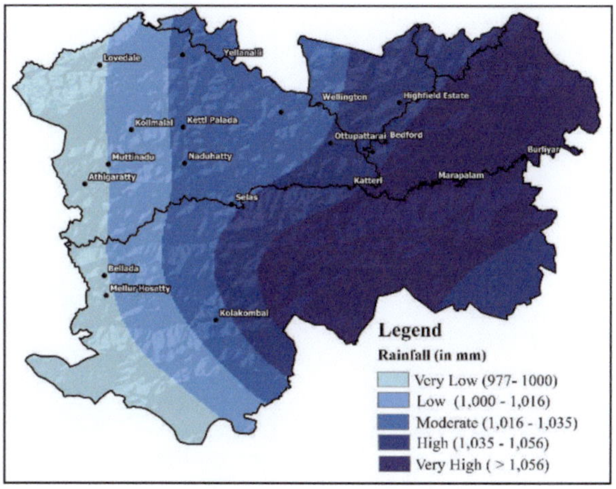

Fig. 5 Mean annual rainfall

rainfall ranges between 977 and 1073 mm. The spatial distribution of rainfall (Table 1, Fig. 5) shows that the very high rainfall class starts from the eastern portion of this study area and it gradually decreased along the western portions with moderate, low, and very low in the northern-western part of this study area.

5.2 *Drainage Density*

In mountainous areas, drainage density serves as an indirect indicator of groundwater conditions, which play a crucial role in landslide activity (Chingkhei et al. 2013). High drainage density tends to have a more fragmented and eroded landscape, leading to higher runoff and erosion rates (Table 2, Fig. 6). Drainage density is sparsely distributed in the Coonoor taluk region. A very high drainage density value of a very high drainage density value of 8.39 is considered to have a lower infiltration potential zone, and a low drainage density value of 36.11% is considered to have a high infiltration potential zone present in this study area.

Areas with very low drainage density, such as Kolakombai and parts of Mellur Hosatty, are less likely to experience water accumulation and thus have a lower

Table 2 Drainage density

Fid	Classes	Area (sq.km)	Area (%)
1	Very low (< 0.90)	83.05	36.11
2	Low (0.90–1.80)	90.29	39.24
3	Moderate (1.80–2.70)	48.2	20.8
4	High (2.70–3.60)	8.39	3.63
5	Very high (> 3.60)	0.51	0.22

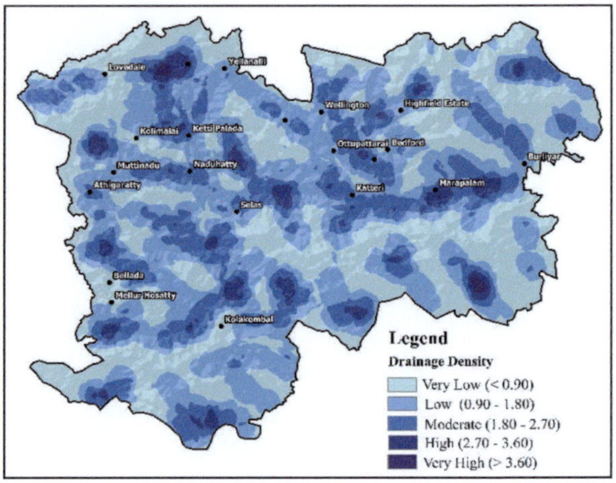

Fig. 6 Drainage density

risk of landslides. Low drainage density regions, including Athigaratty and Bellada, also exhibit reduced susceptibility to landslides. Moderate drainage density zones like Nadhuhatty and Wellington, have a balanced risk of water accumulation and landslide occurrence. High drainage density areas, such as Ketti Palada and Bedford, are highly prone to landslide due to higher water flow and soil saturation. The very high drainage density regions, like Marapalam and Burliyar, are at the greatest risk of landslides, as these areas are likely to experience significant water accumulation, increasing the likelihood of soil instability and landslide.

5.3 *Lineament Density*

Lineaments, representing geological features such as fractures, rock cleavages, and faults, are significant in building pore pressure, further contributing to landslide occurrences (Saha et al. 2023). Based on the lineament density values, the study area has been divided into five zones. Very low, Low, Moderate, High and Very High Lineament densities are high in the western part of the Coonoor taluk (Table 3, Fig. 7). Very low and low class covers an area 82.4% and high class covers 1.70%

Table 3 Lineament density

Fid	Classes	Area (sq.km)	Area (%)
1	Very low (< 0.39)	158.41	68.8
2	Low (0.39–0.78)	31.47	13.6
3	Moderate (0.78–1.18)	35.1	15.25
4	High (1.18–1.57)	3.95	1.7
5	Very high (> 1.57)	1.51	0.65

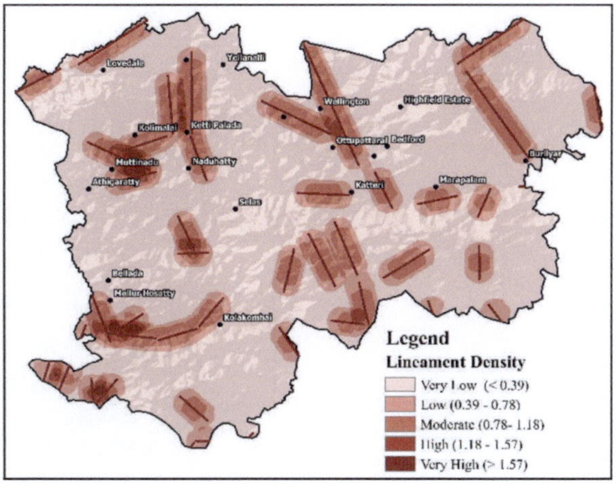

Fig. 7 Lineament density

of the study region. Northern part of the study area showcases significant geological features, indicating fault lines and potential zones of structural weakness. Notable lineaments can also be observed near the areas of Hulical and Kattery, providing valuable insights into the region's geotectonic landscape.

5.4 *Elevation*

Elevation directly impacts weathering rates, vegetation cover, and rainfall intensity, all of which govern landslide dynamics (Sato and Shuin 2025). The study area composes diverse elevation profile that significantly influences its climate, vegetation, and land use. Elevation in this region ranges from 418 m to 2406 m above mean sea level (Table 4, Fig. 8). This variation creates a unique environment where tea plantations flourish, and diverse flora and fauna thrive. Higher altitudes experience cooler climate and receive more rainfall, making them ideal for growing high-quality tea plants. The rugged terrain and steep slopes, make certain areas prone to landslides, necessitating careful land management.

Table 4 Elevation

1BFid	2BClasses	3BArea (sq.km)	4BArea (%)
1	418–979 m	7B18.78	8B8.15
2	979–1314 m	11B24.42	12B10.58
3	1314–1595 m	15B46.06	16B19.96
B4	1595–1829 m	19B65.59	20B28.55
5	1829–2406 m	23B59.07	24B25.6
6	B > 2406 m	27B16.52	28B7.16

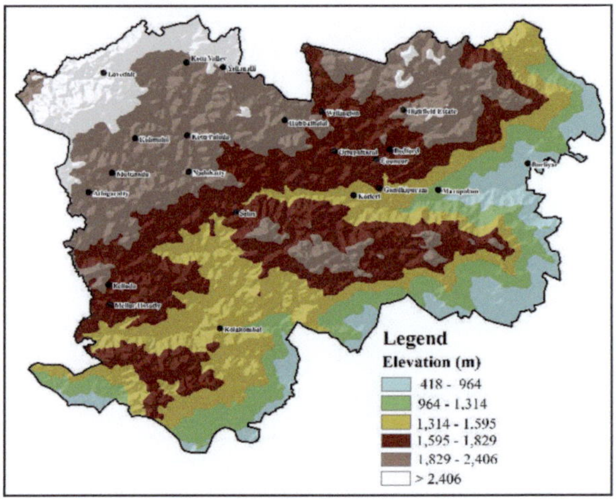

Fig. 8 Elevation

5.5 *Aspect*

Aspect, or the direction a slope faces, is often linked to landslide. It provides insights into bedrock structure and influences sun exposure, which affects vegetation and evapotranspiration (Budianta et al., 2022). Aspect highlights various locations within Coonoor Taluk, such as Wellington, Highfield Estate, and Kotagiri, among others (Table 5, Fig. 9). Aspect is crucial for understanding the slope direction, which has implications for various geographic and environmental studies, including landslide susceptibility mapping.

5.6 *Slope*

Slope gradient is another critical factor in landslide susceptibility, as steeper slopes have a higher probability of failure and mass movement (Lee and Min 2001). For this study slope also chosen as an important parameter and slope has been classified into eight categories namely Flat to Nearly Flat (0–8), Very Gentle (8–15), Gentle

Table 5 Aspect (in degrees)

Fid	Classes	Area (sq.km)	Area (%)
1	North (337.5–22.5)	22.4	9.7
2	North East (22.5–67.5)	27	11.7
3	East (67.5–112.5)	29.7	12.9
4	South East (112.5–157.5)	39.8	17.3
5	South (157.5–202.5)	47.5	20.6
6	South West (202.5–247.5)	32	13.9
7	West (247.5–292.5)	16.3	7.1
8	North West (292.5–337.5)	15.5	6.7

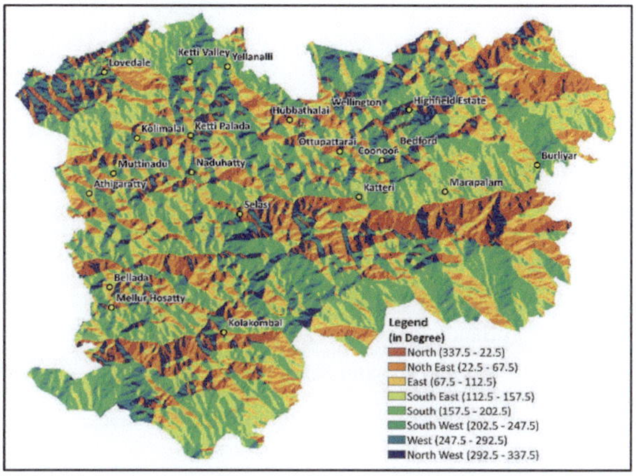

Fig. 9 Aspect

Slope (15–21), Moderately Slope (21–28), Moderately Steep Slope (28–36), Steep Slope (36–46) and Very Steep Slope (> 46) (Table 6, Fig. 10). The Coonoor taluk has 13.95% regions comes under Level to nearly level and very-gentle slope dominate the study area. Moderate steep slope and Steep slope categories are observed in the South and Eastern part of this study area, which covers area of 18.17%.

Table 6 Slope (in percent)

Fid	Classes	Area (sq.km)	Area (%)
1	Flat to nearly level (0–8)	32.13	13.95
2	Very gentle slope (8–15)	55.31	24.01
3	Gentle slope (15–21)	54.8	23.79
4	Moderate slope (21–28)	B41.75	18.12
5	Moderate steep slope (28–36)	27.15	11.78
6	Steep slope (36–46)	14.71	B6.39
7	Very steep slope (> 46)	4.55	1.96

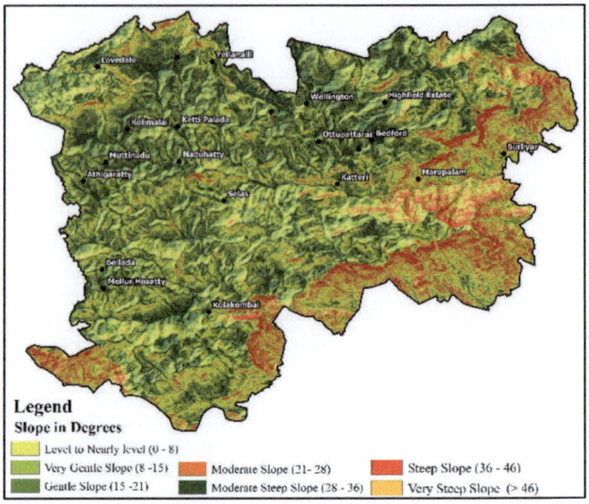

Fig. 10 Slope

Table 7 Soil texture

Fid	Classes	Area (sq.km)	Area (%)
1	Sandy clay loam with sandy clay	153.55	66.71
2	Sand clay loam	45.06	19.59
3	Sandy loam	8.5	3.69
4	Rock with sandy clay loam	23.33	10.01

5.7 *Soil Texture*

Soil texture regulates water movement through penetration, infiltration rates, and water retention capacity. It also affects soil's physical and chemical processes, with clay-rich soils having larger surface areas compared to those dominated by sand and silt (Budianta et al. 2022; Nsengiyumva and Valentino, 2020). Soil texture affects the level of soil permeability and also the level of vulnerability to landslides in the study area. Soil texture is one of the triggering factors of landslide, in the study area soil texture is categorised into four classes such as, sandy clay loam with sandy clay, sandy clay loam, sandy loam and rock with sandy clay loam (Table 7, Fig. 11). Sandy clay loam with sandy clay covers the maximum area of 153.55 sq.km (66.71%).

5.8 *Land Use and Land Cover*

Land use and land cover (LULC) reflect both natural and human-induced changes that directly or indirectly influence slope stability (Quevedo et al. 2023). Landslides are

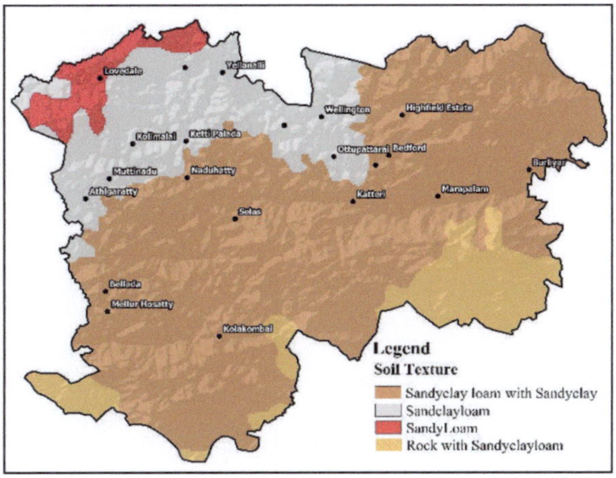

Fig. 11 Soil texture

less likely in areas with vegetation because plant roots provide mechanical support, which enhances the shearing strength of soil (Chen et al., 2019). It is also observed that landslide frequency is substantially lower in areas covered by forests and shrubs. According to Abedin et al. 2020, built-up areas are one of the most landslide-prone land use types. This is due to the modifications made to mountainous terrain for various development activities. LULC is a crucial factor in landslide occurrences, especially in hilly areas. In the study area, forests cover the majority of the land, while agricultural fields and built-up areas with well-connected roads are primarily found in the northern region (Table 8).

Human interference is high in the northern part, but it is less in the southern part due to its thick vegetation. In Coonoor taluk, water bodies cover an area less than 0.20 sq.km, which is dominantly found in the northern part. Vegetation is the dominant class, which covers an area of 208.23 sq.km (Fig. 12). Agriculture class covers an area of 6.45 Sq.km, which is partially distributed in the study area. The fallow land is identified in the western and eastern part of the study area, which covers 7.53 sq.km of area. The Settlement class is observed in the eastern and central part, which covers an area of 3.39%.

Table 8 Land use and land cover

Fid	Classes	Area (sq.km)	Area (%)
1	Waterbody	0.2	0.08
2	Settlement	8.03	3.39
3	Forest	124.2	54
4	Fallow land	7.53	3.36
5	Dense forest	84.03	36.37
6	Agriculture land	6.45	2.8

Fig. 12 LULC

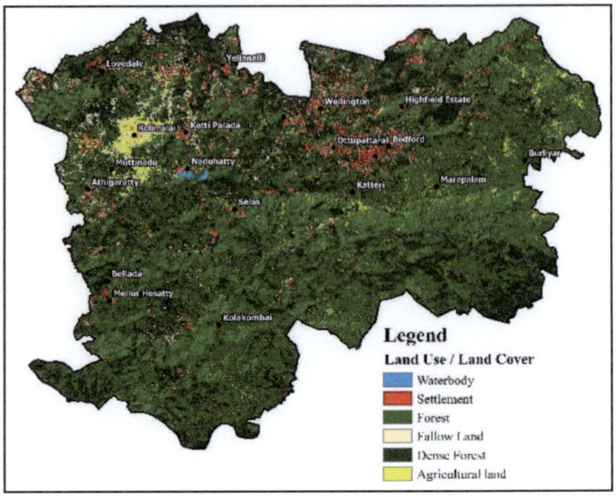

5.9 Random Forest (RF) for Landslide Susceptibility

Random Forest is a machine learning model that is able to generate a large number of decision trees to interpret the spatial relationships in landslide. Random forest produces numerous decision trees during the training time, and generate classes based on the classification or regression individual trees. In a regression algorithm, the dependent variable is estimated using the average of results, whereas in a classification algorithm, decision trees are constructed to determine the class. The Random Forest model does not require any prior assumptions to explain the relationships between dependent and independent variables, making it an effective tool for analysing hierarchical interactions and nonlinear patterns in large datasets. One of its key advantages is its ability to rank factors by importance while efficiently handling high-dimensional datasets.

5.10 Landslide Susceptibility Analysis

Landslide susceptibility zones identified and mapped into five classes (Table 9, Fig. 13). Moderate class occupies the largest area, covering 53.45 sq.km or 23.21% of the total area. High class covers 41.54 sq.km, making up 18.23% of the area, while very high class covers the smallest area of 36.03 sq.km (15.43%). Compared to moderate class, both high and very high classes cover a smaller area, indicating that regions with higher susceptibility to landslides are less extensive. The distribution shows a gradual decrease in area as susceptibility increases from moderate to very high.

Villages with very low landslide susceptibility include Kolakombai and parts of Mellur Hosatty, marked in green. Low susceptibility zones, marked in light green,

Table 9 Landslide susceptibility

Fid	Classes	Area (sq.km)	Area (%)
1	Very low	48.52	21.05
2	Low	50.9	22.08
3	Moderate	53.45	23.21
4	High	41.54	18.23
5	Very high	36.03	15.43

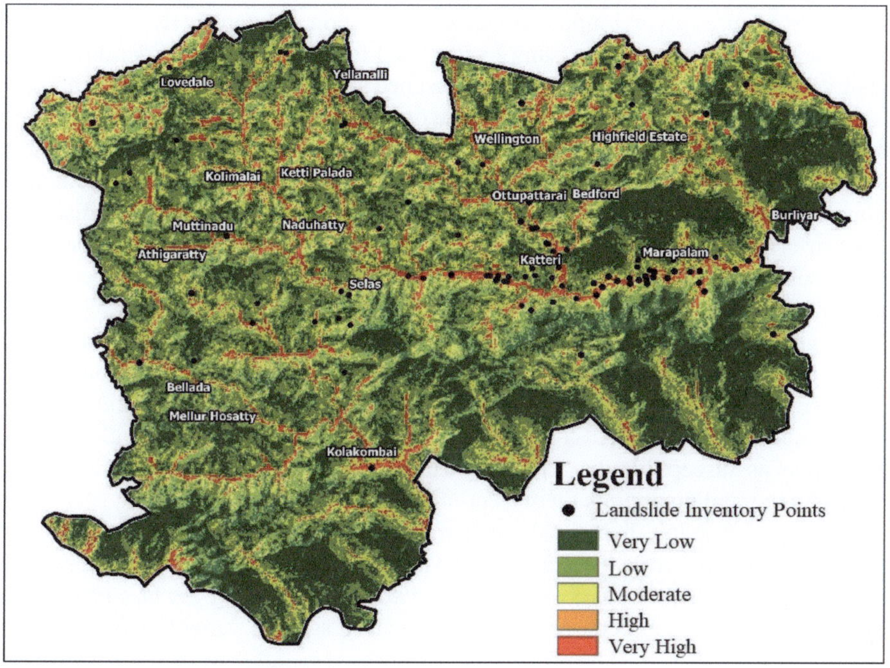

Fig. 13 Landslide susceptibility

encompass Athigaratty and Bellada. Moderate susceptibility areas, marked in yellow, include Nadhuhatty and Wellington. High susceptibility zones, marked in orange, are Ketti Palada and Bedford. Areas with very high susceptibility, marked in red, are predominantly around Yellanalli, Marapalam, and Burliyar. The map also shows landslide inventory points as black dots, indicating areas where landslides have previously occurred, and the road network within the taluk. The susceptibility map highlights the varying degrees of risk across the region, essential for planning and disaster management in Coonoor Taluk.

The landslide susceptibility map was overlaid by the inventory data points of landslides prepared by Geological Survey of India (GSI) to validate the accuracy of the assessment (Fig. 13). It is observed that more than 90% the inventory data overlaps with the susceptibility map produced. However, there are few patches marked highly susceptible for landslide have no intersection with inventory data. This states that these patches must be studied carefully to explain the factors that resist the favourable conditions of landslide.

5.11 Statistical Evaluation of Training Data and Results

Beyond generating classification results, it is essential to statistically evaluate them. This stage requires several inputs, including the region of interest (ROI), study area delimitation, training samples, sample sets used for training and validation, occurrence samples (used in training), non-occurrence samples (used in training), and the landslide susceptibility surface, which is the final product of classification. Analysing the characteristics of the events used to develop the model is crucial, as the probability surface reflects conditions specific to past occurrences. This allows for predicting future events under similar conditions. Additionally, while there is a known disparity in the sizes of occurrence and non-occurrence classes within the study area, the accuracy assessment maintains an equal proportion of occurrence and non-occurrence samples (Fig. 14).

A higher number of non-occurrence samples placed in areas previously identified as not susceptible to landslides could misleadingly inflate accuracy. Conversely, allocating non-occurrence samples in landslide-prone areas would artificially lower the area under the curve (AUC) value. This occurs because the model would classify these areas as susceptible due to their attributes being similar to occurrence areas, yet random sampling assigns them as non-occurrence.

The box-plot graph displays the minimum and maximum observed values, the median, the first quartile (which separates the lowest 25% of the samples), and the third quartile (which marks the upper 25%). By comparing the box-plot graph for occurrence and non-occurrence samples with a specific variable from the data

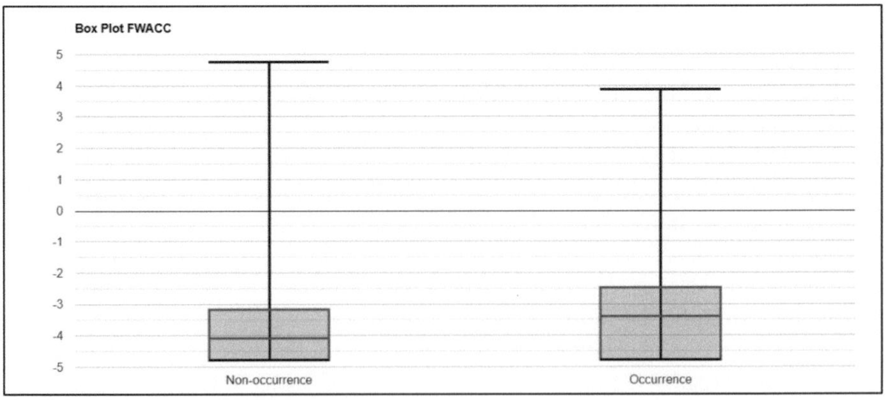

Fig. 14 Box-plot graph

cube, we can assess whether the variable effectively distinguishes between the two sample groups. This analysis is crucial for evaluating the dataset used to train the model, allowing for the refinement of explanatory variables to enhance accuracy. The evaluation of the training samples received from the variables contained in the data cube, and evaluation of the accuracy assessment samples received from the landslide susceptibility surface, and evaluation of the landslide susceptibility surface for the study area are important for the delineation of landslide scars. A histogram evaluates the frequency distribution of training sample values for variables in the data cube. The x-axis represents variable values, while the y-axis shows sample frequency (Figs. 15 and 16). This method helps assess how well variables differentiate samples, ensuring the dataset is effective for the research objective while identifying potential errors or outliers. Another evaluation approach examines how samples align with the probability surface—ideally, occurrence samples fall in high susceptibility zones, while non-occurrence samples remain in lowrisk areas.

Landslide models often mix sample sets due to random sampling. Besides assessing training accuracy, this statistical evaluation highlights model strengths and weaknesses. Here, the x-axis represents landslide susceptibility (%) and the y-axis shows the number of pixels per susceptibility range (Fig. 17). In this study, all four approaches were applied. After processing in GEE, data was downloaded and analysed using Excel for customization and visualization.

The provided graph illustrates the distribution of pixels based on landslide susceptibility rates within a specific area. The horizontal axis represents the landslide susceptibility rate, ranging from 0.0 to 1.0, while the vertical axis indicates the number of pixels, with a maximum count of 9 pixels. From the graph, it is evident that the landslide susceptibility rates are primarily concentrated between 0.0 and 0.5. There is a noticeable peak at a susceptibility rate of around 0.4, with 8 pixels falling into this category. This indicates that a significant portion of the area analysed exhibits a moderate landslide susceptibility. In contrast, very few pixels are observed

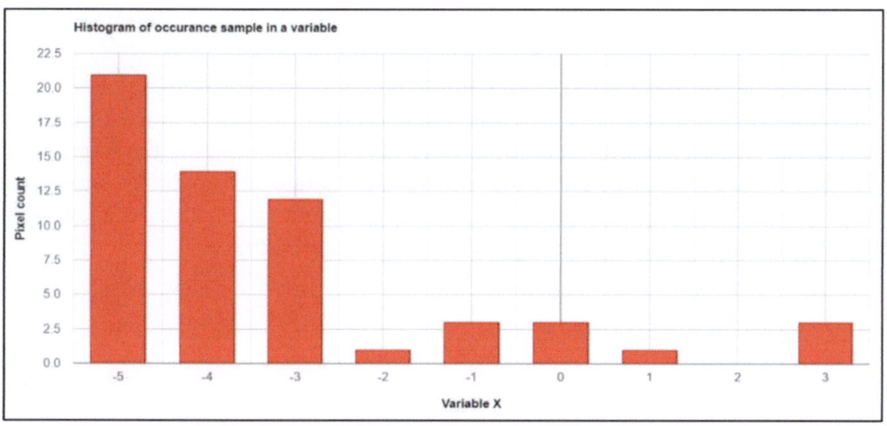

Fig. 15 Histogram of occurrence sample in a variable

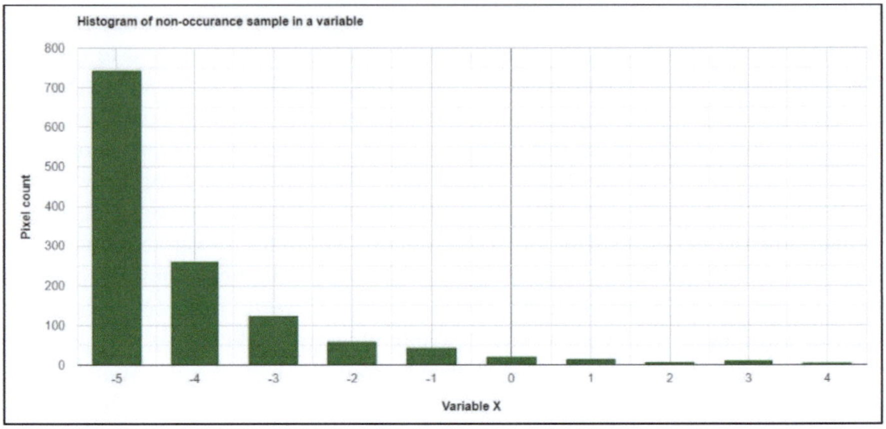

Fig. 16 Histogram of non-occurrence sample in a variable

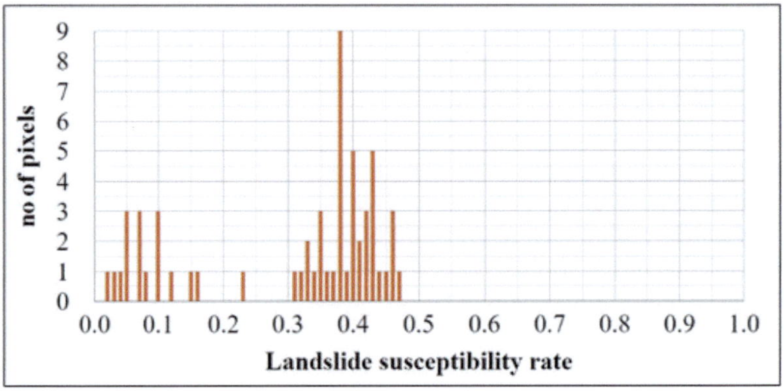

Fig. 17 Landslide susceptibility rate

beyond the 0.5 susceptibility rate, suggesting that high-risk zones are relatively rare within the studied region. Overall, this distribution highlights the varying degrees of landslide risk across different segments of the area, emphasizing the importance of targeted mitigation strategies for zones with moderate susceptibility rates. This analysis provides crucial insights for land use planning and disaster management efforts aimed at minimizing the impact of potential landslides.

6 Evaluation of Model Training Accuracy

Accuracy evaluation is a crucial step in model training, measuring the classifier's ability to differentiate samples and the information gained through modelling. A sample set (30% of total samples) assesses training accuracy and final susceptibility mapping. The ROC curve analyses true positive (TP) and false positive (FP) rates for each probability level. TPR is calculated as TP/(TP + FN), while FPR is FP/(FP + TP). The area under the curve (AUC) simplifies ROC analysis by averaging true positive and false negative rates across all thresholds.

An AUC of 1 indicates a perfect fit, while 0.5 represents randomness, and values below 0.5 indicate unrealistic models. Landslides are often concentrated in small areas, sometimes covering less than 1% of the study region (Fig. 18). When training uses occurrence samples from a single event, even with 30% reserved for accuracy assessment, spatial and temporal correlation can inflate the model's fit, leading to misleading interpretations. To predict future landslides, independent samples from another event should be used. The ROC curve method was applied to assess training performance, with the red curve representing the model's performance. The AUC of 0.859 indicates good predictive ability, distinguishing between susceptible and non-susceptible areas. The red curve above the diagonal dashed line confirms better-than-random performance. The shaded area around the curve represents the confidence interval, estimating model uncertainty. Overall, the ROC curve and AUC value suggest that the landslide susceptibility model for Coonoor Taluk is effective for risk assessment and mitigation planning.

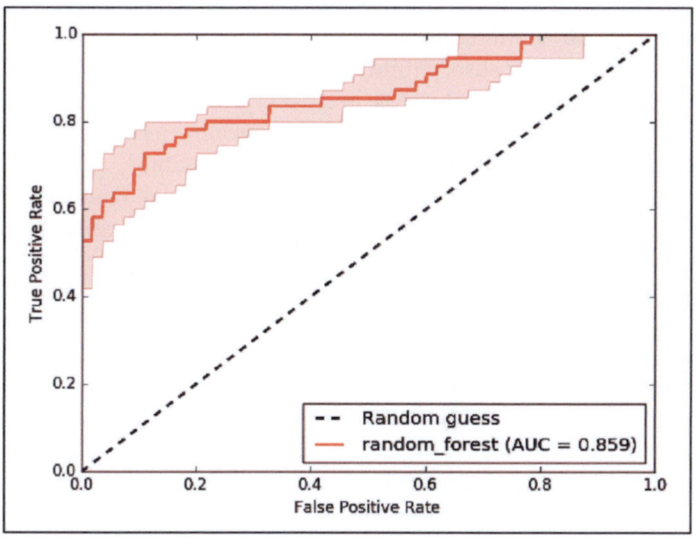

Fig. 18 ROC curve with AUC value

7 Conclusion

Landslides are among the most common geological hazards in hilly regions, causing significant financial and human losses worldwide. This has led to the adoption of management and containment strategies in crisis management and development programs. Landslides are influenced by various triggering factors, and identifying these factors helps pinpoint high-risk areas. Such assessments are crucial for implementing effective mitigation strategies. Various methods—quantitative, qualitative, and hybrid—have been developed for landslide susceptibility analysis, each with its own strengths and limitations. Advances in computing and the use of artificial intelligence have significantly improved the accuracy of susceptibility mapping.

This study applies Google Earth Engine (GEE) and the Random Forest algorithm to assess landslide susceptibility in Coonoor Taluk. The assessment integrates desk studies and remote sensing datasets to build an input database and identify key susceptibility factors. The Random Forest model estimates impact coefficients and classifies susceptibility levels, while GIS tools enhance data analysis, including inventory mapping and hazard planning. The study classifies landslide susceptibility into five categories: very low (21.05%), low (22.08%), moderate (23.21%), high (18.20%), and very high (15.63%). High-susceptibility zones are concentrated along the Coonoor-Mettupalayam highway and in eastern regions like Singara, Kodamalai, Adderaly, and Burliyar. Low and very low susceptibility zones are sparsely distributed. Moderate and highrisk areas employ concrete and gabion retaining walls for mitigation.

Heavy rainfall contributed to numerous landslides, saturating soil and triggering mass movements that displaced soil and vegetation. The affected areas suffered severe damage, with over 60 fatalities and thousands displaced. The landslide scars from these events were used for susceptibility assessment. Retaining structures such as concrete and gabion walls play a key role in erosion prevention and landscape protection. The model achieved an accuracy of 0.859, indicating a strong ability to differentiate susceptible and non-susceptible areas. The susceptibility map is valuable for infrastructure planning, settlements, and land use management. Future improvements could incorporate more detailed historical data and additional variables to enhance risk assessment accuracy.

Although the study achieved high accuracy regarding the assessment of susceptible areas for landslide, there are several patches of high-risk areas with no history of landslides. Inversely, the study also revealed few patches of low landslide susceptible areas with records of landslides in the past. This indicates, there is still a gap in assessing the landslides accurately. Specifically, one of the major factors that could be added in the associated studies is the human activities; inclusion of traces of human activities in the hilly areas could improve the accuracy of such studies.

References

Abedin J, Rabby YW, Hasan I, Akter H (2020) An investigation of the characteristics, causes, and consequences of June 13, 2017, landslides in Rangamati District Bangladesh. Geoenvironmental Disasters 7, 1–19. https://doi.org/10.1186/s40677-020-00161-z

Achu AL, Aju CD, Di Napoli M, Prakash P, Gopinath G, Shaji E, Chandra V (2023) Machine-learning based landslide susceptibility modelling with emphasis on uncertainty analysis. Geosci. Front. 14, 1–14. https://doi.org/10.1016/j.gsf.2023.101657

Aleotti P, Chowdhury R (1999) Landslide hazard assessment: summary review and new perspectives. Bull Eng Geol Env 58(1):21–44

Alimohammadlou Y, Najafi A, Yalcin A (2013) Landslide process and impacts: a proposed classification method. CATENA 104:219–232. https://doi.org/10.1016/j.catena.2012.11.013

Anbalagan R (1992) Landslide hazard evaluation and zonation mapping in mountainous terrain. Eng. Geol. 32, 269–277. https://doi.org/10.1016/0013-7952(92)90053-2

Ballabio C, Sterlacchini S (2012) Support Vector Machines for Landslide Susceptibility Mapping: The Staffora River Basin Case Study, Italy. Math. Geosci. 44, 47–70. https://doi.org/10.1007/s11004-011-9379-9

Bera A, Mukhopadhyay BP, Das D (2019) Landslide hazard zonation mapping using multi-criteria analysis with the help of GIS techniques: a case study from Eastern Himalayas, Namchi, South Sikkim. Nat. Hazards 96, 935–959. https://doi.org/10.1007/s11069-019-03580-w

Budianta W, Ohta H, Takemura J (2022) The Effect of Clay-Soil on Landslide: Case Study from Central Java, Indonesia, in: IOP Conference Series: Earth and Environmental Science. pp. 1–6. https://doi.org/10.1088/1755-1315/1091/1/012012

Chen L, Guo Z, Yin K, Pikha Shrestha D, Jin S (2019) The influence of land use and land cover change on landslide susceptibility: A case study in Zhushan Town, Xuan'en County (Hubei, China). Nat. Hazards Earth Syst. Sci. 19, 2207–2228. https://doi.org/10.5194/nhess-19-2207-2019

Chingkhei RK, Shiroyleima A, Singh LR, Kumar A (2013) Landslide Hazard Zonation in NH-1A in Kashmir Himalaya, India. Int. J. Geosci. 4, 1501–1508. https://doi.org/10.4236/ijg.2013.410147

Costanzo D, Rotigliano E, Irigaray C, Jiménez-Perálvarez JD, Chacón J (2012) Factors selection in landslide susceptibility modelling on large scale following the gis matrix method: Application to the river Beiro basin (Spain). Nat. Hazards Earth Syst. Sci. 12, 327–340. https://doi.org/10.5194/nhess-12-327-2012

Dahal RK, Hasegawa S, Nonomura A, Yamanaka M, Masuda T, Nishino K (2008) GIS-based weights-of-evidence modelling of rainfall-induced landslides in small catchments for landslide susceptibility mapping. Environ Geol 54(2):311–324. https://doi.org/10.1007/s00254-007-0818-3

Eskandarinejad A, Nazari R, Nikoo MR, Arellano D, Pezeshk S, Ghasemi SH (2025) A comprehensive review of geotechnical implications of floods and water-driven disasters. Sci. Total Environ. 985, 179731. https://doi.org/10.1016/j.scitotenv.2025.179731

Guo J, Meng F, Guo J (2024) Gradual Failure of a Rainfall-Induced Creep-Type Landslide and an Application of Improved Integrated Monitoring System: A Case Study. Sensors 24. https://doi.org/10.3390/s24227409

He L, Wu X, He Z, Xue D, Luo F, Bai W, Kang G, Chen X, Zhang Y (2023) Susceptibility Assessment of Landslides in the Loess Plateau Based on Machine Learning Models: A Case Study of Xining City. Sustain. 15, 1–18. https://doi.org/10.3390/su152014761

Highland LM, Bobrowsky P (2008) The landslide handbook—a guide to understanding landslides. U.S. Geological Survey Circular 1325

Hong H (2023) Assessing landslide susceptibility based on hybrid Best-first decision tree with ensemble learning model. Ecol. Indic. 147, 1–16. https://doi.org/10.1016/j.ecolind.2023.109968

Jaiswal P, van Westen CJ, Jetten V (2010) Quantitative landslide hazard assessment along a transportation corridor in southern India. Eng. Geol. 116, 236–250. https://doi.org/10.1016/j.enggeo.2010.09.005

Kadavi PR, Lee CW, Lee S (2018) Application of ensemble-based machine learning models to landslide susceptibility mapping. Remote Sens. 10, 1–18. https://doi.org/10.3390/rs10081252

Kalantar B, Pradhan B, Amir Naghibi S, Motevalli A, Mansor S (2018) Assessment of the effects of training data selection on the landslide susceptibility mapping: a comparison between support vector machine (SVM), logistic regression (LR) and artificial neural networks (ANN). Geomatics, Nat. Hazards Risk 9, 49–69. https://doi.org/10.1080/19475705.2017.1407368

Kamran KV, Feizizadeh B, Khorrami B, Ebadi Y (2021) A comparative approach of support vector machine kernel functions for GIS-based landslide susceptibility mapping. Appl. Geomatics 13, 837–851. https://doi.org/10.1007/s12518-021-00393-0

Lee S, Hong SM, Jung HS (2017) A support vector machine for landslide susceptibility mapping in Gangwon Province, Korea. Sustainability 9, 1–15. https://doi.org/10.3390/su9010048

Lee S, Min K (2001) Statistical analysis of landslide susceptibility at Yongin, Korea. Environ Geol 40:1095–1113. https://doi.org/10.1007/s002540100310

Nawazuzzoha M, Rashid MM, Iqbal F, Suheb Bhardwaj A, Shakeel A, Arabameri A, Naqvi HR (2025) Advanced Machine Learning Models for Landslide Susceptibility Mapping in South Sikkim Himalayas, India. Geol. J. 1–17. https://doi.org/10.1002/gj.5234

Nsengiyumva JB, Valentino R (2020) Predicting landslide susceptibility and risks using GIS-based machine learning simulations, case of upper Nyabarongo catchment. Geomatics, Nat. Hazards Risk 11, 1250–1277. https://doi.org/10.1080/19475705.2020.1785555

Othman AN, Naim WM, M W, S, N (2012) GIS Based Multi-Criteria Decision Making for Landslide Hazard Zonation. Procedia - Soc. Behav. Sci. 35, 595–602. https://doi.org/10.1016/j.sbspro.2012.02.126

Pacheco Quevedo R, Velastegui-Montoya A, Montalván-Burbano N, Morante-Carballo F, Korup O, Daleles Rennó C (2023) Land use and land cover as a conditioning factor in landslide susceptibility: a literature review. Landslides 20, 967–982. https://doi.org/10.1007/s10346-022-02020-4

Panchal S, Shrivastava AK (2022) Landslide hazard assessment using analytic hierarchy process (AHP): A case study of National Highway 5 in India. Ain Shams Eng. J. 13, 1–11. https://doi.org/10.1016/j.asej.2021.10.021

Pham BT, Jaafari A, Prakash I, Bui DT (2019) A novel hybrid intelligent model of support vector machines and the MultiBoost ensemble for landslide susceptibility modeling. Bull. Eng. Geol. Environ. 78, 2865–2886. https://doi.org/10.1007/s10064-018-1281-y

Pourghasemi HR, Jirandeh AG, Pradhan B, Xu C, Gokceoglu C (2013) Landslide susceptibility mapping using support vector machine and GIS at the Golestan province, Iran. J. Earth Syst. Sci. 122, 349–369. https://doi.org/10.1007/s12040-013-0282-2

Pradhan B (2010) Landslide susceptibility mapping of a catchment area using frequency ratio, fuzzy logic and multivariate logistic regression approaches. J. Indian Soc. Remote Sens. 38, 301–320. https://doi.org/10.1007/s12524-010-0020-z

Pradhan B, Lee S, Buchroithner MF (2009) Use of geospatial data and fuzzy algebraic operators to landslide-hazard mapping. Appl. Geomatics 1, 3–15. https://doi.org/10.1007/s12518-009-0001-5

Prakash S (2011) Historical Records of Socio-economically Significant Landslides in India. J. South Asia Disaster Stud. 4, 177–204. https://doi.org/https://www.researchgate.net/publication/272905239_Historical_Records_of_Socio-economically_Significant_Landslides_in_India

Quecedo M, Pastor M, Herreros MI, Merodo JAF (2004) Numerical modelling of the propagation of fast landslides using the finite element method. Int. J. Numer. Methods Eng. 59, 755–794. https://doi.org/10.1002/nme.841

Ramzan M, Cui P, Ualiyeva D, Bazai NA, Baig MA (2025) Optimizing landslide susceptibility mapping under climatic variability: A hybrid approach. J. Rock Mech. Geotech. Eng. 1–24. https://doi.org/10.1016/j.jrmge.2025.07.012

Saha S, Majumdar P, Bera, B (2023) Deep learning and benchmark machine learning based landslide susceptibility investigation, Garhwal Himalaya (India). Quat. Sci. Adv. 10. https://doi.org/10.1016/j.qsa.2023.100075

Saha S, Sarkar R, Roy J, Hembram TK, Acharya S, Thapa G, Drukpa D (2021) Measuring landslide vulnerability status of Chukha, Bhutan using deep learning algorithms. Sci. Rep. 11, 1–23. https://doi.org/10.1038/s41598-021-95978-5

Sato T, Shuin Y (2025) Variation in the frequency and characteristics of landslides in response to changes in forest cover and rainfall in Japan over the last century: A literature review. Catena 249, 1–13. https://doi.org/10.1016/j.catena.2024.108639

Shahabi H, Ahmadi R, Alizadeh M, Hashim M, Al-Ansari M, Shirzadi A, Wolf ID, Ariffin EH (2023) Landslide Susceptibility Mapping in a Mountainous Area Using Machine Learning Algorithms Himan. Remote Sens. 15, 1–18. https:// doi.org/10.3390/rs15123112

Tran VA, Khuc TD, Truong XQ, Nguyen AB, Phi TT (2024) Application of potential machine learning models in landslide susceptibility assessment: A case study of Van Yen district, Yen Bai province, Vietnam. Quat. Sci. Adv. 14, 100181. https://doi.org/10.1016/j.qsa.2024.100181

Tyagi A, Kamal Tiwari R, James N (2022) A review on spatial, temporal and magnitude prediction of landslide hazard. J. Asian Earth Sci. X 7, 1–19. https://doi.org/10.1016/j.jaesx.2022.100099

Ullah I, Aslam B, Hassan S, Ahmad I, Tariq A, Qin S (2022) An Integrated Approach of Machine Learning, Remote Sensing, and GIS Data for the Landslide Susceptibility Mapping. Land 11, 1–20. https://doi.org/10.3390/land11081265

Xie W, Nie W, Saffari P, Robledo LF, Descote PY, Jian W (2021) Landslide hazard assessment based on Bayesian optimization–support vector machine in Nanping City, China. Nat. Hazards 109, 931–948. https://doi.org/10.1007/s11069-021-04862-y

Zhou X, Wen H, Zhang Y, Xu J, Zhang W (2021) Landslide susceptibility mapping using hybrid random forest with GeoDetector and RFE for factor optimization. Geosci. Front. 12, 1–19. https://doi.org/10.1016/j.gsf.2021.101211

Environmental Prediction and Resilience through Geospatial Intelligence

Forecasting Drought from Satellite Imagery: Comparing Stochastic, Classical Machine Learning, and Deep Learning Approaches

Mesfin F. Woldmariam

1 Introduction

Many factors and phenomena like internal displacement, conflict, disease outbreak, and drought are the ones that call humanitarian action. Drought is the most frequently happening factor that make many Ethiopians to be food shortage (Weissman 2003). Aid agencies and key stakeholders in this sector lack reliable and accurate data for planning their scarce resources. The comprehensive study by Mera (2018) indicates Ethiopia has faced drought year after year causing social and economic problems on its people. During drought period millions of people fall in short of food, water, contact diseases, and millions of livestock die. This resulted in government and NGOs to provide emergency food, water, and health aids to both humans and livestock. In most of the cases, the number of aid seekers is massive and this affects government's and aid agencies' planning, budgeting, and mobilizing resources while attempting to respond.

Aid agencies need to change their current approach from being reactive to anticipative approaches (Arendt-Cassetta 2021). For such anticipative approaches, new technologies can support by enabling earlier, faster and relatively more effective and efficient humanitarian action. In this regard, AI tools, mobile apps, chat boats, social Medias, unmanned aerial vehicles, and recently earth observation have got wide usage and help to speed up the assessment, mapping and monitoring of aid seekers.

Advancement in the above technologies, deep learning, predictive analytics, machine learning, and statistical models enable aid agencies greatly in their day to day decision making. These technologies and techniques demonstrated to be successful to forecast the likely hood and features of humanitarian actions related to pandemics, famines, natural disasters, and refugee movements (Hernandez and

M. F. Woldmariam (✉)
College of Business and Economic, Addis Ababa University, Addis Ababa, Ethiopia
e-mail: mesfin.fikre@aau.edu.et

M. O. Sarif and A. Sharifi (eds.), *Advanced Geospatial Intelligence and AI for Environmental Resilience and Sustainable Development*, Advances in Geographic Information Science, https://doi.org/10.1007/978-3-032-03714-5_16

Roberts 2020). Integration latest AI techniques in the form of predictive analytics are also getting more attentions by aid agencies in their prediction activities. Accurate advance prediction enables informed decision making as well as preparation in terms of relief finance, supplies and personnel.

Beside the use of classical machine learning techniques, there are some studies that explored the use of deep learning methods for predicting wealth in sub-Saharan African countries (Jean et al. 2016) as well as non-African countries (Head et al. 2017) with better predictive accuracy. In both studies, their deep learning models achieved R^2 ranging from 0.5–0.75. Specifically, they reported their model has an R^2 values like: for Haiti 0.51; for Malawi 0.55; for Tanzania 0.57; for Nepal 0.64; for Nigeria 0.68; for Uganda 0.69; and for Rwanda 0.75. Similarly, this study focuses on two key tasks: (1) to explore drought the trends of 55 wordas (counties) and make comparisons among the different counties (wordas) and (2) develop predictive models for each of the study wordas.

2 Recent Developments

Reviews of the literature about the use of predictive analytics reveal that many organizations are already using them. For example, (1) https://www.wfp.org is using predictive analytics to track hunger and food security in real time, (2) UNICEF,[1] uses predictive analytics to streamline waste collection in refugee areas in Jordan, (3) JMIR,[2] uses predictive analytics to identify if children can default from their immunization visits. Table 1 summarizes example of how organizations are using predictive analytics in different domain areas.

As technology is developing, many humanitarian action researchers have started exploring the use of uncommon datasets like mobile phone, satellite imagery, and social media posts. The uses of such dataset become normal due to developments in techniques and algorithms of machine learning. It is "a method, as well as a scientific discipline, used to program computers using data instead of explicit instructions" (Bernard 2021a). It involves learning patterns from datasets so that better decisions can be made.

In line with this, the joint study by UC Berkeley's Center for Effective Global Action and Facebook's Data for Good has explored the use of satellite imagery to estimate poverty or wealth of regions (counties). In their study they used publicly available survey data including high-resolution satellite imagery, topographic maps, mobile network data, and connectivity data. In the past these dataset weren't in use by economists or other economic researchers. Rather these researchers were using survey data which is expensive to collect, out dated (usually collected once

[1] "Smart Refugee Camps": applying the best of IoT and ICT for better camp management | by UNICEF Jordan | Medium.

[2] JMIR Public Health and Surveillance—Using Predictive Analytics to Identify Children at High Risk of Defaulting From a Routine Immunization Program: Feasibility Study.

Table 1 Sample application areas of AI based predictive analytics

Example Application	Brief Description
File (unicef.org)	To identify persistently poor households. Even though household economy analysis (HEA) has better accuracy to identify food insecure individuals (Schnitzer 2019), it is found to be expensive and subject to biasness
About—crisis computing (qcri.org)	UN OCHA and UNICEF demonstrated successful use of machine learning techniques (NLP and computer vision techniques) for humanitarian response based on analysis of text, image, video etc. contents posted on social medias
Refugee and Migration analytics[3]	Predictive analytics developed by IBM that enable to determine where people are going and how they are getting there
Water Monitoring in Turkana and Wajir[4]	A dashboard system that allows utility managers to monitor water systems for supply, demand, and functionality in real-time
Platform for real-time impact and situation monitoring—perism/vampire[5]	"Perism is a climate risk monitoring system that integrates geospatial data on hazards such as droughts, floods, tropical storms, and earthquakes, with socioeconomic vulnerability information to inform disaster risk reduction and social assistance programs"
Uber for Waste: Using predictive analytics to streamline waste collection in the camps[6]	"A waste tank nearing capacity smells bad; an overflowing one has health implications. By following the concept of how Uber works, refugees could report on a full tank via their phones, and trucks would be dispatched to do the collection. The software's algorithm uses the laser meter reading to calculate a "fill rate" and predict when the tank will be full again"
Homepage—Tala	This company developed an innovative credit scoring solution from mobile phone datasets and be able to generate credit scores for the previously unbanked

in 10 years), and subject to biased and subjective inputs. For the same survey item (questions) respondents from different locations (wordas) can give different response which increases subjectivity. But, the development in earth observation and remote sensing fields, now researchers and scientists can use satellite imagery to analyze

[3] How to Predict the Next Refugee Crisis (businessinsider.com).

[4] Oxfam (sensorinsight.io).

[5] PRISM | WFP Innovation.

[6] "Smart Refugee Camps": applying the best of IoT and ICT for better camp management | by UNICEF Jordan | Medium.

population densities, quality and availability of roads infrastructure, which would suggest poverty or wealth of locations or wordas and hence better humanitarian targeting.

Based on their study, a region with fewer smartphones and Wi-Fi connectivity routers would suggest less relative wealth compared to regions with more of these infrastructures. Added to this, looking into mobile phone datasets like call duration, volume of mobile internet usage, and kinds of website individuals usually browse are good indicators of wealth. These datasets can be combined with remote sensing imagery to train machine learning model that predicts the absolute and relative wealth of each location (counties). Such insightful discoveries from uncommon dataset can help to make objective and unbiased decisions by government and aid agencies.

By identifying the dense concentration of night time light from satellite images, researchers from Stanford predicted and analyzed poverty level of different places. Similarly, researchers from Berkeley University and Qatar Computing Research Institute (QCRI) used telecommunication datasets from mobile network operators and analyzed poverty level of areas or individuals[7] that can inform aid agencies. Besides, there are many studies that demonstrate the use of satellite imagery for cash aid distribution, distribute COVID-19 aid. For example, an NGO called GiveDirectly use satellite images to find the poorest people in the world-and send them cash.[8] Following this, many scholars emerged to explore the use of satellite imagery for assessing and estimating poverty/wealth, example (Head et al. 2017; Htet et al. 2021; Zixi 2021; Min et al. 2022).

By using nighttime satellites images, (Henderson et al. 2012) elaborated the importance of such uncommon datasets. According to them, regions emitting high levels of artificial light tended to have higher economic output. But when it comes to developing countries (Jean et al. 2016) showed that night-lights alone do poor job of differentiating between poor and ultra-poor regions in sub-Saharan Africa. Countries in this region do rarely have nighttime satellite images. As a supplementary approach, considering mobile phone data is valuable (Steele et al. 2017).

Besides the commonly used classic machine learning models, other forms of AI technologies are becoming a hot topic in humanitarian activities. For example: AI can help (1) where to build schools by identifying locations which don't have schools, (2) to understand what a student needs to learn (Squirrel AI) initiative in china, (3) to learn and recommend where is the best to relocate refugees by the AI for Good Foundation, (4) to track missing children by the TrackChild portal in India, and (5) to diagnose those in an emergency like heart arrest by Danish based company (Jonathan 2019), to forecast forced displacement in Africa (Burkina Faso, Mali, Niger, and Nigeria) by Danish Refugee Council (Leila), to optimize service delivery, to predict conflict, facilitate job matching, and personalized learning by International Rescue Committee in refugee setup, to assist Venusian migrants know their immigration rights by Norwegian Refugee Council, to analyze image and videos if there is any

---

[7] *Joe Bush, How AI is Being Used in Poverty Map, 17 October 2019* https://www.electronicspecifier.com/products/artificial-intelligence/how-ai-is-being-used-to-map-poverty.

[8] How Google and GiveDirectly are using AI mapping to get cash to Hurric (fastcompany.com).

violation of human rights,[9] and to help get more accurate data/information about Syrian conflict by the Carter Center. Even though the list will increase even as we write this paper, one can easily understand the diverse and innovative application of AI and related technologies in humanitarian context.

Coming back to our issue of predicting poverty, there are scholarly works that start combining uncommon sources and datasets. This becomes possible thanks to development of AI and techniques therein. For example, by using deep learning techniques (Jean et al. 2016; Pandey et al. 2018), combined Open Street Map (OSM) and demographic health survey (DHS) by US government agency is becoming common, they are able to predict poverty more accurately and efficiently. By integrating different datasets, better models can be developed that out performs the currently used manual operations (Quinn et al. 2018). Besides predictive analytics and supervised machine learning techniques, it is also possible to cluster locations based on their level of poverty.[10] Such clustering of locations based on their level of poverty can assist to design better aid programs.

In summary, we learned that there is a growing application and use of satellite imagery. Specifically for road quality assessment (Oshri et al. 2018), to map poverty (Kuc and Chormański 2019), to map humanitarian action (Logar et al. 2020), for wealth estimation (Ledesma et al. 2020), for land use land cover (Yang and Liu 2005), to predict household income (Xie et al. 2016), to map flooded areas (Nemni et al. 2020), to discover risk areas of the epidemic disease (Traore et al. 2017). As the result, besides machine learning and AI researchers there is a trend that poverty and development economists start using remote sensing data for their wealth/poverty research and analysis. For them, such datasets has three advantages (1) access to data where it is difficult to access by other means, (2) high spatial resolution; and (3) wide geographic area coverage (Donaldson and Storeygard 2016).

3 Machine Learning and Drought Modeling

Several approaches were used for drought forecasting. These include meteorological drought indices, hydrological drought indices, crop yield models that relied on historical data, remote sensing techniques that used satellite imagery, aerial photography, and expert judgment that involves combining multiple sources of data and information (Belayneh and Adamowski 2013). These approaches were used in conjunction with traditional methods for predicting droughts, such as indigenous knowledge systems, statistical methods, and others. Together, these approaches provided valuable insights into the onset and severity of drought, helping experts and farmers

---

[9] https://www.microsoft.com/en-us/ai/ai-for-humanitarian-action-projects?activetab=pivot1%3aprimaryr5

[10] rs-anderson/Clustering-Ward-Level-Poverty-Using-Satellite-Imagery: Combing satellite imagery and machine learning methods to cluster ward-level poverty in Gauteng, South Africa. (github.com).

take proactive measures to mitigate its impact on crops, water resources, and livelihoods (Tyagi et al. 2022). Despite their usefulness, these approaches were limited in capturing the complex and non-linear relationships between meteorological variables and drought conditions.

Drought time series datasets are usually nonlinear as well as instable (Bouaziz et al. 2021), which makes it challenging to develop accurate models using traditional statistical techniques. As a result, these approaches often fail to capture the true dynamics of drought, leading to inaccurate predictions and inadequate mitigation efforts. This is where machine learning models come in, as they are better equipped to handle complex and non-linear relationships, making them a promising tool for improving drought forecasting and management (Felsche and Ludwig 2021). The advent of machine learning has provided new opportunities for improving drought forecasting accuracy by capturing the complex relationships in large datasets.

4 Statistical, Machine Learning, and Deep Learning Techniques

Even though, the goal of this paper is not to compare these techniques, understanding about when to use which technique as well as their strength and limitation is important.

Statistical techniques dominate the field of forecasting and research issues thereof. But, recently machine learning methods have been proposed in the academic literature as better alternatives to statistical methods. Depending on the nature of datasets, there are studies which found superiority of statistical models over machine learning models (Makridakis et al. 2018; Kumar 2022). But contrary to this, study by Cerqueira et al. (2019) found the reverse. According to the latter scholars, when sample sizes are small, statistical models might outperform machine learning models. Besides sample sizes, existence of strong/weak signals do affect model performance. Machine learning models are more likely suit for large sample sizes with strong signals while statistical models suit for the opposite.[11] The latter are also suitable to model trends and seasonality feature of time series datasets.

As the consequence of drought is massive, it needs accurate forecasting/perdition. Forecasting is one of the first steps to prepare for efficient and effective humanitarian actions. Different modeling techniques like statistical, machine learning and deep learning models have their own benefits and limitations. (Dikshit et al. 2021; Gyaneshwar et al. 2023; Docheshmeh Gorgij et al. 2022; Kaur and Sood 2020; Xu and Mo 2019; Kumer et al. 2023) found that deep learning can outperform other techniques and hence to mitigate impacts of drought.

In this study, we use time series satellite images which are complex and difficult to model using traditional linear time series prediction approaches, especially due to

[11] Time series forecasting: classical methods versus machine learning/artificial intelligence—Cross Validated (stackexchange.com), retrieved on Sep 6, 2023.

nonlinearity and instability nature of such datasets. For datasets of this nature, the use of deep learning is valuable. (Ornella et al. 2019) utilized satellite imagery data and computed normalized difference vegetation index (NDVI) values to develop a CNN model for predicting drought or no drought conditions. Their CNN model has an accuracy of 96%.

Study by Marj and Meijerink (2011) also used time series satellite images to predict agricultural drought using neural network. The result is fair, R^2 of 70%. In a similar study (Jalili et al. 2013) constructed a neural network model that forecasts drought with an accuracy of 90%. But one possible shortcoming is that they predicted for nationwide. Perhaps, had they forecast for smaller geographic area, their model could have predicted with more accuracy. Study by Ghazaryan et al. (2020) also used time series satellite images to study agricultural droughts by studying NDVI. Their model has acceptable accuracy of R^2 of 70%. All these studies show machine learning and deep learning techniques become more efficient for drought prediction than conventional stochastic models, (Nandgude et al. 2023) and (Chaudhari et al. 2021). They have used deep learning techniques and found prediction accuracy of R^2 of 97% while they report low performance for other models.

There are various image processing and analysis approaches for drought assessment. Typical approach might be, analysis of Normalized Difference Vegetation Index (NDVI), Enhanced Normalized Vegetation Index (ENVI), Standardized Precipitation Index (SPI), Palmer Drought Severity Index (PDSI) etc., (Patel et al. 2012) and (Rembold et al. 2015). Processing these dataset is hard in normal computing infrastructure. But, these days thanks to Google Earth Engine, processing satellite imagery become simple. This technology is a cloud-based platform that provides access to a large collection of satellite images as well as capabilities for image processing and analysis (Ejaz et al. 2023) and (Aksoy et al. 2019).

The PDSI, a complete indicator that considers data on precipitation, temperature, soil moisture, and other variables, is not employed by majority of the above studies. But it is one of the best approach to give a measurement of the severity of drought over a longer time frame (often months to years). Comparing the PDSI to the SPI and other remote sensing-derived drought indices, it is stated that PDSI is better suited for longer-term drought prediction (Zhang et al. 2022).

For modeling, it is learned that specialized deep learning types, Long-short term memory (LSTM) are proved to give better prediction accuracy (Dikshit et al. 2021; Xu et al. 2022) and Convolutional Neural Nets (CNN) are also proved to give better drought prediction accuracy (Dhyani and Pandya 2021). As the name signify LSTM model can help to forecast for both short term and long term lead times. Hence, modeling drought with one of these special deep learning models can help to have better forecasting accuracies. Gated Neural networks (G-CNN) are also found to outperform classical machine learning as well as stochastic techniques in predicting droughts (Mei et al. 2022). Table 2 gives summarized comparison of these models.

Table 2 Comparison of statistical, ML, and DL techniques source (Bouaziz et al. 2021)

Statistical	Machine learning	Deep learning
Focus on linear relationship among dependent and independent variables	Can focus on nonlinear relationship among dependent and independent variables. It also applies uniform weightage across time steps	Can focus on nonlinear relationship among dependent and independent variables. It can also retain past information and apply decreasing weightage across time steps
Can use single or many indicators	Can be used for data cleaning like downscaling, screening of indicators, and estimating indices	Can be used for data cleaning like downscaling, screening of indicators, and estimating indices
Unable to deal with non-linearity of data and has issue of multicollinearity	Can deal with nonlinear relationship and issue of multicollinearity	Can deal with nonlinear relationships, multicollinearity, non-stationary, and over fitting lag component
	Unable to deal with non-stationary and over fitting lag components	Explainable DL can generate results that align with the knowledge of physical based model and improve prediction

5 Materials and Methods

In our bigger project study we report the development and comparison of different modeling approaches. We explored and compared performance of (1) stochastic models, (2) classical machine learning models, and (deep learning models- namely fully connected neural nets as well as gated convolutional networks). But, in this paper we only report performance of stochastic models, while giving some explanations for the rest. We have detailed report for stochastic models in other paper. In or study we used 40 years' time series satellite images of 55 geographic areas (wordas) from five Ethiopian drought affected regions. The identified wordas are known for their severe and repeated drought for many years. We selected 10 wordas from Afar region, 17 from Amhara 17, 10 from Oromia, 10 from SNNPR, and 9 wordas from Somalia region.

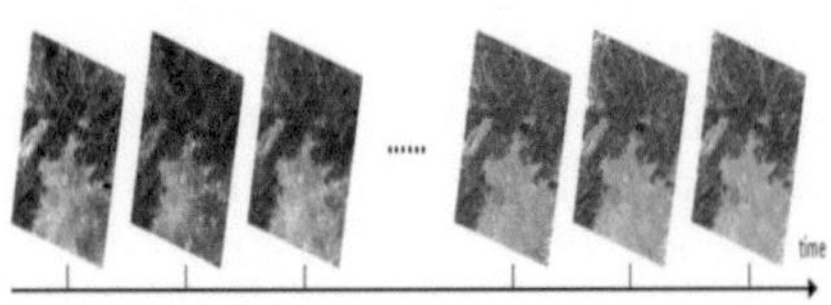

Time series imagery datasets

The datasets are collected by Abatzoglou (2013) and available through Google Earth Engine platform. This dataset contains drought indices derived from the 4-km daily Gridded Surface Meteorological (GRIDMET) dataset

PSDI values of study sites are extracted by using Google Earth Engine platform. These values of the respective research sites can be compared regarding how drought trended overtime. It is an effective measurement index of drought on agricultural

Table 3 Standard PDSI values and interpretations

PDSI value	PDSI category
Above 4.00	Extreme wet spell
3.00 to 3.99	Severe wet spell
2.00 to 2.99	Moderate wet spell
1.00 to 1.99	Mild wet spell
0.50 to 0.99	Incipient wet spell
0.49 to − 0.49	Normal
− 0.50 to − 0.99	Incipient drought
− 1.00 to − 1.99	Mild drought
− 2.00 to − 2.99	Moderate drought
− 3.00 to − 3.99	Severe drought
Below − 4.00	Extreme drought

production, and hence can be considered as drought monitoring tools as well as to give insight to develop contingency plans. Each worda has equal number of images (datasets), monthly images from 1983–2022 years. Table 3 shows possible PSDI values and interpretations. Detailed computations and formulas are available in our other paper and we deliberately left out here.

6 Data Preparation and Modeling

6.1 Modeling with Classical Machine Learning Approaches

In time series past events affect current, this in turn affects future values. As stated above, the PDSI values are time series but they are not structured in such a way for classical machine learning input --- > output forms. Thus, we need to transform our dataset into supervised machine learning problem forms that has input--- > output relationships. In order to predict the t + 1 value of a sequence, say t_i, it is intuitive that previous values of time x_{i-1} are more important than values of time x_{i-2}, which is more important than values at time x_{i-3}, and so on. The Markov model of order 1 takes this important relationship into consideration. As we see from Fig. 1, this approach can be extended by adding more past values to predict the future. For example, in a Markov model of order 2, each element is conditioned on its previous two elements.

In our other paper (not discussed in this paper), we tried to fit the PDSI values into stochastic models and found that almost all the 55 locations were better fitted into ARMA model, with the order of (1, 1), (1, 2), or (1, 3). Taking this information into consideration, we experimented on different machine learning models with lag of 1, 2, 3, and 4. Every time, we changed this lag time (t), the performance of the

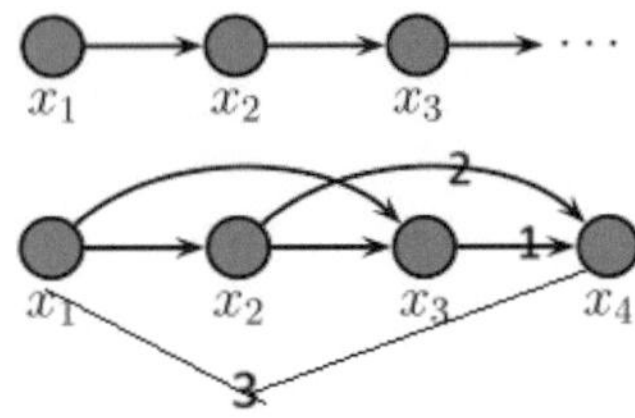

Fig.1 Interdependence of time series dataset values

model is different. After many such experiments, almost all models developed for each wordas revealed going back by two months give more prediction accuracy for models. Here is a sample:

Sample time series values	– 0.62	– 0.84	– 0.67	0.54	0.88	– 1.97	– 2.83	etc.

Once arranged into machine learning input-- output form, it looks like: $\{\{-0.62, -0.84 \to -0.67, \{-0.84, -0.67\} \to 0.54, \{-0.67, 0.54\} \to -0.88, \{0.54, 0.88\}, \to -0.88, \{0.88, -0.88\} \to -1.97, \{-0.88, -1.97\} \to -2.88$, for order two chains. This means, when we assume two past time values affects current values.

Then 90% of the dataset is assigned for training while the remaining 10% is used for testing purpose. By using Wolfram Mathematica 13.2, the machine learning model was trained and evaluated. The following paragraph gives some information about the models.

```
PredictorFunction[ Input type: {Numerical, Numerical}
                   Method: LinearRegression
                   Number of training examples: 425 ]
```

As, can be seen from the above screen shot, for all 55 wordas, Liner Regression Algorithm was picked automatically to be the best algorithm. In terms of performance, each wordas has different values ranging from 42% R^2 (for Tenta worda of Amhara Region) to 95% Boloso Worda of SNNR. As can be seen from Fig. 2, except two wordas (Tenta of Amhara and Dubti of Afar), for the rest wordas, R^2 is above 50%. Actually the majority of them have above 65%, which is impressive.

When comparing R^2 values of wordas across the 5 regions, Oromia, SNNPR, and Somali regions demonstrates higher prediction accuracy than Afar and Amhara. This means, PDSI drought values forms easier learnable patterns for wordas in the three regions than those in the 2 regions. As can be seen, datasets of high performing machine learning models (models of Boloso and Hudet) appear to form some kind of pattern that made the algorithms to learn the pattern easily. As the datasets of low performing models (Tenta and Dubti) are relatively appear to be scattered, the algorithm didn't catch existing patterns good enough and hence low value of R^2. The dataset gives interesting insights. Tenta datasets are kind of scattered with minimal pattern for learning while Boloso datasets form good pattern and hence good model performance. Table 4 gives summary measurement values for all the machine learning models experimented in this study for all the 55 wordas.

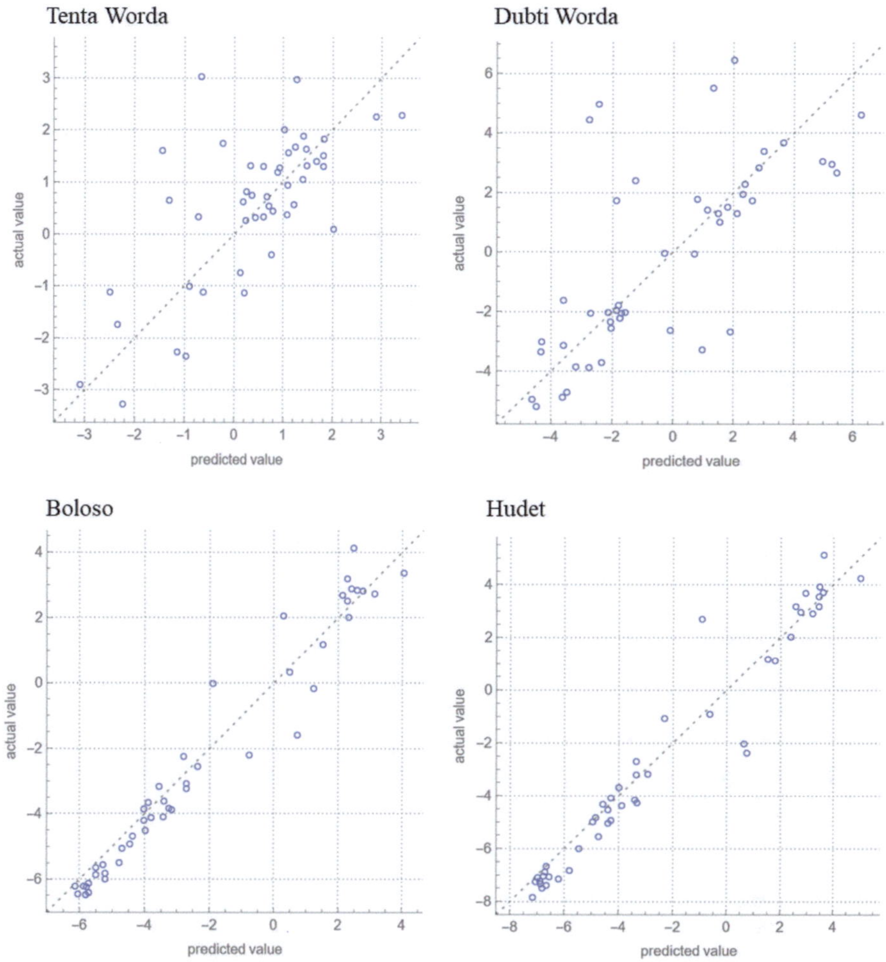

Fig. 2 Accuracy of ML model for two wordas (as a sample)

6.2 *Modeling with Fully Connected Neural Network Architecture*

Neural nets involve inputs (X_s), outputs (y), transfer functions (f), weights (w_s), and bias/error. Weights and bias are called learnable parameters. The inputs can be combined linearly or nonlinearly. Figure 3 depicts typical network architectures. There are different types of neural networks, like fully connected networks, convolutional networks, and recurrent networks. These network types are good at solving certain problem types.

In fully connected network architecture, "neurons are arranged by layers, and every neuron of a given layer sends its output to every neuron of the next layer".

Table 4 R^2 and standard deviations summary of best performing models of study sites

			Conventional ML Models			Fully connected Neural Network Models			Recurrent Nets	
Region	Worda	Stochastic Model	Standard deviation	RSquared	Lag period	Lag Period	RSquared	Standard Deviatio	RSquared	Standard D
Afar	1. Dalol	ARMA (1,2)	1.5624	0.665296	2	3	0.968	0.41	0.993	0.19
	2. Berhale	ARMA (1,2)	2.18756	0.681073			0.945	0.48	0.992	0.18
	3. Koneba	ARMA (1,2)	1.65212	0.73246			0.955	0.44	0.985	0.25
	4. Chifra	ARMA (1,1)	2.25146	0.6173			0.963	0.39	0.99	0.24
	5. Teru	ARMA (1,3)	2.02143	0.801348			0.956	0.43	0.975	0.35
	6. EliDar	ARMA (1, 2)	1.12626	0.537143			0.976	0.36	0.985	0.29
	7. Garani	ARMA (1,3)	1.99059	0.554241			0.964	0.44	0.99	0.23
	8. Erepti	ARMA (1, 2)	1.9518	0.746088			0.973	0.35	0.981	0.009
	9. Dubti	**ARMA (1,1)**	**2.2801**	**0.47**			**0.973**	**0.38**	**0.992**	**0.20**
	10. Malie			**0.716**			**0.970**	**0.39**	**0.989**	**0.24**
Amhara	Beyeda	ARMA (1, 3)	1.11705	0.765199			0.978	0.33	0.988	0.26
	Dahana	ARMA (1, 2)	1.03189	0.678325			0.970	0.37	0.974	0.34
	East Belesa	ARMA (1, 1)	1.83301	0.603141			0.932	0.57	0.983	0.29
	Ibnat	ARMA (1, 2)	1.37123	0.713847			0.936	0.56	0.992	0.19

(continued)

Table 4 (continued)

			Conventional ML Models			Fully connected Neural Network Models			Recurrent Nets	
Region	Worda	Stochastic Model	Standard deviation	RSquared	Lag period	Lag Period	RSquared	Standard Deviatio	RSquared	Standard D
	Gidan	ARMA (1, 3)	1.1144	0.652945			0.942	0.59	0.988	0.27
	Legambo	ARMA (1, 2)	0.673645	0.622436			0.937	0.55	0.989	0.24
	MenzGMider	ARMA (1, 2)	0.783823	0.868625			0.914	0.65	0.994	0.17
	Metema	ARMA (1, 3)	1.27689	0.895238			0.959	0.64	0.985	0.39
	RayaKobo	ARMA (1, 3)	2.26035	0.541942			0.947	0.56	0.978	0.36
	WereBabu	ARMA (1, 3)	2.52112	0.529704			0.938	0.63	0.989	0.27
	GubaLafto	ARMA (1, 2)	1.1632	0.782719			0.928	0.64	0.988	0.26
	Habru	ARMA (1, 3)	2.46612	0.527976			0.930	0.66	0.99	0.25
	Jama	ARMA (1, 3)	0.779705	0.592452			0.935	0.54	0.989	0.22
	Meket	ARMA (1, 1)	1.55282	0.58501			0.931	0.57	0.992	0.19

(continued)

Table 4 (continued)

			Conventional ML Models			Fully connected Neural Network Models			Recurrent Nets	
Region	Worda	Stochastic Model	Standard deviation	RSquared	Lag period	Lag Period	RSquared	Standard Deviatio	RSquared	Standard D
	Simada	ARMA (1, 1)	0.854633	0.656779			0.932	0.55	0.992	0.19
	Tenta	**ARMA (1, 2)**	**1.09487**	**0.419508**			**0.941**	**0.54**	**0.996**	**0.13**
	Wadla	ARMA (1, 3)	0.926855	0.710305			0.938	0.54	0.995	0.16
Oromia	Moyale	ARMA (1, 3)	0.551311	0.939923			0.980	0.38	0.977	0.43
	Guchi	ARMA (1,2)	0.591427	0.918131			0.955	0.57	0.985	0.36
	Gelana	ARMA (1, 3)	0.506452	0.934691			0.954	0.54	0.981	0.35
	Kercha	*ARMA (1, 3)*	*0.983209*	*0.944031*			*0.966*	*0.51*	*0.99*	*0.28*
	BuleHora	ARMA (1, 3)	1.04704	0.933785			0.964	0.55	0.99	0.30
	Girawa	ARMA (1, 2)	1.05635	0.867617			0.955	0.57	0.988	0.29
	Bedeno	ARMA (1, 2)	0.844796	0.871509			0.952	0.56	0.987	0.29
	HambelaWamena	ARMA (1, 3)	0.77342	0.949258			0.970	0.44	0.991	0.24

(continued)

Table 4 (continued)

			Conventional ML Models			Fully connected Neural Network Models			Recurrent Nets	
Region	Worda	Stochastic Model	Standard deviation	RSquared	Lag period	Lag Period	RSquared	Standard Deviatio	RSquared	Standard D
	Deder	ARMA (1,3)	0.709802	0.841436			0.929	0.68	0.982	0.34
	Haromaya	ARMA (1, 2)	0.608902	0.883906			0.927	0.681	0.978	0.37
SNNR	Amaro	ARMA (1, 3)	0.55672	0.938739			0.984	0.33	0.982	0.13
	Derashe	ARMA (1, 3)	0.591427	0.918131			0.948	0.64	0.991	0.26
	SouthAri	ARMA (1, 3)	0.506452	0.934691			0.960	0.56	0.991	0,27
	Kochere	*ARMA (1, 3)*	*0.983115*	*0.944042*			*0.947*	*0.56*	*0.986*	*0.28*
	Yirgachefe	ARMA (1, 3)	1.04713	0.933773			0.943	0.56	0.989	0.25
	Kena	ARMA (1, 3)	1.05635	0.867617			0.956	0.56	0.989	0.28
	Alle	ARMA (1, 3)	0.831571	0.875501			0.964	0.54	0.992	0.26
	Boloso	*ARMA (1, 2)*	*0.77342*	*0.949258*			*0.937*	*0.60*	*0.99*	*0.24*
	Silti	ARMA (1, 3)	0.709802	0.841436			0.943	0.53	0.987	0.25

(continued)

Table 4 (continued)

			Conventional ML Models			Fully connected Neural Network Models			Recurrent Nets	
Region	Worda	Stochastic Model	Standard deviation	RSquared	Lag period	Lag Period	RSquared	Standard Deviatio	RSquared	Standard D
	Malie	ARMA (1, 3)	0.608933	0.883894,						
Somali	Galadi	ARMA (1, 1)	0.551311	0.939923			0.949	0.55	0.977	0.37
	Babille	ARMA (1, 1)	0.592419	0.917856			0.929	0.64	0.99	0.23
	Bokh	ARMA (1, 1)	0.506452	0.934691			0.919	0.67	0.987	0.26
	Hudet	*ARMA (1, 2)*	*0.983115*	*0.944042*			*0.939*	*0.60*	*0.985*	*0.30*
	Tulu	ARMA (1, 2)	1.04704	0.933785			0.928	0.67	0.989	0.25
	Filtu	ARMA (1, 2)	1.13693	0.846649			0.948	0.54	0.98	0.34
	Danbal	ARMA (1, 1)	0.844796	0.871509			0.926	0.64	0.986	0.28
	Harshin	ARMA (1, 1)	0.77342	0.949258			0.916	9.72	0.986	0.29
	Awbar	ARMA (1, 1)	0.709802	0.841436			0.926	0.72	0.991	0.25

(Lag for ML and Deep Learning Models is determined by experimenting on **+ *1*** period on automatically selected stochastic model orders)

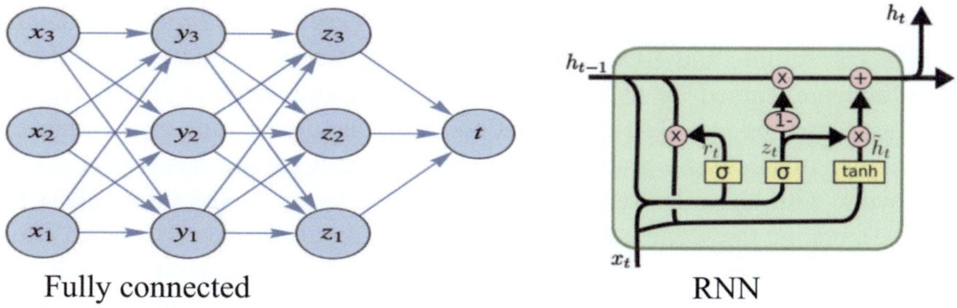

Fig. 3 Schematic architecture of fully connected and recurrent nets

Before 2017, this network architecture hasn't taken over the classic machine learning methods on structured problems. CNN and RNN architectures still use the concept of layers, but their connectivity is quite different from multilayer perceptron (Villalobos Alva and Villalobos Alva 2021).

Deep learning neural net is powerful to learn complex interactions among variables in a dataset, like images, texts, as well as time series datasets. The complex interactions among previous (past) values can affect current time values. This characteristic makes time series data to be kind of unstructured and hence better captured by deep learning methods. As can be seen by Fig. 4, in Wolfram Language neural networks can be defined by chaining layers one after the other and bundling the whole layer into a function called NetChain[{Layer1, 2, ...Layer$_n$}] and if the nets are complex, the layers can be interconnected by using a function called NetGraph[]. The use of many layers each with its own functions can form deep learning. This is where neural nets outperform classical machine learning approach (Villalobos Alva and Villalobos Alva 2021).

Accordingly, we analyzed the drought index dataset by using the fully connected neural network architecture. The best performing architecture has four linear chains of layers that perform linear transformations and two layers that perform nonlinear transformations. To minimize the cost, MeanSquareLossLayer[] function is used. Figure 4 shows how the different components of neural net (linear, nonlinear, and activation functions) are interconnected. Network training is optimized by tuning these components or parameters, until the minimum overall cost is achieved.

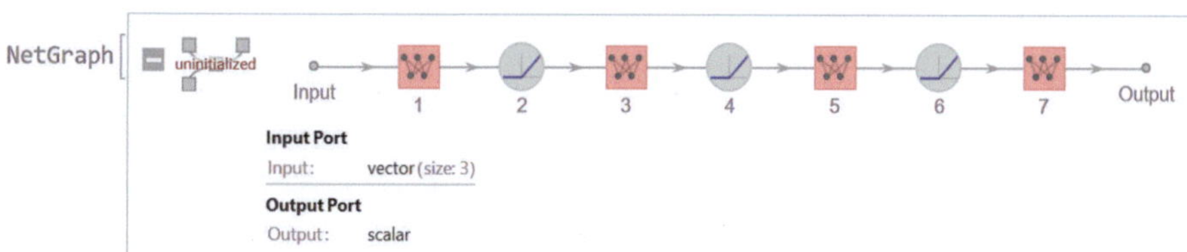

Fig. 4 Deep learning net components and their interconnection

Like classical machine learning, for deep learning nets we also divided the dataset into (90% for training and 10% for testing) purpose. Network models of all the 55 study sites were trained by using 64 batch size, learning rates of 0.001, ADAM optimization method, and L2Regularization techniques (to avoid) over fitting are used. The use of stochastic gradient descent has advantages like faster processing, easy to work with large datasets (as it feeds in smaller batch sizes than all at the same time), behind the scene it also adds an implicit regularization which can allow for better generalization. As of the early 2020s, just about all neural networks are trained with the stochastic gradient descent method. This means, "for each iteration of the optimization process, we need to compute the derivatives of the cost (a.k.a. the gradient) with respect to each parameter for the current set of parameters", (Bernard 2021b) (Fig. 5).

When we compare the performance of fully connected neural networks of the 55 location against the respective classical machine learning models, we can see (from Table 4) there is significant difference between them. Even those places like (Tenta and Dubti with low performance accuracy of 41.9% and 47% R^2in classical machine learning techniques) have got an accuracy of 97% and 94% of R^2 in fully connected neural nets. This is a clear demonstration that neural networks are superior to classical machine learning models. Table 4 gives a summary of model performances

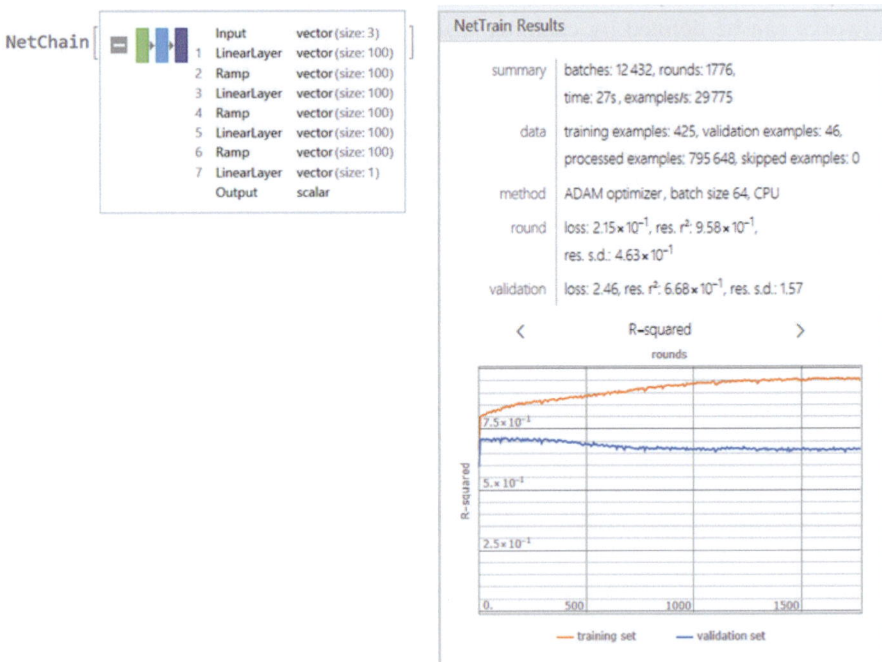

Fig. 5 Trained deep learning model and its performance for dalol worda (sample)

of stochastic models (not discussed in this paper), classical machine learning, fully connected nets, as well as gated recurrent neural nets.

6.3 Deep Learning: Gated Recurrent Nets

Besides classical machine learning and fully connected nets, we also experimented with deep learning neural networks, specifically gated neural networks. These networks are considered to be a better fit for time series datasets. Unlike other nets, recurrent networks can go many time periods back and consider its impact on current or future values. This assertion is witnessed in this paper as well. With respect to performance, recurrent nets out performed both fully connected neural nets and classical machine learning models. Table 4 gives summary and comparative information about the models (Fig. 6).

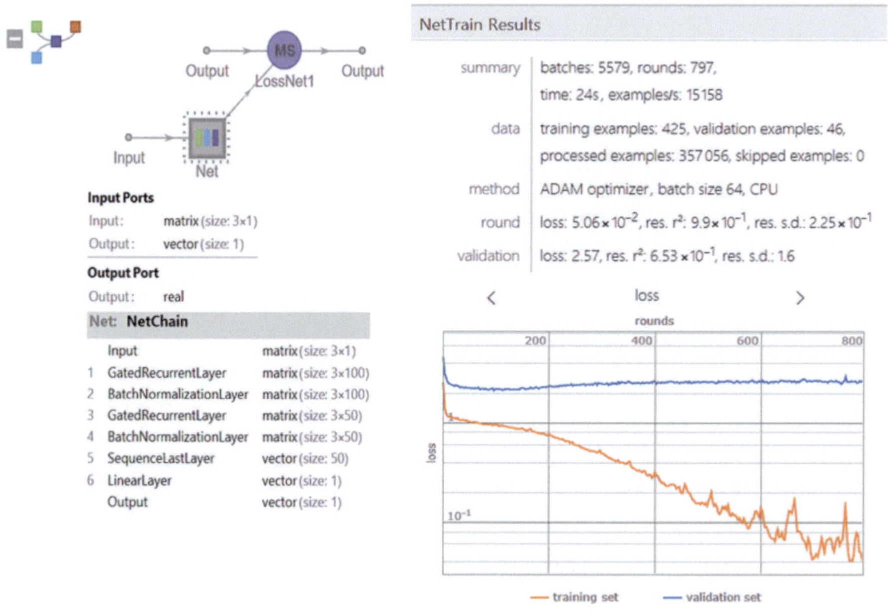

Fig. 6 Trained model and its performance for dalol worda (sample)

7 Conclusion and Limitation

The present paper found that deep learning model has the most accuracy to predict drought indices for all 55 study sites. Compared to classical machine learning and fully connected deep learning models, gated recurrent neural networks have better performances. Thus, by deploying such models on drought index datasets, it is possible to have almost accurate insights and anticipations about locations and communities who might be victims of drought and make informed decisions and planning. Making such information transparent to stakeholders can also reduce the problem of misdirected aid and wrong targeted beneficiaries that is motivated by biased politicians and their cadres.

Acknowledgements We would like to express our gratitude to scholarship funders—International Development Research Centre (IDRC) and Swedish International Development Cooperation Agency (SIDA), under scholarship programme—Artificial Intelligence for Development (AI4D) Africa, and scholarship fund manager—Africa Centre for Technology Studies (ACTS).

References

Abatzoglou JT (2013) Development of gridded surface meteorological data for ecological applications and modeling. Int J Climatol 33(1):121–131

Aksoy S, Gorucu O, Sertel E (2019) Drought monitoring using MODIS derived indices and Google earth engine platform. In: 2019 8th International conference on agro-geoinformatics (agro-geoinformatics), pp 1–6. IEEE

Arendt-Cassetta L (2021) From digital promise to frontline practice: new and emerging technologies in humanitarian action. OCHA, New York

Belayneh A, Adamowski J (2013) Drought forecasting using new machine learning methods. J Water Land Dev 18(9):3–12

Bernard E (2021) PP 2. Introduction to Machine Learning: Copyright © 2021 by Wolfram Media, Inc.

Bernard E (2021) Introduction to machine learning. Copy right Wolfram Media, Inc.

Bouaziz M, Medhioub E, Csaplovisc E (2021) A machine learning model for drought tracking and forecasting using remote precipitation data and a standardized precipitation index from arid regions. J Arid Environ 189:104478

Cerqueira V, Torgo L, Soares C (2019) Machine learning vs statistical methods for time series forecasting: Size matters. arXiv preprint arXiv:1909.13316

Chaudhari S, Sardar V, Rahul DS, Chandan M, Shivakale MS, Harini KR (2021) Performance analysis of CNN, alexNet and VGGNet models for drought prediction using satellite images. In: 2021 Asian conference on innovation in technology (ASIANCON), pp 1–6. IEEE

Dhyani Y, Pandya RJ (2021) Deep learning oriented satellite remote sensing for drought and prediction in agriculture. In: 2021 IEEE 18th India council international conference (INDICON), pp 1–5. IEEE

Dikshit A, Pradhan B, Alamri AM (2021) Long lead time drought forecasting using lagged climate variables and a stacked long short-term memory model. Sci Total Environ 755:142638

Docheshmeh Gorgij A, Alizamir M, Kisi O, Elshafie A (2022) Drought modeling by standard precipitation index (SPI) in a semi-arid climate using deep learning method: long short-term memory. Neural Comput Appl 1–18

Donaldson D, Storeygard A (2016) The view from above: applications of satellite data in economics. J Econ Perspect 30(4):171–198
Ejaz I, Li W, Naseer MA, Li Y, Qin W, Farooq M, Yu K (2023) Detection of combined frost and drought stress in wheat using hyperspectral and chlorophyll fluorescence imaging. Environ Technol Innov 30:103051
Felsche E, Ludwig R (2021) Applying machine learning for drought prediction in a perfect model framework using data from a large ensemble of climate simulations. Nat Hazard 21(12):3679–3691
Ghazaryan G, Dubovyk O, Graw V, Kussul N, Schellberg J (2020) Local-scale agricultural drought monitoring with satellite-based multi-sensor time-series. Giscience Remote Sens 57(5):704–718
Gyaneshwar A, Mishra A, Chadha U, Raj Vincent PD, Rajinikanth V, Pattukandan Ganapathy G, Srinivasan K (2023) A contemporary review on deep learning models for drought prediction. Sustainability 15(7):6160
Head A, Manguin M, Tran N, Blumenstock JE (2017) Can human development be measured with satellite imagery? Ictd 17:16–19
Henderson JV, Storeygard A, Weil DN (2012) Measuring economic growth from outer space. Am Econ Rev 102(2):994–1028
Hernandez K, Roberts T (2020) Predictive analytics in humanitarian action: a preliminary mapping and analysis
Htet NL, Kongprawechnon W, Thajchayapong S, Isshiki T (2021) Machine learning approach with multiple open-source data for mapping and prediction of poverty in Myanmar. In: 2021 18th International conference on electrical engineering/electronics, computer, telecommunications and information technology (ECTI-CON), pp 1041–1045. IEEE
Jalili M, Gharibshah J, Ghavami SM, Beheshtifar M, Farshi R (2013) Nationwide prediction of drought conditions in Iran based on remote sensing data. IEEE Trans Comput 63(1):90–101
Jean N, Burke M, Xie M, Alampay Davis WM, Lobell DB, Ermon S (2016) Combining satellite imagery and machine learning to predict poverty. Science 353(6301):790–794
Hatch Jonathan (2019). https://www.bond.org.uk/person/jonathan-hatch/
Kaur A, Sood SK (2020) Deep learning based drought assessment and prediction framework. Eco Inform 57:101067
Kuc G, Chormański J (2019) Sentinel-2 imagery for mapping and monitoring imperviousness in urban areas. Int Arch Photogramm Remote Sens Spat Inf Sci 42:43–47
Kumar S (2022) Statistical vs machine learning vs deep learning modeling for time series forecasting. In: Statistical vs machine learning vs deep learning modeling for time series forecasting—Sumit's diary (reachsumit.com)
Kumer SA, Gogu LB, Mohan E, Maloji S, Natarajan B, Sambasivam G, Tyagi VB (2023) Track and noise separation based on the universal codebook and enhanced speech recognition using hybrid deep learning method. IEEE Access 11:120707–120720
Ledesma C, Garonita OL, Flores LJ, Tingzon I, Dalisay D (2020) Interpretable poverty mapping using social media data, satellite images, and geospatial information. arXiv preprint arXiv:2011.13563
Topic Leila https://reliefweb.int/report/world/ai-humanitarian-sector
Logar T, Bullock J, Nemni E, Bromley L, Quinn JA, Luengo-Oroz M (2020) PulseSatellite: a tool using human-AI feedback loops for satellite image analysis in humanitarian contexts. In: Proceedings of the AAAI conference on artificial intelligence, vol 34, no 09, pp 13628–13629
Makridakis S, Spiliotis E, Assimakopoulos V (2018) Statistical and machine learning forecasting methods: concerns and ways forward. PLoS ONE 13(3):e0194889
Marj AF, Meijerink AM (2011) Agricultural drought forecasting using satellite images, climate indices and artificial neural network. Int J Remote Sens 32(24):9707–9719
Mei P, Liu J, Liu C, Liu J (2022) A deep learning model and its application to predict the monthly MCI drought index in the Yunnan Province of China. Atmosphere 13(12):1951
Mera GA (2018) Drought and its impacts in Ethiopia. Weather Clim Extrem 22(June):24–35

Min PP, Gan YW, Hamzah SNB, Ong TS, Sayeed MS (2022) Poverty prediction using machine learning approach. J Southwest Jiaotong Univ 57(1)

Nandgude N, Singh TP, Nandgude S, Tiwari M (2023) Drought prediction: a comprehensive review of different drought prediction models and adopted technologies. Sustainability 15(15):11684

Nemni E, Bullock J, Belabbes S, Bromley L (2020) Fully convolutional neural network for rapid flood segmentation in synthetic aperture radar imagery. Remote Sensing 12(16):2532

Ornella L, Kruseman G, Crossa J (2019) Satellite data and supervised learning to prevent impact of drought on crop production: meteorological drought. In: Drought-detection and solutions. IntechOpen

Oshri B, Hu A, Adelson P, Chen X, Dupas P, Weinstein J, Ermon S (2018) Infrastructure quality assessment in Africa using satellite imagery and deep learning. In: Proceedings of the 24th ACM SIGKDD international conference on knowledge discovery & data mining, pp 616–625

Pandey SM, Agarwal T, Krishnan NC (2018) Multi-task deep learning for predicting poverty from satellite images. In: Thirty-Second AAAI conference on artificial intelligence

Patel NR, Parida BR, Venus V, Saha SK, Dadhwal VK (2012) Analysis of agricultural drought using vegetation temperature condition index (VTCI) from Terra/MODIS satellite data. Environ Monit Assess 184:7153–7163

Quinn JA, Nyhan MM, Navarro C, Coluccia D, Bromley L, Luengo-Oroz M (2018) Humanitarian applications of machine learning with remote-sensing data: review and case study in refugee settlement mapping. Philos Trans R Soc A: Math, Phys Eng Sci 376(2128):20170363

Rembold F, Meroni M, Urbano F, Royer A, Atzberger C, Lemoine G, Haesen D (2015) Remote sensing time series analysis for crop monitoring with the SPIRITS software: new functionalities and use examples. Front Environ Sci 3:46

Schnitzer P (2019) How to target households in adaptive social protection systems? Evidence from humanitarian and development approaches in Niger. J Dev Stud 55(sup1):75–90

Steele JE, Sundsøy PR, Pezzulo C, Alegana VA, Bird TJ, Blumenstock J, Bengtsson L (2017) Mapping poverty using mobile phone and satellite data. J R Soc Interface 14(127):20160690

Traore BB, Kamsu-Foguem B, Tangara F (2017) Data mining techniques on satellite images for discovery of risk areas. Expert Syst Appl 72:443–456

Tyagi S, Zhang X, Saraswat D, Sahany S, Mishra SK, Niyogi D (2022) Flash drought: review of concept, prediction and the potential for machine learning, deep learning methods. Earth's Futur 10(11):e2022EF002723

Villalobos Alva J, Villalobos Alva J (2021) Neural networks with the wolfram language. In: Beginning mathematica and wolfram for data science: applications in data analysis, machine learning, and neural networks, pp 331–373

Weissman F The food emergency in Ethiopia: what the drought conceals. Available at FW, 2003, Famine en Ethiopie. Ce que cache la sécheresse _UK, art, p.7_ (msf-crash.org)

Xie M, Jean N, Burke M, Lobell D, Ermon S (2016) Transfer learning from deep features for remote sensing and poverty mapping. In: Proceedings of the AAAI conference on artificial intelligence, vol 30, no. 1

Xu D, Zhang Q, Ding Y, Zhang D (2022) Application of a hybrid ARIMA-LSTM model based on the SPEI for drought forecasting. Environ Sci Pollut Res 29(3):4128–4144

Xu L, Mo KC (2019) A preliminary study of deep learning based drought forecast. In: 44th NOAA annual climate diagnostics and prediction workshop. NOAA, Durham, NC, pp 63–66

Yang X, Liu Z (2005) Using satellite imagery and GIS for land-use and land-cover change mapping in an estuarine watershed. Int J Remote Sens 26(23):5275–5296

Zhang Q, Shi R, Xu CY, Sun P, Yu H, Zhao J (2022) Multisource data-based integrated drought monitoring index: model development and application. J Hydrol 615:128644

Zixi H (2021) Poverty prediction through machine learning. In: 2021 2nd international conference on e-commerce and internet technology (ECIT), pp 314–324. IEEE

Smart Geospatial Intelligence and AI for a Resilient Environment: A Review

Arshad Husain and Subhojit Bhowmick

1 Introduction

The environment faces critical problems which include severe environmental situations such as global warming alongside biodiversity destruction and tree loss and unpredictable disasters. Technology advances are needed in greater numbers so environmental challenges intensify while creating pressure for sustainable development and enhanced operational resilience. GEOINT technology combined with AI functions provides optimal solutions to solve environmental issues (Akkaya and Demirel 2005). Through their combination with AI-driven analytic methods and remote sensing programs and satellite data algorithms these technologies reveal vital environmental understanding and assessment findings in addition to resource management expertise. Through geospatial intelligence practitioners gather spatial information which they analyze then map to interpret environmental processes that are complex in nature. GEOINT becomes more accurate at processing large datasets and identifying trends when AI capabilities are added to its function (Corbane et al. 2015). The essential capability allows people to prepare for changing climates and plan disasters and protect natural resources (Migallón et al. 2022). AI-controlled analytical models speed up real-time geospatial operations enabling governments and research institutions and organizational entities to base their decisions on data to minimize environmental hazards and advance sustainability goals (Abbasi and Hanandeh 2016). The chapter evaluates modern applications of geospatial intelligence and artificial intelligence to enhance environmental resilience and sustainable development. The section identifies significant technological progress together with

A. Husain (✉) · S. Bhowmick
Bennett University, Greater Noida, Uttar Pradesh, India
e-mail: arshad.husain18@gmail.com

M. O. Sarif and A. Sharifi (eds.), *Advanced Geospatial Intelligence and AI for Environmental Resilience and Sustainable Development*, Advances in Geographic Information Science, https://doi.org/10.1007/978-3-032-03714-5_17

the problems and new directions in this discipline. Society will create successful environmental protection plans and sustainable programs by adopting these innovative approaches.

2 Related Work

Several scientific studies and technological advancements demonstrate that artificial intelligence geospatial analytics successfully manages the environmental issues (Chojer et al. 2022). The scientists have used machine learning and deep learning systems to analyze the satellite images that contributes to monitoring climate change, tracking deforestation, and disaster response programs (Corbane et al. 2015). The effective early detection of wildfires using AI-based models derived from satellites has demonstrated an excellent capacity to minimize response delay periods while reducing the impacts of wildfires. AI together with remote sensing technologies has produced a set of deforestation monitoring tools that provide necessary information for conservation (Oruganti et al. 2023). With precision farming applications using Artificial Intelligence, agriculture functions with greater efficiency in water management along with soil health assessment and yield prediction, which forms sustainable systems for food production (Manapragada et al. 2025). AI-driven geospatial analytics empower cities to make the most of their data to create sustainable infrastructure and optimize transportation systems, thereby reducing carbon emissions in cities (Migallón et al. 2022). Real-time environmental monitoring has become more efficient due to AI-supported IoT devices and has, therefore, facilitated proactivity against water and air pollution (Galaz et al. 2021). However, many issues continue to face AI-driven geospatial applications such as problems related to data acquisition as well as system biases towards automated systems and issues related to ethical usage (Brahart and See 2004).

Researcher refinement of these technologies occurs in a continuous process to improve their capacity for serving all demographics. The analysis presents essential field research to build substantial understanding required for environmental sustainability research on AI-enhanced geospatial intelligence.

3 AI and Geospatial Intelligence (GEOINT)

Geospatial intelligence underwent fundamental changes in fundamental cartographic practices through integrated operations between satellite images coupled with Geographic Information Systems (GIS) together with GPS technology and real-time spatial analytical systems (Corbane et al. 2015). The environmental awareness from GEOINT helps decision making for disaster situations and sustainable city planning. The basic function of GEOINT provided defense needs for military until civilian adaptations increased at a rapid pace during the past several decades. The modern applications utilizing GEOINT technology support environmental projects for observation purposes alongside public health analysis and infrastructure planning decisions and climate adaptation initiatives.

3.1 The Role of AI in GEOINT

Artificial intelligence systems have transformed how data collection and information processing operations in geospatial data teams function. ML and DL algorithms, along with neural networks, collaborate to create a system of advanced algorithms that enhance GEOINT operations (Fig. 1) by following various pathways (Biu et al. 2024).

Artificial Intelligence technology accelerates the speed and precision of geospatial intelligence, enabling large-scale analysis and offering effective solutions (Table 1) for complex environmental problems (Waykar and Yambal 2025; Coorey 2018).

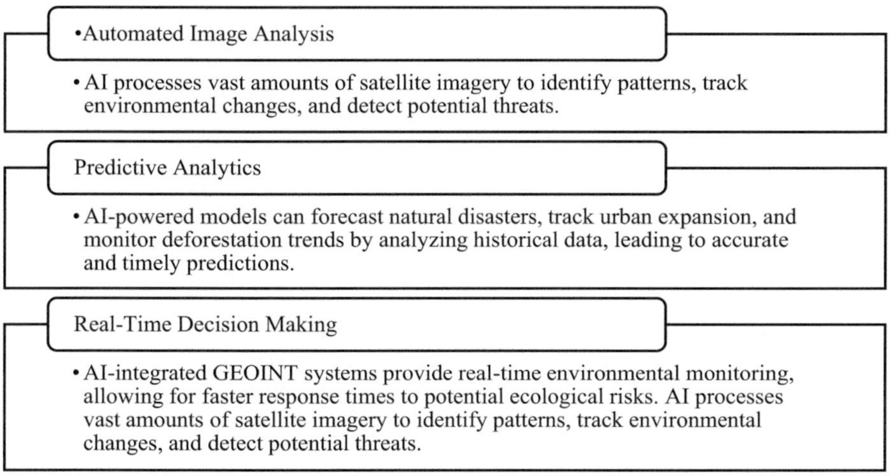

Fig. 1 AI applications in environmental monitoring, including image analysis, predictive analytics, and real-time decision-making

Table 1 Summary of key research in AI-driven geospatial intelligence for environmental resilience (Ref)

Research area	Key studies and contributions	Technologies used	Impact on environmental resilience	References
Climate change monitoring	Studies on AI-driven climate modeling using geospatial data to track temperature shifts and CO_2 emissions	Machine learning (ML), GIS, remote sensing	Improved accuracy in climate projections and risk assessments	Alotaibi and Nassif (2024), Reichstein et al. (2019), Hawchar et al. (2020)
	Study focused AI to enhance soil and land management practices on the ability of forests to prevent climate change	Soil-land inference model (SoLIM) combining GIS to improve soil mapping. ML algorithms, ERA5-Land	Improving the resolution and accuracy of soil map	Zhu et al. (2001), Forzieri et al. (2022)
Disaster prediction and management	Modern disaster management systems integrate theory, practice, software, and hardware, relying on spatial data for risk assessment and post-disaster updates	ML, GIS and remote sensing, multi-sensor systems	Faster disaster response and mitigation strategies	Kemper and Kemper (2020), Gevaert et al. (2021)
Flood susceptibility	The bagging-LMT model, combining bagging ensemble and Logistic Model Tree (LMT), was developed to map flood susceptibility in the Haraz watershed, northern Iran, using a spatial database with eleven flood conditioning factors	ML algorithms	Evaluation and comparison with four benchmark models showed that Bagging-LMT outperformed them, making it a promising tool for sustainable flood management	Chapi et al. (2017)

(continued)

Table 1 (continued)

Research area	Key studies and contributions	Technologies used	Impact on environmental resilience	References
Costal management	The integration of AI algorithms with GIS offers a proactive approach to coastal management by predicting erosion risks and enhancing marine infrastructure resilience	GIS, AI, IoT, Remote sensing	This article explores AI-driven erosion prediction, IoT-based real-time monitoring, and GIS visualization to support adaptive planning, infrastructure maintenance, and sustainable coastal development	Samaei and Ghahfarrokhi (2023), Ricketts (1992)
Biodiversity conservation	AI-powered species tracking using drones and satellite data for deforestation detection. Projects like global forest watch	AI-powered drones, IoT, GIS, Remote sensing	Real-time monitoring of deforestation and endangered species	Shivaprakash et al. (2022), Raihan (2023)
Smart cities and urban planning	Research on AI-driven urban design to optimize traffic flow, reduce pollution, and improve energy efficiency	AI-based traffic models, GIS, IoT sensors	More sustainable urban environments with reduced carbon emissions	Kalusivalingam et al. (2021), Rojek et al. (2024), Yadav et al. (2024)

(continued)

Table 1 (continued)

Research area	Key studies and contributions	Technologies used	Impact on environmental resilience	References
Agriculture and food security	AI-enhanced precision farming for soil monitoring, irrigation management, and pest detection	AI Algorithms, GIS, Remote sensing	Increased agricultural efficiency and sustainable food production	Pandey and Mishra (2024), Rugji et al. (2024), Chavan et al. (2024), Behairy et al. (2023)
Renewable energy optimization	AI-powered solar and wind energy site selection models. Grid optimization studies	Geospatial AI, Smart grids, Predictive analytics	Enhanced renewable energy planning and reduced energy wastage	Shivaprakash et al. (2022), Raihan (2023), Kalusivalingam et al. (2021), Rojek et al. (2024), Yadav et al. (2024), Pandey and Mishra (2024), Rugji et al. (2024), Chavan et al. (2024), Behairy et al. (2023), Pant (2025), Resch et al. (2014)

4 Applications of Geospatial Intelligence in Environmental Resilience

4.1 Climate Monitoring and Adaptation

Climate change, resulting from elevated atmospheric carbon dioxide levels that contribute to global temperature rise, presents an important risk to environmental and socio-economic systems around the world. Historical data indicates that atmospheric CO2 concentrations rose from 284 ppm in 1832 to 397 ppm in 2013, primarily due to fossil fuel consumption and change in land uses pattern, which highlights the anthropogenic roots of this global problem (Hawchar et al. 2020; Bagheri et al. 2023).

The remote sensing and GIS are the essential technologies to analyses climate change by providing timely, comprehensive and precise data on climate patterns, variations, vegetation dynamics and land uses pattern. Recent research demonstrates a shift towards multi-source datasets, integrating ground-based observations, UAV imagery, AI-generated predictions, and climate reanalysis datasets such as CHIRPS, Princeton, and ERA5 to enhance the accuracy of climate modelling. Advanced machine learning (ML) and artificial intelligence (AI) models, such as Support Vector Machines (SVMs), Random Forest (RF), Artificial Neural Networks (ANNs), Analytical Hierarchy Process-Multi Criteria Decision Making (AHP-MCDM), and deep learning methodologies, significantly enhance climate risk assessments. The

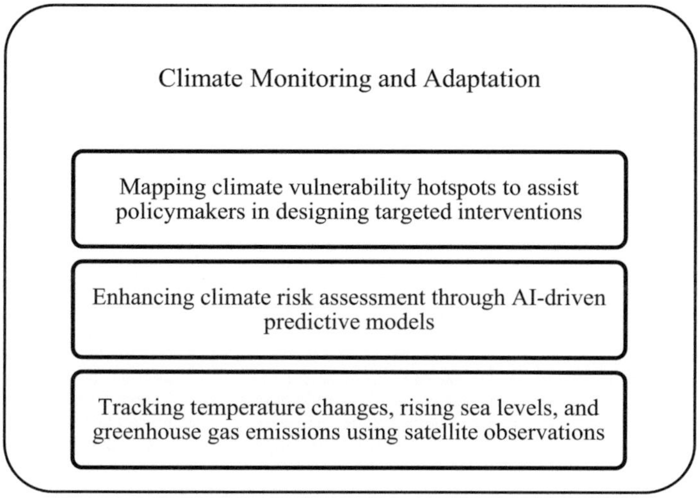

Fig. 2 Climate monitoring and application

AI and GIS integrated in climate monitoring and mitigations which are categorised in Fig. 2 (Sa'adi et al. 2024; Pour et al. 2023; Kemarau et al. 2025).

4.2 Disaster Risk Reduction and Management

Climate change and the growing urban population enhance the susceptibility of individuals and infrastructure to disasters. In 2021, disasters impacted about 98.4 million individuals, resulting in more over 15,000 fatalities and an estimated economic loss of USD 171.3 billion. The geospatial data is playing key role to map affected population and infrastructures in emergency time. The remote sensing images from drones and satellites can be used to analyse pre-disaster and post disaster mitigation options. The AI-algorithms rapidly and accurately process the data and develop risk models to give better distribution of aid (Gevaert et al. 2021; Chapi et al. 2017; Gonzalez et al. 2016).

AI and GEOINT are transforming disaster management and the applications in different areas described in Fig. 3.

4.3 Biodiversity and Ecosystem Monitoring

About 30% of the Earth's terrestrial area is comprised of forests, rendering them the most dominant ecosystems on land and habitat to 90% of the planet's species.

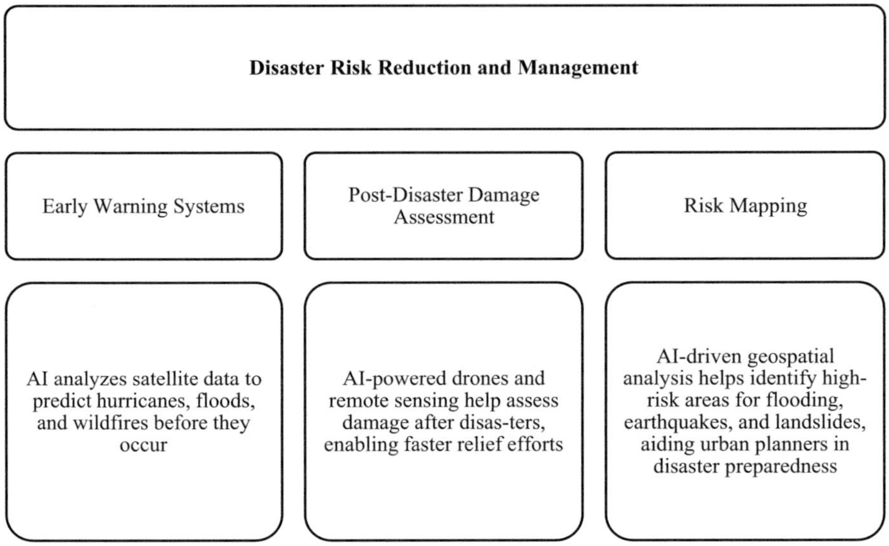

Fig. 3 Disaster risk reduction and management

The monitoring of biodiversity is a necessary sector where AI and ML with integrating GIS have been increasingly utilised remarkably (Fig. 4). AI/ML methodologies, especially those related to picture recognition and data processing, are employed to monitor biodiversity, evaluate ecosystem health, and track changes in species distributions and habitat conditions. These strategies provide a comprehensive and economical solution to conventional observation techniques, which frequently depend on human collecting information and have restrictions in scope and coverage.

The Maxent algorithm, a machine learning model for spices distribution, widely used for biodiversity monitoring purpose. The other model such as Genetic Algorithm for Rule-set Production (GARP) used for predicting species ranges based on environmental variables. The deep learning algorithm use data from camera images for wildlife monitoring. Unmanned aerial vehicles (UAVs) with artificial intelligence methodologies help track objects, and infrared imaging helps observe wildlife populations in forest ecosystems.

These technologies include satellite imagery, unmanned aerial systems, and marine robots provide information about the coral reef ecosystem, including its extent, reef type, geomorphological zonation, reef substrate, and benthic structure. In recent years, drones and small autonomous surface vehicles have been employed for the mapping and monitoring of coral reefs in shallow regions (less than 20 m in depth) (Gonzalez et al. 2016; Alotaibi and Nassif 2024; Piñeros et al. 2024).

Forests, in opposition to agricultural systems, exhibit a dynamic nature that requires effective management, especially as many nations are currently pursuing strategies to achieve transformative change to meet their nationally determined

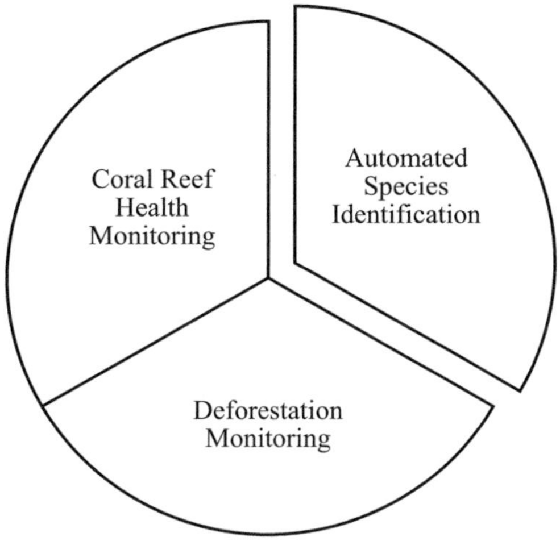

Fig. 4 AI applications in environment conservation

contributions (NDCs). Technology may successfully overcome the challenges of conventional data collection and analysis methods, thereby significantly contributing to the improvement of forest management and conservation. Access to extensive and high-quality datasets, Internet-of-Things (IoT) network infrastructure, advanced technologies such as high-resolution cameras, satellite systems, sensors, drones, and unmanned aerial vehicles (UAVs), along with computational resources and storage, are critical prerequisites for the development and implementation of AI across various domains. The necessity to utilise a combination of these criteria has propelled research on AI applications across many domains of biodiversity conservation and forestry. These categories encompass forest inventory, identification of illegal wildlife, wood trafficking, and logging (Raihan 2023; Pramudya et al. 2023).

4.4 Sustainable Water Resource Management

Water scarcity is a growing concern on all continents, and the supply of clean, drinkable water. The issue is exacerbated by irresponsible resource utilisation and conventional inefficiencies. Approximately 2.4 billion individuals reside in water-stressed nations, with 800 million inhabiting regions classified as experiencing high to critical water shortages in 2020 (Nations 2023).

AI methods, such as artificial neural networks (ANN), back propagation neural networks (BPANN), feed-forward neural networks (FFANN), multi-layer perceptron (MLP), and Bayesian neural networks (BNN), are employed to model groundwater quality, utilising their selected model design within the designated study area (Fig. 5).

Furthermore, significant potential exists for identifying groundwater pollution sources and optimising monitoring networks through integrating GIS and AI (Kontos et al. 2022; Jayakumar et al. 2024).

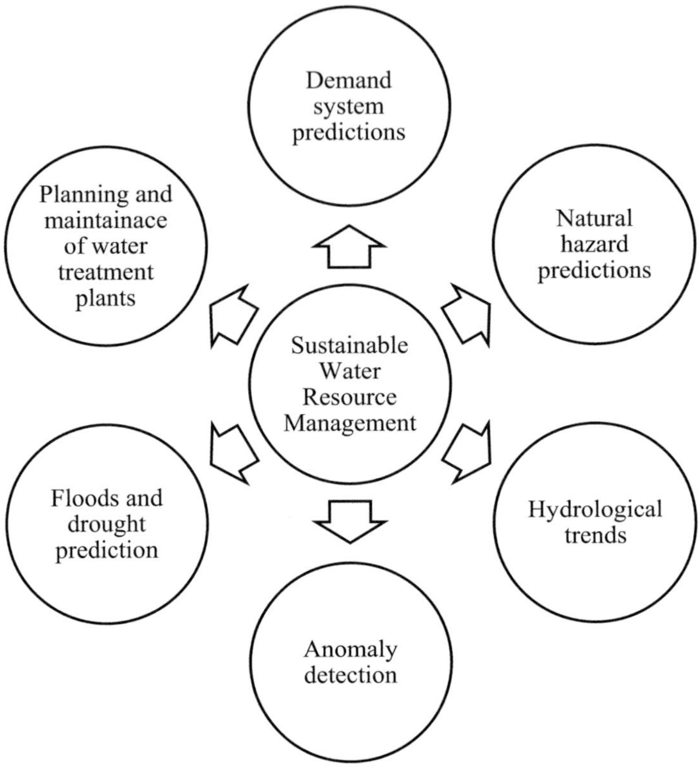

Fig. 5 Sustainable water resource management

5 Applications of Geospatial Intelligence in Environmental Resilience

5.1 AI for Smart Cities and Sustainable Urban Planning

The worldwide urban population is anticipated to increase to 66% or 70% by 2050 (Szpilko et al. 2023). There are increasing concerns regarding the environmental, administrative, and safety risks. Smart cities combine several technologies, including the Internet of Things, blockchain, artificial intelligence, machine learning, and deep reinforcement learning, to deliver holistic solutions.

The massive data generated from different sectors used to enhance the accuracy and precision of the model which strengthening automation. The different sectors in which the AI and the geospatial intelligence contribute is categorized in Fig. 6 (Szpilko et al. 2023).

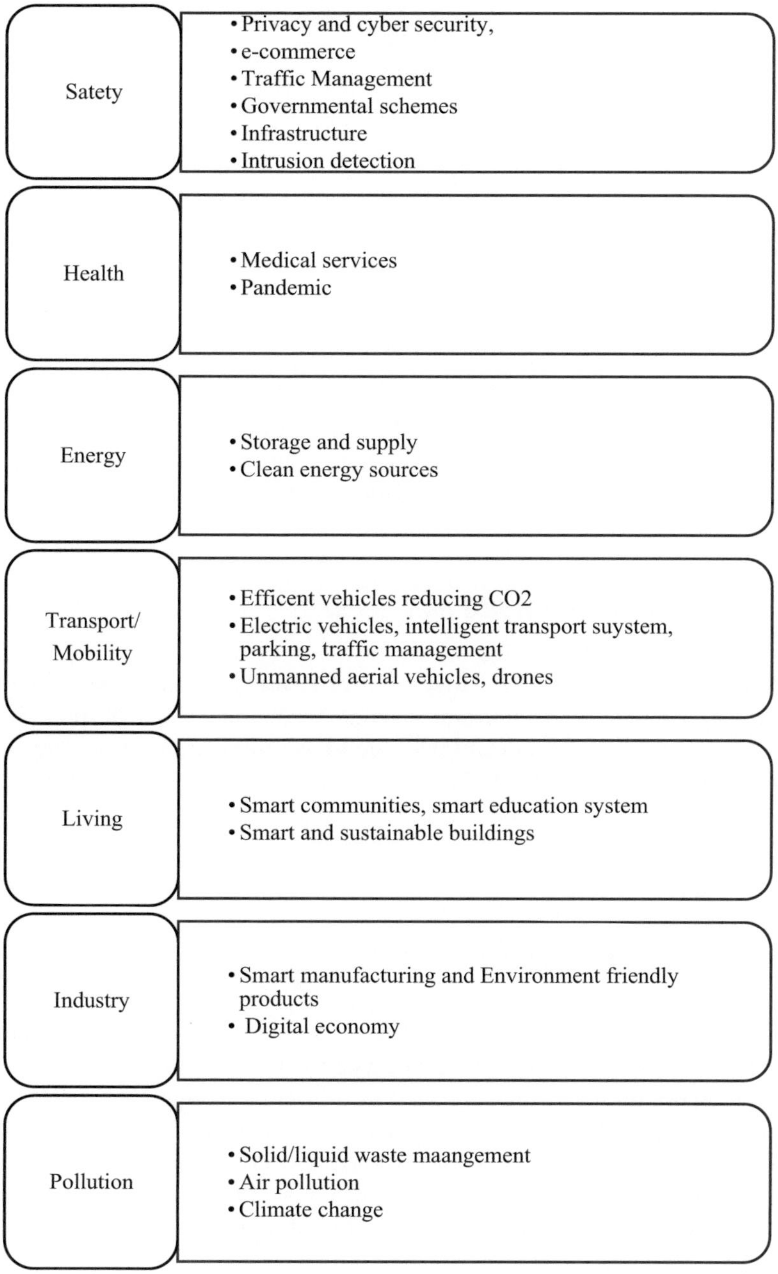

Fig. 6 AI contribution in different sectors

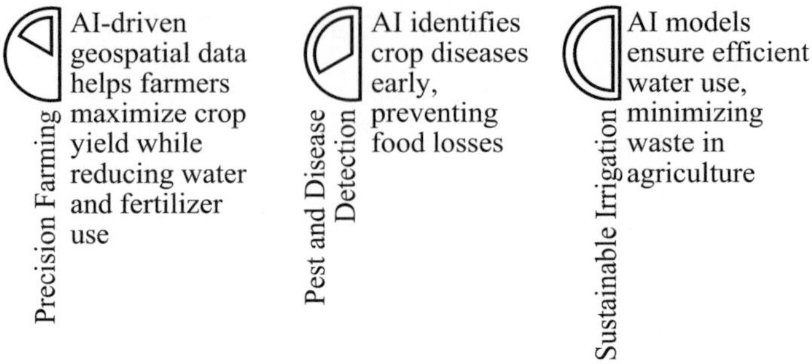

Fig. 7 AI contribution in agriculture and food security

The AI is used in different area for safety in smart cities to provide better living quality, medical facilities, fastest transportation, safety of lives and privacy, efficient and uninterrupted energy supply and pollution control in urban areas.

5.2 AI in Agriculture and Food Security

The global population is projected to reach over 10 billion by 2050, necessitating an increase in agricultural production amidst modest economic growth of approximately 50% compared to 2013. AI and GEOINT are transforming agriculture (Fig. 7).

These innovations support sustainable food production, helping to ensure global food security while minimizing environmental impact (Kutyauripo et al. 2023; Ranganathan et al. 2018).

5.3 Renewable Energy Optimization

The incorporation of artificial intelligence (AI) technology has enhanced the reliability, efficiency, and sustainability of renewable energy systems.

The incorporation of AI has emerged as a possible pathway for optimisation. The integration of renewable energy technology and artificial intelligence offers a ground-breaking method that possesses considerable potential for improving system efficiency, reliability, and sustainability (Ukoba et al. 2024).

AI enhances the adoption and efficiency of renewable energy sources by:

i. Identifying Ideal Solar and Wind Energy Locations: AI analyzes geospatial data to determine optimal sites for renewable energy projects.
ii. Smart Grid Optimization: AI-driven models improve energy distribution, reducing waste and inefficiency.

5.4 *Reducing Carbon Footprint with AI*

Artificial intelligence (AI) stands out from the plethora of alternatives due to its exceptional ability to decrease environmental impact and streamline processes in a variety of industries. AI thus sets itself apart from the plethora of possibilities. Modern technology is said to be essential for efficient resource management, energy optimisation, emissions reduction, and the incorporation of artificial intelligence into environmental sustainability (Roy and Mitra 2020).

AI-powered tools help in:

i. Tracking and Reducing Carbon Emissions: AI models analyze geospatial data to monitor emissions and recommend sustainable alternatives.
ii. Carbon Credit Monitoring: AI assists in ensuring compliance with environmental regulations and policies.

By integrating AI into geospatial intelligence, governments, businesses, and communities can take more effective steps toward a low-carbon, sustainable future.

6 Conclusion

AI operating with GEOINT delivers an innovative solution to build environmental sustainability together with resilience development. The combination of live spatial data with predictive analytics and automated systems enables leaders plus scientists and industrial sectors to acquire instruments for addressing major environmental problems. Organizations use artificial intelligence-powered geospatial intelligence systems to enhance their climate change adaptability along with their disaster preparedness capabilities and biodiversity protection and resource utilization optimization. AI technology plays an essential role in city development together with food production ministries and renewable energy transition to create sustainable future operations.

The entire marketplace will remain inaccessible to AI-powered GEOINT tools despite their available benefits. The principal concerns involve protecting data privacy as well as artificial intelligence discriminatory tendencies as well as problems regarding digital equality and the consequences of computing resource usage on the environment. Complete resolutions to these challenges need collaborative effort between policy makers and researchers and technology leaders who must join forces with community members. Technical deployment of geospatial intelligence systems needs both ethical artificial intelligence practices alongside strategic data management frameworks to ensure equitable system distribution.

Future advancement of geospatial intelligence systems through Artificial Intelligence technology will guide global sustainability planning activities. Satellite imaging advances alongside real-time data processing and AI algorithms improve environmental monitoring abilities which enables prompt prevention procedures.

Progress through international technological advancement allows us to integrate AI and GEOINT capabilities for creating sustainable planet development which benefits upcoming generations.

References

Abbasi M, El Hanandeh A (2016) Forecasting municipal solid waste generation using artificial intelligence modelling approaches. Waste Manage 56:13–22

Abrahart RJ, See L (2000) Comparing neural network and autoregressive moving average techniques for the provision of continuous river flow forecasts in two contrasting catchments. Hydrol. Process. 14, 2157–2172

Alotaibi E, Nassif N (2024) Artificial intelligence in environmental monitoring: in-depth analysis. Discov Artif Intell 4(1):84

Bagheri M, Ibrahim ZZ, Mansor S, Abd Manaf L, Akhir MF, Talaat WIAW, Wolf ID (2023) Hazard assessment and modeling of erosion and sea level rise under global climate change conditions for coastal city management. Natural Hazards Rev 24(1):04022038

Biu PW, Oliha JS, Obi OC (2024) The evolving role of geospatial intelligence in enhancing urban security: A review of applications and outcomes. GSC Adv. Res. Rev. 18, 215–224. https://doi.org/10.30574/gscarr.2024.18.2.0058

El Behairy RA, El Arwash HM, El Baroudy AA, Ibrahim MM, Mohamed ES, Rebouh NY, Shokr MS (2023) Artificial intelligence integrated GIS for land suitability assessment of wheat crop growth in arid zones to sustain food security. Agronomy 13(5):1281

Chapi K, Singh VP, Shirzadi A, Shahabi H, Bui DT, Pham BT, Khosravi K (2017) A novel hybrid artificial intelligence approach for flood susceptibility assessment. Environ Model Softw 95:229–245

Chavan L, Kumbhalkar R, Agrawal P, Rakesh N, Parvathi R, Shahid M (2024) AI in agriculture: yield optimization and food security. In: 2024 4th international conference on technological advancements in computational sciences (ICTACS). IEEE, pp 536–541

Chojer H, Branco PTBS, Martins FG, Alvim-Ferraz MCM, Sousa SIV (2022) Can data reliability of low-cost sensor devices for indoor air particulate matter monitoring be improved?—An approach using machine learning. Atmos Environ 286:119251

Coorey RS (2018) The evolution of geospatial intelligence. In: Australian contributions to strategic and military geography, pp 143–151

Corbane C, Lang S, Pipkins K, Alleaume S, Deshayes M, García Millán VE, Strasser T, Vanden Borre J, Spanhove T, Förster M (2015) Remote sensing for mapping natural habitats and their conservation status—new opportunities and challenges. Int J Appl Earth Obs Geoinf 37:7–16

Forzieri G, Dakos V, McDowell NG, Ramdane A, Cescatti A (2022) Emerging signals of declining forest resilience under climate change. Nature 608(7923):534–539

Galaz V, Centeno MA, Callahan PW, Causevic A, Patterson T, Brass I, Baum S, Farber D, Fischer J, Garcia D, McPhearson T, Jimenez D, King B, Larcey P, Levy K (2021) Artificial intelligence, systemic risks, and sustainability. Technol Soc 67:101741

Gevaert CM, Carman M, Rosman B, Georgiadou Y, Soden R (2021) Fairness and accountability of AI in disaster risk management: opportunities and challenges. Patterns 2(11)

Gonzalez LF, Montes GA, Puig E, Johnson S, Mengersen K, Gaston KJ (2016) Unmanned aerial vehicles (UAVs) and artificial intelligence revolutionizing wildlife monitoring and conservation. Sensors 16(1):97

Hawchar L, Naughton O, Nolan P, Stewart MG, Ryan PC (2020) A GIS-based framework for high-level climate change risk assessment of critical infrastructure. Clim Risk Manage 29:100235

Jayakumar D, Bouhoula A, Al-Zubari WK (2024) Unlocking the potential of artificial intelligence for sustainable water management focusing operational applications. Water 16(22)

Kalusivalingam AK, Sharma A, Patel N, Singh V (2021) Enhancing smart city development with AI: leveraging machine learning algorithms and IoT-driven data analytics. Int J AI and ML 2(3)

Kemarau RA, Suab SA, Eboy OV, Sa'adi Z, Echoh DU, Sakawi Z (2025) Integrative approaches in remote sensing and GIS for assessing climate change impacts across Malaysian ecosystems and societies. Sustainability 17(4):1344

Kemper H, Kemper G (2020) Sensor fusion, GIS and AI technologies for disaster management. Int Arch Photogramm Remote Sens Spat Inf Sci 43:1677–1683

Kontos YN, Kassandros T, Perifanos K, Karampasis M, Katsifarakis KL, Karatzas K (2022) Machine learning for groundwater pollution source identification and monitoring network optimization. Neural Comput Appl 34(22):19515–19545

Kutyauripo I, Rushambwa M, Chiwazi L (2023) Artificial intelligence applications in the agrifood sectors. J Agric Food Res 11:100502

Manapragada NVSK, Mandelmilch M, Roitberg E, Kizel F, Natanian J (2025) Remote sensing for environmentally responsive urban built environment: A review of tools, methods and gaps. Remote Sens. Appl. Soc. Environ. 38, 101529. https://doi.org/10.1016/j.rsase.2025.101529

Migallón V, Navarro-González FJ, Penadés H, Penadés J, Villacampa Y (2024) Designing a Parallel Nonlinear Model for Predicting Nitrogen Oxide Emissions. AIP Conf. Proc. 3094, 25–27. https://doi.org/10.1063/5.0210140

Nations U (2023) The sustainable development goals report 2023: special edition, towards a rescue plan for people and planet. In: United Nations, New York, NY, USA

Oruganti RK, Biji AP, Lanuyanger T, Show PL, Sriariyanun M, Upadhyayula VKK, Gadhamshetty V, Bhattacharyya D (2023) Artificial intelligence and machine learning tools for high-performance microalgal wastewater treatment and algal biorefinery: a critical review. Sci Total Environ 876:162797

Pandey DK, Mishra R (2024) Towards sustainable agriculture: harnessing AI for global food security. Artif Intell Agric

Pant P (2025) AI for renewable energy grid management and storage. In: Cases on AI-driven solutions to environmental challenges. IGI Global Scientific Publishing, pp 319–354

Piñeros VJ, Reveles-Espinoza AM, Monroy JA (2024) From remote sensing to artificial intelligence in coral reef monitoring. Machines 12(10):693

Pour SM, Masoumi AM, Dujardin NS (2023) Towards a technology-driven adaptive decision support system for integrated pavement and maintenance strategies (TDADSS-IPM): focus on risk assessment framework for climate change adaptation. Transp Res Procedia 72:4468–4475

Pramudya FS, Bong L, Rolling E, Awirya AA, Gunawan AAS (2023) Forest loss analysis and calculation with geospatial artificial intelligence: a case study of Papua Province. Procedia Comput Sci 216:346–355

Raihan A (2023) Artificial intelligence and machine learning applications in forest management and biodiversity conservation. Nat Resour Conserv Res 6(2):3825

Ranganathan J, Waite R, Searchinger T, Hanson C (2018) How to sustainably feed 10 billion people by 2050, in 21 charts. World Resources Institute

Reichstein M, Camps-Valls G, Stevens B, Jung M, Denzler J, Carvalhais N, Prabhat F (2019) Deep learning and process understanding for data-driven Earth system science. Nature 566(7743):195–204

Resch B, Sagl G, Törnros T, Bachmaier A, Eggers JB, Herkel S, Gündra H (2014) GIS-based planning and modeling for renewable energy: challenges and future research avenues. ISPRS Int J Geo-Inf 3(2):662–692

Ricketts PJ (1992) Current approaches in geographic information systems for coastal management. Mar Pollut Bull 25(1–4):82–87

Rojek I, Mikołajewski D, Dorożyński J, Dostatni E, Mreła A (2024) An ML-based solution in the transformation towards a sustainable smart city. Appl Sci 14(18):8288

Roy S, Mitra M (2020) AI Driven strategies for carbon footprint reduction-a scholarly exploration. Int J Curr Res Sci Eng Technol 10:2250–1770

Rugji J, Erol Z, Taşçı F, Musa L, Hamadani A, Gündemir MG, Siddiqui SA (2024) Utilization of AI–reshaping the future of food safety, agriculture and food security–a critical review. In: Critical reviews in food science and nutrition. Springer, Heidelberg, pp 1–45

Sa'adi Z, Al-Suwaiyan MS, Yaseen ZM, Tan ML, Goliatt L, Heddam S, Shafik SS (2024) Observed and future shifts in climate zone of Borneo based on CMIP6 models. J Environ Manage 360:121087

Samaei SR, Ghahfarrokhi MA (2023) AI-enhanced GIS solutions for sustainable coastal management: navigating erosion prediction and infrastructure resilience. In: 2nd International conference on creative achievements of architecture, urban planning, civil engineering, and environment in the sustainable development of the Middle East

Shivaprakash KN, Swami N, Mysorekar S, Arora R, Gangadharan A, Vohra K, Kiesecker JM (2022) Potential for artificial intelligence (AI) and machine learning (ML) applications in biodiversity conservation, managing forests, and related services in India. Sustainability 14(12):7154

Szpilko D, Naharro FJ, Lăzăroiu G, Nica E, de la Torre Gallegos A (2023) Artificial intelligence in the smart city—a literature review. Eng Manag Prod Serv 15(4):53–75

Ukoba K, Olatunji KO, Adeoye E, Jen TC, Madyira DM (2024) Optimizing renewable energy systems through artificial intelligence: review and future prospects. Energy Environ 0958305X:241256293

Waykar YA, Yambal S (2025) AI and geospatial intelligence: unlocking new possibilities for the future. In: Harnessing AI in geospatial technology for environmental monitoring and management, pp 1–28. IGI Global Scientific Publishing

Yadav AK, Sharma AK, Patel M, Rathore NS (2025) Intelligent Urban Futures: Exploring AI and Machine Learning Applications in Smart Cities. In: Goar, V., Kuri, M., Kumar, R., Senjyu, T. (eds) Advances in Information Communication Technology and Computing. AICTC 2024. Lecture Notes in Networks and Systems, vol 1075. Springer, Singapore. https://doi.org/10.1007/978-981-97-6106-7_20

Yao D, Liu JJ (2020) Competitive pricing of mixed retail and e-tail distribution channels. Omega 33 (2005) 235–247

Zhu AX, Hudson B, Burt J, Lubich K, Simonson D (2001) Soil mapping using GIS, expert knowledge, and fuzzy logic. Soil Sci Soc Am J

Uncovering Air Pollution Risks in Indian Scenario with Explainable AI: A Case Study Perspective

Shallu Sharma, Jagriti Saini, and Sumit Kumar

1 Introduction

With the swift developments in transportation and industrialization, energy requirements across the globe are rapidly increasing, resulting in the extensive release of pollutants and exhaust gases along with the massive burning of large amounts of oil, coal, and bio-mass fuels (Wu et al. 2024). Every day, a variety of harmful gases such as carbon dioxide (CO_2), carbon monoxide (CO), ozone (O_3), and particulate matter (PM) are produced during fossil fuel consumption (Chen et al. 2017). This direct emission of hazardous components into the atmosphere becomes the primary source of air pollution. Over the years, air pollution has become a global issue and has a detrimental impact on the environment as well as human health. The issues are more prevalent in low and middle-income countries where the majority of populations are still using biomass fuels and other traditional methods for cooking and heating. In Indian cities, the major source of pollution is industrial activities whereas in rural areas, the burning of biomass fuel is the main contributor to degraded air quality (Munsif et al. 2021; Kansiime et al. 2022). Other contributors to contaminated air quality include vehicles and crop burning. The statistical aspects reveal industrial activities cause 50% of the air pollution in India and vehicles are the second biggest contributors with 27%. Furthermore, crop burning leads to 17% of air pollution in

S. Sharma
School of Computer Science Engineering and Technology, Bennett University, Greater Noida, Uttar Pradesh, India

J. Saini
Eternal RESTEM, Mandi, Himachal Pradesh, India

S. Kumar (✉)
Department of Electronics and Communication, Jain (Deemed to be University), Bengaluru, Karnataka, India
e-mail: kumarsumit8@gmail.com; sumit.k@jainuniversity.ac.in

M. O. Sarif and A. Sharifi (eds.), *Advanced Geospatial Intelligence and AI for Environmental Resilience and Sustainable Development*, Advances in Geographic Information Science, https://doi.org/10.1007/978-3-032-03714-5_18

rural and few urban areas; however, the remaining degradation in the quality of breathing air occurs due to domestic cooking activities that involve multiple unclean cooking methods (Natarajan et al. 2024). Long-term and short-term exposure to air pollution can lead to critical consequences for the ecosystems, public health, and climate. Reports reveal that it greatly contributes to the occurrence of critical health issues such as stroke, heart issues, cancer, and breathing problems (Apte and Salvi 2016; Alahmad et al. 2023). Prolonged exposure to PM with a diameter less than 2.5 μm leads to a rise in relative risk of cardiovascular disease by 1–3% (Rahman et al. 2024). The critical consequences related to air pollution not only pose adverse impacts on human health but affect the economy as well.

As per estimates provided by the World Health Organization (WHO), China and India report almost 7 million annual fatalities due to air pollution; this number is expected to rise to 9 million by the year 2030 (Rahman et al. 2024). Reports reveal that 9 out of 10 people across the world are breathing in highly contaminated air. Statistics show that in a survey of the most polluted countries, India ranked 8th among 131 countries in 2022. The Air Quality Index (AQI) is a standard tool for measuring the quality of air and it provides information about the health impacts of air pollution. AQI can be used to help people get updates about any possible threats related to the degraded air quality in the environment. It is a numerical measure that helps to quantify the quality of air at a specific location including the impact of multiple pollutants. The highly polluted country (Chad) is reported to have an AQI of 169 whereas the AQI of India was reported to be 144 (Natarajan et al. 2024). The AQI is a dimensionless factor that can be categorized into six different classes following the guidelines from the Environmental Protection Agency (EPA) in the United States. The mathematical formulation of AQI Score calculation can be provided as follows using Eq. 1 (Natarajan et al. 2024).

$$AQI = \frac{I_{high} - I_{low}}{C_{high} - C_{low}}(C - I_{low}) + I_{low} \tag{1}$$

In this equation, C indicates pollutant concentration, C_{low} indicates the value less than normal concentration of pollutants, whereas C_{high} represents the value above normal pollutant concentration. The index breakpoint with reference to C_{low} and C_{high} is further represented with I_{low} and I_{high}, respectively. Level 1 of AQI is represented as 'good' with a range between 0 and 50; level 2 is 'satisfactory' with a range of 51–100. The third level is 'moderate' with a range fixed into 101–200; whereas 4th level is named 'poor' with an AQI score of 201–300. The AQI range of 301–400 is defined as 'very poor' for breathing and above 400 is indicated as 'severe'. Since critical air pollution levels can pose serious harm to human health, it is essential to predict AQI to protect public health as well as to avail advance insights about air quality conditions so that necessary proactive measures can be taken. There is a range of pollutants that are used to calculate AQI; the list includes sulfur dioxide (SO_2), nitrogen dioxide (NO_2), CO, CO_2, O_3, $PM_{2.5}$, PM_{10}, benzene, and ammonia (NH_3). However, the adequate selection of pollutants depends on the particular objective and numerous variables, including measurement techniques, data accessibility, and monitoring frequency. High AQI indicates severe contamination in the air that can pose an adverse impact on human health. Therefore, timely forecasting of such critical scenarios can help in improving the overall outcomes for the health and well-being of human beings.

A timely understanding of future trends can help agencies and governments plan effective traffic management, environmental policies, industrial layout, and urban as well as rural air quality. Several researchers in the past have used various machine learning, ensemble learning, and deep learning algorithms for forecasting air quality conditions. Adequate selection of methods can help in accurate decision-making to mitigate contamination and avoid critical scenarios with public health. This study focuses on the design of a predictive modeling technique using Geospatial data from different Indian cities. The dataset used in this study has been collected using sensor networks installed at different stations in various Indian cities. The authors used explainable AI for evaluating air pollution scenarios that can help to protect public health & well-being. Various data pre-processing and cleaning methods are deployed in the process to make field data ready for the forecasting algorithm. The main contribution of this study is provided below:

- It has been demonstrated in this study that addressing missing values significantly improves model performance, with R^2 scores increasing by 10–20% on average across test and validation sets.
- It is showcased that tree-based ensemble models (Random Forest, XGBoost, LightGBM) consistently outperformed others, achieving over 94% test accuracy.
- Identified $PM_{2.5}$, CO, and PM_{10} as the most critical predictors across all models.
- Provided a comprehensive comparison of different regression techniques, aiding in selecting the most suitable model based on data structure and preprocessing strategies.

The remaining article is presented as follows. Detailed information about recent advancements in air pollution assessment and management is provided in the 'Recent Developments' section. Furthermore, the 'Materials and Methods' section provides a detailed description of the dataset used and the methods used for forecasting AQI conditions using field data. The detailed analysis and interpretation of findings is provided in the 'Results and Discussions' section; it also highlights the experimental setup, performance metrics and comparative analysis of the proposed methodology. Ultimately, the 'Conclusion' section highlights the advantages, limitations, and possible future implications of this study.

2 Recent Developments

Since air pollution management has been a major concern for governments across the world, several researchers have made contributions in this area. Recent developments show the use of IoT and AI-based algorithms to assess and forecast critical scenarios in air pollution management (Sharma et al. 2022; Saini et al. 2020). The authors

(Liu et al. 2019) developed a regression model for predicting AQI using random forest regression and support vector regression. They also tested the initial relationships between various air indicators such as total NO_x, and $PM_{2.5}$ along with AQI. They assessed model performances using regression analysis metrics such as root mean squared error (RMSE), correlation coefficient (R), and coefficient of determination (R^2). Lin et al. (2020) used a neuro-fuzzy network to predict air quality from historical time-series data. The authors in this study designed fuzzy clusters along with specific fuzzy rules that are applied over a four-layer fuzzy neural network. Furthermore, the steepest descent backpropagation and particle swarm optimization algorithms are used to train the network. The proposed system was capable enough to provide forecasts for air pollutant concentrations with automatic extraction of feature variables. Another study (Phruksahiran 2021) presented an improvement in AQI prediction using a geographically weighted predictor (GWP) methodology. The researchers in this paper utilized ensemble prediction algorithms in combination with machine learning algorithms while using additional predictor variables for hourly analysis. The model was trained on the Bangkok air quality dataset and the comparative analysis shows better performance in comparison to existing studies in literature.

The researchers (Mani et al. 2022) used machine learning techniques to forecast AQI influenced by several individual pollutants such as PM_{10}, $PM_{2.5}$, SO_2, O_3, CO, and NO_2. Initially, they used Multi multi-linear regression with time series analysis and linear regression modeling to predict AQI; the model was trained using sensor data collected from the Central Pollution Control Board in Chennai city of India. Furthermore, they designed an Auto-Regressive Integrated Moving Average (ARIMA) method to perform time series modeling for analyzing trends of AQI in the future. In order to forecast AQI conditions using the previous year's data, researchers (Ravindiran et al. 2023) tested multiple machine learning and ensemble learning models including XGBoost, Adaboost, Catboost, Random Forest, and LightGBM. The dataset used in this study was collected from Vishakhapatnam city in India while focusing on 10 meteorological parameters and 12 air pollutants. Comparative analysis reveals that the Catboost model outperformed all other models. Furthermore, the authors (Natarajan et al. 2024) proposed an optimized method for forecasting AQI in major cities of India. This study focuses on the combination of the Decision Tree and Grey Wolf Optimization (GWO) algorithm to predict critical AQI conditions using datasets collected from the major cities of India. Moreover, several studies have explored the major emission sources, sensor networks for measuring pollutant concentrations, and the significance of comprehensive assessment of models in urban and rural areas. In summary, the literature survey shows the importance of advanced modeling techniques in handling AQI forecasting tasks. It shows a pathway to using more effective and reliable methodologies for addressing computational complexities, real-time data analysis & forecasting for improved outcomes. Adequate application of advanced methodologies can help policy makers, government agencies and end users make active decisions to mitigate air pollution related challenges.

3 Material and Methods

In this study, the dataset was obtained from Kaggle for AQI prediction, consisting of AQI data from various Indian metropolitan cities from 2015 to 2020 and can be accessed from the weblink: https://www.kaggle.com/datasets/rohanrao/air-quality-data-in-india. This dataset consists of 16 attributes, including data for 12 pollutants (illustrated in Fig. 1), along with information about the 'City' and 'Date' of data collection. Additionally, it includes the corresponding AQI and the AQI_Bucket, which categorizes air quality into six levels: Good, Satisfactory, Moderate, Poor, Very Poor, and Severe, based on AQI ranges. This dataset has been made available publicly by the Central Pollution Control Board, Government of India.

A Box-and-Whisker plot analysis is performed to examine the distribution of data statistically. It helps in identifying the outliers as well as an understanding of the data's variability and range. It elucidates the extent to which extreme values deviate from the majority of the data and provides insights into the underlying patterns and inconsistencies. Therefore, this analysis is critical in guiding decisions regarding the choice of appropriate preprocessing techniques and the selection of models to perform accurate predictions. In Table 1, each pollutant is represented with the following statistical measures i.e., Count, Mean, Standard Deviation, Minimum value in the data, 25% (Q1), 50% (Median), 75% (Q3), Maximum which is representing the total number of valid observations for each pollutant; the average concentration level, variability in the levels of concentrations, minimum concentration value, the lower 25% of the data, the middle concentration value, the higher 25% of the data and the maximum

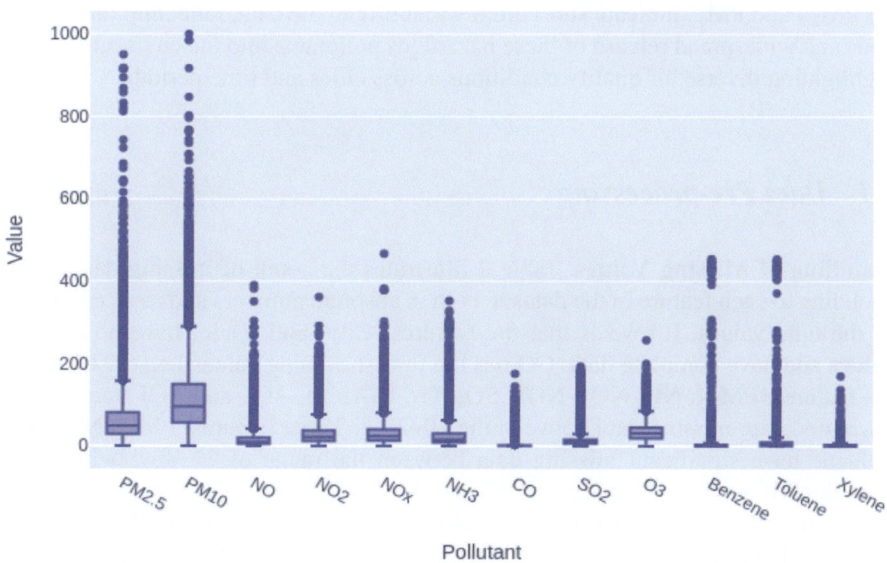

Fig. 1 A box-and-whisker plot represents the distributions of pollutants' in AQI_Data

Table 1 Descriptive statistical analysis of pollutants' concentration data

Index	Count	Mean	Std	Min	25%	50%	75%	Max
$PM_{2.5}$	24,933.0	67.450	64.661	0.04	28.82	48.57	80.59	949.99
PM_{10}	18,391.0	118.127	90.605	0.01	56.25	95.68	149.745	1000.0
NO	25,949.0	17.574	22.785	0.02	5.63	9.89	19.95	390.68
NO_2	25,946.0	28.560	24.474	0.01	11.75	21.69	37.62	362.21
NO_x	25,346.0	32.309	31.646	0.0	12.82	23.52	40.1275	467.63
NH_3	19,203.0	23.483	25.684	0.01	8.58	15.85	30.02	352.89
CO	27,472.0	2.248	6.962	0.0	0.51	0.89	1.45	175.81
SO_2	25,677.0	14.531	18.133	0.01	5.67	9.16	15.22	193.86
O_3	25,509.0	34.491	21.694	0.01	18.86	30.84	45.57	257.73
Benzene	23,908.0	3.2808	15.811	0.0	0.12	1.07	3.08	455.03
Toluene	21,490.0	8.700	19.969	0.0	0.6	2.97	9.15	454.85
Xylene	11,422.0	3.070	6.323	0.0	0.14	0.98	3.35	170.37

concentration value, respectively. It has been observed that the pollutants $PM_{2.5}$, CO, SO_2, and O_3 show good coverage while Xylene shows the least coverage within the dataset in terms of count. Outliers have also been noticed, especially in the recorded values for pollutants like $PM_{2.5}$, PM_{10}, and Benzene. While the data appears highly imbalanced and exhibits the presence of extreme values, these cannot be categorically assumed as outliers due to the alarming rise in pollutant concentrations, which significantly deteriorates air quality. Additionally, the large standard deviations observed for $PM_{2.5}$ and PM_{10} indicate substantial variability in the data, reflecting the continuous and widespread release of these hazardous pollutants into the environment and highlighting diverse air quality conditions across cities and time periods.

3.1 Data Pre-processing

Handling of Missing Values. Table 2 illustrates the count of missing data corresponding to each feature in the dataset, both in absolute numbers and as a percentage of the total values. It reveals that the features 'City' and 'Date' have no missing values and have complete data. CO has the fewest missing values i.e., 6.97% while the features: $PM_{2.5}$, NO, NO_2, NOx, SO_2, O_3, Benzene, AQI, and AQI Bucket each have moderate missing data between the 10–20%. Three features, PM_{10}, NH_3, and Toluene have significant missing data between the range of 21–40% which is a considerable portion of the dataset. Xylene is found as the extremely high missing data with 61.32% of the total values. Missing a significant portion of the Xylene makes it challenging to include this feature in the analysis as imputation may introduce bias. Hence, Xylene has been dropped in the development of this predictive model. To handle missing data for AQI_Bucket and AQI, the rows with missing values have also been removed. However, the interpolation method has been applied to the features with moderate missing data based on the pattern of nearby data points.

Feature Normalization. This is a very crucial step in the application of machine learning while dealing with the data which brings all the considered features to a comparable scale. Since AQI data consists of attributes with different scales therefore normalizing these attributes will aid in preventing features with larger values from dominating the learning process and misleading the model. Feature normalization enhances the performance of the machine learning models by improving stability and convergence. Here, the AQI Data is standardized by a mean of zero and a standard deviation of one.

Feature Selection. In order to select the most significant features from the AQI dataset, a heatmap of Pearson correlation coefficients is generated to analyze relationships between various pollutants and AQI. In Fig. 2, the magnitude and direction of correlation are represented by the dark to light color, ranging from − 1 to 1. Here, dark shades close to -1 are used to represent a strong negative correlation while light shades close to + 1 represent a strong positive correlation. It has been observed from Fig. 2 that Carbon Monoxide (CO) has the highest correlation i.e., 0.67 with AQI. It indicates that the increased level of CO considerably impacts air quality degradation. Particulate matter i.e., $PM_{2.5}$ and PM_{10} are major contributors to poor air quality, showing a high correlation of 0.66 and 0.53 with AQI in respective manner. Sulfur oxides, Nitrogen oxides, and dioxides mainly from industrial and vehicular emissions contribute significantly to air pollution with a correlation factor of 0.48,

Table 2 Missing values for features in the AQI _Data

Features	Missing values count	Missing values % of total values
City	0	0.00
Date	0	0.00
$PM_{2.5}$	4598	15.57
PM_{10}	11,140	37.72
NO	3582	12.12
NO_2	3585	12.13
NO_x	4185	14.17
NH_3	10,328	34.97
CO	2059	6.97
SO_2	3854	13.05
O_3	4022	13.61
Benzene	5623	19.04
Toluene	8041	27.22
Xylene	18,109	61.32
AQI	4681	15.85
AQI_Bucket	4681	15.85

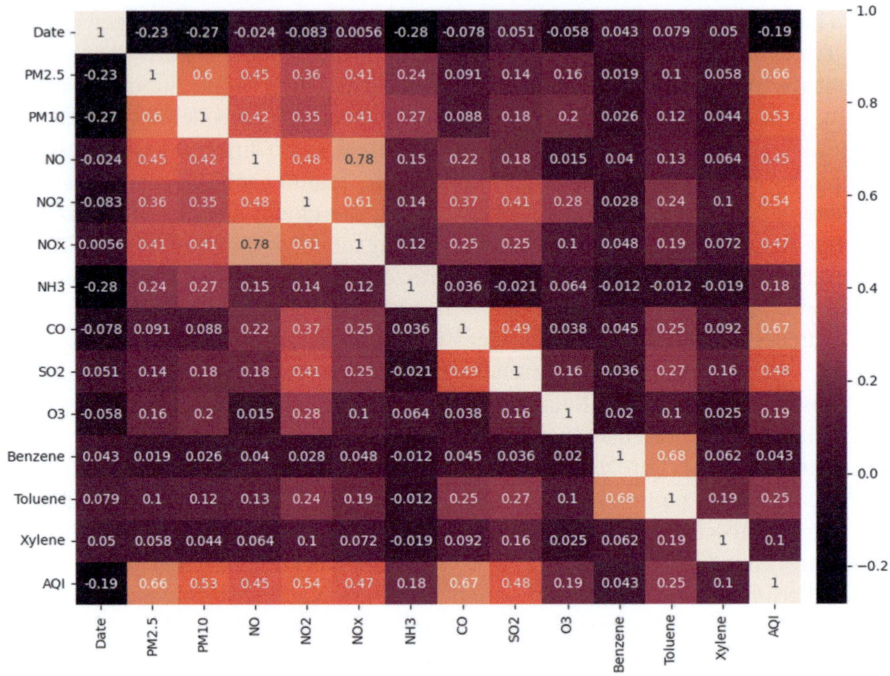

Fig. 2 A Pearson correlation matrix, represents the intensity and direction of correlation among all features of AQI_data

0.47, and 0.54, respectively. Ozone and Ammonia (NH_3) indicate a moderate level of correlation with AQI i.e., 0.19 and 0.18, respectively. Benzene, Toluene, and Xylene do not contribute significantly to AQI prediction as these pollutants have weak correlation with a factor of 0.043, 0.25, and 0.10, respectively. Therefore, pollutants with weak correlation have not been considered further in AQI prediction.

3.2 Training Protocol

This section outlines the structured approach followed to train and optimize the machine learning model. To perform this work, a variety of regression models including linear, kernel-based, tree-based, and ensemble models have been considered for AQI prediction. Initially, the models are set up with predefined or default hyperparameters. The models are trained using the AQI dataset, wherein, 80% of the data is used for training, 10% for validation, and 10% for testing of the models. Furthermore, the R^2 score (coefficient of determination) is used to evaluate the performance of regression models. It measures how well the model's predictions align with the actual target values. The mathematical formula, used to compute the R^2 score is as follows:

Where, SS_{res} = Sum of squared residuals or error between actual and predicted values.

SS_{tot} = Total sum of squares (variance in actual data).

If the R^2 score is equal to 1, represents a perfect model, equal to 0 represents no variance in the data, while less than 0 depicts that the model is a poor fit.

3.3 *Training Models*

Since the selection of an appropriate model depends on various factors such as dataset size, feature relationships, and computational capacity of the utilized systems, therefore, a concise but comprehensive comparison of all the considered regression models i.e., linear models, regularized models, tree-based models, and boosting methods with their advantages and disadvantages are provided in tabular form in this section, (see Table 3).

4 Results and Discussion

In this section, the results obtained from the implementation of regression models over the AQI data are represented in Tables 4, 5, and 6. Since we considered the R^2 Score as a performance evaluation metric, therefore, a comparative analysis of the R^2 Score before handling missing values, after handling missing values without feature selection, and after performing feature selection along with the handling of missing values in the AQI dataset is shown. From the analysis, a significant improvement in model performance has been observed after handling the missing data and feature selection.

Before handling of missing values in the AQI dataset, all linear models (Linear Regression, Ridge, Lasso, SGD, and Elastic Net) exhibited a moderate R^2 Score (66–77%) which improved to 80–85% after treating the problem of missing values. It indicates that the stability and generalization ability of the model gets enhanced after handling missing values. SVR represented a relatively very low R^2 Score (64–65%) due to the extreme sensitivity of SVR towards missing and non-normalized data. It demonstrated significant improvement to 78–82% due to the fullness of the data. It has been analyzed from the result analysis that Ensemble learning models (Gradient Boosting, XGBoost, Random Forest, LightGBM) performed notably better than linear models and SVR, even before the handling of missing values. Among ensemble learning models, Random Forest and XGboost showed a strong generalization ability however, their performance has been improved further from 97.86% to 98.55% (train) and 93.43% to 98.03% (train), respectively.

Table 3 Regression models' description with their advantages and disadvantages

Models	Description	Advantages	Disadvantages	Best suited for	References
Linear Regression	Models a linear relationship between input and output using a straight-line equation	Simple, interpretable, computationally efficient	Assumes linearity, sensitive to outliers	Simple problems where relationships are linear	James et al. (2023); Hastie et al. (2009)
Ridge Regression	Linear regression with L2 regularization to reduce overfitting	Handles multicollinearity, and prevents overfitting	Does not perform feature selection (coefficients shrink but don't become zero)	High-dimensional data with correlated features	Hoerl et al. (n.d.)
SGD Regressor	Uses stochastic gradient descent (SGD) for optimizing linear regression iteratively	Works well for large datasets, fast and scalable	Requires careful tuning of hyperparameters, sensitive to feature scaling	Large-scale datasets where batch learning is infeasible	Bottou (2010)
Elastic Net	Combines L1 (Lasso) and L2 (Ridge) regularization for better generalization	Improves performance in cases with highly correlated features and prevents over-regularization seen in Lasso	Computationally expensive than Lasso or Ridge alone and requires hyperparameter tuning to balance L1 and L2 penalties	Handles multicollinearity, and performs feature selection	Zou and Hastie (2005)
Lasso Regression	Linear regression with L1 regularization, shrinking some coefficients to zero for feature selection	Useful for sparse models where only a few features contribute significantly and prevent overfitting	Not ideal for datasets where all features contribute because it forces some coefficients to zero	Automatic feature selection reduces model complexity	Tibshirani (1996)

(continued)

Table 3 (continued)

Models	Description	Advantages	Disadvantages	Best suited for	References
Support Vector Regressor (SVR)	Uses Support Vector Machines (SVMs) for regression by fitting a hyperplane with minimal error	Works well for non-linear relationships using different kernel functions	Computationally expensive on large datasets, especially with non-linear kernels	Works well for complex, non-linear data	Smola and Schölkopf (2004)
Gradient Boosting Regressor (GBR)	An ensemble technique that builds trees sequentially, optimizing residual errors	High accuracy, robust to outliers, handles non-linearity	Computationally expensive, prone to overfitting if not tuned well	Complex regression problems requiring high accuracy	Friedman (2001)
XGBoost (Extreme Gradient Boosting)	Optimized gradient boosting with regularization, parallelization, and missing value handling	Fast training, highly accurate, handles large datasets well	Requires careful tuning, prone to overfitting	Large-scale datasets, ML competitions, and real-world applications	Chen and Guestrin (2016)
Random Forest Regressor	An ensemble of decision trees trained using bagging (Bootstrap Aggregation)	Reduces variance, robust to overfitting, and handles non-linearity well	Less interpretable than linear models, and slower for very large datasets	General-purpose regression tasks, structured/tabular data	Breiman (2001)
LightGBM (Light Gradient Boosting Machine)	A fast and efficient gradient-boosting model using histogram-based learning	Faster than XGBoost, handles large datasets efficiently	Can overfit small datasets, sensitive to hyperparameters	Large datasets with high-dimensional features require fast training	Ke et al. (2017)

Table 4 R^2 Score before handling missing values

Regression model	R^2 score %		
	Training	Validation	Test
Linear Regression	66.64	67.37	66.76
Ridge	66.64	67.37	66.76
SGD Regressor	66.60	67.38	66.75
Elastic Net	62.69	63.60	62.76
Lasso	66.57	67.37	66.66
SVR	64.01	65.46	64.41
Gradient Boosting Regressor	83.21	81.76	83.31
XG Boost	93.43	84.02	92.21
Random Forest Regressor	97.86	84.73	93.61
Light GBM	88.68	84.58	87.91

Table 5 R^2 Score after handling missing data without feature selection

Regression model	R^2 score %		
	Training	**Validation**	**Test**
Linear Regression	85.36	80.24	82.90
Ridge	85.36	80.24	82.90
SGD Regressor	85.34	80.06	82.84
Elastic Net	79.49	77.34	77.53
Lasso	85.34	80.34	82.89
SVR	84.70	78.64	82.20
Gradient Boosting Regressor	91.75	85.06	88.84
XG Boost	98.03	87.08	94.73
Random Forest Regressor	98.55	86.80	94.77
Light GBM	94.74	86.53	91.58

Table 6 R^2 Score after handling missing data and performing feature selection

Regression model	R^2 score %		
	Training	Validation	Test
Linear Regression	84.01	86.17	82.93
Ridge	84.01	86.17	82.93
SGD Regressor	83.87	86.00	82.85
Elastic Net	78.40	79.65	77.08
Lasso	83.99	86.15	82.90
SVR	83.19	85.88	82.32
Gradient Boosting Regressor	91.29	89.55	90.24
XG Boost	97.77	89.33	95.71
Random Forest Regressor	98.55	90.49	96.17
Light GBM	94.14	90.04	92.79

Table 6 presents the R^2 Score of regression models after handling missing values and feature selection. It has been evident from the results that feature selection impacted the performance of Random Forest Regressor significantly due to which Random Forest surpassed all the considered models with the highest R^2 Score of 96.17% over the test data. However, XGBoost and Gradient Boosting Regressor also provided comparable results to Random Forest when trained using selected features. Linear models underperformed in comparison to tree-based methods when feature selection is performed. The disruption of linear relationships among features due to feature selection may be one of the rationales for lower performance.

These findings underscore the importance of data preprocessing techniques, like missing value handling and normalization, in enhancing model reliability and predictive power for AQI forecasting or similar regression-based tasks.

4.1 Demystifying Regression Models Using Explainable AI

This section discusses the concept of explainable AI in the regression models for more clarity on the performance of considered regression models for AQI data. In this context, SHAP (SHapley Additive exPlanations) and LIME (Local Interpretable Model-

agnostic Explanations) techniques are used to explain regression models. Wherein, SHAP provides local and global interpretability by assigning importance value to each feature and shows how each feature contributes to that prediction. However, LIME creates local approximations of the model.

In Fig. 3, SHAP analysis for all the considered regression models represents insights into feature importance and their impact on predictions. It has been observed from the figure that $PM_{2.5}$, CO, and PM_{10} are the most influential feature across all models, indicating its strong correlation with the target variable. Features including SO_2, NO_2, NO, and O_3, exhibit moderate influence with varying degrees of importance depending upon the model. NH_3 and NOx show the least contribution for almost all the models and represent minimal impact on predictions. Linear regressor shows a well-balanced distribution of feature importance wherein $PM_{2.5}$, CO, and PM_{10} contribute significantly. The ridge regressor is also behaving like a linear regressor but with slightly less variance in feature importance. Low variance in feature importance is caused by regularization which reduces the dominance of a few features. SGD regressor displays higher variation in feature contributions compared to ridge and linear models, still with larger impact of $PM_{2.5}$, PM_{10}, and CO. SVR and lasso regressor follow the similar behavior possessed by linear regressor model but lasso regressor forces some coefficients to zero, consequently reduces the contribution of less significant features like NH_3. ElasticNet balances between lasso and ridge with moderate regularization. Tree-based regression models provided more refined feature rankings compared to linear models. Gradient boosting regressor, XGBoost shows sharper feature importance distribution. Similar to XGBoost, LightGBM provides a more balanced feature distribution. However, the random forest regressor demonstrates the most balanced feature distribution over LightGBM.

5 Perspectives: Study Limitations and Future Implications

This section highlights the limitations and future implications of this study. The authors in this case study considered a limited AQI dataset provided by the Indian Pollution Control Board, but the trained models still lack generalization due to differences in geographical condition of a region. Despite handling missing values, potential inconsistencies, sensor errors, and biases in data collection are other challenging factors that will affect model performance. Imbalance in the distribution of pollution levels in different regions along with extreme values of pollutants in the data sometimes leads to confusion with outliers is one of the potential challenges in handling AQI data.

Future implications of this study may include the development of real-time air quality monitoring data with adaptive models that dynamically update based on incoming observation. Aggregation of AQI data with satellite imagery, meteorological factors, and traffic data may also enhance predictive accuracy and robustness. Implementing Explainable AI (XAI) techniques i.e., LIME can help in understanding how specific pollutants contribute to AQI fluctuations. Optimization of models for deployment in IoT-based smart city frameworks can also be a future aspect of this study which may enable real-time AQI prediction for urban planning and policy-making. Comparing current models with deep learning approaches such as LSTMs (Long Short-Term Memory Networks) or Transformer-based models could help in capturing complex temporal dependencies in air quality trends.

6 Conclusion

This study concludes that $PM_{2.5}$, CO, and PM_{10} are the most influencing features across all models for predicting the target variable, therefore, ensuring a high correlation with the target variable. On the other side, Benzene, Toluene, and Xylene are found as the least influencing features which do not contribute significantly to AQI prediction as these pollutants have a weak correlation with AQI data. Further, it has been observed that handling missing values in the data has a profound impact on the predictive performance of regression models. While all models benefited from this preprocessing step, tree-based ensemble models such as Random Forest, XGBoost, and LightGBM remained the top performers, achieving over 94% test accuracy. Simpler models like Linear Regression, Ridge, and Lasso showed significant improvement as well, making them more viable for predictive analysis post-preprocessing. Handling missing data enhances model generalization, as reflected in

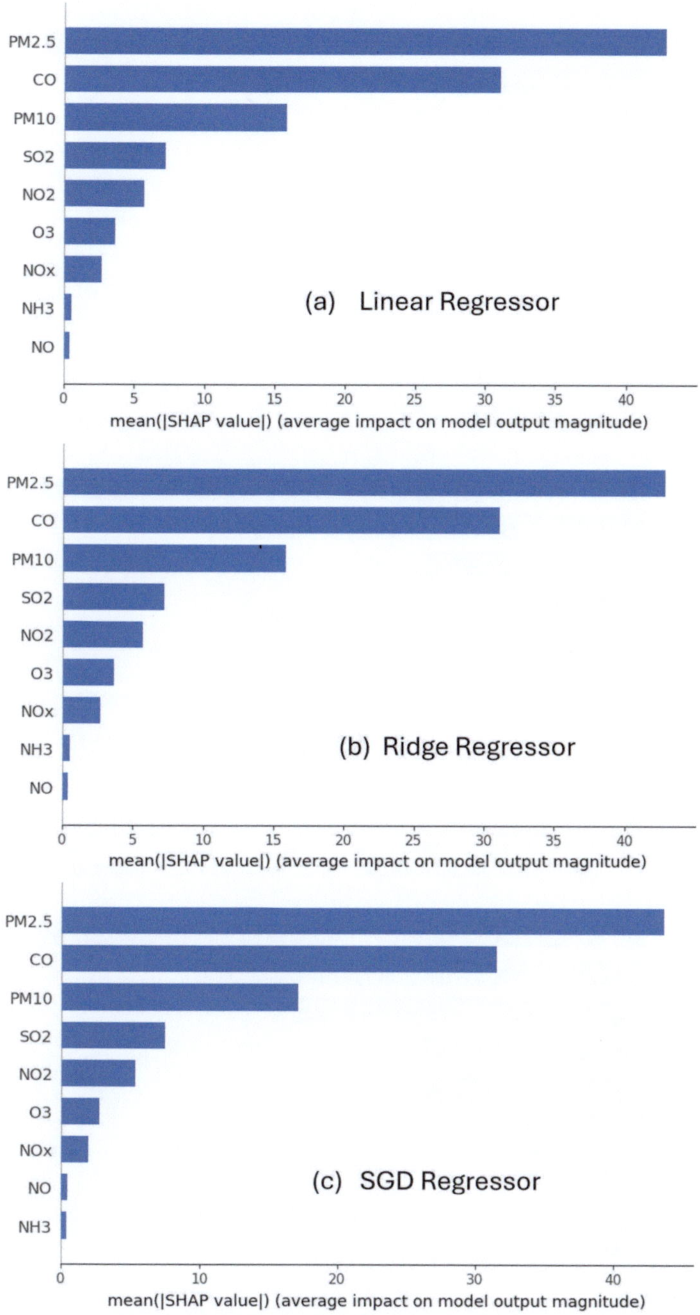

Fig. 3 SHAP analysis, representing the impact of each feature on the prediction of the model **a** Linear Regressor, **b** Ridge Regressor, **c** SGD Regressor, **d** SVR, **e** Lasso Regressor, **f** ElasticNet, **g** Gradient Boost, **h** XG Boost, **i** Light GBM, **j** Random Forest Regressor

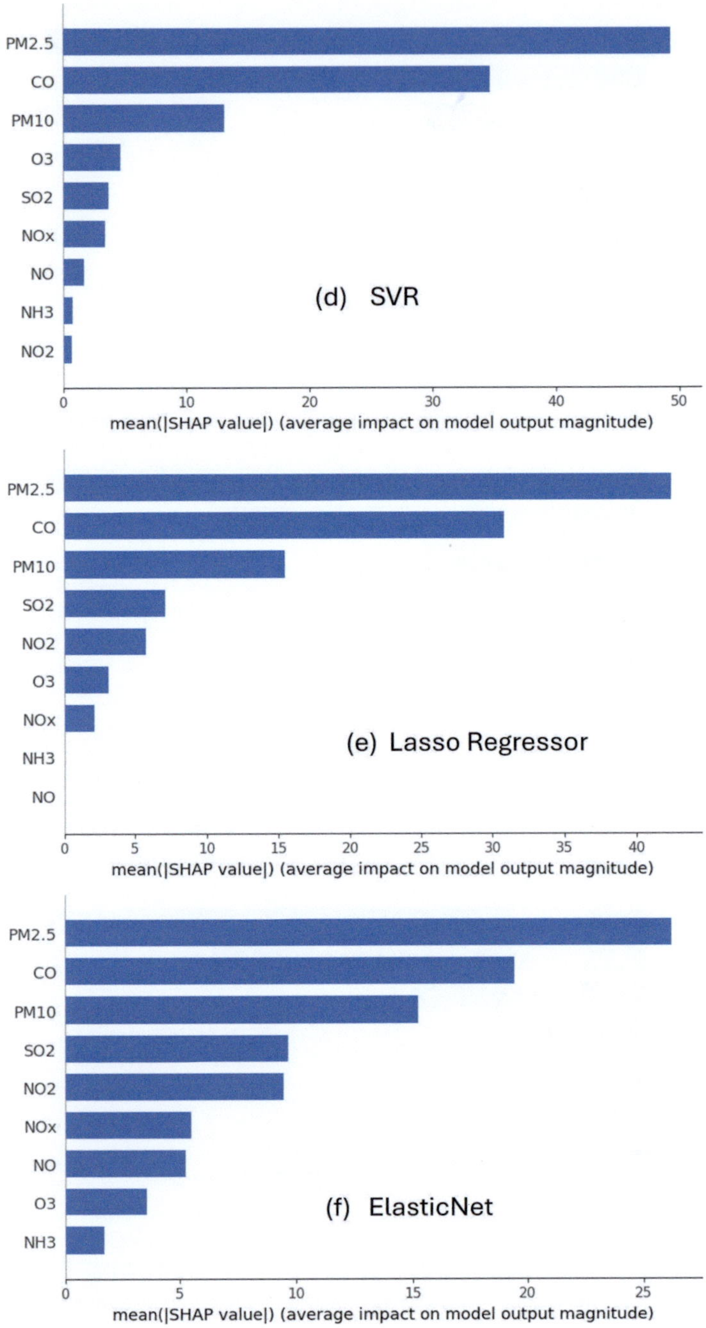

Fig. 3 (continued)

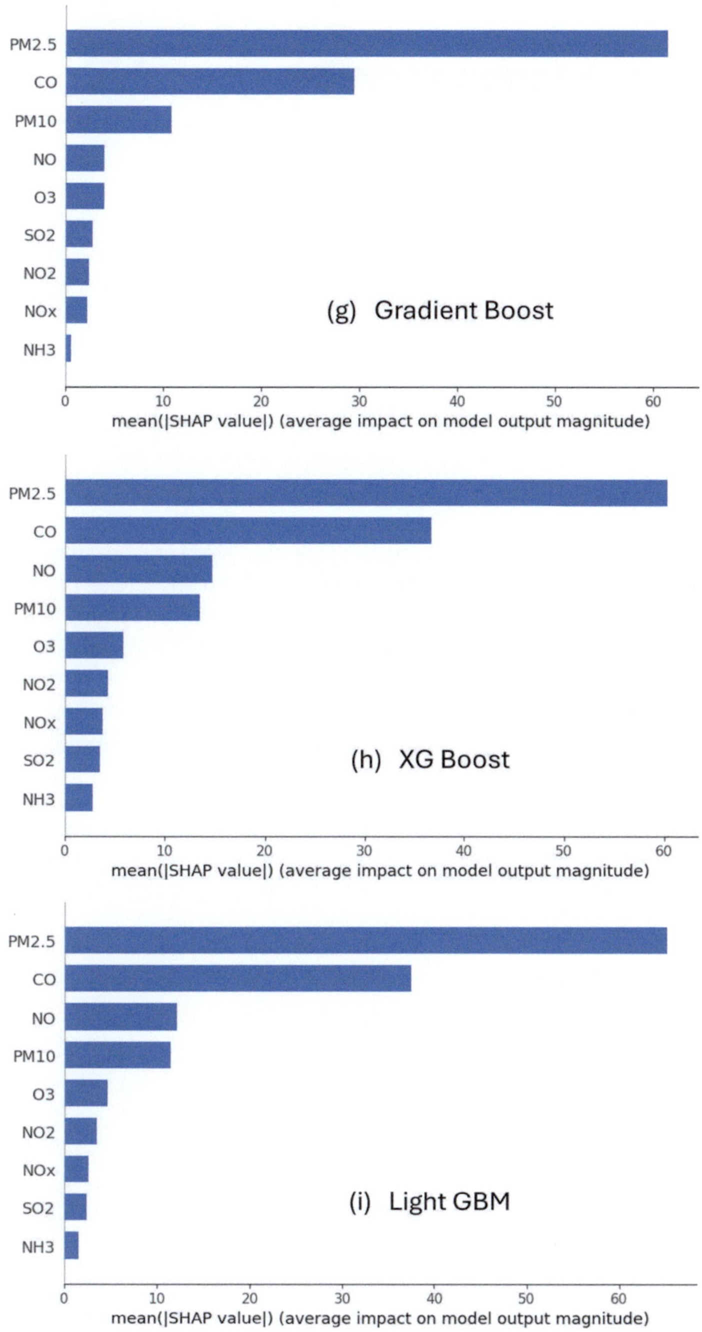

Fig. 3 (continued)

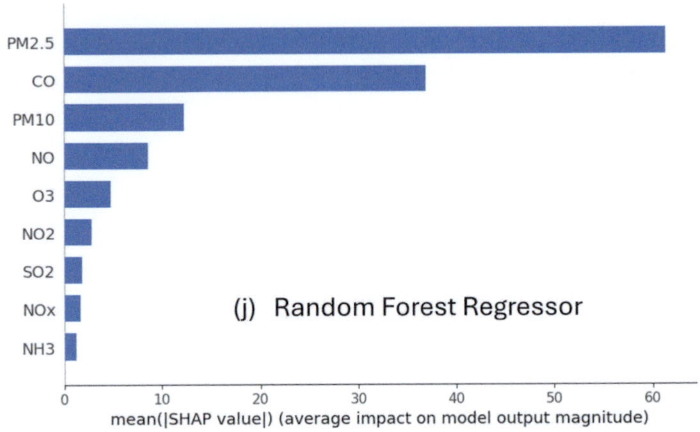

Fig. 3 (continued)

improved validation and test scores. Feature selection also showed a drastic improvement in the predictive performance of the models. The differences in improvement across models highlight the varying sensitivity of machine learning algorithms to missing data, emphasizing the need for careful preprocessing. Tree-based models (XGBoost, Random Forest) offer better interpretability and feature differentiation compared to linear models. Gradient Boosting methods provide sharper insights into feature impact, making them preferable for performance-driven applications.

Acknowledgements We would like to express our sincere gratitude to the School of Computer Science and Engineering, Bennett University, Greater Noida and the Department of Electronics and Communication Engineering, Jain (Deemed-to-be-University) Gobal Campus, Bengaluru for providing the necessary resources and research infrastructure that facilitated this study.

AI Disclosure There is no use of AI in this study.

Conflict of Interest Disclosure There are no conflicts of interest in this study.

References

Alahmad B, Khraishah H, Althalji K et al (2023) Connections between air pollution, climate change, and cardiovascular health. Can J Cardiol 39:1182–1190. https://doi.org/10.1016/j.cjca.2023.03.025

Apte K, Salvi S (2016) Household air pollution and its effects on health. F1000Research 5:2593. https://doi.org/10.12688/f1000research.7552.1

Bottou L (2010) Large-scale machine learning with stochastic gradient descent. In: Lechevallier Y, Saporta G (eds) Proceedings of COMPSTAT'2010. Physica-Verlag HD, Heidelberg, pp 177–186. https://doi.org/10.1007/978-3-7908-2604-3_16

Breiman L (2001) Random forests. Mach Learn 45:5–32. https://doi.org/10.1023/A:1010933404324

Chen J, Li C, Ristovski Z et al (2017) A review of biomass burning: emissions and impacts on air quality, health and climate in China. Sci Total Environ 579:1000–1034. https://doi.org/10.1016/j.scitotenv.2016.11.025

Chen T, Guestrin C (2016) XGBoost: a scalable tree boosting system. In: Proceedings of the 22nd ACM SIGKDD international conference on knowledge discovery and data mining. Association for Computing Machinery, New York, NY, USA, pp 785–794. https://doi.org/10.1145/2939672.2939785

Friedman JH (2001) Greedy function approximation: a gradient boosting machine. Ann Stat 29:1189–1232. https://doi.org/10.1214/aos/1013203451

Hastie T, Tibshirani R, Friedman J (2009) The elements of statistical learning. Springer, New York, NY. https://doi.org/10.1007/978-0-387-84858-7

Hoerl AE, Kennard RW (n.d.) Ridge regression: biased estimation for nonorthogonal problems. https://doi.org/10.2307/1267351

James G, Witten D, Hastie T, et al (2023) An introduction to statistical learning: with applications in Python 112(1). Springer. https://doi.org/10.1007/978-1-4614-7138-7_1

Kansiime WK, Mugambe RK, Atusingwize E et al (2022) Use of biomass fuels predicts indoor particulate matter and carbon monoxide concentrations; evidence from an informal urban settlement in Fort Portal city, Uganda. BMC Public Health 22:1723. https://doi.org/10.1186/s12889-022-14015-w

Ke G, Meng Q, Finley T et al (2017) LightGBM: a highly efficient gradient boosting decision tree. In: Advances in neural information processing systems. Curran Associates, Inc. https://doi.org/10.5555/3294996.3295074

Lin Y-C, Lee S-J, Ouyang C-S, Wu C-H (2020) Air quality prediction by neuro-fuzzy modeling approach. Appl Soft Comput 86:105898. https://doi.org/10.1016/j.asoc.2019.105898

Liu H, Li Q, Yu D, Gu Y (2019) Air Quality index and air pollutant concentration prediction based on machine learning algorithms. Appl Sci 9:4069. https://doi.org/10.3390/app9194069

Mani G, Viswanadhapalli JK, S AA (2022) Prediction and forecasting of air quality index in Chennai using regression and ARIMA time series models. J Eng Res 10:179–194. https://doi.org/10.36909/jer.10253

Munsif R, Zubair M, Aziz A, Zafar MN (2021) Industrial air emission pollution: potential sources and sustainable mitigation. IntechOpen. https://doi.org/10.5772/intechopen.93104

Natarajan SK, Shanmurthy P, Arockiam D et al (2024) Optimized machine learning model for air quality index prediction in major cities in India. Sci Rep 14:6795. https://doi.org/10.1038/s41598-024-54807-1

Phruksahiran N (2021) Improvement of air quality index prediction using geographically weighted predictor methodology. Urban Climate 38:100890. https://doi.org/10.1016/j.uclim.2021.100890

Rahman MdM, Nayeem MdEH, Ahmed MdS et al (2024) AirNet: predictive machine learning model for air quality forecasting using web interface. Environ Syst Res 13:44. https://doi.org/10.1186/s40068-024-00378-z

Ravindiran G, Hayder G, Kanagarathinam K et al (2023) Air quality prediction by machine learning models: a predictive study on the indian coastal city of Visakhapatnam. Chemosphere 338:139518. https://doi.org/10.1016/j.chemosphere.2023.139518

Saini J, Dutta M, Marques G (2020) A comprehensive review on indoor air quality monitoring systems for enhanced public health. Sustain Environ Res 30:6. https://doi.org/10.1186/s42834-020-0047-y

Sharma M, Kumar N, Sharma S et al (2022) Assessment of fine particulate matter for Port City of eastern peninsular india using gradient boosting machine learning model. Atmosphere 13:743. https://doi.org/10.3390/atmos13050743

Smola AJ, Schölkopf B (2004) A tutorial on support vector regression. Stat Comput 14:199–222. https://doi.org/10.1023/B:STCO.0000035301.49549.88

Tibshirani R (1996) Regression shrinkage and selection via the Lasso. J Royal Statist Soc Ser B (Methodological) 58:267–288. https://www.jstor.org/stable/2346178

Wu J, Chen X, Li R et al (2024) A novel framework for high resolution air quality index prediction with interpretable artificial intelligence and uncertainties estimation. J Environ Manage 357:120785. https://doi.org/10.1016/j.jenvman.2024.120785

Zou H, Hastie T (2005) Regularization and variable selection via the elastic net. J Royal Stat Soc Ser B (Statistical Methodology) 67:301–320. https://doi.org/10.1111/j.1467-9868.2005.00503.x

MIX
Papier aus verantwortungsvollen Quellen
Paper from responsible sources
FSC® C105338

FSC
www.fsc.org

If you have any concerns about our products,
you can contact us on
ProductSafety@springernature.com

In case Publisher is established outside the EU,
the EU authorized representative is:
Springer Nature Customer Service Center GmbH
Europaplatz 3, 69115 Heidelberg, Germany

Printed by Libri Plureos GmbH
in Hamburg, Germany